T0189638

IFIP Advances in Information and Communication Technology 423

Editor-in-Chief

A. Joe Turner, Seneca, SC, USA

Editorial Board

Foundations of Computer Science
Mike Hinchey, Lero, Limerick, Ireland

Software: Theory and Practice
Michael Goedicke, University of Duisburg-Essen, Germany

Education
Arthur Tatnall, Victoria University, Melbourne, Australia

Information Technology Applications
Ronald Waxman, EDA Standards Consulting, Beachwood, OH, USA

Communication Systems
Guy Leduc, Université de Liège, Belgium

System Modeling and Optimization
Jacques Henry, Université de Bordeaux, France

Information Systems
Jan Pries-Heje, Roskilde University, Denmark

ICT and Society
Jackie Phahlamohlaka, CSIR, Pretoria, South Africa

Computer Systems Technology
Paolo Prinetto, Politecnico di Torino, Italy

Security and Privacy Protection in Information Processing Systems
Kai Rannenberg, Goethe University Frankfurt, Germany

Artificial Intelligence
Tharam Dillon, Curtin University, Bentley, Australia

Human-Computer Interaction
Annelise Mark Pejtersen, Center of Cognitive Systems Engineering, Denmark

Entertainment Computing
Ryohei Nakatsu, National University of Singapore

IFIP – The International Federation for Information Processing

IFIP was founded in 1960 under the auspices of UNESCO, following the First World Computer Congress held in Paris the previous year. An umbrella organization for societies working in information processing, IFIP's aim is two-fold: to support information processing within its member countries and to encourage technology transfer to developing nations. As its mission statement clearly states,

> IFIP's mission is to be the leading, truly international, apolitical organization which encourages and assists in the development, exploitation and application of information technology for the benefit of all people.

IFIP is a non-profitmaking organization, run almost solely by 2500 volunteers. It operates through a number of technical committees, which organize events and publications. IFIP's events range from an international congress to local seminars, but the most important are:

- The IFIP World Computer Congress, held every second year;
- Open conferences;
- Working conferences.

The flagship event is the IFIP World Computer Congress, at which both invited and contributed papers are presented. Contributed papers are rigorously refereed and the rejection rate is high.

As with the Congress, participation in the open conferences is open to all and papers may be invited or submitted. Again, submitted papers are stringently refereed.

The working conferences are structured differently. They are usually run by a working group and attendance is small and by invitation only. Their purpose is to create an atmosphere conducive to innovation and development. Refereeing is also rigorous and papers are subjected to extensive group discussion.

Publications arising from IFIP events vary. The papers presented at the IFIP World Computer Congress and at open conferences are published as conference proceedings, while the results of the working conferences are often published as collections of selected and edited papers.

Any national society whose primary activity is about information processing may apply to become a full member of IFIP, although full membership is restricted to one society per country. Full members are entitled to vote at the annual General Assembly, National societies preferring a less committed involvement may apply for associate or corresponding membership. Associate members enjoy the same benefits as full members, but without voting rights. Corresponding members are not represented in IFIP bodies. Affiliated membership is open to non-national societies, and individual and honorary membership schemes are also offered.

Luis M. Camarinha-Matos   Nuno S. Barrento
Ricardo Mendonça (Eds.)

Technological Innovation for Collective Awareness Systems

5th IFIP WG 5.5/SOCOLNET Doctoral Conference on
Computing, Electrical and Industrial Systems, DoCEIS 2014
Costa de Caparica, Portugal, April 7-9, 2014
Proceedings

 Springer

Volume Editors

Luis M. Camarinha-Matos
Nuno S. Barrento
Ricardo Mendonça
Universidade Nova de Lisboa
FCT - Department of Electrical Engineering
Campus de Caparica, 2829-516 Monte Caparica, Portugal
E-mail: cam@uninova.pt
nsbarrento@refer.pt
r.mendonca@campus.fct.unl.pt

ISSN 1868-4238   e-ISSN 1868-422X
ISBN 978-3-662-52564-7   e-ISBN 978-3-642-54734-8 (e-Book)
DOI 10.1007/978-3-642-54734-8
Springer Heidelberg New York Dordrecht London

Typesetting: Camera-ready by author, data conversion by Scientific Publishing Services, Chennai, India

Printed on acid-free paper

Springer is part of Springer Science+Business Media (www.springer.com)

Preface

This proceeding book, which collects relevant research results produced in engineering doctoral programs, focuses on socio-technical systems capable of extensive sensing and multi-source, multi-modal information processing; harnessing collective intelligence for promoting innovation; and taking good, informed, and sustainability-oriented decisions: collective awareness systems. These systems leverage the ubiquitous computing and "network effect" by combining open social media, distributed knowledge creation, and data acquisition from real environments ("Internet of Things"), thus linking objects, people, and knowledge in order to foster new forms of social and business innovation.

Although typical PhD students are not experienced researchers, but rather in the process of learning how to do research, observation of worldwide publications shows that a high number of technologically innovative ideas are produced in the early careers of researchers. The DoCEIS series of doctoral conferences on Computing, Electrical and Industrial Systems aims at creating a space for sharing and discussing ideas and results from doctoral research in these inter-related areas of engineering. Innovative ideas and hypothesis can be better enhanced when presented and discussed in an encouraging and open environment. DoCEIS aims to provide such an environment, releasing PhD students from the pressure of presenting their propositions in more formal contexts.

The fifth edition of DoCEIS, which was sponsored by SOCOLNET and IFIP, attracted a considerable number of paper submissions from a large number of PhD students (and their supervisors) from 22 countries. This book comprises the works selected by the International Program Committee for inclusion in the main program and covers a wide spectrum of topics, ranging from collaborative networks to microelectronics. As such, research results and on-going work are presented, illustrated, and discussed in areas such as:

- Collaborative Networks
- Computational Systems and Human-Computer Interfaces
- Self-organizing Manufacturing Systems
- Manufacturing and Supervision
- Robotics and Mechatronics
- Embedded Systems and Petri Nets
- Energy Systems and Smart Grid
- Monitoring and Optimization in Energy
- Electronics and Telecommunications

Focusing on the main theme of the conference, and as a gluing element, all authors were asked to explicitly indicate the (potential) contribution of their work to the collective awareness systems. The idea was not to "deviate students' attention" from their core research. The core of each paper was aimed to be defined around the PhD research topic and the innovative contributions that such research brings to each specific area. Nevertheless, it was also anticipated, and confirmed by the submissions, that virtually any research topic in this broad engineering area could either benefit from a collective awareness systems perspective, or being a direct contributor with models, approaches, and technologies for further development of such systems.

On the other hand, researchers are increasingly requested to be able to "position" their research in a wider scope and establish links with other disciplines. More and more funding agencies are requiring research proposals to include an element of multi-disciplinarity. Therefore, this "exercise" requested by DoCEIS can be seen as part of the process of acquiring such skills, which are mandatory in the profession of a PhD.

We expect that this book will provide readers with an inspiring set of promising ideas and new challenges, presented in a multi-disciplinary context, and that by their diversity these results will trigger and motivate richer research and development directions.

We would like to thank all the authors for their contributions. We also appreciate the efforts and dedication of the DoCEIS Program Committee members who both helped with the selection of articles and contributed with valuable comments to improve their quality.

February 2014

Luis M. Camarinha-Matos
Nuno Silvério Barrento
Ricardo Mendonça

Organization

 5th IFIP/SOCOLNET Doctoral Conference on COMPUTING, ELECTRICAL AND INDUSTRIAL SYSTEMS

2014 Costa de Caparica, Portugal, April 7–9, 2014

Conference and Program Chair

Luis M. Camarinha-Matos, Portugal

Organizing Committee Co-chairs

Luis Gomes, Portugal
João Goes, Portugal
João Martins, Portugal

International Program Committee

Andy Adamatzky, UK
Marian Adamski, Poland
José Júlio Alferes, Portugal
Josué Álvarez-Borrego, Mexico
Carlos Henggeler Antunes, Portugal
Helder Araujo, Portugal
Amir Assadi, USA
José Barata, Portugal
Fernando Maciel Barbosa, Portugal
Olga Battai-Guschinskaya, France
Marko Beko, Portugal
Luis Bernardo, Portugal
Nik Bessis, UK
Vedran Bilas, Croatia
Xavier Boucher, France
Erik Bruun, Denmark
Giuseppe Buja, Italy
Teodoro Calonge Cano, Spain
Luis M. Camarinha-Matos, Portugal
António Cardoso, Portugal

João Catalão, Portugal
Wojciech Cellary, Poland
Naoufel Cheikhrouhou, Switzerland
Alok Choudhary, UK
Fernando J. Coito, Portugal
Luis Correia, Portugal
Luis Cruz, Portugal
Ed Curry, Ireland
Jorge Dias, Portugal
Rolf Drechsler, Germany
Pedro Encarnação, Portugal
Ip-Shing Fan, UK
Florin G. Filip, Romania
Maria Helena Fino, Portugal
José M. Fonseca, Portugal
Fausto P. Garcia, Spain
Paulo Gil, Portugal
João Goes, Portugal
Luis Gomes, Portugal
Antoni Grau, Spain

Michael Huebner, Germany
Tomasz Janowski, Macau
Ricardo Jardim-Gonçalves, Portugal
Hans-Jörg Kreowski, Germany
Paulo Leitão, Portugal
J. Tenreiro Machado, Portugal
Veljko Malbasa, Serbia
João Martins, Portugal
Paulo Miyagi, Brazil
Ángel Molina, Spain
Jörg Müller, Germany
Ferrante Neri, UK
Horacio Neto, Portugal
Rui Neves-Silva, Portugal
Henrique O'Neill, Portugal
Luis Oliveira, Portugal
Manuel D. Ortigueira, Portugal
Angel Ortiz, Spain
Gordana Ostojic, Serbia
Peter Palensky, Austria
Luis Palma, Portugal
Nuno Paulino, Portugal

Carlos Eduardo Pereira, Brazil
Willy Picard, Poland
Paulo Pinto, Portugal
Armando Pires, Portugal
Ricardo Rabelo, Brazil
Rita Ribeiro, Portugal
Juan Rodriguez-Andina, Spain
Enrique Romero, Spain
Jose de la Rosa, Spain
Pierluigi Siano, Italy
Fernando Silva, Portugal
Adolfo Steiger-Garção, Portugal
Sasu Tarkoma, Finland
João Manuel Tavares, Portugal
Klaus-Dieter Thoben, Germany
Stanimir Valtchev, Portugal
Manuela Vieira, Portugal
Ricardo Vigario, Finland
Dmitri Vinnikov, Estonia
Wuqiang Yang, UK

Organizing Committee (PhD Students)

Pedro Arsénio
Nuno Barrento
António Furtado
Ali Gharbali
João Guerreiro
Fábio Januário
Luis Romba Jorge

Rui Lopes
Catarina Lucena
Ricardo Mendonça
Carlos Oliveira
Eduardo Ortigueira
Hugo Serra
Nuno Vilhena

Technical Sponsors

 Society of Collaborative Networks

IFIP WG 5.5 COVE
Co-operation infrastructure for Virtual Enterprises
and electronic business

 IEEE-Industrial Electronics Society

Organizational Sponsors

 UNINOVA

Organized by:

PhD Program on Electrical and Computer Engineering FCT-UNL.

Table of Contents

Part V: Monitoring and Supervision Systems

Part VI: Advances in Manufacturing

Part VII: Human-Computer Interfaces

Part VIII: Robotics and Mechatronics

Part IX: Petri Nets

Part X: Multi-energy Systems

Part XI: Monitoring and Control in Energy

Part XII: Modeling and Simulation in Energy

Part XIII: Optimization Issues in Energy - I

Part XIV: Optimization Issues in Energy - II

Part XV: Operation Issues in Energy - I

Part XVI: Operation Issues in Energy - II

Part XVII: Power Conversion

Part XVIII: Telecommunications

Part XIX: Electronics: Design

Part XX: Electronics: RF Applications

Part XXI: Electronics: Devices

Part I
Introduction

Towards Collective Awareness Systems

Luis M. Camarinha-Matos, João Goes, Luis Gomes, and João Martins

Department of Electrical Engineering, Faculty of Sciences and Technology,
Universidade Nova de Lisboa, 2829-516 Caparica, Portugal
cam@uninova.pt

Abstract. Progress on smart interconnected devices and sensors allow access to large amounts of real time information coming from multiple sources, allowing reaching new levels of awareness and creating opportunities for new applications. Motivating engineering PhD students to look into the potential of this new wave of research is a crucial element in their education. With this aim, the doctoral conference DoCEIS'14 focused on technological innovation for collective awareness systems, challenging the contributors to analyze in which ways their technical and scientific work could contribute to or benefit from this paradigm. The results of this initiative are briefly analyzed in this chapter.

Keywords: Collective Awareness, Internet of Things, Cyber Physical Systems.

1   Introduction

Fast progress on devices with increased computational capabilities and connected to computer networks, namely to Internet, give users access to unprecedentedly high amounts of real time information coming from a large diversity of distributed sources. This situation greatly extends our capability of being aware of the surrounding environment and expands the reach-ability of our perception horizon.

The multi-dimensional nature of the accessible information sources and their growing pervasiveness opens the opportunity for new levels of awareness, and thus the development of new applications, virtually in all domains of human activity.

In this context, a large variety of research challenges emerge, ranging from devices development, communication mechanisms and protocols, to high-level information interpretation, and novel services development. The very nature of the involved areas, combining the physical and the cyber worlds, requires a multi-disciplinary approach, in which the competencies typically covered by the disciplines of Electrical and Computer Engineering play a relevant role.

Considering that a substantial amount of technological innovation results from the research works of engineering PhD students, it is important to call the attention of students in this area, which typically tend to focus on a specific research topic, for the potential of the "interconnected information sources" and the role they can play in the innovation process. The DoCEIS'14 (5th Doctoral Conference on Computing, Electrical and Industrial Systems) was thus organized with this mission and some of the results are summarized in this paper.

L.M. Camarinha-Matos et al. (Eds.): DoCEIS 2014, IFIP AICT 423, pp. 3–10, 2014.
© IFIP International Federation for Information Processing 2014

2   Related Concepts and Trends

The term collective awareness represents a concept that has been evolving since the 19th century [1], originally related to the works of the French sociologist Émile Durkheim, and referring to the common norms, values, and beliefs shared by the members of a community. With the developments of the information and communication technologies, the concept has been revisited and rephrased under the perspectives of human-computer interaction, computer-supported collaborative work, virtual communities, and information management systems. One example reflecting this evolution is the definition provided by Daassi and Favier [1]: "*Collective awareness refers to a common and shared vision of the whole team's context which allows members to coordinate implicitly their activities and behaviors through communication*".

Depending on the adopted perspective, different definitions have been proposed, which might be seen as different types of awareness. In this way, some authors have attempted to also classify the different types of awareness. For instance, Daassi et al. [2] consider: Activity awareness, Process awareness, Availability awareness, Informal awareness, Workspace awareness, Social awareness, and Group-structural awareness.

Progress on web-based technologies, social networks, wireless sensor networks, and the proliferation of smart devices connected to Internet, are inspiring other views and extending the concept of awareness to other directions. For instance, the FutureICT project [3] elaborated a research roadmap for what they call *Planetary Nervous System*, which addresses awareness issues at the world-scale. The roadmap addresses the needed scientific and technological developments based on three pillars: social sensing, social mining, and the idea of trust networks and privacy-aware social mining. In the framework of the European Research Programs FP7 and Horizon 2020, a particular focus on using collective awareness to solve big societal problems has emerged under the term "Collective Awareness Platforms for Sustainability and Social Innovation". Such platforms are defined as "*ICT systems leveraging the network effect … for gathering and making use of open data, by combining social media, distributed knowledge creation, and Internet of Things. They are expected to support environmentally aware, grassroots processes and practices to share knowledge; to achieve changes in lifestyle, production and consumption patterns; and set up more participatory democratic processes*" [4].

Also as a result of recent technological developments, a number of other related terms have emerged to represent partially overlapping perspectives or focused application contexts. Some examples:

- **Context Awareness**: A term frequently associated to mobile smart devices and regarded as the capability to both sense and react based on their environment and circumstances of events. Examples of characteristics of a context include location (where), time (when), identity (who), and activity (what). Other features / information can nevertheless be considered. The notion has been generalized in context-aware computing to the capability of understanding the *context of use* [5]. A survey of context-aware systems can be found in [6].

- *Ambient Intelligence*: A concept that represents electronic-enhanced environments, which are sensitive and responsive to the presence of people [7]. Therefore it builds upon the notions of pervasive computing, embedded systems, context awareness, and human-centric computer interaction. One of the relevant application areas is the so-called ambient assisted living (AAL), which uses technology to assist elderly. In these environments technology becomes invisibly embedded in our natural surroundings, present whenever we need it.

- *Collective Intelligence*: A notion representing the shared intelligence of groups, which emerges as a result of interactions (e.g. collaboration and competition) among group members. It is a form of distributed intelligence and implies some form of collective awareness (members' awareness of each other, awareness of the environment or task at hands and its progress, etc.). Internet developments, allowing for shared information and rapid information flows, organization and operation of communities, contributes to the emergence of many examples of collective intelligence (web-enabled collective intelligence) [8].

- *Big data*: A term that refers to massive data sets and stream computing that due to their large size and complexity are beyond the capabilities of traditional databases and software techniques. The expansion of sensing capabilities, sensor networks, smart devices, and other sources, generating huge amounts of data, motivates the importance of the topic. Interest in big data has emerged in science (e.g. medicine, astronomy and spatial exploration), business (e.g. understanding market trends, social media analytics, energy management in smart grid), security (e.g. analyzing surveillance data, identifying hidden networks), and many other [9]. Context awareness or collective awareness may require the adoption of techniques being developed for big data.

- *Internet of Things* (IoT): Understood as "*a dynamic global network infrastructure with self-configuring capabilities based on standard and interoperable communication protocols where physical and virtual "things" have identities, physical attributes, and virtual personalities and use intelligent interfaces, and are seamlessly integrated into the information network*" [10]. In this context, a "thing" can be understood as a real/physical or digital/virtual entity that exists and moves in space and time and is capable of being identified [11]. A large class of objects connectable to Internet constitutes (smart) sensors or (smart) devices able to provide status information, thus an important enabler for collective awareness.

- *Internet of Events*. Less popular than IoT, it corresponds to a perspective of the IoT that puts the emphasis on time dependency and discrete events handling [12]. As such, events modeling and management, time critical reactivity, and process modeling and supervision are the relevant issues here.

- *Sensing Enterprise*: A notion introduced by the FInES (Future Internet Enterprise Systems) cluster of projects [13] to refer to an enterprise anticipating future decisions by using multi-dimensional information captured through physical and virtual objects and providing added value information to enhance its global context awareness. In other words, this notion particularly focuses on enriching enterprises' context awareness through intelligent, interconnected and interoperable smart components and devices to power enterprise systems, making them responsive to events in real time and aiming at reaching seamless transformation of (raw) data to (tailored) information and (experienced) knowledge.

Each one of these concepts share a number of facets with the notion of collective awareness, namely sensing / acquisition of multi-source distributed data, interpretation / mining of these data, structured communities or ecosystems, and some forms of collaboration.

When the term "system" is added to the picture, it typically refers to a kind of cyber-physical system which not only considers the interlinking of the physical and cyber worlds, but also supports a community of interacting entities, helps integrating and analysing rich multimodal information sets towards reaching a common and shared vision of the surrounding environment and situations, and facilitates collaboration among these entities. Such system is thus an instrument for achieving collective awareness.

3   Example Contributions

Collective Awareness Systems (CAS) involve a set of topics that are particularly relevant for Electrical and Computer Engineering (ECE) researchers and professionals. In the last decades, the scope of ECE has expanded so widely that it risks some "fragmentation". Most professionals (and students) focus on a specialization sub-field, rarely mastering a comprehensive view of the whole field. However, most contemporary challenges in our society require the capability to address problems from a multi-disciplinary perspective. Since Collective Awareness Systems require a "strong dialogue" among the various sub-fields of electrical and computer engineering (and other areas), it represents a particularly interesting subject to bring them together.

Under this assumption, a challenge was presented to DoCEIS'14 conference participants [14], which are doctoral students from various countries and different sub-fields, as summarized in the following alternative questions:

  − *In which aspects your research can contribute to the development of the Collective Awareness Systems?*
or
  − *In which aspects your area of work could be affected / influenced in the future by the development in Collective Awareness Systems?*

As a result, all contributors made an effort to analyze the relationship between their specific research work and the *Collective Awareness Systems*. Among the accepted papers there is an almost balanced distribution between those that can benefit from CAS adoption and those that contribute to the development of support technologies and models for CAS [14], as summarized respectively in Fig. 1 and Fig. 2. The figures also include the percentage of contributions in each specific sub-field.

Energy management

[62.2%]

- Better energy management / informed decision making through **effective customer involvement**
  - Defining energy consumption & production patterns
  - Active role of customers in energy market & better reaction to market fluctuation
  - Empowering and motivating citizens to make informed decisions as consumers towards collective environmentally sustainable behavior
  - Shifting demand peaks to off-peak periods
  - Facilitate collective decisions regarding renewable energy sources
- Effective **support for smart grids**, which strongly depend on data
  - Use of real-time data from home smart appliances to better energy monitoring / management
  - Collect data on power quality, behavioral patterns of devices connected to the grid
  - Fast detection of power outages, mitigation of faulty power quality
  - Help managing smart multi-energy systems through more precise information on consumption
  - Better interation between smart grid and electric vehicles - bidirectional flows of energy and information
- Facilitate **optimization** approaches
  - Minimize energy production with pollutant fossil fuels and maximize production based on renewable sources
  - Facilitate coping with non-linear loads
  - Optimize placing switches on the grid based on higher information awareness
  - Optimal scheduling based on dynamic prices
  - Best bidding approaches in energy markets

Manufacturing

[24.3%]

- Improve **self-organization and adaptation capabilities** of manufacturing systems
  - Creating collective awareness on shopt floor logistics, facilitating self-configuration abilities
  - Facilitate fault tolerance / fault mitigation through reconfiguration
  - Collecting data on products and processes to better cope with volatile business opportunities
- Allow **collaborative supervision and control** systems
  - Bio-inspired view of manufacturing systems as collectives of autonomous intelligent entities that interact to achieve common goals
  - Muti-agent based control systems, allowing agents to make informed local decisions
  - Intelligent shopfloor entities able to reason and make decisions according to contextual information, leading to collective behavior
  - Collaboration among sub-systems in disperse manufacturing systems
- Faciliate **interactions between different engineering areas**
  - Interactions between product design and production / assembly lines
  - Increase awareness on organizational management through competencies modeling

Other areas

[13.5%]

- Facilitate **collaboration in enterprise networks**
  - Allow informed distribute decision making
  - Support more effective negotiation processes
- **Optimization** of resources and processes in farm management
- **Tracking** mobile equipments / robot **positioning**
- **Road safety**: improving communication between vehicles

Affected by / benefiting from Collective Awareness Systems

Fig. 1. DoCEIS'14: Benefiting from Collective Awareness Systems

In terms of areas that can benefit from a collective awareness systems approach, energy management is the most represented in terms of submissions to the conference. This is certainly a result of the current challenges faced by society regarding the development of sustainable energy solutions. The materialization of smart grids strongly relies on ICT and sensing / measuring capabilities in order to shift the emphasis from the traditional "predict and supply" model to a more flexible and responsive "demand-based" model [15]. The full realization of this idea requires highly distributed cyber-physical functionalities, but also the involvement of the customers in close (collaborative) interaction with providers, a process that should go far beyond the traditional client-supplier relationship. The concept of collective awareness can help reaching a shared vision of long-term sustainability and facilitate more effective use of resources.

Another area indicated as major potential beneficiary is manufacturing. Achieving truly agile manufacturing systems, able to dynamically adjust to market dynamics, requires novel approaches that take the shop-floor as a collaborative ecosystem of (progressively more) intelligent and autonomous machines / resources. The implementation of self-organizing / adaptive capabilities, and collaborative supervision systems, requires reaching a high-level of collective awareness and even collective intelligence among the members of this ecosystem. On the other hand, there is a need for stronger synergies among the various engineering areas involved in product design, manufacturing system design, manufacturing system deployment, manufacturing system operation, etc., which can also benefit from the conceptual insights of collective awareness systems.

Although less represented in this conference, several other areas can benefit from CAS, e.g. collaborative enterprise networks, logistics, road safety, etc.

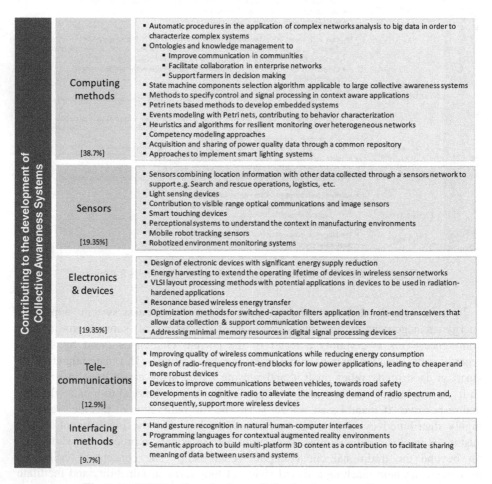

Fig. 2. DoCEIS'14: Contributing to Collective Awareness Systems

In terms of conceptual and technological contributions to the development of CAS, DoCEIS'14 includes a vast number of elements, ranging from the hardware level (electronic devices, sensors, telecommunication devices and systems), to software (interfacing and computing methods and models), including specific approaches for embedded systems design.

Perhaps the weakest part is the development of intelligent functionalities to reach high levels of awareness and collaboration, as well as proper consideration of safety and privacy issues.

On the other hand, many of these developments are designed in the context of vertical / specialized applications. An urgent need is thus to develop more generic platforms and components (de-verticalization) that can be applied across various domains. On the methodological side, there is also a need to invest more on engineering methods for design, deployment and operation of such systems.

Given the scope of DoCEIS, the mentioned contributions are naturally biased by an engineering perspective. The development of advanced collective awareness systems would, however, require the involvement of other disciplines.

4   Concluding Remarks

The conceptual framework and developments of collective awareness systems are likely to have a strong impact in many sectors of society. Some of these impacts are already visible, but their importance is likely to grow in the coming years.

The needed technological components and approaches to connect the physical and the cyber worlds, as well as required modelling and intelligent functionalities, open important expansion opportunities for the area of Electrical and Computer Engineering. This is clearly reflected in the contributions to the DoCEIS'14 doctoral conference.

Acknowledgments. This work was supported in part by the GloNet project (FP7 programme) funded by the European Commission.

References

1. Daassi, M., Favier, M.: Groupware and team aware. In: Dasgupta, S. (ed.) Encyclopedia of Virtual Communities and Technologies, pp. 228–231. Idea Group Inc. (2006)
2. Daassi, M., Daassi, C., Favier, M.: Integrating visualization techniques in groupware interfaces. In: Dasgupta, S. (ed.) Encyclopedia of Virtual Communities and Technologies, pp. 279–284. Idea Group Inc. (2006)
3. Giannotti, F., Pedreschi, D., Pentland, A., Lukowicz, P., Kossmann, D., Crowley, J., Helbing, D.: A planetary nervous system for social mining and collective awareness. The European Physical Journal Special Topics 214, 49–75 (2012)
4. Sestini, F.: Collective awareness platforms: Engines for sustainability and ethics. IEEE Technology and Society Magazine 2012, 54–62 (2012)

5. Schmidt, A.: Context-Aware Computing: Context-Awareness, Context-Aware User Interfaces, and Implicit Interaction. In: Soegaard, M., Dam, R.F. (eds.) The Encyclopedia of Human-Computer Interaction, 2nd edn., The Interaction Design Foundation, Aarhus (2013), http://www.interaction-design.org/encyclopedia/context-aware_computing.html

6. Baldauf, M., Dustdar, S., Rosenberg, F.: A survey of context-aware systems. Int. J. Ad Hoc and Ubiquitous Computing 2(4), 263–277 (2007)

7. Cooky, D., Augustoz, J., Jakkula, V.: Ambient Intelligence: Technologies, Applications, and Opportunities. Pervasive and Mobile Computing 5(4), 277–298 (2009)

8. Malone, T.W., Laubacher, R., Dellarocas, C.: The collective intelligence genome. MIT SLOAN Management Review 51(3), 21–31 (2010)

9. McLelland, C.: Big data: An overview. ZDnet (October 2013), http://www.zdnet.com/big-data-an-overview-7000020785/

10. Sundmaeker, H., Guillemin, P., Friess, P., Woelfflé, S. (eds.): Vision and Challenges for Realising the Internet of Things. CERP-IoT, European Commission (2010)

11. Camarinha-Matos, L.M., Goes, J., Gomes, L., Martins, J.: Contributing to the Internet of Things. In: Camarinha-Matos, L.M., Tomic, S., Graça, P. (eds.) DoCEIS 2013. IFIP AICT, vol. 394, pp. 3–12. Springer, Heidelberg (2013)

12. Ortner, E., Schneider, T.: Temporal and Modal Logic Based Event Languages for the Development of Reactive Application Systems. In: Proc. 1st International Workshop on Complex Event Processing for the Future Internet, Vienna, Austria, September 28-30 (2008)

13. FInES, Embarking on New Orientations Towards Horizon, Position Paper, European Commission (2013), http://www.fines-cluster.eu/jm/Documents/Download-document/409-FInES-Horizon-2020_Position-Paper-v2.0_final.html

14. Camarinha-Matos, L.M., Barrento, N., Mendonça, R. (eds.): Technological Innovation for Collaborative Awareness Systems. IFIP AICT, vol. 423. Springer, Heidelberg (2014)

15. Pitt, J., Bourazeri, A., Nowak, A., Roszczynska-Kurasinska, M., Rychwalska, A., Santiago, I.R., Sanchez, M.L., Florea, M., Sanduleac, M.: Transforming Big Data into Collective Awareness. Computer 46(6), 40–45 (2013)

Part II
Collaborative Networks

Negotiation Support in Collaborative Services Design

Ana Inês Oliveira[1,2] and Luis M. Camarinha-Matos[1,2]

[1] CTS, Uninova, 2829-516 Caparica, Portugal
[2] Faculdade de Ciências e Tecnologia, Universidade Nova de Lisboa, Campus da Caparica,
Quinta da Torre, 2829-516 Monte Caparica, Portugal
{aio,cam}@uninova.pt

Abstract. Due to the continuous changes in markets and society, companies and organizations have to constantly adapt themselves to maintain competitiveness. One of the options is to strategically join their competencies to rapidly respond to a business or collaboration opportunity through a goal-oriented network. Nevertheless, as market demands are continuously and fast evolving, besides taking advantage of clustering themselves, companies and organizations have to find better solutions to meet the customer needs. In this context, this paper presents an approach based on a negotiation support environment that enhances the design of new business services under a collaborative perspective, so that the customer's requirements can be better fulfilled.

Keywords: Collaborative Network, Negotiation Support Environment, Business Services, Co-Design, Service Co-Creation Network, Service Design.

1 Introduction

During the last decade, in manufacturing and service industries, collaboration among small and medium enterprises (SMEs) has focused on competencies and resources sharing as an approach to both create new competitive environments, as well as achieve agility to rapidly answer market demands. Working in collaboration implies on one hand sharing the opportunities and gained profits, but on the other hand also sharing the risks and losses, which in turn increases the survival chances of SMEs. Moreover, one market trend is to have more and more customizable products, so companies are forced to change their paradigm from mass-produced products towards highly customized products and ultimately one-of-a-kind products [1]. This trend is reflected in the term mass-customization that refers to products and services which meet the needs/choices of each individual customer with regards to the variety of different product features.

In GloNet project [2] the design, development and deployment of an agile virtual enterprise environment for networks of SMEs involved in highly customized and service-enhanced products through end-to-end collaboration with customers and local suppliers (co-creation of products and business services) is envisaged [1]. Collaboration among manufacturers, customers and local suppliers can be improved if a methodological approach is followed to design new business services to enhance the

L.M. Camarinha-Matos et al. (Eds.): DoCEIS 2014, IFIP AICT 423, pp. 13–20, 2014.
© IFIP International Federation for Information Processing 2014

existing products. This leads to the notion of co-design, where all relevant stakeholders are collectively involved in the design process of a new service [3]. In addition, to support the network of partners (a virtual organization, VO) in achieving agreements during the design of a new business service, it is desirable to have a negotiation environment that facilitates collaboration, and decreases the risks and amount of time spent in this process. The achieved agreements will be the basis for the implementation of the designed service [4, 5].

Considering this background and a virtual organization breeding environment (VBE) [6] context, that supports and fosters the creation of dynamic VOs, one relevant research question that emerges is:

How can an electronic negotiation support environment increase the agility in the creation process of successful dynamic virtual organizations?

One important motivation for this work is that by contextualizing the design of a new business service in a collaborative environment, it can also use the same negotiation support environment mechanisms to reach agreements, as the ones that are used for the VO creation process in the VBE environment. Therefore, in order to give an answer to this assumption, this paper presents a service co-design negotiation support system that is being designed and developed in the scope of the GloNet project and part of a PhD research work. The paper is then organized as follows: Section 2 identifies the relationship to collective awareness systems; Section 3 gives a brief overview of related literature namely on business services and services composition, the GloNet co-creation network characteristics, and service design; Section 4 illustrates the service co-design negotiation support system; and Section 5 presents the conclusions and future work.

2   Relationship to Collective Awareness Systems

Collective awareness systems refer to systems that intend to interrelate several sources of data and provide users with relevant data so that they can take better informed and sustainable decisions [7]. These systems harness concepts from other fields such as: the Internet of Things for data collection; Social Networks for interactions; and Wikis for the coproduction of new knowledge.

As previously mentioned, in the manufacturing and service industry, the trend is to move towards a global networked economy, being therefore important to build a collective network context awareness so that it is possible to share business, knowledge and collaboration. With this support it is possible to build alliances to develop creativity [8]. Considering the co-design of new business services where different and diverse stakeholders (including the customer) intervene in the process to achieve new business services design, the collective awareness systems appear with relevance once they can take more accurate and innovative decisions based on the environmental collected data.

3   Context and Related Areas

To contextualize the service co-design negotiation support, this section gives a brief overview of related literature.

Business Services and Composite Business Services. A business service refers to an organized set of added value activities from a business perspective [9], considering issues such as the delivery conditions, service level agreements, period, availability, etc. [1]. It corresponds to the service provided to the customer, and can be implemented by manual services and/or software services whose flow that can be modeled in different business process. Business services delivered to the customer can be composed of several atomic business services. The service providers of such business services together can form a virtual organization to deliver the composite business through a new entity that is the service integrator (that acts as the service provider of the composite business service) [10]. In GloNet, the design of new business services can include this composition of services.

GloNet Co-creation Network. One of the relevant business scenarios identified in GloNet is aimed at providing an environment that supports and promotes the collaborative design of business services. It includes the aspects of mass customization as well as the emergence of new products and solutions to identified needs, through collaboration between manufacturers, customer and members of the customer's community [1]. Therefore, this network (that is a VO) aims at co-design and co-innovate new value-added services for products. During the life-time of a certain product, several service co-creation networks might be created depending on the number of times new promising business ideas come up for new services [11]. Also, an analysis of the IT implementation requirements of those business applications can be performed [12]. This network shall be based on a collaboration environment that helps designing and providing business services based on innovation, knowledge and customer orientation, through collaboration between the different stakeholders (manufacturers, customer and members of the customer's community - open innovation approach).

Service Design. It is an interdisciplinary area that connects all relevant stakeholders in the design of services through some methodological approach [13]. Service design aims at designing user-oriented services making them useful, effective and different from existing ones. It potentiates co-creation between the different users of a service, and the providers [14]. In the different service design activities, especially in co-creation, not all stakeholders need to be involved at the same time; they can be involved just in the specific moments or situation in which they take part [3], which applies to the co-creation and co-design case of GloNet. There are a number of methods and tools for service design that have been emerging [15]; some of which can be found in http://www.servicedesigntools.org/. Nevertheless, some of these *tools* are just manual methods to organize a collaborative process, others use software tools, but no integrated environment is available, neither any integration between service design and service delivery environments is available. GloNet project is not focused on designing business services but rather on creating a collaborative environment where new multi-stakeholder services can emerge and be provided.

4   Research Contribution and Innovation

Negotiation in collaboration plays an important role as it can provide a support environment for the formation of VOs [6, 16, 17], and essentially improve the entire process of establishing a VO agreement that can lead to the governing rules and principles of the consortium during the operation phase. When designing new business services, there is already a consortium (acting like a VO) with the aim of reaching agreements on the design requirements of the new business service. Therefore, a negotiation support environment can contribute to boost the participation in the business service design, namely if the decision making process is fundamentally made by human actors. This is done by providing them with support functionalities, specifically: basic collaboration, data storage, editing, templates management, notary and traceability. As such, there are similarities to a negotiation support environment for the VO formation phase as already supported in the current PhD research work [4, 5, 18]. As a result, the current work on negotiation support for the design of new business services also supports the adopted research hypothesis:

The process of creating dynamic virtual organizations can become more agile if an appropriate electronic negotiation wizard environment is established with the necessary soft modeling characteristics to structure and conduct the entire negotiation process, making it traceable, reducing the collaboration risks, and managing the participants' expectations. Moreover, the negotiation environment should be customizable according to different collaboration levels, either in terms of commitment or in terms of duration.

Nevertheless, the negotiation support environment for the design of new business services, represents a different collaboration level and improves the entire process of reaching agreements on the design of new business services so that they can be properly developed and/or implemented. To design new business services, a service design methodology has been adopted [18]. As mentioned, a number of methods and tools for service design can be found in different disciplines and for different purposes [15]. Nevertheless, most of the tools are simply manual methods to organize collaboration processes. In GloNet, the aim is to use a service design methodology and adapt it to a philosophy of co-design where new business services can be designed in a collaborative environment with multi-stakeholders and customer involvement. The main steps of the approach are summarized in Table 1.

Table 1. Service Design Methodology in co-creation teams

Service Design steps	Description
1 Identify needed service	Brainstorming exercise involving an analysis of the needs and characteristics of the customer.
2 Design touchpoints diagram	To identify user interaction points with the service.
3 Design blueprint diagram	To describe the nature and the characteristics of the service interaction in enough detail to verify, implement and maintain it. It includes: temporal order, timings, and line of visibility (denoting what the customer sees and *back-office*).
4 Storyboard / storytelling	A tool derived from the cinematographic tradition; it is the representation of use cases through a series of drawings or textual description, put together in a narrative sequence that illustrates a sequence of events such as a customer journey.
5 Service prototyping	Involving the selection, assembly and integration of the various service components (atomic services).

For this purpose, it is therefore important to identify: who are the participants; the touchpoints with the customer; and how the participants can share information and documentation between them. Table 2 below addresses some of the relevant characteristics of service design and how they are related to GloNet co-design.

Table 2. Service Design Methodology in co-creation teams

	Service Design relevant characteristics	*Relevance for co-creation teams*
Participants	Service design assumes the involvement of various participants from different backgrounds and specially the interaction with the customer.	Co-creation of a new service is expected to involve a temporary collaborative network (VO), including different stakeholders, from geographically dispersed manufacturers, to the providers and supporting institutions close to the customer. The customer is also an active part.
TouchPoints	In service design it is particularly relevant to identify the customer journey in the process of receiving the service, and thus the points of interaction with the service provider.	Aiming user-centered services and being the customer an active part of co-design, it is very important to consider his interactions with the service, namely the moments and places that he gets into direct contact with the service.
Sharing	Service design methods, even if not supported by software tools, a shared space where all participants can visualize the progression of the design process is assumed.	Collaborative environment where the involved participants can interact in the design and creation processes and reach agreements.

Service Co-design Negotiation Support System. The *Services Co-Design Negotiation Support (CoDeN)* system is intended to provide a collaborative environment for design of new business services where the various involved participants can reach agreements on what is decided. The involved participants (including the customer) in this process are defined a priori. Similar to a negotiation support system for VO creation [18], this system is also intended to generate an agreement that represents the reached consensus. Nevertheless, unlike the VO formation negotiation, here there are no free negotiation topics. Instead, the consensus that has to be reached based on the service design methodology that serves as a guide for the negotiation. Examples of available templates can be found in [19]. Similar to other VOs, the entire process is conducted by the VO Planner that acts as the co-creation team mediator.

In Fig. 1, the main flow and interactions of the system are illustrated. The thick arrows represent the strong negotiation that is required in the co-design process and the numbers correspond to the steps of the approach summarized in Table 2. Also, this system directly interacts with: the negotiation support for agreement establishment that allows clients to exchange information with warranty of authenticity and validity as well as providing a safe repository for saving and requesting documentation.

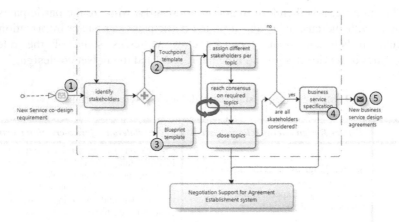

Fig. 1. Services co-design negotiation support system interactions

Below there is an adapted i* Rationale Strategic model where the involved actors as well as their dependency objectives with the system are illustrated. Within the boundaries of the *Services Co-Design Negotiation Support* system the respective tasks and sub-tasks are presented. The services that directly interact with the involved actors and other related systems are also depicted.

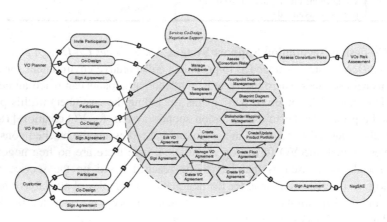

Fig. 2. Adapted i* Rationale Strategic Model for the Services Co-Design Negotiation Support system

Discussion of Results. The work presented in this paper is the result of a service co-design negotiation support system specification aimed at supporting the co-creation team mediator in the negotiation process of a new business service design. The support environment to achieve agreements on the business service design is based on software system that assists the human decision making and guides the generation of the final agreement (based on specific templates), and stores the participants' received input. In this way, the process becomes structured and traceable. Moreover, the intended environment is based on the same mechanism already specified for negotiation during a VO formation phase, meaning that it has to be adaptable to different types of collaboration. Through the interaction with other systems, namely: *(i)* support for agreement establishment, it is also possible to guarantee a degree of *authenticity*; and *(ii)* VO risks assessment is possible to reduce the risks in collaboration and increase adaptation to unexpected events because partners expectations are also considered. Through the on-going PhD research and participation in the European research project GloNet, the topic of new business service design gains more significance. Therefore, the test bed and validation of the current work is being considered in this context.

5   Conclusions and Further Work

To compete in the global market, where customer's requirements are becoming more specific and customizable, it is important that companies maintain efficiency and the sustainability level, so they tend to collaborate in order to accomplish those requirements of almost one-of-a kind products or services. One factor that is important is the agility that these companies can have to interoperate and co-work with each other in a co-creation environment. In order that the co-creation or co-design of a new business service becomes more feasible, it is important to have an advanced service to support the existing team, so that the negotiation process to establish agreements can be achieved. This paper summarizes an on-going research work that tries to fulfill the above requirements: the specification of the service co-design negotiation support system. As some of the concepts and preliminary results have already been positively validated, it is now foreseen to achieve a more advanced environment that comprehends most of the described areas with the needed adaptations to support the aimed negotiation support for dynamic VOs creation and new business services design. The emphasis of the current work is on the relations and interactions of the several sub-systems that together provide an advanced consortia formation support, namely: *(i)* the basic services for VO creation that are related to the order characterization and partners search; *(ii)* the base negotiation support with negotiation templates management and support for agreement establishment; and *(iii)* functionalities to reduce potential risks in collaboration. The ongoing validation process is intended to consist in peer validation and supported by the EU project GloNet.

Acknowledgments. This work has been supported by the *Collaborative Networks and Distributed Industrial Systems* Research Group of Uninova and partly by the GloNet project (FP7 programme) funded by the European Commission.

References

1. Camarinha-Matos, L., et al.: Collaborative Business Scenarios in a Service-Enhanced Products Ecosystem. In: Collaborative Networks in the Internet of Services, pp. 13–25. Springer (2012)
2. Camarinha-Matos, L.M., Afsarmanesh, H., Koelmel, B.: Collaborative networks in support of service-enhanced products. In: Camarinha-Matos, L.M., Pereira-Klen, A., Afsarmanesh, H. (eds.) PRO-VE 2011. IFIP AICT, vol. 362, pp. 95–104. Springer, Heidelberg (2011)
3. Sanders, E.B.-N., Stappers, P.J.: Co-creation and the new landscapes of design. Co-design 4(1), 5–18 (2008)
4. Oliveira, A.I., Camarinha-Matos, L.M., Pouly, M.: Agreement negotiation support in virtual organisation creation–an illustrative case. Production Planning & Control 21(2), 160–180 (2010)
5. Oliveira, A.I., Camarinha-Matos, L.M.: Negotiation Support and Risk Reduction in Collaborative Networks. In: Camarinha-Matos, L.M., Tomic, S., Graça, P. (eds.) DoCEIS 2013. IFIP AICT, vol. 394, pp. 15–24. Springer, Heidelberg (2013)
6. Camarinha-Matos, L., Afsarmanesh, H.: A framework for virtual organization creation in a breeding environment. Annual Reviews in Control 31(1), 119–135 (2007)
7. Sestini, F.: Collective Awareness Platforms: Engines for Sustainability and Ethics. IEEE Technology and Society Magazine 31(4), 54–62 (2012)
8. Di Maio, P., Ure, J.: An open conceptual framework for operationalising collective awareness and social sensing. In: Proceedings of the 3rd International Conference on Web Intelligence, Mining and Semantics. ACM (2013)
9. Brentani, U.: Innovative versus incremental new business services: different keys for achieving success. Journal of Product Innovation Management 18(3), 169–187 (2001)
10. Camarinha-Matos, L.M., et al.: Collaborative Business Services Provision. In: ICEIS 2013 – 15th Int. Conf. on Enterprise Information Systems, Angers, France (2013)
11. Camarinha, L.M., et al.: Supporting product-servicing networks. In: IESM 2013 – 5th Int Conf. on Industrial Engineering and Systems Management. IEEE Explore, Rabat (2013)
12. Yang, S., Zhang, S., Kong, L.: Research on the cooperative decision of business application needs in IT governance. In: 2012 IEEE 16th International Conference on Computer Supported Cooperative Work in Design (CSCWD). IEEE (2012)
13. Mager, B., Sung, T.: Special issue editorial: Designing for services. International Journal of Design 5(2), 1–3 (2011)
14. Sandberg, F.: Co-creating collaborative food service opportunities through work context maps. In: Proceedings of 3rd Service Design and Service Innovation Conference, ServDes. Linköping Electronic Conference Proceedings, vol. 67. Linköping University Electronic Press, Linköping (2012)
15. Wild, P.J.: Review of Service Design Methods. In: IPAS Project Deliverable I15.6. University of Cambridge, Cambridge (2009)
16. Stelmach, M., et al.: Distributed contract negotiation system for virtual organizations. Procedia Computer Science 4, 2206–2215 (2011)
17. Picard, W., Rabelo, R.J.: Engagement in collaborative networks. Production Planning & Control: The Management of Operations 21(2), 101–102 (2010)
18. Oliveira, A.I., Camarinha-Matos, L.M.: Electronic Negotiation Support Environment in Collaborative Networks. In: Camarinha-Matos, L.M., Shahamatnia, E., Nunes, G. (eds.) DoCEIS 2012. IFIP AICT, vol. 372, pp. 21–32. Springer, Heidelberg (2012)
19. Namahn, Window, Y.: Service Design Toolkit (2013),
   http://www.servicedesigntoolkit.org/ (November 07, 2013)

Research on Collaborative Processes in Non-Hierarchical Manufacturing Networks

Beatriz Andrés and Raul Poler

Research Centre on Production management and Engineering (CIGIP),
Universitat Politècnica de València (UPV), Plaza Ferràndiz y Carbonell, 2, 03801 Alcoy, Spain
{beaana,rpoler}@cigip.upv.es

Abstract. Collaborative networks research has increased over the last years due to the advantages experienced by the enterprises that establish collaboration. In this paper, a set of relevant collaborative processes are identified in the literature and analysed considering their specific application in non-hierarchical manufacturing networks (NHN). Besides, collaboration within networked partners is enhanced through designing a roadmap to deal with the migration path towards collaborative processes establishment. Currently, the research aim is focused on modelling the network and the collaborative processes, established within the partners, through a mathematical model that identifies the collaboration objects. In this context, experts will be able to promote the establishment of collaboration, analyse the processes performed within a network and consider their redesign/design towards collaboration. The model would identify the objects of collaboration, in order to be analysed and thus serve as a tool to improve the enterprises' performance and, consequently, the network performance.

Keywords: collaborative processes, non-hierarchical networks, collaborative networks, SMEs.

1 Introduction

The interest increase on collaborative processes establishment, within partners of the same network, has resulted in an increment on the emergence of the number of network topologies [1], [2]. Amongst all the network topologies this research is focused on collaborative non-hierarchical manufacturing networks (NHN) [3]. Unlike hierarchical networks (HN), based on centralised approaches of decision making in which one partner possesses all the power, NHN are characterised by the establishment of collaborative processes with decentralised decision making models (DDM). The establishment of collaborative DDM in NHN implies that all the network partners are autonomous; all decisional independent units are collaboratively involved in the management of the network processes and integrated with different degrees of collaboration. Involved partners equally enjoy power sharing and status, and no individual partner leads the network [4].

For enterprises whose nature does not imply a hierarchy in the network structure, conforming decentralised and collaborative structures, that is NHN, provide important

L.M. Camarinha-Matos et al. (Eds.): DoCEIS 2014, IFIP AICT 423, pp. 21–28, 2014.

benefits [5]. These benefits are led to improve the network competitiveness, innovation, partners' adaptability, customers' satisfaction and inefficient processes elimination [6]. DDM, in which the NHN are based on, improves each network node commitment as regards to the overall goal of the network while improving communication, collaboration and flows among nodes. Furthermore, NHN, as equally considers all network partners, helps SMEs to position in the global market [7]. NHN are characterised by long term partnerships with close collaboration. Accordingly, collaborative DDM changes the way how processes are executed in a network, implying an evolution towards collaboration, in which the exchanges of information are to be done in an interoperable way and the business processes are jointly performed. Thus, this paper particularly focuses on collaborative NHN.

1.1 Motivation and Research Question Formulation: Collaborative Processes in Non-Hierarchical Networks (NHN)

Research in collaborative non-hierarchical manufacturing networks is motivated, on the one hand, by the call funded by the European Commission, "FP7-NMP-2008-SMALL-2" (Activity code *NMP-2008-3.3-1: Supply chain integration and real-time decision making in non-hierarchical manufacturing networks*) [8]. On the other hand, the Intelligent Manufacturing Systems initiative (iNet-IMS) has encouraged to (i) analyse the needs that arise from relationships between SMEs belonging to NHN, (ii) analyse the technology innovation trends to support DDM, (iii) analyse standards for information exchange to support collaborative processes and (iv) define a framework for collaboration in NHN [7]. Moreover, the growing interest of enterprises on establishing collaboration has led to study the establishment of collaborative processes. Therefore, the development of a Collaborative Framework (proposing models, guidelines and tools) to handle with the barriers SMEs can encounter when establish collaborative processes, is considered the research line followed by the thesis. Taking into account the above said, the research questions are hereafter raised:

(i)   Identify, through the literature review, relevant collaborative processes that networked SMEs perform [6].
(ii) Analyse the collaborative processes treatment in order to determine how the approaches provided in the literature deal with the NHN characteristics and therefore can be applied to support the establishment of collaborative processes within NHN partners [3], [6].
(iii) Recognise the non-covered processes, as regards the solutions that do not treat the collaborative process in specific contexts of collaborative and decentralised NHN [3], [9].
(iv) Provide solutions in the NHN context to fill the literature gaps of non-covered collaborative processes. In order to fill these gaps, research on new approaches to model collaboration is to be considered from the NHN perspective [10], [11].
(v)   Build a Collaborative Framework containing a set of models (M), guidelines (G) and tools (T) to support SMEs on their participation in collaborative processes within decentralised NHN structures. The Collaborative Framework consists of a set of building blocks characterised by the type of collaborative process and the type of solutions (M, G and T) to cope with the process [6].

(vi) Develop a roadmap to support SMEs on the migration path from non-collaborative NHN towards collaborative NHN [12].

(vii) Model the collaborative network, NHN, and the objects identified to deal with the collaboration and formally conceptualise the collaborative processes performed. The formal model will serve researchers as a tool to analyse the degree of collaboration set up within groups of networked partners.

Once identified the research questions, the paper is devoted to identify the relationship between the research developed within the paper and the *Collective Awareness Systems* (section 2). Besides, the literature review results as regards collaborative processes are shown (section 3). The research contribution and innovation is proposed (section 4) and research results are discussed (section 5). Finally, conclusions and future research lines are identified (section 6).

2   Relationship to Collective Awareness Systems

Collective Awareness Platforms for Sustainability and Social Innovation (CAPS) are ICT systems leveraging the emerging "network effect" by combining open online social media, distributed knowledge creation and data from real environments in order to create awareness of problems and possible solutions requesting collective efforts enabling new forms of social innovation [13].

In collaborative NHN contexts CAPS can be adopted to support distributed enterprises to collaboratively act. CAPS can give the clue to achieve a situation in which independent partners establish collaborative processes and handle with the encountered barriers in a sustainable way. The decentralised decision making perspective of NHN requires greater exchanges of knowledge and information, and commitment of all companies without losing their authority in decision making. In the light of this, CAPS would provide a proper platform to support (i) the partners' information access in real-time and easily understandable, (ii) the consideration of both SMEs and the network, (iii) the access to environmental models and simulations to deal with collaboration and (iv) the collaboration promotion within the partners.

The main aim proposed in this paper is to achieve sustainable NHN, characterised by DDM through establishing sustainable collaborative processes. For these reasons, CAPS in NHN are to be considered to integrate all decisional units, perform greater exchanges of information, share responsibilities, involve all the SMEs in the network decision-making, tackle problems jointly and take into account the objectives of all the partners by equally considering all the networked nodes.

3   Literature Review: Collaborative Processes

The literature review carried out collects the diverse knowledge on the research arena of collaborative processes, both in HN and NHN contexts. The works analysed so far (1981-2012) reveal the existence of an extensive literature concerning collaborative processes. The literature review was carried out considering the most relevant processes to establish collaboration; the processes were reviewed considering the solution's approaches used to deal with them: models, guidelines and tools.

Afterwards, an analysis was completed in order to identify into which extent the treatment given in the literature was appropriate to be applied from the NHN perspective; that is the extent into which the approaches to support collaborative processes can be employed in decentralised and collaborative networks. Three classification criterion are used: (i) *NHN* when most of the contributions to deal with the collaborative processes are designed from the NHN perspective (ii) *HN → NHN* when most of the contributions are designed form the HN perspective but can be adapted to the collaborative and decentralised features that characterise the NHN and (iii) *HN* when most of the contributions in the literature are designed for HN.

The processes identified are classified according to the three decision making levels, strategic, tactical and operational and according to the solutions degree of application (NHN, HN→NHN and HN) (table 1).

Conforming to the classification criteria, the existence of some processes that are not specifically treated form the NHN perspective are identified.

The main conclusion deduced from the literature review is led to consider that there are a set of processes that still do not have a complete treatment in the literature due to the approaches to support the collaborative processes are not provided form the NHN perspective. These processes are identified considering those belonging to the group of *HN* and *HN →NHN* (see table 1).

Table 1. Collaborative Processes and Degree of Application [3], [6]

	STRATEGIC		TACTICAL		OPERATIONAL	
	(1) Network Design	HN→NHN	(1) Forecast Demand	NHN	(1) Scheduling	HN→NHN
	(2) Decision System Design	NHN	(2) Operational Planning	HN→NHN	(2) OPP	NHN
	(3) Partners Selection	HN→NHN	(3) Replenishment	HN→NHN	(3) Lotsizing Negotiation	HN
Collaborative Processes	(4) Strategy Alignment	HN	(4) Performance Management	NHN	(4) Inventory Management.	HN→NHN
	(5) Partners Coordination and Integration	HN→NHN	(5) Knowledge Management.	HN→NHN	(5) Information Exchange Management	NHN
	(6) Product Design	HN→NHN	(6) Uncertainty Management.	HN	(6) Process Connection	HN→NHN
	(7) PMS Design	NHN	(7) Negotiation Contracts among partners	HN→NHN	(7) Interoperability	NHN
	(8) Coordination Mechanisms Design	HN→NHN	(8) Share costs/profits	HN		
			(9) Coordination Mechanisms Management	HN→NHN		

4   Research Contribution and Innovation in Collaborative Non-Hierarchical Manufacturing Networks

According to the literature reviewed, a set of non-covered processes are identified as regards the collaborative NHN perspective. Therefore, the main aim is to design solutions to fill the gaps in those processes whose degrees of application are classified as *HN* and *HN→NHN*. The processes classified with the **HN** degree of application are the *strategy alignment, uncertainty management, share costs/profits* and *lotsizing negotiation*. An example is considered with a solution proposal for the *share costs/profits* process. The methodology, *Share Profits in Non-Hierarchical Networks* (SP-NHN), is designed to ensure equitable sharing among networked partners to foster collaborative and trust behaviours and deal with the gap of collaboratively *share costs/profits* [10]. The research also considers the processes classified in **HN→NHN**, in order to adapt the solutions provided in the literature and procure suitable solutions from the NHN perspective. An example can be found with the *operational planning* process, considered a complex activity in collaborative NHN due to the agreements and standardised processes demanded. In the light of this, a solution is adapted, from the literature considering the novel *Supply Chain Agent-based modelling Methodology that supports a Collaborative Planning Approach* (SCAMM-CPA), in order to handle with the problems associated when the *operational planning* is performed in NHN under a collaborative perspective [11].

Taking into account the literature reviewed and admitting that many factors and conditions may cause threats, specifically in SMEs, when establishing collaborative processes [3, 9] in decentralised NHN; a roadmap, to support the new SMEs challenges, is provided. The importance on collaboration within the networks under study encourages the development a roadmap to deal with the evolution from non-collaborative scenarios towards the establishment of collaborative processes among the SMEs decided to participate in decentralised and collaborative NHN. The roadmap, *NHNmap*, consists of a set of ten phases structured into four main areas (i) collaboration establishment, (ii) performance evaluation, (iii) solutions' proposal to overcome possible barriers appearing when collaborative processes are established, and (iv) information and technology systems to efficiently manage the decentralised decision making models that characterise the NHN [12].

An additional research contribution is identified with the modelling of collaborative networked processes, as well as modelling the NHN. This research line is the early stages of development.

The modelling research line is motivated due to, currently, the models proposed in the literature are based on individually model specific collaborative processes, such as interoperability, collaborative product development or knowledge management, amongst others. Therefore, a gap is found to provide a general only model that could be valid for all the identified collaborative processes. Thus, regardless the process to be modelled an only model could be used through identifying different objects that take part in each of the collaborative process. In the light of this, a new research line is proposed to design a collaborative model and identify the objects participating in the establishment of collaborative partnerships. The model is to be generally designed in order to be used by the vast majority of the collaborative processes previously defined in the literature reviewed [6]. The model is to be developed to support researchers on the formal conceptualisation of the collaborative processes, giving them an insight of (i) how to analyse the processes and measure the collaboration (ii) how to design a collaborative process, if this does not already exist or (iii) how to redesign a process, if this has not been already executed from a collaborative perspective, in order to globally improve the network performance and individually improve the enterprises' performance. An example can be encountered in [4], that models the network, the relations established within the networked partners, and the links between them to determine where the partners' transactions lead. A model is proposed to allow researchers to identify the power degree of each networked partner and therefore determine the power distribution. Once the network, the partners and the partners' relations are modelled, *Markov Chains* are used to compute the power distribution. Modelling the power and therefore the relationships of the network nodes allows to better consider the networked partners' relationships and obtain more sustainable and balanced networks.

5   Discussion of Results and Critical View

Taking into account the results obtained from the literature reviewed, it is evidenced that a high percentage of the approaches developed in the collaboration research field are mostly designed from the HN perspective, so do not have a full application on collaborative processes established by SMEs belonging to NHN. The lower complexity on HN treatment makes them more studied. Contrarily, research in collaborative NHN environments is less widespread due to researchers have to deal with companies that could be part of several production networks at the same time; what motivates the study of these networks. In the light of this, the research focuses on the creation and management of non-hierarchical manufacturing networks and the proposal of supporting approaches for SMEs to establish collaborative processes in networks characterised by DDM, in order to simultaneously deal with both the enterprises' objectives the and the global objective defined for the network.

Considering the outcomes from the literature review and the results based on the iNet-IMS initiative, relevant collaborative processes and barriers associated to SMEs, when participating in NHN, are identified. A Collaborative Framework is built as a set of solutions classified into models, guidelines and tools to cope with the SMEs requirements when participating in collaborative NHN. The research developed provides, researchers and practitioners, an approach to handle with the changes needed to be performed by the SMEs in order to efficiently achieve the desired future state of establishing collaborative processes in NHN, and, therefore, get the benefits derived from the collaboration.

Furthermore, the design of a collaborative model allows to (i) model the network and the collaborative processes, (ii) quantitatively analyse the collaborative relationships, (iii) determine how the network system is, AS-IS and (iv) propose a collection of tools and guidelines to cope with collaborative barriers so as to improve the establishment of collaborative processes within the involved parts of the NHN. A quantitative analysis on the established collaborative processes is needed to identify the most relevant processes performed and those that are lagging behind but indeed are important to be performed. The quantification will determine how the system is and how important is to make improvements throughout the network.

Dealing with the NHN and the barriers associated with the collaborative processes establishment is a laborious task due to the added difficulty to individually consider each of the companies with its objectives, strategies and particularities. Moreover the existence of conflicting objectives appearing, due to companies belong to more than one network, are to be taken into account. This research work is a step forward in the study of real networks consisting of autonomous SMEs and deals with the next generation of manufacturing enterprises embedded in global environments characterised by multi-lateral collaborations.

6   Conclusions and Further Work

Research in NHN has been launched only few years ago therefore there is a long way of work to cover in the future. This research is focused on identifying those processes that can be collaboratively established within a NHN and modelling the NHN in order have a current view on the processes performed in a NHN, with the main aim to improve the partners collaboration through solutions procurement. The Collaborative Framework for NHN will give researchers a tool to, once identified the barriers when collaborative processes are established, overcome the appearing weakness through the application of models, guidelines and tools provided.

The high interest on the topic under research and the wide variety on the approaches to deal with collaboration, leads the research contributions to: (i) summarise the existing knowledge regarding the establishment of collaborative processes within networks, specifically in NHN; (ii) provide a roadmap to overcome the possible barriers appearing when SMEs decide to participate in collaborative NHN and (iii) design a formal model defining the collaboration objects to allow researchers to specifically model the collaborative processes and the network.

Future research lines are led to provide solutions amongst all the processes that are not treated form the NHN perspective; for processes with **HN** degree of application solutions will be designed; for processes with **HN→NHN** degrees of application existent solutions in HN would be adapted to be applied in NHN contexts. The results obtained so far reveal the need of providing a collaborative framework to cope with those processes that are treated in the literature without considering NHN features. The work developed so far has provided a set of solutions to overcome the collaborative processes form the NHN perspective [10], [11], [12]. This research begins a series of solution proposals to build a Collaborative Framework for NHN, designing new models, guidelines and tools to support the establishment of collaborative processes in the networks under study. Future work is led to complete table 1 and achieve, for all the processes, **NHN** degrees of application.

The standardisation of the collaborative processes established within the networked SMEs is to be done in order to consider the information exchange systems in an interoperable way and the negotiation mechanisms to regulate the collaboration within the network.

Considering the aforementioned, future work is led to (i) identify or design models guidelines and tools to deal with collaborative processes in NHN contexts, (ii) consolidate collaborative processes across a network through Information Systems and Technologies (SI / IT), (iii) promote collaboration through the implementation of the roadmap, (iv) validate the proposed Collaborative Framework through its application to various industrial pilots and (v) design a formal model for collaboration and implement it in real networks.

References

1. Camarinha-Matos, L.M., Afsarmanesh, H.: Collaborative Networks: A New Scientific Discipline. Journal of Intelligent Manufacturing 16(4), 439–452 (2005)
2. Camarinha-Matos, L.M., Afsarmanesh, H., Galeano, N., Molina, A.: Collaborative Networked Organizations – Concepts and Practice in Manufacturing Enterprises. Computers and Industrial Engineering 57(1), 46–60 (2009)
3. Andres, B., Poler, R.: Relevant problems in collaborative processes of non-hierarchical manufacturing networks. J. Industrial Engineering and Management 6(3), 723–773 (2013)
4. Andrés, B., Poler, R.: A Method to Quantify the Power Distribution in Collaborative Non-hierarchical Networks. In: Camarinha-Matos, L.M., Scherer, R.J. (eds.) PRO-VE 2013. IFIP AICT, vol. 408, pp. 660–669. Springer, Heidelberg (2013)
5. Stadtler, H.: A framework for collaborative planning and state-of-the-art. Or Spectrum 31(1), 5–30 (2009)
6. Andres, B.: Collaborative Processes Analysis in Non-Hierarchical Manufacturing Networks. In: Ortiz, A. (ed.) 6th International Conference on Interoperability for Enterprise Systems and Applications. Doctoral Symposium, pp. 13–22. Editorial Universitat Politècnica de València (2012)
7. Poler, R., Carneiro, L.M., Jasinski, T., Zolghadri, M., Pedrazzoli, P.: Intelligent Non-hierarchical Manufacturing Networks, 448 p. Iste, Wisley (2013)
8. European Commission. Work Programme. Cooperation Theme 4 Nanosciences, Nanotechnologies, Materials and New Production Technologies – NMP (2008), http://ec.europa.eu/research/participants/portal/download?docId=22687
9. Andrés, B., Poler, R.: Methodology to Identify SMEs Needs of Internationalised and Collaborative Networks. In: Emmanouilidis, C., Taisch, M., Kiritsis, D. (eds.) Advances in Production Management Systems, Part II. IFIP AICT, vol. 398, pp. 463–470. Springer, Heidelberg (2013)
10. Andres, B., Poler, R.: A Methodology to Share Profits and Costs in Non-Hierarchical Networks. In: Prado-Prado, J.C., García-Arca, J. (eds.) Annals of Industrial Engineering, Springer, London (in press, 2014)
11. Andres, B., Poler, R., Herández, J.E.: An operational planning solution for SME's in collaborative and non-hierarchical networks. In: Hernández, J.E., Liu, S., Delibǎic', B., Zaraté, P., Dargam, F., Ribeiro, R. (eds.) Logic of Programs 1983. LNCS, vol. 164, pp. 46–56. Springer, Heidelberg (1984)
12. Andrés, B., Poler, R.: A Roadmap Focused on SMEs Decided to Participate in Collaborative Non-Hierarchical Networks. In: Camarinha-Matos, L.M., Xu, L., Afsarmanesh, H. (eds.) Collaborative Networks in the Internet of Services. IFIP AICT, vol. 380, pp. 397–407. Springer, Heidelberg (2012)
13. Collective Awareness Platforms for Sustainability and Social Innovation, https://ec.europa.eu/digital-agenda/en/collective-awareness-platforms

A Knowledge Management Framework
to Support Online Communities Creation

Catarina Lucena[1,2], João Sarraipa[1,2], and Ricardo Jardim-Goncalves[1,2]

[1] Departamento de Engenharia Electrotécnica, Faculdade de Ciências e Tecnologia, FCT,
Universidade Nova de Lisboa, 2829-516 Caparica, Portugal
[2] Centre of Technology and Systems, CTS, UNINOVA, 2829-516 Caparica, Portugal
`{cml,jfss,rg}@uninova.pt`

Abstract. Nowadays, there are constant increases of people that actively produce and consume online content. They use various tools to create and share information with others, namely: blogs; forum; photo & video sharing and social networking sites. As a consequence the development of services or mechanisms for the acquisition of knowledge from Internet users is currently observed as crucial. However, most of such knowledge is often unstructured and is only available in a tacit form. Thus, this paper presents a framework to facilitate knowledge management to support communities' establishment. Its main objective is to build a community knowledge base from cyclical conversion of individual to organizational knowledge representations creation, and consequently, to allow specific reasoning and decision-making.

Keywords: Knowledge Management, Collective Awareness, Ontology.

1 Introduction

Social Web is a set of relationships that link together people over the Web [1]. Social Web encompasses how social software is designed and developed in order to support and foster social interactions [2]. This social software tools typically handle the capturing, storing and presentation of communication and focus on establishing and maintaining a connection among users [3]. Social Web focuses on human communication, and as consequence the vast majority of the information is in human readable format only. Thus software programs cannot understand and process this information, and much of the potential of the Web remains unusable. This leads to the necessity of having an approach or solution to concretize the semantic web concept.

The original *Scientific American* article on Semantic Web concept appeared on 2001 [4]. It described the evolution of a Web that consisted in a large amount of documents for humans to read and that included data and information for computers to manipulate [5]. Semantic Web can be seen as a set of technologies that help knowledge sharing across the web between different applications. Ontologies based technologies play a prominent role in the Semantic Web [6]. They make possible the widespread publication of machine understandable data, opening opportunities to automatic knowledge reasoning able to potentiate the generation of context awareness systems.

L.M. Camarinha-Matos et al. (Eds.): DoCEIS 2014, IFIP AICT 423, pp. 29–36, 2014.

In this paper an initial assessment related to the necessity of gathering knowledge from social web applications is made. This gathering and storing of knowledge in a formalized structure will make possible to software applications to enrich the knowledge retrieved from individuals, adjusting it to specific objectives enabling its systems to have the notion of a context. This will contribute, for instance, to help users to find others to interact with about a specific purpose, which will facilitate to online communities creation and also to enable them to have contextual awareness able to make decisions supported by the knowledge handled. Thus, based on this, it is then proposed an ontology-based framework to contribute to such online communities creation. Afterwards, it is presented an architecture based in an application scenario of the proposed framework, followed by some conclusions and future work statements.

2   Relationship to Collective Awareness Systems

In today's internet, many applications arise in the context of "Web 2.0", a concept popularized by O'Reilly Media [7]. Web 2.0 focuses in collaboration of users and sharing information between them. On the one hand, Web 2.0 focuses on enhance creativity, information sharing and collaboration among users. On the other hand, Semantic Web refers to the intelligent interaction among systems and applications by deploying ontologies, semantic annotation of Web content, and reasoning. Its ultimate goal is to make data understandable to computers [8]. Together they contribute for software applications with knowledge from social software. This leads to the Web 3.0 appearance, where Semantic Web technologies are integrated into Web 2.0 technologies, or powering, large-scale Web applications [9].

Folksonomies are inventive Web 2.0 tools, which aggregates both web 2.0 and semantic web characteristics being great for categorizing documents and resources in a collaborative way [10]. They arise when a large number of people are interested in particular information and are encouraged to describe it [5]. The term was originally defined by Thomas Vander Wal [11] as the result of personal free tagging of information and objects (anything with a URL) for one's own retrieval. The person consuming the information does the act of tagging.

By other hand, ontologies specify a conceptualization of a domain in terms of concepts, attributes, and relations [12]. Consequently, semantic web and its applications rely heavily on formal ontologies to structure data for comprehensive and transportable machine understanding. In addition, the Semantic Web's success is then dependent on the quality of its underlying ontologies [13], [14].

As mentioned before, social software normally shows limitations related to use or in the formalization of user's knowledge. For that reason a new framework to gather and formalize knowledge from social software, allowing reasoning and support to decision making is proposed (see Figure 1). This framework uses ontologies and folksonomy concept principles to enable knowledge management features, able to facilitate the achievement of collective awareness. Thus, this proposed framework represents the author's research, which contributes itself to collective awareness systems establishment. Collective awareness refers to a common and shared vision of the whole team's or community context that allows members to coordinate implicitly their activities and behaviors through communication [15]. Making decisions with awareness will reduce the effort to coordinate tasks and resources by providing a context in which to interpret utterances and to anticipate actions [16]. This is also an objective that the framework intends to contribute to, where collective awareness is supported by the knowledge acquisition features from social networks through specific reasoning over that knowledge. This facet will facilitate communities' coordination by helping them in decision-making. Decision-making can be regarded as the cognitive process resulting in the selection of a specific resource or action among several alternative scenarios. Every decision making process produces a final choice [17].

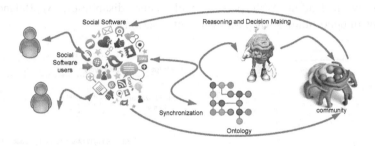

Fig. 1. Framework for Knowledge Management to support online communities creation

As can be seen in Figure 1, the proposed framework is composed by four modules: 1) Social Software; 2) Synchronization; 3) Ontology; and 4) Reasoning and Decision Making. The input of the framework is the Social Software users and the output is the community in general, which in this case raises some awareness using the knowledge represented in the system. The first module is social software, which is characterized by being collaboratively handled by social web users. This module can be a blog, wiki, forum, etc. However, as mentioned before, the content of social software is human readable only. This means that it is not formalized to facilitate its computerized use. For that reason, the synchronization module has incorporated some changes to that social software in order to transform its content to machine interpretable information. This synchronization is made between the knowledge represented in the used social software and the one in the ontology.

The role of the ontology in the proposed framework is to hold the gathered knowledge about a domain in a structured way, with a specific purpose to facilitate reasoning over that knowledge. Such reasoning is made by the module Reasoning and Decision Making, which is able to supply personalized or specific knowledge, representing to communities an access to structured contextual information. Contextual information is used by humans when communicate with each other. By improving computers access to such context, we increase the richness of communication human-computer and make it possible to produce more useful computational services [18]. Thus, these communities can, between other options, use the retrieved knowledge, and consequently to facilitate the interaction with other users with same interests, experience, etc.

3   ALTER-NATIVA's Knowledge Management Architecture

The proposed approach can be applied to various domains. In this case it is described the application of the framework to develop a Knowledge Management architecture for the ALTER-NATIVA community (Figure 2).

ALTER-NATIVA is one ALFA III project. ALFA is a program of cooperation between Higher Education Institutions of the European Union and Latin America. ALFA III retains the original objective of the previous phases of the ALFA Program, that is, to promote higher Education in Latin America as a means to contribute to the economic and social development of the region [19]. The main goals of the project were to: 1) Give education for everyone; 2) Give an environment of formation to professors when leading with persons with some disabilities; 3) Balance the inequalities of opportunities when accessing information.

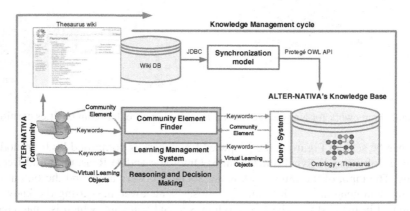

Fig. 2. ALTER-NATIVA's Knowledge Management architecture

In this architecture a complete Knowledge Management cycle was implemented. The beginning of the cycle or the input of the architecture is the knowledge of the ALTER-NATIVA community. This knowledge is processed and consumed by the community, which then uses it to re-feed this cycle with more knowledge. The gather of knowledge from the community is focused on the creation of a Thesaurus to represent the information related to the concepts/keywords (lexicon) of four distinct areas, namely Literature, Mathematics, Science and Teaching.

The Social Software used in the developed architecture was the Wikipedia. Wikipedia is an open encyclopedia that is collaboratively edited by its users. This software provides so called categories that are used to classify articles and other pages. The process of assigning categories to Wikipedia articles is a kind of collaborative tagging like in a folksonomy, but in this case, it is related to the categorization of concepts based on its definitions. One particularity is that the resulting category system is a thesaurus. A Thesaurus can be represented by a set of classes organized as a taxonomy representing domain reference concepts with associated meanings about domain in a semantic related structure. However, despite this strong usability, Wikipedia is not fully prepared to be integrated with other information systems. For that reason, in ALTER-NATIVA were implemented and used two extensions to allow a structured insertion of knowledge. The MediaWiki community developed the CategoryTree extension. It facilitates an organized introduction of concepts in a hierarchical form, representing the wanted concept tree. The other extension was developed to avoid the insertion of data with HTML codification (e.g. in the middle of tags), which resulted in an easy and coherent knowledge insertion form for the community users. This extension is composed by four textboxes (Figure 3 a)): in the first textbox it is inserted the father of the concept (category of which this concept is member); in the second it is inserted the definition of the concept; in the third the creation date; and in the forth the author/authors identification.

The synchronization module of this architecture uses Wikipedia Data Base to detect any changes in the log of information that could have occurred in the Wiki Thesaurus since its last run. JDBC (Java Database Connectivity) is used to querying Wiki Thesaurus database. If any change occurred, it then updates the Knowledge Base's Thesaurus accordingly. The OWL API provides classes and methods to load and save OWL files, to query and manipulate OWL data models. This module runs periodically thanks to a "cron job" implemented in the Wikipedia's server. Thus, changes made by the users are constantly updated establishing in this way, a dynamic knowledge handling feature since its gathering from web users till its availability to the community.

The establishment of a Knowledge Base arises from the necessity to formalize the knowledge related to the ALTER-NATIVA's domain and allowing its communities to actively participate in its maintenance. The ALTER-NATIVA's Knowledge Base is split in two distinct parts: the "Ontology" and the "Thesaurus", which are connected by the relation "has keywords". This relation helps to characterize the resources of this community e.g. Virtual Learning Objects (VLO) making use of the knowledge gathered by the architecture. The Ontology is dedicated to represent the Knowledge of this community's resource. It main objective is to facilitate resources representation (categorization), searching and recommendation. Resources categorization uses the lexicon represented in the thesaurus. Each resource can be related to a set of concepts of the thesaurus.

The resources searching functionality use specific characteristics of the resource e.g. level of expertise of a professor, and the concepts that characterize them, which then can be used to find a professor with a pre-determined profile. The resources' recommendation functionality is an advanced feature that automatically records specific patterns usage information, as an example, a professor of Mathematics normally has interest in differential calculus. Thus, next time a professor of this kind of profile enters the system, the system automatically recommends to him a set of resources. It could be, for instance, a specific type of VLO related to differential calculus.

The ALTER-NATIVA Knowledge Base is able to support on the suggestion of users to interact with, based on common interests and levels of expertise potentiating communities' creation. This is done by the advanced services of the Community Element Finder of the architecture (Figure 2). Other feature is the recommendation of VLO to which is done by the advanced services of the module Learning Management System. Whether the contextual information is formal, seed by the keywords and associated meaning or by the ontology, the expertise of the platform users is available for the implemented architecture be able to recommend with some context awareness. This could be reached by associating for instance the levels of expertise of users to the collected by associating the number of times that the user consults a specific topic. Then, by specific inference it is possible for the system to be aware of some context. One professor with a specific profile normally has interests or needs about a particular thing (e.g. VLO).

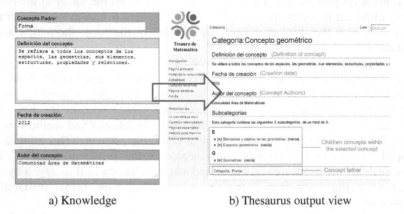

a) Knowledge   b) Thesaurus output view

Fig. 3. Knowledge gathering in the wiki insertion form

4   Demonstration Scenario

The ALTER-NATIVA's project main goal is to define curricular guidelines with technological support for higher education in the areas of language, mathematics and science with specific focus in supporting people in context of diversity. Thus, one of the main scenarios of this project pursue the idea of supporting teachers from different disciplines, many of them from pedagogical careers, with accessible tools that would facilitate them in the creation of working groups for discussion and for the generation of specific VLOs supported by a recommendation system [20]. As a consequence, the importance of developing services or mechanisms to gather knowledge of the domain experts was observed. Since in this scenario, professors are the main actors and better know how to characterize this project's resources: the Thesaurus for each of the ALTER-NATIVA project areas (literature, mathematics and science areas) gathered the knowledge directly from the professors.

An example of a contribution to the thesaurus creation is shown in Figure 3. An expert inserts the information corresponding to de concept "Concepto geométrico" in the form of Figure 3 a). When the user submits it, the output is the web page of Figure 3 b). After having the thesaurus with substantial contents, it is possible trough Community Element Finder services, search for specific profile of people. As an example, it is possible to search for professors that are experiment in "Geometric concepts" ("Concepto geométrico"). Another similar contribution in the possibility of search for a specific VLO characterized for keywords using the Learning Management System, as an example a VLO, which has the keyword "Geometric concepts". The system is also able of advanced recommendations based on the historical information of users. Taking as an example, if a user consults several times a VLO, is most likely that he/she wants to consult other ones with similar characterization in terms of keywords. The same kind of recommendation is also made when searching for a user to interact with. If a user wants to develop a VLO characterized by a set of keywords, a possible suggestion of user to interact are other users that use to develop Virtual Learning Objects with the same keywords characterization.

5 Conclusions

The proposed framework establishes a set of components with specific characteristics to guide on the creation of software platforms able to support online communities' creation. Its main characteristic is related to its capability of gathering and formalization of knowledge from its users to afterwards use it for specific reasoning services. Due to the use of such knowledge, these services aggregate a collective view of a domain knowledge. Depending on its design, such services are able to provide its related systems with specific reasoning and Decision-making ability.

In the example of ALTER-NATIVA project the services were designed to recommend specific resources in relation to a determined profile of user. Thus they give some support to what is defined as "context awareness". Depending on the profile (context) of the user, the system is able (aware) to suggest the most appropriate resources to such kind of people, which will help them in their (online) community creation.

As future work, the authors want to integrate ontology learning in the framework. The concept Ontology learning represents methods and techniques used to build, semi-automatically or automatically, ontology from scratch, enriching, or adapting an existing ontology using several sources [21]. Thus, he authors' intend to create a prototype to allow an autonomous gather of knowledge from social software, which capable afterwards used to automatically build its related ontology. One example of a step in execution/implementation is to consider Wikipedia categories as ontological classes and the corresponding pages as instantiations of the knowledge base. This would allow a higher dynamic conversion from the individuals to community knowledge.

References

[1] Appelquist, D., Brickley, D., Carvahlo, M., Iannella, R., Passant, A., Perey, C., Story, H.: A Standards-based, Open and Privacy-aware Social Web. W3C Incubator Group Report, W3C (December 2010)

[2] Porter, J.: Designing for the social web. New Riders Press (2008)

[3] Allen, C.: Tracing the Evolution of Social Software (2004)

[4] Berners-Lee, T., Hendler, J., Lassila, O.: The Semantic Web. Sci. Am. 284(5), 34–43 (2001)

[5] Shadbolt, N., Berners-Lee, T., Hall, W.: The Semantic Web Revisited. IEEE Intell. Syst. 21(3), 96–101 (2006)

[6] Taye, M.M.: Understanding Semantic Web and Ontologies: Theory and Applications. CoRR, vol. abs/1006.4 (2010)

[7] O'Reilly, T.: What is Web 2.0: Design Patterns and Business Models for the Next Generation of Software (2005)

[8] Devedzic, V., Gasevic, D. (eds.): Web 2.0 & Semantic Web, vol. 6. Springer (2009)

[9] Hendler, J.: Web 3.0 Emerging. IEEE Comput. 42(1), 111–113 (2009)

[10] Kaminski, J.: EDITORIAL Folksonomies boost Web 2.0 Functionality 13(2), 1–8 (2009)

[11] Vander Wal, T.: Folksonomies definition and wikipedia (2005),
http://www.vanderwal.net/random/entrysel.php?blog=1750
(accessed: October 16, 2013)

[12] Fensel, D.: Ontologies: A Silver Bullet for Knowledge Management and Electronic Commerce, 2nd edn. Springer-Verlag New York, Inc., New York (2003)

[13] Maedche, A., Staab, S.: Ontology Learning for the Semantic Web. IEEE Intell. Syst. 16(2), 72–79 (2001)

[14] Hazman, M., El-Beltagy, S.R., Rafea, A.: Article: A Survey of Ontology Learning Approaches. Int. J. Comput. Appl. 22(8), 36–43 (2011)

[15] Daassi, M., Favier, M.: Groupware and team aware: Bridging the gap between technologies and human behaviour. Encycl. Virtual Communities Technol. (2005)

[16] Gutwin, C., Greenberg, S., Roseman, M.: Workspace Awareness in Real-Time Distributed Groupware: Framework, Widgets, and Evaluation. In: Proceedings of HCI on People and Computers XI, pp. 281–298 (1996)

[17] Reason, J.: Human Error, 302 p. Cambridge University Press, Cambridge (1990)

[18] Abowd, G.D., Dey, A.K., Brown, P.J., Davies, N., Smith, M., Steggles, P.: Towards a Better Understanding of Context and Context-Awareness. In: Proceedings of the 1st International Symposium on Handheld and Ubiquitous Computing, pp. 304–307 (1999)

[19] European Commision. Development and Cooperation - EUROPEAID,
http://ec.europa.eu/europeaid/where/latin-america/
regional-cooperation/alfa/publications_en.html
(accessed: October 17, 2013)

[20] Sarraipa, J., Baldiris, S., Fabregat, R., Jardim-Gonçalves, R.: Knowledge Representation in Support of Adaptable eLearning Services for All. In: Procedia CS, vol. 14, pp. 391–402 (2012)

[21] Gómez-Pérez, A., Manzano-Macho, D.: A survey of ontology learning methods and techniques (2003)

Part III
Computational Systems

Analysis of Complex Data by Means
of Complex Networks

Massimiliano Zanin[1,2,3], Ernestina Menasalvas[2], Stefano Boccaletti[4],
and Pedro A. Sousa[1]

[1] Faculdade de Ciências e Tecnologia, Departamento de Engenharia Electrotécnica,
Universidade Nova de Lisboa, 2829-516 Caparica, Portugal
[2] Center for Biomedical Technology, Universidad Politécnica de Madrid,
28223 Pozuelo de Alarcón, Madrid, Spain
[3] Innaxis Foundation & Research Institute, José Ortega y Gasset 20, 28006, Madrid, Spain
[4] CNR - Institute of Complex Systems, Via Madonna del Piano 10, 50019 Sesto Fiorentino,
Florence, Italy
m.zanin@campus.fct.unl.pt, ernestina.menasalvas@upm.es,
stefano.boccaletti@fi.isc.cnr.it, pas@holos.pt

Abstract. In the ever-increasing availability of massive data sets describing
complex systems, *i.e.* systems composed of a plethora of elements interacting in
a non-linear way, *complex networks* have emerged as powerful tools for
characterizing these structures of interactions in a mathematical way. In this
contribution, we explore how different Data Mining techniques can be adapted
to improve such characterization. Specifically, we here describe novel
techniques for optimizing network representations of different data sets;
automatize the extraction of relevant topological metrics, and using such
metrics toward the synthesis of high-level knowledge. The validity and
usefulness of such approach is demonstrated through the analysis of medical
data sets describing groups of control subjects and patients. Finally, the
application of these techniques to other social and technological problems is
discussed.

Keywords: Complex systems, complex networks, data mining.

1   Introduction

Networks are all around us: from a social point of view, when we are ourselves, as
individuals, the units of a network of social relationships of different kinds [1]; but we
are also the result of networks of biochemical reactions, and of electro-chemical
interactions between neurons [2]. Furthermore, our world around us is organized in
networks, from physical transportation networks [3] up to virtual information webs
[4]. While the mathematical formulation of networks as mathematical objects started
in 1736 when the Swiss mathematician Leonhard Euler published the solution of the
Königsberg bridge problem, graph representations can be found back in 980 AD [5].
Only in recent years, thanks to the increasing capacity of computation centers on the
one side, and availability of public data sets on the other, complex network analysis

L.M. Camarinha-Matos et al. (Eds.): DoCEIS 2014, IFIP AICT 423, pp. 39–46, 2014.

has witnessed a revolution, which has yielded a vast theoretical and applied body of research. Interested readers may refer to Refs. [6], [7], [8], [9] for further information.

In spite of this evolution, several research questions have still to be tackled, and new ones appear when novel applications of network theory are proposed. Among these, the PhD Thesis "Complex Networks and Data Mining: Toward a new perspective for the understanding of Complex Systems" [10] proposes the use of data mining techniques to solve the following four research questions: (i) how to pre-select relevant features for minimizing the cost of network reconstruction, (ii) how to design new network reconstruction techniques, (iii) how to use network representations to improve data mining tasks, and (iv) how to assess and optimize the significance of a network representation.

As will be elaborated in Section 2, the problem posed in the fourth research question is of utmost relevance when the system under analysis is a *Collective Awareness System* (CAS). In this contribution, we review the methodology developed inside [10] to deal with this problem, which allows automatizing the process of obtaining the best network representation of a given system. This guarantees that the highest quantity of information is extracted from the system, thus maximizing the knowledge gained from it.

2   Relationship to Collective Awareness Systems

Under the umbrella of the 'FuturICT' FET Flagship Pilot Project, the European research community has already analyzed the implications and requirements of a collective awareness system, called *Planetary Nervous System* (PNS) [11]. Among the expected benefits of such world-scale sensory system, the PNS would be able to record and mine the digital footsteps created by human activity, as well as to unveil the knowledge hidden in such social big data, thus allowing addressing some fundamental questions about social dynamics. Nevertheless, implementing such system will require overcoming several challenges: some of them of a technical nature and expected to be solved in the next years, as for instance increased computational capabilities, others requiring a change in the data processing paradigm. Specifically, new algorithms for finding patterns in large sets of data will be required, with two specific targets: *i)* handle fragmented, low-level and incomplete data, and *ii)* adapt to the specific characteristics of these data sets, including their networked multi-dimensional nature and semantic richness [11].

Within this context, a natural solution has been already identified in the complex network theory, as it provides a rigorous framework for the study of structures created by relationships between the elements of a complex system [12]. Potential applications include the analysis of the dynamics of social systems, *e.g.* the diffusion of opinions [13] or of diseases [14]; the analysis of mobility patterns, by modeling pairs of origin – destination locations as links; or semantic text analysis, for creating structured taxonomies over texts [15].

If one is to apply complex network analysis to data coming from CAS like the PNS, whose size is expected to exceed the capacity of human analysts, principles and methods should be found to ensure a way for an automated knowledge extraction. The methodology presented in this contribution aims at providing an automatic procedure for obtaining the best network representation of a given system. Traditionally such optimization step has mainly been performed by means of the expert judgment of the researcher: yet, it is unfeasible to manually optimize the network representation of a CAS, due to the quantity of information it encodes. Thus, we expect an important added value when the methodology here presented is positioned between the data gathering, and the human-based network analysis phases of CAS management.

3   Proposed Methodology

As previously introduced, this Section proposes a novel way of optimizing the network representation of a complex system, *e.g.* a CAS, by means of data mining techniques, as developed in [10]. Such methodology will be presented in Section 3.2: before that, Section 3.1 will introduce the reader to the different ways of constructing a network representation starting from a raw data set.

3.1   Network Reconstruction Frameworks

As a first step in the analysis of a complex system, it is necessary to create a network representation of it; this, in turn, requires two steps: map each element of the system into a node of the network, and assess the existence of a relationship between pairs of nodes.

When relationships between the system elements are defined upon a physical support, their identification is a straightforward task, and the researcher only needs to map them into the network representation. For instance, one may consider the air transportation network: when airports are represented by nodes, links are naturally established between pairs of them if at least one direct flight is connecting these two airports [3].

In the absence of such relationships, links can still be built, provided a vector of *observables*, *i.e.* of measurements representing some properties of the system, can be associated to each node. In this case, each link represents the presence of a *functional relationship* between the data corresponding to that pair of nodes, and the resulting networks are called *functional networks*. For instance, if one is to analyze the structure of a stock market, each stock may be represented by a node, with pairs of them connected whenever there is a significant correlation in their price evolution through time [16]. Fig. 1 reports a simple example, in which a network is created by calculating Person's linear correlation between the evolutions of four U.S. stock prices.

It is important to notice that the requirement of having a vector of observable for each node precludes the use of functional representations for systems whose elements are characterized by a single value. Examples include tissues and organic sample analysis, like spectrography; genetic expression levels of individuals, without evolution through time [17]; biomedical analyses, *e.g.* the study of brain oxygen consumption by means of neuroimaging techniques [18]; or social network analyses, when just a snapshot of users characteristics is available [13]. To overcome this

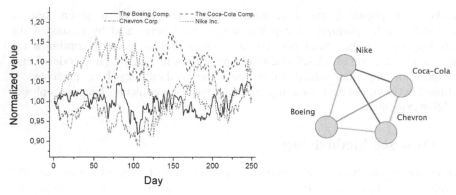

Fig. 1. Example of a *functional network* reconstruction. (Left) Time evolution of four stocks composing the Dow Jones Industrial Average index; each time series corresponds the evolution of prices from 1st January to 31st December 2012, and values are normalized to 1.0 in the first day. (Right) Resulting network, where each node is a stock, and links are weighted according to the Pearson's linear correlation between the corresponding time series; green (red) shades indicate positive (negative) correlations.

limitation, a novel method was recently proposed, which allows treating collections of isolated, possibly heterogeneous, scalars, *e.g.* sets of biomedical tests, as networked systems. The method yields a network where each node represents an observable, and links codify the distance between a pair of observables and a model of their typical relationship within the studied population [19], [20].

3.2   Network Optimization and Analysis

Whether the network representation is assembled by mapping physical connections, by constructing a *functional* representation, or by using the technique proposed in Refs. [19], [20], two further steps are required: *i)* transform the *fully-connected weighted network* (as the one depicted in Fig. 1) into a *structured unweighted network*, and *ii)* extract a set of metrics describing some topological characteristics. It is worth noticing that both steps are characterized by some level of arbitrariness. While binarizing the network, it is necessary to define a threshold, such that links with a weight lower than this reference value are deleted. Also, among the large group of available topological metrics, the researcher has to choose the one he / she considers being relevant for describing the system under study.

A new methodology has been proposed for addressing these two issues, based on the application of data mining techniques [21]. By starting from an external classification as ground truth, *e.g.* control subjects and patients suffering from some disease, it is possible to use the output of a data mining classification task as a proxy for the relevance of the network representation under study. This yields criteria for an optimal network representation with respect to a given problem.

Following the approach proposed in Ref. [21], instead of applying a single pre-determined threshold τ, such that links whose weight is lower than τ are deleted, a set of thresholds $T = \{\tau_1, \tau_2, \ldots\}$ is applied, covering the whole range of applicable thresholds. Furthermore, a large set of measures M is extracted from each network,

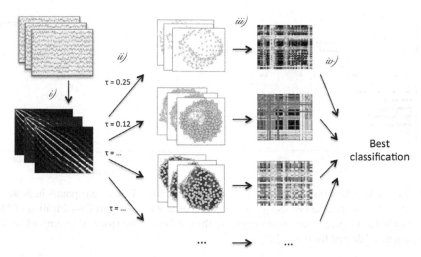

Fig. 2. Optimizing network reconstruction. Following the creation of weighted cliques (step *i*), these are transformed into a set of unweighted adjacency matrices by dint of different thresholds (step *ii*); a set of features is extracted for each network (step *iii*), and this is used as input for a data mining task (step *iv*). Finally, the best classification is used to choose the most relevant threshold and topological metrics. Adapted from Ref. [21].

including the most relevant macro-, meso- and micro-scale topological features of a complex network (see Ref. [8] for a review of applicable metrics). At the end of the process, the initial raw data are therefore converted into a large set of measures, representing a wide sample of the possible analyses that may be performed from a complex network perspective.

Once the raw data have been transformed into a large set of topological metrics, the problem faced by the researcher is the identification of the optimal subset of metrics for describing the system. Here we propose the use of a data mining classification task for automatizing this process. Specifically, for each threshold τ_i, and for each pair (or triplet) of metrics, subjects are classified; the percentage of subjects correctly classified is then used as a proxy of the relevance of such set of parameters. Indeed, if a good classification is achieved, the considered parameters and network metrics correctly represent the structural differences between the two classes of subjects. Thus, the best classification corresponds to both the best set of metrics and to the corresponding best threshold. Fig. 2 proposes a graphical representation of this process, where information flows from the left (raw data and weighted fully-connected networks) to the right (final classification).

4   An Application to Mild Cognitive Impairment

To demonstrate the validity of the proposed approach, we here consider a set of magneto-encephalographic data (MEG), and identify the features that better differentiate healthy subjects from patients suffering from *Mild Cognitive Impairment*

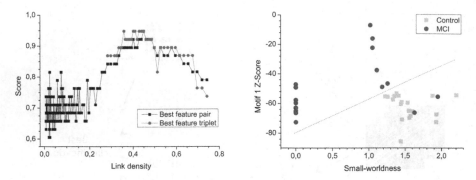

Fig. 3. (Left) Classification score as a function of link density; black (red) points indicate the best classification score obtained using pairs (triplets) of features. (Right) Classification of MCI and healthy subjects; green (red) points represent the position in the space of features of healthy (MCI) patients. Adapted from Ref. [20].

(MCI). MCI is a disease, considered a prodromal stage of *Alzheimer's*, characterized by cognitive impairments beyond those expected based on the age of the patient, but which are not significant enough to interfere with their daily activities. The data set comprises recordings from nineteen patients and nineteen healthy volunteers during a modified Sternberg's letter-probe task, which requires participants to firstly memorize a set of five letters presented on a computer screen, for then pressing a button when a member of the previous set is detected.

Following the methodology proposed in Section 3.2, 178 networks have been created for each subject, corresponding to the number of different thresholds considered. From each one of these networks, 72 different topological metrics have been calculated. A classification task was ultimately performed for each pair and triplet of considered features, using a Support Vector Machine algorithm [22]. Fig. 3 (Left) reports the precision (percentage of correctly classified subjects) corresponding to the most representative pair (triplet, in red) of features, as a function of the link density obtained by applying different thresholds. Classification was also attempted with other algorithms, including Naive Bayes and neural networks [23], producing qualitatively comparable results.

Several relevant conclusions can be derived from Fig. 3. Firstly, the best classification rate (95%) is obtained for sufficiently low threshold values, *i.e.* including a great quantity of links inside the analysis. Specifically, the maximum score corresponds to including about 40% of the links. Remarkably, the functional brain network literature typically considers networks with a 5% link density [24]. The increase in the number of links, as suggested by the proposed methodology, has a major consequence: allowing a better consideration of meso-scale structures, *e.g.* of motifs, that is specific connectivity patterns formed by 3 nodes [17]. Furthermore, results corresponding to low link densities are much more unstable, as demonstrated by the leftmost part of the plot in Fig. 3. Clearly, the addition, or deletion, of a few links has a major effect in the topology, changing the meaning of all metrics calculated on the top of it. Therefore, these results invite to reconsider many studies made in the Literature about functional brain network reconstruction, and validate the hypothesis that a data mining approach can improve the understanding of complex systems.

5   Conclusion and Discussion

In conclusion, in this contribution we have described and reviewed how the application of data mining techniques can be used to improve and optimize the reconstruction of complex networks, representing for instance Collective Awareness Systems. In turn, the resulting networks can be used to extract knowledge about the topological properties of the corresponding systems, in a way that goes beyond the capacity of classical data mining. Such advantages come at a cost: due to the high number of analyses required, *e.g.* the extraction of several topological metrics for different threshold values, there is an important increase in the computational cost, especially when compared with standard data mining algorithms.

Beyond the proposed biomedical example, such methodology can be applied in any scenario in which a complex network representation is expected to be relevant. Thus, this includes the analysis of any system whose dynamics is defined by the relationships between its elements: from social networks created by interacting individuals, up to technological networks, as communication or transportation systems. Furthermore, such elements may not be homogeneous, or interactions may develop through different channels – what is known as a *multi-layer* (of *multiplex*) network [25], [26]. For all of this, the approach here presented is expected to be of relevance for future applications of complex network techniques to the field of Collective Awareness System.

Acknowledgments. The authors acknowledge the computational resources, facilities and assistance provided by the Centro computazionale di RicErca sui Sistemi COmplessi (CRESCO) of the Italian National Agency for New Technologies, Energy and Sustainable Economic Development (ENEA), and of the Universidad Politécnica de Madrid's CeSViMa (Madrid Supercomputing and Visualization Center).

References

1. Knoke, D., Yang, S.: Social Network Analysis. Sage (2008)
2. Bullmore, E., Sporns, O.: Complex brain networks: graph theoretical analysis of structural and functional systems. Nature Reviews Neuroscience 10, 186–198 (2009)
3. Zanin, M., Lillo, F.: Modelling the air transport with complex networks: A short review. The European Physical Journal Special Topics 215, 5–21 (2013)
4. Albert, R., Jeong, H., Barabási, A.L.: Error and attack tolerance of complex networks. Nature 406, 378–382 (2000)
5. Strano, E., Zanin, M., Estrada, E., Lillo, F.: Spatially embedded socio-technical complex networks. The European Physical Journal Special Topics 215, 1–4 (2013)
6. Albert, R., Barabási, A.L.: Statistical mechanics of complex networks. Reviews of Modern Physics 74 (2002)
7. Boccaletti, S., Latora, V., Moreno, Y., Chavez, M., Hwang, D.U.: Complex networks: Structure and dynamics. Physics Reports 424, 175–308 (2006)
8. Costa, L.D.F., Rodrigues, F.A., Travieso, G., Villas Boas, P.R.: Characterization of complex networks: A survey of measurements. Advances in Physics 56, 167–242 (2007)

9. Costa, L.D.F., Oliveira Jr, O.N., Travieso, G., Rodrigues, F.A., Villas Boas, P.R., Antiqueira, L., Viana, M.P., Correa Rocha, L.E.: Analyzing and modeling real-world phenomena with complex networks: a survey of applications. Advances in Physics 60, 329–412 (2011)
10. Zanin, M.: Complex Networks and Data Mining: Toward a new perspective for the understanding of Complex Systems. PhD Thesis (2014)
11. Giannotti, F., Pedreschi, D., Pentland, A., Lukowicz, P., Kossmann, D., Crowley, J., Helbing, D.: A planetary nervous system for social mining and collective awareness. The European Physical Journal Special Topics 214, 49–75 (2012)
12. Havlin, S., Kenett, D.Y., Ben-Jacob, E., Bunde, A., Cohen, R., Hermann, H., Kantelhardt, J.W., Kertész, J., Kirkpatrick, S., Kurths, J., Portugali, J., Solomon, S.: Challenges in network science: Applications to infrastructures, climate, social systems and economics. The European Physical Journal Special Topics 214, 273–293 (2012)
13. Castellano, C., Fortunato, S., Loreto, V.: Statistical physics of social dynamics. Reviews of Modern Physics 81 (2009)
14. Pastor-Satorras, R., Vespignani, A.: Epidemic dynamics and endemic states in complex networks. Physical Review E 63, 066117 (2001)
15. Navigli, R., Velardi, P., Faralli, S.: A graph-based algorithm for inducing lexical taxonomies from scratch. In: Twenty-Second International Joint Conference on Artificial Intelligence-Volume, pp. 1872–1877. AAAI Press (2011)
16. Mantegna, R.N.: Hierarchical structure in financial markets. The European Physical Journal B-Condensed Matter and Complex Systems 11, 193–197 (1999)
17. Milo, R., Shen-Orr, S., Itzkovitz, S., Kashtan, N., Chklovskii, D., Alon, U.: Network motifs: simple building blocks of complex networks. Science 298, 824–827 (2002)
18. Phelps, M.E., Mazziotta, J.C.: Positron emission tomography: human brain function and biochemistry. Science 228, 799–809 (1985)
19. Zanin, M., Boccaletti, S.: Complex networks analysis of obstructive nephropathy data. Chaos: An Interdisciplinary Journal of Nonlinear Science 21, 033103 (2011)
20. Zanin, M., Alcazar, J.M., Carbajosa, J.V., Sousa, P., Papo, D., Menasalvas, E., Boccaletti, S.: Parenclitic networks' representation of data sets. arXiv:1304.1896 (2013)
21. Zanin, M., Sousa, P., Papo, D., Bajo, R., García-Prieto, J., del Pozo, F., Menasalvas, E., Boccaletti, S.: Optimizing functional network representation of multivariate time series. Scientific Reports 2 (2012)
22. Steinwart, I., Christmann, A.: Support vector machines. Springer (2008)
23. Bishop, C.M., Nasrabadi, N.M.: Pattern recognition and machine learning. Springer (2006)
24. Buldú, J.M., Bajo, R., Maestú, F., Castellanos, N., Leyva, I., Gil, P., Sendiña-Nadal, I., Almendral, J.A., Nevado, A.: del-Pozo, F., Boccaletti, S.: Reorganization of functional networks in mild cognitive impairment. PLoS One 6, e19584 (2011)
25. Kivelä, M., Arenas, A., Barthelemy, M., Gleeson, J.P., Moreno, Y., Porter, M.A.: Multilayer Networks. arXiv:1309.7233 [physics.soc-ph] (2013)
26. Cardillo, A., Gómez-Gardeñes, J., Zanin, M., Romance, M., Papo, D., del Pozo, F., Boccaletti, S.: Emergence of network features from multiplexity. Scientific Reports 3 (2013)

A Conceptual Model of Farm Management Information System for Decision Support

George Burlacu[1], Ruben Costa[2], Joao Sarraipa[2], Ricardo Jardim-Goncalves[2], and Dan Popescu[1]

[1] University "Politehnica" of Bucharest,
Faculty of Automatic Control and Computer Science, Bucharest, Romania
[2] Centre of Technology and Systems, CTS, UNINOVA, 2829-516 Caparica, Portugal
{burlacu_george85,dan_popescu_2002}@yahoo.com,
{rddc,jfss,rg}@uninova.pt

Abstract. In a today economy, it is crucial to have systems able to handle information with precision. In addition, it is also important to apply technological innovations in the various domains, with the objective to modernize and transform them to become more competitive. In this paper, a conceptual framework for a Farm Management Information System (FMIS) is presented. It is focused in the different ways of using the information coming from various sources as sensors to assist farmers in decision making of agriculture business.

Keywords: Ontology, Management Information System, Data Acquisition.

1   Introduction

Precision agriculture is a modern method to make agriculture, which refers to optimizing the production through the fusion of traditional mechanized agriculture procedures with new technologies such as monitoring systems, command & control systems, geographical location systems and support information systems. These optimizations have the aim of choosing the right time for culture seeding, at the right place and monitoring the culture throughout the growth period, depending on the various parameters.

Precision agriculture uses various tools for collecting information such as soil quality sampling, remote field sensing and yield monitoring. Optimization focuses on increasing yields, reducing cultivation costs and minimizing environmental impacts [1]. The challenges that precision agriculture tries to overcome are:

- Automatic collection of the related information about the physical structure and chemical composition of the soil;
- Development of a well-structured database, which should be integrated in GIS (Geographical Information Systems);
- Development of intelligent agricultural machinery for farm working with high degree of spatial accuracy;
- Nutrients distribution in variable doses required for uniform plants growth, in order to compensate the unevenness of soil characteristics;

L.M. Camarinha-Matos et al. (Eds.): DoCEIS 2014, IFIP AICT 423, pp. 47–54, 2014.

- Pesticides applications have to take into consideration the nature of existing pests and weeds in crops [2];
- Making crop productivity maps in real time based on flow sensors mounted in combine hopper that harvested grain [3];
- Making electrical conductivity distribution maps of soil. These maps provide an overview of water distribution and concentration in soil [4].

Traditional agriculture uses a variety of techniques to improve land quality in order to make it suitable for planting. This includes using animal manure and digging water-channels for field's irrigation. Modern agriculture uses technologies for plant breeding and pesticides and fertilizers optimization.

2   Relationship to Collective Awareness Systems

Collective Awareness Systems are designed to support the environment, by sharing knowledge, best practices and processes, in order to achieve changes in lifestyle, in terms of production and consumption [5].

In this context, the paper propose a framework for a Farm Management Information System, which is a system able to manage the information with precision, having the purpose to optimize both resources and processes that occurs in the farm, using sensors systems and other types of technologies for monitoring the land.

Furthermore, the paper present an ontology that has been developed with the aim of representing the knowledge used for understanding the concepts and relations regarding the Farm Management Information System. Also, the ontology offers support to farmers in decision-making regarding the land and crop management.

3   Related Work

A Farm Management Information System (FMIS) is a management information system designed to assist agricultural farmers to perform various tasks ranging from operational planning, implementation and documentation for assessment of performed field work.

Sørensen proposed a conceptual model to develop a FMIS (Fig. 1a)). The main actor in this picture is the farm manager who needs to manage the overall agricultural crop production. The managerial demands are caused not only by the internal farm activity of production, but also by external entities like government, customers, universities, who increase the pressure on the agricultural sector to change the methods of production, whether the quality, the price or the technological improvements is considered [6]. The proposed FMIS aims to support management tasks and real-time decision making, as well as compliance management by automating data acquisition, contextualization operations taking into account external parameters (e.g., regulations, Best Management Practices (BMP), market information, etc.). In addition, the structure of FMIS should allow the connection with external systems (e.g., market, financial, administration, etc.).

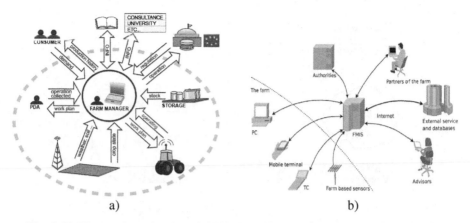

Fig. 1. FMIS: a) A Conceptual model [6]; b) Architecture from the user viewpoint [7]

The automatically acquired data provide complete monitoring of operational activities. The farm manager Information initiates the both internal and external filtering, taking into account the planned operational activities. Based on this, an execution plan can be generated and sent to the executants (e.g. equipment, staff or supply service that are designed to perform the operation).

In addition to the conceptual model proposed by Sørensen, Payman Salami and Hojat Ahmadi [7] defined a system architecture (Fig. 1b)) focused in how the user of the system should understand it. The communication is represented by arrows, which are intentionally left vague, meaning that they do not have a specific protocol or content of the communication. The end user does not need to know or even care how the communication is handled between the various systems, only that it occurs and that it is possible.

In Romania, agriculture plays an important role in our daily lives; many families produce the necessary revenues existence in this way. In recent years, this sector has increasingly presented more interest and attracted many development funds. Even so, many farmers choose to work the land in the traditional way. Their motivation is mainly related to the financial aspect necessary to implement a FMIS.

4 Farm Management Information System Framework

The new agricultural technologies such as Variable Rate Technologies (VRT), Remote Sensing, GIS and Global Positioning System (GPS), in addition to the developments in modelling and simulation of crop production, provide numerous opportunities for the development of Precision Agriculture. In Fig. 2 is represented a new conceptual model of FMIS. Unlike the conceptual models proposed by Sørensen, Salami and Ahmadi, this framework focuses on the way it performs data acquisition, mapping, making spatial distribution maps, productivity maps and profit maps. The acquired information is stored on a platform together with the information from the various external web services; the system purposes some possibilities from its knowledge base, through which the farmer can take a decision on land or crops.

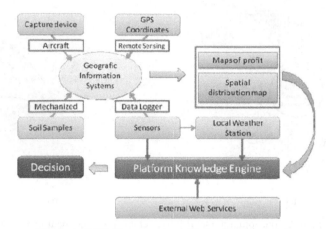

Fig. 2. A FMIS Conceptual model for decision support

The purpose of Web services is to enable communication, interoperability and data exchange between applications. Through these services the farmer has access to: new technology, resources (human, materials), information from the market (competitors, prices, forecasts), information from ministries and the European Union, etc. The GIS Software's are developed for viewing and interpreting the collected data, having special options for embedding geographic latitude and longitude. Based on this data, the farmer makes spatial distribution maps, taking into account certain parameters (location of interest points, distances between certain parts with reference to whole land areas or only to certain areas of interest) [8].

Data obtained through remote acquisition methods play an important role in building databases of geographical data within a GIS. Remote acquisition implies the existence of a capture device, attached to a platform (usually aircraft) [9]. Most of data in the current systems of agricultural information management are obtained through the interpretation of aerial photographs. Using a GIS, the farmer can analyze and interpret only digital type data. So the geographic information that does not meet this requirement must be digitized. Digitization is the process through which a physical copy of a map or a plan is converted into a digital environment using programs for objects representation in 2D or 3D. Digitization can be done both by hand and by using semi-automat devices that are connected to a computer and can transmit the coordinates of a point on a 2D surface.

Topographic information from GPS have various roles: fields definition in terms of geographical boundaries, monitoring the routes of agricultural machinery in the field, to observe their position in real time, mapping the soil parameters, crops mapping, maximizing the yield per hectare and monitoring the spread of specific disease in crops. Precision spatial dimension that provides the GPS system has been manifested in the classical form of mechanized agriculture, while the development of new technologies and equipment tend to exploit to the fullest the potential acquired [10]. Presently, a large amount of data from field operations are collected by agricultural machines and transmitted using various data storage and transmission media [11]. The classical sampling, which was done manually by collecting soil samples from different areas of land, at random or after a specific pattern, has moved in a mechanized collection by using modern agricultural machinery. This practice is known as "sampling" [12]. Sampling is used to obtain an overview distribution of soil chemical properties and/or to check for any abnormalities in normal chemical properties of soil. Also physical sizes (soil electro conductivity, soil pH, concentrations of nitrates, potassium, sodium) are measured directly in the field by specific laboratory values methods. Also, measuring the position of these points is recorded using GPS [13].Data from the mapping process are used to estimate the production of agricultural crops, in the entire area or specific areas of interest. In this estimation process, data from the history of agricultural land are taken into account and analyzed. A profit map can be created using inputs records in crops and outputs in harvest. So, a farmer can determine what areas of the field aren't profitable.

Farmer can also make crop productivity maps, using sensors that measure and record in real time the harvested crops volume. Various parameters like quantity per hectare and flow are measured. In this way, crop productivity maps are a valuable solution in farm management. Spatial distribution maps of the relevant parameters are ranked based on several criteria (visualization, physical support, technical type mapping, type of variables, type of representation, etc.). One advantage of using spatial distribution maps is the avoidance of the surplus or deficit in natural or mineral fertilizer (nitrogen, phosphorus, potassium, magnesium, nitrates). Automated VRT allow farmers to vary inputs, such as fertilizers, pesticides and seeding rates [14]. Varying input rates aims at either increasing yields or reducing costs, depending on the farmer's goal for the production system. Auto guidance systems, on the other hand, assist equipment operators to drive through the fields so that efforts could be focused on other important tasks. Therefore, these technological tools help to reduce redundancy, labour costs and to save operation times.

The sensors are used to determine soil properties (ground temperature/air temperature, humidity), hydric stress, and degree of crop disease (using the reflection of light on the leaves to determine their level of chlorophyll). Also, the sensors are used for measuring various parameters like electrical conductivity, high concentrations of potassium, nitrates, sodium, pH, or can be used to measure the soil and crops properties. The need for increase the production and simultaneously the efforts for saving resources make the sensor systems the best-allied tool. Their function requires the physical storage of data, using devices known as "data loggers", which are equipped with non-volatile memory. Data transmission is done either by cable (wired) or by radio signals (wireless) [15].

Using remote sensing sensors (aerial or satellite sensors) lands maps and cultures maps are obtained. The "aerial photographer" is an optical sensor that observes the variations in soil colour, crop development and land boundaries. Images obtained by satellite or by an aircraft can provide maps of vegetative indices, which reflect also the plant health, the soil conditions or the crops status. The obtained database is more accurate (contains a large number of measured values and parameters), thereby it increases the quality of the agricultural production estimations and decisions.

In Fig. 3 different use of type of sensors that can be found in farm are presented. The role of Context allocator is to adapt the nodes behaviour in order to divide the workload based on operating or network conditions. A sensor node is a node in a wireless sensor network. This node gathers sensory information and communicates with other connected nodes in the network and is capable to perform some processes [16]. FMIS can use sensors in three different ways: competitive; collaborative; and individual.

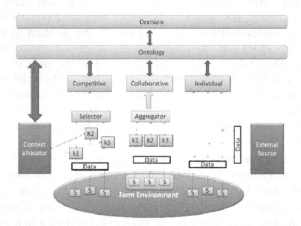

Fig. 3. Different type of sensors used in farm

The first case represents the competitive level of sensors. In this case, if the system needs to wait for the operator's intervention, maybe it becomes too slow, and thus the decision-making capability can give large benefits. So using the context allocation, new tasks are allocated to suitable nodes. In the second case, the sensors collaborate forming a network. Wireless Sensor Networks (WSNs) have been deployed as a cost-effective communications technology, which allows the acquisition and transmission of different data from the crop to final users. These WSNs consists on non-intrusive communication devices of small size, to which one or more precision sensors for data collection are adapted. Once the information arrives at their operators, it is further processed and studied in order to make an appropriate decision. In the third case, using for example moisture sensors and data coming from the weather station (external source), the farmer can set a daily schedule pre-programmed for soil irrigation, according to a soil moisture level, calibrated by the farmer. A soil moisture sensor will be connected to each zone, in the relevant regions in order to assess the humidity for the entire area. Thus, the water consumption used for this purpose is minimized. If the soil moisture, at one time, is already higher than the reference level, irrigation is not done in that area and so it saves water.

4.1 Farm Management Information System Ontology

The ontology has been developed with the aim of representing the knowledge that can be used for understanding the concepts and relations regarding the FMIS, with the purpose of supporting farmers in decision-making regarding the land and crop management. The ontology was built through Protégé editor, using the OWL Full version, mainly because of its high expressiveness capability. Fig. 4 a) illustrates an excerpt of the defined ontology representing a FMIS aligned with its conceptual model (Fig. 2). The ontology intends to answer different questions about FMIS features. As an example, Fig. 4b) illustrates how the ontology represents sensors characteristics. The types of sensors used in FMIS described in [2], are represented as instances of Sensors class [1], which provides data for GIS [3], able to measure: concentration, electro-conductivity, moisture and temperature [4].

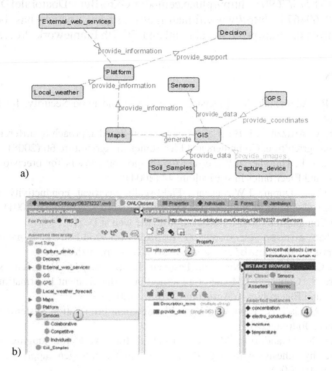

Fig. 4. FMIS Ontology: a) structure excerpt; b) sensors characterization

5   Conclusions and Future Work

In Romania, after many years of decline, reviving the agricultural activities has become a priority. In this context, precision agriculture supported by management information systems of the farm activities represents a viable and effective solution for its modernisation. In this paper, it is proposed a FMIS for precision agriculture, which is focused on functions to handle: land mappings, internal and external data collection, monitoring parameters, and data processing for decision support. Through such type of features, the farmer is able to see information about new technology, available resources (human, materials), information from the market (competitors, prices, forecasts), information coming from ministries and European Union, so it can take a better profitable decision. FMIS related knowledge is represented in a designed ontology to facilitate formal data acquisition from various monitoring, control, analysis and tracking systems for the referred precision farming implementation.

In the future authors intend to develop a platform that will integrate this ontology and external services; business process modelling between farmer, suppliers and potential clients using Business Process Model and Notation (BPMN); to then effectively test the proposed FMIS in pilot farms.

Acknowledgments. This article results has been developed within the EXPERT POSDRU/107/1.5/S/76903 (http://plone.cempdi.pub.ro/BurseDoctoraleID76903) and FITMAN N° 604674 (http://www.fitman-fi.eu) projects, which has been funded respectively from the European Social Fund and EC 7th Framework Programme.

References

1. Gebbers, R., Adamchuk, V.: Precision Agriculture and Food Security. In: Science, vol. 327 (2010)
2. Tellaeche, A., Artizzu, X.P.B., et al.: A new vision-based approach to differential spraying in precision agriculture. Computers and Electronics in Agriculture 60 (2008)
3. Adamchuk, V.I., Hummel, J.W., et al.: On-the-go soil sensors for precision agriculture. Computers and Electronics in Agriculture 44 (2004)
4. Johnson, C.K., Doran, J.W., et al.: Field-scale electrical conductivity mapping for delineating soil condition. Soil Science Society of America Journal 65 (2001)
5. European Commission (2013), Retrieved from the web at Dez13:
   http://ec.europa.eu/digital-agenda/en/collective-awareness-platforms-sustainability-and-social-innovation
6. Sørensen, C., Fountas, S., Nash, E., Pesonen, L., Bochtis, D., Pedersen, S., Basso, B., Blackmore, S.: Conceptual Model of a Future Farm Management Information System. In: Computers and Electronics in Agriculture, vol. 72 (2010)
7. Salami, P., Ahmadi, H.: Review of Farm Management Information Systems (FMIS). New York Science Journal (2010)
8. Bachmaier, M., Gandorfer, M.: A conceptual framework for judging the precision agriculture hypothesis with regard to site-specific nitrogen application. Precision Agriculture 10 (2008)
9. Wang, N., Zhang, N.Q., et al.: Wireless sensors in agriculture and food industry - Recent development and future perspective. Computers and Electronics in Agriculture 50 (2006)
10. Lamba, D., Fraziera, W.P., et al.: Improving pathways to adoption: Putting the right P's in precision agriculture. Computers and Electronics in Agriculture 61 (2007)
11. Nikkilä, R.: Farm Management Information System Architecture for Precision Agriculture. Department of Computer Science and Engineering, Laboratory of Computer and Information Science. Helsinki University of technology, Espoo, Finland (2007)
12. Schnug, E., Panten, K., et al.: Sampling and nutrient recommendations - The future. In: Int. Soil and Plant Analysis Symp., Minneapolis, Minnesota, Marcel Dekker Inc. (1997)
13. Adamchuk, V.I., Morgan, M.T., et al.: An automated sampling system for measuring soil pH. Trans. Sensors (2011)
14. Srivastava, S.: Space Inputs for Precision Agriculture: Scope for Prototype Experiments in the Diverse Indian Agro-Ecosystems. In: Map Asia 2002, Report, 7 - 9, Bangkok, Thailand, 1–4 (August 2002) (accessed February 4, 2011)
15. Ruiz-Garcia, L., Lunadei, L., et al.: A Review of Wireless Sensor Technologies and Applications in Agriculture and Food Industry: SoA and Current Trends, Sensors (2009)
16. Dargie, W., Poellabauer, C.: Fundamentals of wireless sensor networks: theory and practice. John Wiley and Sons (2010) ISBN 978-0-470-99765-9

Ontology Transformation of Enterprise Architecture Models

Marzieh Bakhshadeh[1], André Morais[1], Artur Caetano, and José Borbinha[1,2]

[1] Information Systems Group, INESC-ID, Rua Alves Redol 9, 1000-029 Lisboa, Portugal.
[2] Instituto Superior Técnico, Technical University of Lisbon, Avenida Rovisco Pais 1,
1049-001 Lisboa, Portugal
{marzieh.bakhshadeh,andre.coutinho,artur.caetano,
jlb}@ist.utl.pt

Abstract. Enterprise architecture supports the analysis and design of business-oriented systems through the creation of complementary perspectives from multiple viewpoints over the business, information systems and technological infrastructure, enabling communication between stakeholders. However, enterprise architecture modelling languages lack representation schemas that support the computable assessment of its models. This paper applies model transformation to address this issue. The proposed approach translates models specified using ArchiMate into OWL. The resulting ontological representation is therefore computable, allowing for the analysis of the consistency and completeness of the enterprise architecture models. The applicability of the approach is shown through a case study.

Keywords: enterprise architecture, model transformation, ontology, ArchiMate, OWL.

1 Introduction

Enterprise architecture is defined by Lankhorst as [1] "a coherent whole of principles, methods, and models that are used in the design and realisation of an enterprise's organisational structure, business processes, information systems, and infrastructure". Enterprise architecture languages are usually specified without formal semantics or representation schemas that facilitate analysing its models. For instance, TOGAF is specified using a combination of textual descriptions and object-oriented models [2] and the BPMN metamodel [3] lacks a formal semantics that hinders checking the correctness of models [4]. The same issue occurs with the object-oriented metamodel of the ArchiMate language [5]. ArchiMate lacks a formal representation schema that could help to facilitate the analysing of its models.

There has been a growing interest in Ontology Engineering during the last decade [6]. The most widely used definition characterizes ontologies as "formal, explicit specification of a shared conceptualization" [7]. Ontologies are becoming more important, by their role on the so called Semantic Web [8] [9] [10]. In different fields such as software engineering [11] [12] [13]. The use of ontologies and associated

L.M. Camarinha-Matos et al. (Eds.): DoCEIS 2014, IFIP AICT 423, pp. 55–62, 2014.

techniques in EA is increasing, with proposals of EA based on ontologies for improving the models and their semantics, as witnessed in [14] [15] [16] [17] [18] [19].

In the past few years a number of different approaches for evaluating EA models have been described. In [20], and [21], the authors have implemented a tool for analysing EA models, which guides the creation of enterprise information system scenarios in the form of enterprise architecture models and generates quantitative assessments of the scenarios as they evolve. In [22], the authors propose an uniform approach for capturing quality attribute requirements and analysing system and software architecture. However, such approach did not considered the business architecture [23].

This paper presents an application of model transformation to translate the standard ArchiMate enterprise architecture language into the OWL ontology language [24]. OWL is the latest and the most complex ontology language presented by W3C. With OWL ArchiMate metamodel can be represented with ontologies. Ontologies can assist architects by depicting all the consequences of their model. Formal ontology machineries also help architects to view and understand the implicit consequences of explicit statements and can help to ensure that a model is consistent. Logical reasoning can be applied to check the inconsistencies on the models and for inferring different dependencies between different elements of enterprise architecture models.

In this sense, we try to contribute to the following:

- Given Archimate metamodel described in a single XML document. How can we transform XML to OWL;
- Define a set of techniques on how to specify enterprise architecture models in order to improved meta-model conformance verification of models, through the verification of logical inconsistencies present in models;

The paper is organized as follows: in Section 2 the relationship of our topic to collective awareness systems is described. Section 3, the ArchiMate meta-model is introduced and the transformation process of the ArchiMate metamodel and ArchiMate models are presented. After that, in Section 4, a validation based on reasoning using a predefined set of competency questions on a specific use case has been performed. Finally, in Section 5 we conclude remarks drawn from the work and our perspectives for further research.

2   Collective Awareness Systems

Collective awareness is a critical part of collaboration within communities; especially computer-mediated communities [25] in our case, collective awareness can be achieved by analyzing the stakeholders view points, as such, enterprise architects need to conceive views from the viewpoint of the stakeholders and to address their concerns and requirements. However, the existence of semantic gaps between architects and stakeholders may produce conceptual misalignments which can negatively affect the architecture.

3   Transformation Process

3.1   ArchiMate

ArchiMate is an open and independent modelling language, from the Open Group, for enterprise architecture that is supported by different tool vendors and consulting firms [5].

The ArchiMate framework organizes its metamodel in a three by three matrix: the rows capture the enterprise domain layers (business, application, and technology), and the columns capture cross layer aspects (active structure, behaviour and passive structure).

3.2   Transformation of the ArchiMate Metamodel to OWL

The transformation process uses 1) an OWL representation of the ArchiMate metamodel and 2) OWL representations of ArchiMate models.

A set of transformation rules maps each element from the ArchiMate metamodel into the corresponding OWL representation. Each concept is transformed into an OWL Class. Relationships are transformed as OWL Object Properties. ArchiMate metamodel are described in a single XML document represented in Listing 1. Table 1 shows the Transformation Rules for the ArchiMate metamodel. In Table 2 the Archi XML representation element and respective OWL Class is depicted, next in Table 3 Archi XML representation element relations and respective OWL ObjectProperties is shown.

```xml
<?xml version="1.0" encoding="UTF-8"?>
<!-- ArchiMate 2.0 relationship rules -->
<relationships version="2.0">
    <elements>
        <source element="BusinessActor">
            <target element="BusinessActor" relations="cfgostu"
/>
            <target element="BusinessRole" relations="fiotu" />
                              .
                              .
                              .
            <target element="WorkPackage" relations="o" />
            <target element="Deliverable" relations="o" />
            <target element="Plateau" relations="o" />
            <target element="Gap" relations="cgos" />
        </source>
    </elements>
</relationships>
```

Listing 1. XML File structure

Table 1. Transformation Rules for the ArchiMate metamodel

XML	OWL
source element	Class
target element	Class
Relationships	object properties

Table 2. Archi XML representation element and respective OWL Class

XML	OWL
<source element="BusinessActor">	<Declaration> <Class IRI=#BusinessActor"/> </Declaration>

Table 3. Archi XML representation element relations and respective OWL ObjectProperties

XML	OWL
<source element="BusinessActor"> <target element="BusinessObject" relations="ao"/> <target element="Contract" relations="ao" /> </source>	<Class IRI=#BusinessActor"/> <ObjectAllValuesFrom> <ObjectProperty IRI="#accesses"/> <ObjectUnionOf> <Class IRI="#BusinessObject"> <Class IRI="#Contract"/> </ObjectUnionOf> </ObjectAllValuesFrom>

After the concepts and relationships are represented as classes and object properties, the ontology constraints still need to be included. Restrictions were added to the properties: Inverse ObjectProperties and SuperObjectProperties axioms were added to the OWL ontology, so that derived relationships can be extracted through the use of reasoners. Moreover, two SuperObjectProperty chains were created for modeling dependencies between different elements. The dependsDown ObjectProperty is thus a SuperObjectProperty of the aggregation, composition, assignment, usage, and realization ObjectProperties that resulted from the conversion of the ArchiMate relations, while the dependsUp ObjectProperty fills the same purpose for the counterpart InverseObjectProperties.

3.3   Transformation of ArchiMate Models

The ArchiMate models, i.e. the instances of the metamodel, are converted to OWL using a converter that converts a CSV file, generated using the Archi tool [26] plug-in that export such file, in representation of an ArchiMate model to OWL. Figure 1 shows an OWL representation of the ArchiMate metamodel, along with classes and object properties.

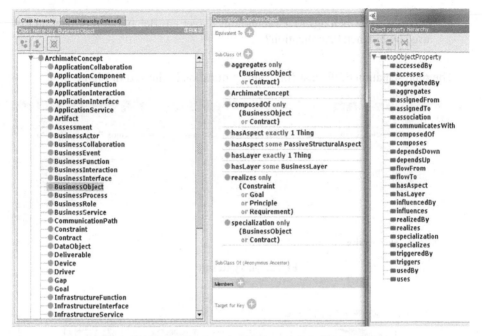

Fig. 1. Business Function class and corresponding properties

3.4   Reasoning

OWL language is based on Description Logics (DL), which is one of the main representational facilities of OWL. OWL- DL enables taking advantage of existing inference and querying mechanisms to analyses the models and assessing their consistency. A Description Logic reasoner performs various inferring services [27].

In the case of ArchiMate, it is possible to take advantage of reasoning in order to check the inconsistencies on the models and for inferring different dependencies between different elements of enterprise models.

4   Case Study

ArchiSurance is a fictitious insurance company used throughout the ArchiMate 2.0 as a case. The scenario was converted to the OWL to be represented as the instances of the ArchiMate metamodel; this example in this section is used to illustrate the capabilities of reasoning, by validating the correctness of the ontology. A set of predefined competency questions were used in order to validate ontology [28]. One of the competency questions defined to validate the integrated ontology is :

- What Services belongs to a Technology Layer are Behavioural Aspect and is used by Financial Application?

The representation of the last competency question is showed in Figure 2.

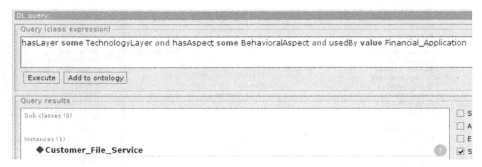

Fig. 2. Query Result

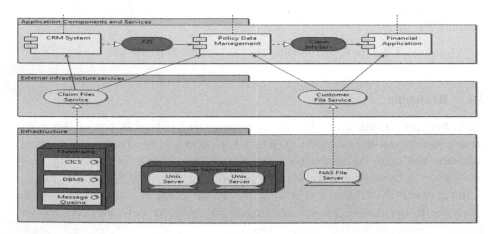

Fig. 3. Archinsurance Layered View

The model below in Fig. 3 depicts a high-level view of the application components and infrastructure in the Archinsurance scenario.

5   Conclusions and Future Work

This paper presents an application of model transformation to translate models described with the standard ArchiMate enterprise architecture language into the OWL ontology language. The resulting artefact is an ArchiMate ontology. This ontology was validating by a set of competency question in a use case. The OWL language gives a formal representation of the ArchiMate metamodel and enables reasoning to be applied to enterprise architecture models. As it was proven, ontologies assist to cover the semantic behind enterprise architecture concepts. In this sense, we presented some techniques that can be used for specification and analysis of models, allowing a better understanding of them. Future works will focus on the application of this approach to new scenarios in order to discover the analysis possibilities, considering the usage of different reasoning and querying techniques.

References

1. Lankhorst, M.: Enterprise architecture at work: Modelling, communication and analysis. Springer (2013)
2. Group, T.: TOGAF(2009)
3. Allweyer, T.: BPMN 2.0. BoD (2010)
4. Ye, J., Sun, S., Song, W., Wen, L.: Formal Semantics of BPMN Process Models Using YAML. In: Intelligent Information Technology Application, pp. 70–74 (2008)
5. Group, T.: Archimate 2.0 Specification (2012)
6. Jenz, D.: Business Process Ontologies: Speeding up Business Process Implementation (2003)
7. Gruber, T.: A Translation Approach to Portable Ontology Specifications. Knownledge Acquisitions 5, 199–220 (1993)
8. Berners-Lee, T.: The semantic web. Scientific American (2001)
9. Shadbolt, N.: The semantic web revisited (2006)
10. Hitzler, P.: Foundations of semantic web technologies. Chapman and Hall/CRC (2011)
11. Gašević, D.: Model driven architecture and ontology development. Springer (2006)
12. Pan, J.: Ontology-Driven Software Development. Springer (2013)
13. Happel, H.-J.: Applications of ontologies in software engineering. In: Proc. of Workshop on Sematic Web Enabled Software Engineering (2006)
14. Uschold, M.: The enterprise ontology (1998)
15. Geerts, G.: An ontological analysis of the economic primitives of the extended-REA enterprise information architecture (2002)
16. Kang, D.: An ontology-based enterprise architecture (2010)
17. Wagner, G.: Ontologies and Rules for Enterprise Modeling and Simulation (2011)
18. Azevedo, C., Almeida, J., Sinderen, M., Quartel, D., Guizzardi, G.: An ontology-based semantics for the motivation extension to archimate. In: 15th IEEE International Enterprise Distributed Object Computing Conference, pp. 25–34 (2011)
19. Almeida, J., Guizzardi, G.: An ontological analysis of the notion of community in the RM-ODP enterprise language. Computer Standards & Interfaces, 257–268 (2013)
20. Johnson, P., E.: A tool for enterprise architecture analysis (2007)
21. Buschle, M., J.: A tool for enterprise architecture analysis using the prm formalism (2011)

22. SoS, A.: A uniform approach for system of systems architecture (2009)
23. Gagliardi, J.: A workshop on analysis and evaluation (2010)
24. McGuinness, D., van Harmelen, F.: OWL Web Language Overview. W3CRecommendation 10, 1–19 (2004)
25. Favier, M.: Developing a Measure of Collective Awareness in Virtual Teams (2007)
26. Archi - Archimate Modelling. In: Overview, http://archi.cetis.ac.uk (accessed December 05, 2012)
27. Protégé. In: what is protégé?, http://protege.stanford.edu/ (accessed December 05, 2012)
28. Fox, M., Gruninger, M.: Enterprise Modeling (1998)

Multi-platform Semantic Representation
of Interactive 3D Content

Jakub Flotyński and Krzysztof Walczak

Poznań University of Economics,
Niepodległości 10, 61-875 Poznań, Poland
{flotynski,walczak}@kti.ue.poznan.pl
http://www.kti.ue.poznan.pl

Abstract. In this paper, a semantic approach to building multi-platform 3D content is proposed. The presented solution is intended to enable flexible and efficient creation of 3D presentations covering a wide range of target platforms – visualisation tools, content representation languages and programming libraries. Referring to the semantics of particular content elements can facilitate conceptual knowledge-based content creation at arbitrarily high levels of abstraction, and it can improve indexing, searching and analysis of 3D content in a variety of application domains on the web.

Keywords: 3D web, Semantic Web, 3D content, multi-platform, ontology.

1   Introduction

Widespread use of interactive 3D technologies and multimedia platforms, including powerful and omnipresent mobile devices, has been recently enabled by the significant progress in hardware performance, the rapid growth in the available network bandwidth as well as the availability of versatile input-output devices. 3D technologies become increasingly used in various application domains, such as education, training, entertainment and social media, significantly enhancing possibilities of presentation and interaction with multimedia information sources, thus increasing collective awareness of their users.

However, to reach a high number of recipients on the web, support for a diversity of hardware and software systems must be provided. Currently, wide coverage of different hardware and software systems by 3D presentations is typically achieved by providing separate implementations of various 3D content browsers and presentation tools. However, in contrast to the development of individual 3D content browsers and presentation tools for different target systems, compatibility of 3D content representation with diverse popular presentation platforms could improve the reuse of 3D content components and the overall use of 3D content on the web. In such an approach, once 3D content is created, it can be presented using different platforms. Moreover, such approach does not require users to install additional software, but it can leverage well-established 3D content browsers and presentation tools that may already be installed on the users' systems (e.g., Adobe Flash Player or WebGL/X3DOM-compliant web browsers).

L.M. Camarinha-Matos et al. (Eds.): DoCEIS 2014, IFIP AICT 423, pp. 63–72, 2014.
© IFIP International Federation for Information Processing 2014

However, currently, the development of 3D platforms is driven by large industry players in a competitive environment and the issue of multi-platform 3D content presentation is still neglected, resulting in fragmentation of content and technologies. This is an important obstacle preventing the mass use of 3D application interfaces.

The main contribution of this paper is a multi-platform semantic representation of interactive 3D content. The approach permits flexible and efficient creation of 3D content for a variety of target presentation platforms. In the proposed solution, once the structure of 3D content is designed, it can be automatically transformed into different final presentation forms, which are suited to different 3D content presentation platforms. The selection of the target platforms to be used is an arbitrary decision of a system designer. Referring to the semantics of particular 3D content components and the conformance to the well-established Semantic Web standards enables 3D content representation that is independent of particular browsers and presentation tools, permits reflection of complex dependencies and relations between content components, and can facilitate indexing, searching and analysis of 3D content in a variety of web applications.

2 Interactive 3D Content in Collective Awareness Systems

Support for content presentation across different platforms is essential for building collective awareness systems. Several works have been devoted to multi-platform 3D content presentation. In [1], a specific 3D browser plug-in for different web browsers has been described. In [2], an approach to hardware multi-platform 3D content presentation based on MPEG-4 has been proposed. In [3], an approach to multi-platform visualisation of 2D and 3D tourism information has been presented. In [4], an approach to adaptation of 3D content complexity with respect to the available resources has been proposed. In [5], a multi-platform on-line game has been presented. In [6], integrated information spaces combining hypertext and 3D content have been proposed to enable dual-mode user interfaces – embedding 3D scenes in hypertext and immersing hypertextual annotations into 3D scenes – that can be presented on multiple platforms on the web. The aforementioned works cover the development of 3D content presentation tools and environments as well as contextual platform-dependent content adaptation. However, they do not address comprehensive and generic methods of content transformation to improve building of multi-platform 3D content presentations.

Collective awareness systems require sharing the meaning of data and content between users and systems. Numerous works have been devoted to semantic 3D representation of information. In [7], integration of X3D and OWL using scene-independent ontologies and the concept of semantic zones have been proposed. In [8], an ontology for X3D as well as semantic properties for coupling VR scenes with domain knowledge have been described. In [9], 3D content representation based on reusable elements with specific roles has been introduced. In [10], an approach to generating virtual words upon domain ontologies has been considered. In [11], semantic entities in VR applications have been discussed. The aforementioned approaches address different aspects of semantic modelling of 3D content, but they lack general solutions for comprehensive conceptual creation of 3D content with respect to its components, properties and relations, at an arbitrarily high level of semantic abstraction.

3   The Method of Semantic 3D Content Creation

Although several approaches have been proposed for semantic modelling of 3D content, they lack general and comprehensive solutions for flexible creation of 3D representation of information. The following requirements, which have been specified for an approach to 3D content creation, go beyond the current state of the art in semantic modelling of 3D content. First, the approach should enable conceptual declarative modelling of content with discovery of hidden knowledge, which is not specified explicitly, but has impact on the modelled content. Second, it should reduce the effort in content design by enabling modelling of complex content components and properties at arbitrarily chosen levels of abstraction, including both the aspects that are directly related to 3D content and the aspects that are specific for a particular domain (e.g., to facilitate content creation by domain experts who are not IT-specialists). Third, it should be independent of particular hardware and software platforms to enable multi-platform 3D content presentation.

This section provides an overview of a method of semantic creation of interactive 3D content (proposed in [12]), which leverages the model of multi-platform 3D content proposed in the next section. In the presented method, the creation of 3D content is a sequence of partly dependent activities (Fig. 1). Modelling of 3D content in the first three activities – design of a concrete representation of 3D content, mapping the concrete representation to domain-specific concepts, and design of a conceptual representation of 3D content – is performed by a developer and a domain expert, depending on the expertise required in the particular modelling activity. The two following activities of the method – expanding the content representation and building the final content representation – are performed automatically by specific software. The activities precede 3D content presentation, which may be done using various content presentation platforms.

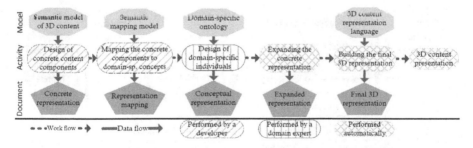

Fig. 1. Semantic creation of 3D content

The design of a concrete semantic representation (activity 1) – provides particular elements of 3D content to enable representation of domain-specific concepts that are further used in activity 3. The result of activity 1 is a concrete semantic representation of 3D content, which is a knowledge base compliant with the platform-independent semantic model of 3D content (PIM – proposed in [13]). The knowledge base incorporates semantic components and properties that are directly related to 3D content, e.g., meshes, groups of objects, materials, viewpoints, events, etc.

The mapping of a concrete 3D content representation (created in activity 1) to domain-specific semantic concepts enables 3D presentation of domain-specific knowledge bases (created in activity 3) [14]. Mapping is performed once for a particular domain-specific ontology and a concrete representation, and it permits the reuse of concrete components and properties for forming 3D representations of various domain-specific individuals (which conform to the domain-specific ontology).

The design of a conceptual semantic representation (activity 3) enables creation of 3D content at a high level of abstraction with a domain-specific ontology. This activity can be performed many times for a particular domain-specific ontology, a concrete representation and a mapping. The following activities of the content creation process can be performed automatically. Expanding the content representation multiplies the concrete components (created in activity 1), which are associated with domain-specific concepts, and assigns them directly to domain-specific individuals (created in activity 3). Building a final 3D representation is a transformation of the expanded representation to its final 3D counterpart, which is encoded using a particular 3D content representation language. This stage of the method is an extension of the approach proposed in [15], and it uses the multi-platform model of 3D content, which is the main contribution of this paper.

4 Multi-platform Model of 3D Content

Although numerous solutions have been proposed for creating semantic representations of 3D content, they do not enable flexible creation of multi-platform 3D content, which can be used for visualisation of various types of information using a multitude of available content presentation tools.

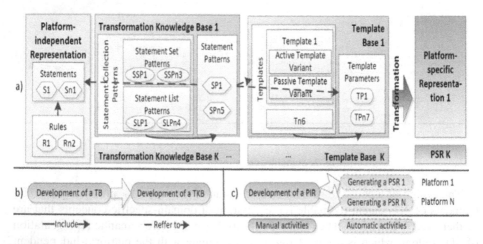

Fig. 2. Multi-platform model of 3D content (a), introducing a new platform to the system (b), and modelling multi-platform 3D content (c)

In this section, a new multi-platform model of 3D content is proposed (Fig. 2a). The model extends the semantic content model (cf. [13,14]) and enables the last activity of the content creation method (cf. [12] and Section 3) – building the final 3D content representation. The model consists of four parts: *platform-independent content representations* (PIRs), *platform-specific content representations* (PSRs), *template bases* (TBs) and *transformation knowledge bases* (TKBs). The model enables transformation of PIRs to PSRs, which may be visualised with diverse 3D content representation languages (e.g., X3D, Java), programming libraries (e.g., Java3D, Away3D) and game engines (e.g., Unity, Unreal), which are determined by the content presentation platforms to be used. Transformation is performed using TKBs and TBs, which are specific to particular presentation platforms. An individual TB and its corresponding TKB are created once every time a new presentation platform is introduced to the system (Fig. 2b). Since a TB and a TKB are added, they may be used for the development of various PSRs that are permitted by the new presentation platform (Fig. 2c). The particular parts of the proposed content model are described in the following subsections.

4.1   Platform-Independent Representations

A *platform-independent representation* (PIR) of 3D content is a knowledge base comprised of platform agnostic semantic elements: individuals – reflecting 3D content components, and properties – reflecting dependencies and relations between the components. Every PIR is compliant with the PIM (cf. [13] and Section 3). A PIR declaratively describes 3D content, as it is a set of semantic *statements* (Ss) and *rules* (Rs), which may be described using, e.g., RDF and SWRL, respectively. Every S is a triple consisting of a *subject*, a *property* and an *object*. Ss are used to describe aspects of 3D content that are related to geometry (e.g., shape), structure (e.g., sub-components), space (e.g., position), appearance (e.g., texture) and simple behaviour (e.g., animations). Every R is an *implication* including a *body* and a *head*, which are conjunctions of Ss. An R is interpreted in the following way: if the *body* is satisfied, then the *head* is also satisfied. Rules are used to describe complex behaviour of 3D content, in particular complex animations and interactions, which may depend on multiple factors, such as time, state of objects, user actions, etc. PIRs can be presented on different platforms when transformed to PSRs with appropriate TBs and TKBs.

4.2   Platform-Specific Representations and Template Bases

A *platform-specific representation* (PSR) of 3D content is a counterpart to a PIR, and it includes 3D objects and 3D scenes, which are encoded using a 3D content representation language. Every PSR is built upon a TB, as it is a combination of *templates* (Ts), which are parameterised fragments of code (e.g., sequences of instructions) of a particular 3D content representation language. Ts may be linked to individual Ss as well as to Ss that are included in Rs of PIRs. For some of the target platforms used, this context of linking may determine the form of the T used. Every T may be given in two *template variants* that differ depending on the context of its use. *Active template variants* (ATVs) are used for the independent Ss (to express logical facts) and the Ss included in the *heads* of Rs (to express logical results), while *passive template variants* (PTVs) are used for the Ss included in the *bodies* of Rs (to express logical conditions). For instance, the S [object pim:shape "cone"] may be transformed to an ATV [object.shape="cone"] or to a PTV [object.shape. isEqual("cone")], which are encoded in an object-oriented language.

To enable flexible transformation of PIRs to PSRs, the granularity of Ts should reflect the semantic granularity of their corresponding Ss – Ts should neither extend nor narrow the meaning of the linked Ss. Composing basic Ss into semantically more expressive Ss is performed in the previous activities of the semantic modelling method presented. *Template parameters* (TPs) enable the reuse of individual Ts in different contexts in combination with other Ts, e.g., by joining or nesting Ts. All PSRs that are based on a common TB and which are presented on a common target platform are generated from PIRs using a common TKB.

4.3   Transformation Knowledge Bases

A *transformation knowledge base* (TKB) links an individual TB to the PIM, thus enabling reflection of semantic Ss by parameterised Ts. A TKB specifies a transformation of PIRs to the corresponding PSRs that are to be presented on a common target platform. Semantic elements contained in a TKB do not influence the modelled 3D content. A TKB consists of *statement patterns* (SPs), each of which matches a group of possible Ss that should be processed collectively. Linking a group of Ss to an SP is performed by semantic generalisation. A generalisation may pertain to the *subject*, the *property* or the *object* of Ss. For instance, a possible generalisation (an SP) of the S [object pim:color "red"] in terms of property, is the SP [object pim:appearanceProperty "red"], while a possible generalisation of the S [object rdf:type pim:Mesh3D] in terms of object, is the SP [object rdf:type pim:GeometricalComponent].

A target PSR, which is generated based on Ss of a PIR, may require an exchange of TPs in a group of different Ts, which are associated with these Ss, and it may require a specific order of Ts. For this purpose, SPs may be gathered into *statement collection patterns* – *statement set patterns* (SSPs), which do not respect the order of the associated Ts, and *statement list patterns* (SLPs), which do respect the order of the associated Ts in the resulting PSR. For instance, while a declarative PIR including the pair of Ss [light pim:intensity "10". light rdf:type pim:DirectionalLight.] does not depend on the order of the Ss, a corresponding imperative PSR may require the following order of the associated Ts (imperative code instructions) [DirectionalLight light = new DirectionalLight(); light.intensity = 10;].

5   Example of a Multi-platform 3D Content Representation

In this section, an example of a multi-platform 3D content representation is discussed. The example 3D scene (Fig. 3) could be used, e.g., in a collective educational system or a virtual museum system. The scene includes a light source and a 3D model of a plough, which has been retrieved from a repository of a virtual museum of agriculture. In the example, the Adobe Flash Player is selected as the target presentation platform. However, in general, arbitrary selected multiple platforms could be used. A PIR of the scene is presented in Listing 1. It uses the PIM (the pim prefix) to describe the 3D content components regardless of any particular target platform. A TB and a TKB, which enable encoding of PIRs with the ActionScript imperative programming language and the Away3D library, are presented in Listings 2 and 3. The listings include only elements that are crucial for the discussion.

The S1 and S2 *statements* (Listing 1) create the scene and set its background with the appropriate colour. The *Ss* are processed according to the SLP1 (Listing 3), which includes two SPs. The S1 is dynamically linked to T1 (by its *property* and *object*) and the S2 is dynamically linked to T2. The T1 is a template of an ActionScript document and it initializes a 3D view, while the T2 sets the background of a scene. The order of the SPs in the SLP1 determines the order of the T1 and T2 in the resulting document (a PSR). The SLP1 requires a new object to be created before any property of this object is set (according to the imperative programming paradigm). If the T linked to the SP1 includes the $actions TP, the T linked to the SP2 is injected into this TP, else the T of the SP2 follows the T of the SP1 in the final PSR. The S3 and S4, which create a light source, are processed in the same order (determined by the SLP1) as the previous *Ss*, independently of their initial order in the PIR.

The plough, which is a virtual museum artefact, is a structural component that consists of the three geometrical components—a box, a wheel and a frame (S5-S7). The SLP2 is applied to the S5-S7—in the resulting PSR, instructions that create new objects in the scene, precede instructions that link these objects by properties (T3). In the example, the resulting PSR (an ActionScript document) includes the Ts in the required order, with the proper TPs specified. The PIR has been created with the Protégé editor. However, a tool for visual semantic modelling can be developed.

Fig. 3. A final 3D content representation

```
S1: scene rdf:type pim:Scene.
S2: scene pim:backgroung "beige".
S3: light pim:intensity "10.0".
S4: light rdf:type pim:DirLight.
S5: plough rdf:type
        pim:StructuralComponent.
S6: plough pim:includes box ,
        wheel , frame.
S7: box, wheel, frame rdf:type
        pim:GeometricalComponent.
```

Listing. 1. A platform-independent representation (PIR) of 3D content

```
01: T1: {
02:    $declarations
03:    public function Main():void {
04:       var $object:View3D = new
              View3D();
05:       $actions } }
06: T2: { $object.backgroundColor =
          $data; }
07: T3: {
          $object1.addChild($object2); }
```

Listing. 2. A platform-specific template base (TB) for an object-oriented language

```
SLP1: { SP1: ?object rdf:type
          pim:ContentComponent.
   SP2: ?object pim:dataProperty
          ?data.
   SP1.$actions == SP2 }
SLP2: { SSP: { SP1: ?object1
          rdf:type pim:ContentComponent.
   SP2: ?object2 rdf:type
          pim:ContentComponent. }
   SP3: ?object1 pim:objectProperty
          ?object2. }
```

Listing. 3. A transformation knowledge base (TKB)

6   Conclusions and Future Works

In this paper, a new approach to building multi-platform representations of 3D content has been proposed. The presented solution has several important advantages in comparison to the available approaches to 3D content presentation. First, it is more convenient for environments, which cover various hardware and software systems, as the development of TBs and TKBs requires less effort than the development of individual content models or content browsers. Second, the possible use of well-established 3D content presentation tools, programming languages and libraries liberates users from the installation of additional software, which can improve the dissemination of 3D content. Third, the use of Semantic Web standards permits conceptual knowledge-based content modelling that refers to hidden information inferred using, e.g., RDF, OWL or SWRL reasoners. Moreover, semantic representation enables more efficient and flexible methods of indexing, searching and analysis of the content regarding complex dependencies and relations between its components, provides methods of describing rules of combining different components and permits description of complex content behaviour.

The following directions of future research are possible. First, the proposed approach needs to be implemented for selected content representation languages (e.g., ActionScript and X3D), and combined with the approach to transformation of 3D content description formats proposed in [15]. Second, the approach should be evaluated and compared to other solutions in terms of both: the possibilities of high-level conceptual content creation and the effort in the implementation of multi-platform 3D content presentations. Third, the complexity of TBs and TKBs for different target languages – imperative and declarative – could be compared. Furthermore, a visual modelling tool supporting the semantic content creation can be developed. Finally, a persistent link between semantic and final content representations can be proposed to provide real-time synchronisation of the content state.

References

1. Mendes, C.M., Drees, D.R., Silva, L., Bellon, O.R.: Interactive 3D visualization of natural and cultural assets. In: Proc. of the 2nd Workshop on eHeritage and Digital Art Preservation, Firenze, Italy, pp. 49–54 (2010)
2. Celakovski, S., Davcev, D.: Multiplatform real-time rendering of MPEG-4 3D scenes with Microsoft XNA. In: ICT Innovations, pp. 337–344. Springer (2010)
3. Almer, A., Schnabel, T., Stelzl, H., Stieg, J., Luley, P.: A tourism information system for rural areas based on a multi-platform concept. In: Proc. of the 6th Int. Conf. on Web and Wireless Geographical Inf. Systems, Hong Kong, China, pp. 31–41 (2006)
4. Tack, K., Lafruit, G., Catthoor, F., Lauwereins, R.: Platform independent optimisation of multi-resolution 3D content to enable universal media access. The Visual Computer 22(8), 577–590 (2006)
5. Han, J., Kang, I.-G., Hyun, C., Woo, J.-S., Eom, Y.-I.: Multi-platform online game design and architecture. In: Costabile, M.F., Paternó, F. (eds.) INTERACT 2005. LNCS, vol. 3585, pp. 1116–1119. Springer, Heidelberg (2005)
6. Jankowski, J., Decker, S.: A dual-mode user interface for accessing 3d content on the worldwide web. In: Proc. of the 21st Int. WWW Conf., Lyon, France, pp. 1047–1056 (2012)
7. Pittarello, F., Faveri, A.: Semantic description of 3D environments: a proposal based on web standards. In: Proceedings of the 11th International Conference on 3D Web Technology, Columbia, MD, USA, pp. 85–95 (2006)
8. Kalogerakis, E., Christodoulakis, S., Moumoutzis, N.: Coupling Ontologies with Graphics Content for Knowledge Driven Visualization. In: VR 2006 Proceedings of the IEEE Conference on Virtual Reality, Alexandria, VA, USA, pp. 43–50 (2006)
9. Walczak, K.: Flex-VR: Configurable 3D Web Applications. In: Proc. of the Int. Conf. on Human System Interaction, pp. 135–140. Kraków (2008) ISBN: 1-4244-1543-8
10. De Troyer, O., Kleinermann, F., Pellens, B., Bille, W.: Conceptual modeling for virtual reality. In: Tutorials, Posters, Panels and Industrial Contributions at the 26th International Conference on Conceptual Modeling, vol. 83, pp. 3–18. Australian Computer Society, Inc. Darlinghurst (2007) ISBN: 978-1-920682-64-4

11. Latoschik, M.E., Frohlich, C.: Semantic Reflection for Intelligent Virtual Environments. In: IEEE Virtual Reality Conference, Charlotte, NC, USA, pp. 305–306 (2007)
12. Flotyński, J., Walczak, K.: Semantic Modelling of Interactive 3D Content. In: Proc. of the 5th Joint Virtual Reality Conference, Paris, France, pp. 41–48 (2013)
13. Flotyński, J., Walczak, K.: Semantic Multi-layered Design of Interactive 3D Presentations. In: Proceedings of the Federated Conference on Computer Science and Information Systems, Kraków, Poland, pp. 541–548 (2013)
14. Flotyński, J., Walczak, K.: Conceptual Semantic Representation of 3D Content. In: Abramowicz, W. (ed.) BIS 2013 Workshop. LNBIP, vol. 160, pp. 244–257. Springer, Heidelberg (2013)
15. Flotyński, J., Dalkowski, J., Walczak, K.: Building multi-platform 3D virtual museum exhibitions with Flex-VR. In: Proceedings of the 18th International Conference on Virtual Systems and Multimedia, Milan, Italy, pp. 391–398 (2012)

Part IV
Self-organizing Manufacturing Systems

Performance Assessment in Self-organising Mechatronic Systems: A First Step towards Understanding the Topology Influence in Complex Behaviours

Pedro Neves[1], Luis Ribeiro[2], Mauro Onori[1], and José Barata[2]

[1] KTH Royal Institute of Technology, Department of Production Engineering,
Stockholm, Sweden
[2] CTS UNINOVA, Dep. de Eng. Electrotecnica, F.C.T. Universidade Nova de Lisboa,
Monte da Caparica, Portugal
{pmsn2,onori}@kth.se, {ldr,jab}@uninova.pt

Abstract. The research and development of self-organising mechatronic systems has been a hot topic in the past 10 years which conducted to very promising results in the close past. The proof of concept attained in IDEAS project [1] that plug&produce can be achieved in these systems opens up new research horizons on the topics of system design, configuration and performance evaluation. These topics need to consider that the systems are no longer static prototypes but instead several distributed components that can be added and removed in runtime. The distribution of modules in the system and their inherent connections will then potentially affect the system's global behaviour. Hence it is vital to understand the impact on performance as the system endures changes that affect its topology. This article presents an exploratory test case that shows that as a system evolves (and the nature of the network of its components changes) the performance of the system is necessarily affected in a specific direction. This performance landscape is necessarily complex and very likely nonlinear. Simulation plays therefore an important role in the study of these systems as a mean to generate data that can be later on used to generate macro level knowledge that may act as a guideline to improve both design and configuration.

Keywords: Multi-Agent Systems, Performance Assessment, Evolvable Production Systems, Self-Organizing Systems, Simulation.

1 Introduction

Assembly lines are flow oriented production systems which were originally built for cost-efficient mass production. Given the current turbulent markets they are now struggling to cope with constant re-design and re-configurations imposed by high customisation and low volumes [2]. As the business paradigm shifts towards an increased customization and personalization the requirements imposed on production change [3], [4]. Traditional systems present optimal solutions for particular forecasted products. However, they fail to present the desired agility to follow volatile market

L.M. Camarinha-Matos et al. (Eds.): DoCEIS 2014, IFIP AICT 423, pp. 75–84, 2014.

needs. This augments the need of systems that are rapidly deployable, reconfigurable and autonomous and that can accommodate the required changes with minimal integration/programming effort and therefore maximum cost-effectiveness [5]. Modern production paradigms emerged in the last two decades aiming at offering responsive and cost effective solutions. Some examples are Bionic Manufacturing Systems [6], Reconfigurable Manufacturing Systems [7], Evolvable Production Systems (EPS) [8], Holonic Manufacturing Systems [9] and Changeable Manufacturing Systems [10]. They all share some core principles such as modularity, structure, heterogeneity, autonomy, interaction and dynamics [11]. All modern paradigms aim at the encapsulation of module functionalities as services to enable their seamless integration in production systems to tackle interoperability between systems and re-usability of legacy equipment. Modern paradigms exhibit a shift from centralized to distributed control architectures where each node of the system is capable of taking autonomous decisions and interacting with the other nodes on the network. This makes them suitable to handle aspects related to dynamic addition/removal of heterogeneous modules and change of products on the fly with little or none reprogramming efforts.

The Manufuture roadmap [12] has pointed out the need of cross-sectorial research to develop adequate IT structures to support scalable and interoperable control systems, plug&play production modules and responsive factories through co-operative and self-organising control systems. Last decade's ICT research has been fruitful leading to architectures and IT middleware that support ready to use intelligence and autonomy at device-level. The most common approaches to implement distributed control automation are Multi-Agent Systems (MAS) and Service-Oriented Architectures (SOA). Given their loosely coupled nature they enable the dynamic composition of complex entities from simple services and support the dynamic plug and unplug of self-contained heterogeneous mechatronic modules [1], [11], [13], [14], [15].

These characteristics are desirable to increase system's responsiveness and tackle disturbances in the shop floor (e.g. machine failures, bottlenecks, volume and product change, etc.) [5], [16]. Nevertheless due to the early stage development of these systems, there is a lack of methods to support their design and configuration and to evaluate their behaviour although some analysis work has been carried out in [17], [18]. The purpose of this paper is therefore show a potential new system analysis path that may contribute to the design and configuration of complex manufacturing systems with a potential of generalization to collective systems. Particularly it provides a preliminary behavioural assessment of an EPS system to unveil the influence of system component's topology in the total make span. The subsequent details are organized as follows: section 2 frames this work under the scope of collective awareness systems; section 3 presents the state of the art of assembly line design, the research gap and the proposed analysis method; in section 4 it is detailed a preliminary experiment that motivates this research; and finally some concluding remarks are offered in section 5.

2   Relation to Collective Awareness Systems

The proliferation of embedded computing and web technology has opened a door for exploring distributed automation systems. In such a context, all entities in the shop floor are to some extent intelligent, able to reason and make decisions dynamically according to their context, forming dynamic loosely coupled networks of cooperative entities that exhibit collective behaviour. The main advantages of such approach are the increase of local autonomy to allow the addition/removal of equipment and improved robustness, fault tolerance and adaptability by empowering self-* properties [1], [19].

The increase of IT infrastructures and IT middleware that support intelligence at device level also enables the collection of huge amounts of data concerning product and process design, logistics, assembly, quality control, scheduling, maintenance, fault detection, etc.; extending largely previous available data and adding more reliability to it [20]. Extracting knowledge from these operations is perceived as a tremendous opportunity to improve productivity and efficiency both at product and system levels which is important to reduce costs. This can contribute to the comprehension of the local-global relations in the system and the development of better system designs that can fully explore the self-organizing essence of modern paradigms. As a natural consequence, this would contribute to the development of collective awareness systems and allow enterprises to target volatile business opportunities and build the fundamental pillars for competitive sustainable innovation.

3   State of the Art and Problem Definition

Assembly line design has as main goal the maximization of the ratio between throughput time and required costs and incorporates all decisions concerning resource planning, sequence planning and system balancing [2]. These activities are often carried out by a human expert based on experience and know-how. Consequently many design alternatives are left unexplored and the quality of the line design will be highly dependent on the competence of designer [21]. Assembly Line Design (ALD) comprises two main sub-topics: Assembly Line Balancing (ALB) and Resource Planning (RP). ALB's more common problems consist in assigning tasks to workstations to satisfy a specific objective function while RP problems, on the other hand, consider the case when more than one type of equipment is available [2]. Considering the state of the art of ALD, one can find a vast number of contributions in literature regarding ALB. Especially regarding Simple Assembly Line Balancing (SALB) which introduces several simplifications and restrictions to the General Assembly Line Balancing (GALB) problem such as: homogeneous products, fixed cycle times, deterministic operation times of tasks and serial layout line. These simplified models often fail to reproduce the reality and therefore are rarely used in practical industrial applications [21], [22]. Given the early stage development of self-organising mechatronic systems little research has been conducted on assembly line design. Nevertheless, the different nature and goals of these systems in comparison to traditional ones enables the identification of some aspects that might demand new methods: 1) modern paradigms do not target the design of prototypic tailored systems for a specific product but instead the design of responsive and cost-effective solution for evolving requirements; 2) the encapsulation of module functionalities as services and their seamless integration in production systems enables different equipment to be plugged and unplugged in runtime without reprogramming; 3) equipment functionality can be very specific or multi-purpose hence the layouts and network structure are fundamental; 4) modules are intelligent and the system behaves as a cooperative network of modules contrary to centralized automation approaches; 5) task assignment is performed dynamically following the current system status and there is no a-priori assignment of product tasks to specific equipment. The above mentioned points justify the need of a method that can present suitable real-life configurations to apply in Self-organizing mechatronic systems for scenarios of constant system evolution to better fit evolving production requirements. The overall behaviour of these systems results from the interactions and self-organization of its constituents which are prime characteristics of complex adaptive systems [23]. Furthermore as suggested in [24] the topology of the network can play major role in the self-organising phenomena present in random networks. The research question addressed by this work is:

- "How can the performance of an assembly system composed by a dynamic network of intelligent collaborative modules be evaluated?"
- "What methods can support the design and configuration of an assembly system composed by a dynamic network of intelligent collaborative modules?"

To address these questions a simulation tool was developed to generate data resulting from distinct system modifications and assess the impact of those changes in the system make span. Given the experimental nature of the work the classical research method was adopted.

4   Experiment Design

4.1   The Simulation Agents

In order to design and implement an experiment to study the influence of network topology on self-organising mechatronic systems the EPS paradigm was considered. A full description of Evolvable Production Systems principles and architectures can be found on [25], [26] and the references therein. The starting point of the current research is the generation of representative data. This data has to be necessarily generated by simulation, taking some basic EPS systems and allowing them to evolve. The test-case presented in this article has the goal of providing a first indication of whether network topological features can be used to characterize the assembly system composed by a network of modules or not. And more important if these features exhibit any relation with traditional performance metrics such as the make span. The preliminary experiment considered is this paper is based in the evolution of a U-shaped serial line. The original layout is transformed by adding more paths between the stations in the system. The simulation environment was developed using the JADE framework [27] and all communications in the platform follow the FIPA-Request protocol [28]. The system has 5 basic agents that can be deployed:

- Station Agent – abstracts a module in the system where specific assembly processes can be performed (corresponds to a node in the network).
- Router Agent – abstracts a diverter and links conveyors together being responsible to route the carriers to different conveyors.
- Conveyor Agent – abstracts a conveyor with specific size and capacity controlling the flow of products.
- Carrier Agent – abstracts a carrier in the system responsible to carry the product and follow its production plan. The carriers queue in the Entry Point until they are allocated to a product, and then follow the requests of the product. When they receive requests from the product they are responsible to find the shortest path using Djikstra Algorithm and inform the product when they reach a location where the process can be executed.
- Entry/Exit Point Agent – these agents are responsible for the entrance and exit of carriers in the system.
- Product Agent – abstracts a product and manages its production plan.

4.2   Experimental Setup

In the following experiment a setup with 6 stations with functionalities glue, stack, screw, pick&place, insert and weld is used. It is considered that all of them require the same time to execute their process (3 seconds). The positions of the stations in the system are fixed and only the paths connecting them change. All the conveyors have the same size (4m) and speed (1m/s). Consequently all the carriers travel at the same speed inside the conveyors and have the same size (0,2m). Conveyors are connected by routers and the transfer time is corresponds to a fixed time of one second. Upon request from the product to find a specific process in the system, the carrier will calculate the shortest path at each step of the move (i.e. at each new node) and send request messages to conveyors, stations and routers in order to go through them. Once it arrives to a station that can execute the requested process it sends an acknowledgment (inform message) to the product agent so it can request the skill execution to the station. One Entry Point agent and one Exit Point agent are also deployed where the carriers can enter and leave the system respectively. The considered base system for the tests is depicted in figure 1 (left). A total of 18 system evolutions (adding successive links) were considered and each one of these system was ran 4 times. Table 2 presents the network features extracted and the achieved make span results. Gephi [29] was used to extract the following system features:

- Average Degree Centrality – the centrality degree of a vertex is given by the total number of vertices adjacent to the vertex.
- Average Closeness Centrality – the closeness degree measures the degree to which a node is near all other nodes in a network. It is the inverse of the sum of the shortest distances between each node and every other node in the network.

- Average Betweenness Centrality – it measures to which extent a node is connected to other nodes that are not connected to each other.
- Average Cluster Coefficient – it corresponds to the measure of degree to which nodes in a graph tend to cluster together.

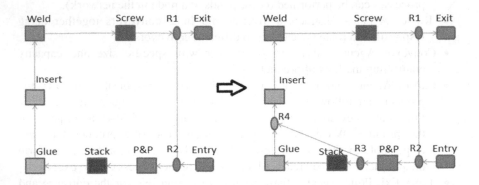

Fig. 1. Example of evolution from Network 1 to Network 2

A total of 12 carriers were deployed which are allocated to the products upon request, and then a total of 80 products from 8 different types (10 of each) were progressively deployed and associated with those carriers (Table 1).

Table 1. Products' workflows deployed

Product	Skill 1	Skill 2	Skill 3	Skill 4	Skill 5
PA1	Pick&Place	Glue	Insert	Weld	Screw
PA2	Stack	Glue	Insert	Screw	Weld
PA3	Insert	Weld	Screw	Glue	Screw
PA4	Stack	Screw	Pick&Place	Glue	Weld
PA5	Pick&Place	Insert	Screw	Weld	Stack
PA6	Glue	Insert	Screw	Weld	Stack
PA7	Insert	Glue	Screw	Weld	Pick&Place
PA8	Weld	Screw	Glue	Stack	Pick&Place

The products "compete" in the platform to be allocated and it is not guaranteed the order they enter the line and therefore results can vary substantially between executions as one can observe in the analysis of results in the next section.

4.3   Preliminary Result's Assessment

The obtained results are depicted in table 2 presenting the average system make span and the network features of each system evolution. It can be observed that as more paths are added in almost all cases the average system make span decreases. Nevertheless, the most significant leaps in performance gain are found in evolution 1→2, 2→3, and 6→7. In the other evolutions the performance is not significantly improved which gives us an indication that exploring some paths can be more rewarding than others. The decrease of make span is somehow expected as the increase of paths results in shortest paths to different destinations and the influence of modules' positions is lowered and the travel time is minimized. Regarding the network features mentioned in section 4.2 one can observe they exhibit a consistent variability as more paths are added. The base system considered is a typical U-shaped serial line where each station has one inbound and one outbound and as more paths are added (more inbounds and outbounds to stations) the centrality degree of nodes increases. With the evolution of the system the average closeness decreases since this feature measures the proximity of a node to all nodes in the network.

Table 2. System features and achieved simulation results

Sys.	AVG degree centrality	AVG closeness centrality	AVG Between. centrality	AVG Cluster Coeff.	AVG Make Span	Standard deviation
1	1	2.81246	12	0	871,4335	30,5132
2	1.125	2.37798	9.25	0	802,8395	26,2787
3	1.25	2.21131	8.25	0	728,3610	4,4376
4	1.375	2.04464	7.25	0	712,9408	20,6743
5	1.625	1.8006	5.8112	0	693,1173	17,0078
6	1.75	1.71725	5.24999	0	696,9508	18,4620
7	1.875	1.675595	5	0	640,8648	16,1297
8	2	1.5774	4.375	0.108	625,1785	11,5933
9	2.125	1.5149	4	0.198	627,0783	11,2726
10	2.375	1.452	3.62501	0.333	618,2798	13,6728
11	2.5	1.411	3.375	0.367	625,6625	3,85056
12	2.625	1.36896	3.125	0.4	622,2423	9,8987
13	2.75	1.3481	3	0.458	620,0778	2,0295
14	2.875	1.3094	2.75	0.479	621,0133	20,7888
15	3.125	1.3304	2.5	0.485	611,1700	8,5325
16	3.25	1.3095	2.375	0.515	618,4785	5,4101
17	3.375	1.2262	2.2504	0.544	615,6785	14,6321
18	3.5	1.20525	2.125	0.56	605,8348	23,6286
19	3.625	1.1845	2	0.577	598,7913	11,4700

The average betweenness decreases with system iteration since it measures to which extent nodes are connected to other nodes that are not connected to each other. This makes sense since the connectivity is improving in the system and almost all nodes will be connected to each other. The average cluster coefficient also increases as it measures how nodes cluster together (triplets of nodes). Regarding the existence of relations between these network features and the make span one can notice that they exist. As the average degree augments the make span is reduced which suggests that the connectivity of the nodes is an important feature to include in the system characterization. A similar conclusion can be deducted by analysing the average closeness of the nodes in the system. When it decreases the make span also decreases since it measures the degree to which a node is near all other nodes in a network. This means that the average closeness can also be potentially considered to characterize the system. The same conclusions can be taken from looking at both average betweenness and average clustering coefficient since they are varying consistently in one direction with the decrease of make span which makes them good candidates for system characterization as well.

This first test-case suggests that there are in fact relations between network features and the system make span that can be further studied. A limiting factor in this experiment was the use of only 12 carriers in all tests. The main goal of this experiment was to have a first indication if network metrics present a consistent variability that can be linked with the overall system performance that results from the interactions and self-organisation of modules. It was possible to conclude that this relation exists and can be further explored. If these relations are fully understood then we can realize the influence of size and topology in the self-organising process and potentially contribute to the generation of better system design and configuration. In highly pluggable systems most of the times the challenge is not to plug more devices but to understand which, where and how to plug these devices and the preliminary test case shows that some system evolutions are more significant than others. Hence the use of machine learning methods will be fundamental as it will enable us to relate all these variables and discover redundancies and dependencies between the features analysed and extract rules that map their real relations to the make span and other performance metrics . Since the relation between performance metrics and networks features is not necessarily linear, Rough Sets present a good candidate technique for this task since they are powerful in the discovery of redundancies and dependencies between the features of objects to be classified and therefore they represent a good approach to classification and rule extraction [30]. Rough set theory provides a mathematical approach using lower and upper approximations to deal with uncertain information and has proved to be useful in exploratory manufacturing applications for fault diagnosis [31], [32].

5   Concluding Remarks

To date most assembly design methods in use target optimized designs from scratch assuming constant the product mix and volumes and trying to match the needed processes in equipment in a repetitive way which is suitable mainly for large volumes and stable production. This research targets, on the other hand highly reconfigurable systems that can evolve autonomously such as the ones developed under the FP7 IDEAS project. One important point is therefore to guide the direction of this evolution. On the other hand data collection from existing system will enable better initial designs. Therefore this research aims and understanding what are the relevant variables, from a system network perspective that can affect its overall performance. The preliminary test case suggests that there are correlations between the network characteristics and its performance. It also shows that certain evolutionary steps are much more relevant than others. The ability to evolve a system in the right direction using the correct "shortcuts" and avoiding mechanical or logical reconfigurations with a marginal gain is extremely important from a cost perspective. Simulation will play a decisive role in the generation of data that can be mined and reused for the assessment of what are the most important features and changes to consider in an EPS system. The present work is only the first step towards a far more ambitious goal. The next steps of this research include the investigation of the suitability of graphs with small-world properties to describe the systems' topology and perform tests by using the generation of random networks and evolutions in the simulation tool. Additionally the performance assessment context will be extended to include more indicators and there will be an incorporation of rule-extraction methods such as rough sets to help us dealing with massive amounts of data and bridge these two dimensions.

References

1. Onori, M., et al.: The IDEAS Project: Plug & Produce at Shop-Floor Level. Assembly Automation 32(2), 4–4 (2012)
2. Boysen, N., Fliedner, M., Scholl, A.: A classification of assembly line balancing problems. European Journal of Operational Research 183(2), 674–693 (2007)
3. Maffei, A.: Characterisation of the Business Models for Innovative, Non-Mature Production Automation Technology, KTH (2012)
4. Hu, S.J., et al.: Assembly system design and operations for product variety. CIRP Annals-Manufacturing Technology 60(2), 715–733 (2011)
5. Onori, M., Barata, J.: Outlook report on the future of European assembly automation. Assembly Automation 30(1), 7–31 (2009)
6. Ueda, K.: A concept for bionic manufacturing systems based on DNA-type information. In: 8th International PROLAMAT Conference on Human Aspects in Computer Integrated Manufacturing, North-Holland Publishing Co. (1992)
7. Koren, Y., et al.: Reconfigurable manufacturing systems. CIRP Annals-Manufacturing Technology 48(2), 527–540 (1999)
8. Onori, M., Barata, J.: Evolvable Production Systems: New domains within mechatronic production equipment. In: IEEE International Symposium on Industrial Electronics (ISIE). IEEE, Bari (2010)
9. Valckenaers, P., et al.: Holonic manufacturing systems. Integrated Computer-Aided Engineering 4(3), 191–201 (1997)
10. Wiendahl, H.P., et al.: Changeable manufacturing-classification, design and operation. CIRP Annals-Manufacturing Technology 56(2), 783–809 (2007)
11. Ribeiro, L., et al.: MAS and SOA: A Case Study Exploring Principles and Technologies to Support Self-Properties in Assembly Systems. In: SARC 2008: Self-Adaptation for Robustness and Cooperation in Holonic Multi-Agent Systems, p. 43 (2008)
12. Jovane, F., Westkamper, E., Williams, D.: The ManuFuture Road: Towards Competitive and Sustainable High-Adding-Value Manufacturing. Springer (2008)

13. Jammes, F., Smit, H.: Service-oriented paradigms in industrial automation. IEEE Transactions on Industrial Informatics 1(1), 62–70 (2005)
14. Shen, W., et al.: Applications of agent-based systems in intelligent manufacturing: An updated review. Advanced Engineering Informatics 20(4), 415–431 (2006)
15. Marik, V., Lazansk, J.: Industrial applications of agent technologies. Control Engineering Practice 15(11), 1364–1380 (2007)
16. Ribeiro, L., Barata, J.: Re-thinking diagnosis for future automation systems: An analysis of current diagnostic practices and their applicability in emerging IT based production paradigms. Computers in Industry 62(7), 639–659 (2011)
17. Ribeiro, L., Rosa, R., Barata, J.: A structural analysis of emerging production systems. In: 10th IEEE International Conference on Industrial Informatics (INDIN). IEEE (2012)
18. Farid, A.: An Axiomatic Design Approach to Non-Assembled Production Path Enumeration in Reconfigurable Manufacturing Systems. In: IEEE International Conference on Systems, Man, and Cybernetics (SMC 2013), Manchester, UK (2013)
19. Ribeiro, L., et al.: Self-organization in automation-the IDEAS pre-demonstrator. In: 37th Annual Conference of the IEEE Industrial Electronics Society (IECON 2011). IEEE, Melbourne (2011)
20. Harding, J.A., Shahbaz, M., Kusiak, A.: Data mining in manufacturing: a review. Journal of Manufacturing Science and Engineering 128, 969 (2006)
21. Michalos, G., Makris, S., Mourtzis, D.: An intelligent search algorithm-based method to derive assembly line design alternatives. International Journal of Computer Integrated Manufacturing 25(3), 211–229 (2012)
22. Boysen, N., Fliedner, M., Scholl, A.: Assembly line balancing: which model to use when? International Journal of Production Economics 111(2), 509–528 (2008)
23. McCarthy, I.P., Rakotobe-Joel, T., Frizelle, G.: Complex systems theory: implications and promises for manufacturing organisations. International Journal of Manufacturing Technology and Management 2(1), 559–579 (2000)
24. Barabási, A.-L., Albert, R.: Emergence of scaling in random networks. Science 286(5439), 509–512 (1999)
25. Ribeiro, L., et al.: Evolvable Production Systems: An Integrated View on Recent Developments. In: 6th International Conference on Digital Enterprise Technology. IEEE, Hong Kong (2009)
26. Onori, M.: Evolvable Assembly Systems - A New Paradigm? In: 33rd International Symposium on Robotics Stockholm (2002)
27. Bellifemine, F.L., Caire, G., Greenwood, D.: Developing multi-agent systems with JADE. Wiley (2007)
28. FIPA, The foundation for intelligent physical agents (2008), http://fipa.org
29. Bastian, M., Heymann, S., Jacomy, M.: Gephi: an open source software for exploring and manipulating networks. In: ICWSM (2009)
30. Pawlak, Z.: Rough set theory and its applications to data analysis. Cybernetics & Systems 29(7), 661–688 (1998)
31. Kusiak, A.: Rough set theory: a data mining tool for semiconductor manufacturing. IEEE Transactions on Electronics Packaging Manufacturing 24(1), 44–50 (2001)
32. Peng, J.-T., Chien, C.-F., Tseng, T.: Rough set theory for data mining for fault diagnosis on distribution feeder. In: IEE Proceedings- Generation, Transmission and Distribution, IET (2004)

Assembly Features Utilization to Support Production System Adaptation

Baha Hasan[1,*], Mauro Onori[2], and Jan Wikander[1]

[1] Department of Machine Design, The Royal Institute of Technology (KTH),
Stockholm, Sweden
[2] Department of Production Engineering, The Royal Institute of Technology (KTH),
Stockholm, Sweden

Abstract. The purpose of this paper is to introduce a proposed methodology to extend the evolvable assembly system (EAS) paradigm for product design by utilizing assembly features in a product. In this paper, assembly features are used to bridge the gap between product design and assembly process by matching features of a part in an assembly to operations of a process in the EAS ontology. This can be achieved by defining and extracting a new set of assembly features called process features, which are features significant to specific and well- defined assembly operations. The extracted assembly features are represented in a proposed model based on product topology. A case-study example is conducted to illustrate the new methodology. A process-feature ontology is proposed as well in order to match the assembly requirements represented by process features with the available processes and skills in the EAS ontology so that adaptation of the production system can be achieved.

Keywords: Adaptation, Evolvable, Assembly, Features, EAS, Process, Skill, Ontology.

1   Introduction

Modern production system paradigms have to cope with several critical issues arising from the need for mass customization, such as short product life-cycles, an increasing number of product variants and frequently changing customer requirements. To overcome these challenges, production systems have to be more adaptive such that they can rapidly respond to the required changes.

Several approaches [1], [2], [3] have been proposed in order to support adaptivity of production systems. One of these approaches is the Evolvable Assembly System (EAS) approach, which was proposed in 2002 and developed during the EUPASS project. EAS aims to cope with unpredictable and changing production requirements by building evolvable capabilities into the production system. As stated by Onori et al. [4], [5], [6] evolvability is not only the ability of system components to adapt to changing requirements, but also a characteristic, which assists the processes in

---

* Corresponding author.

L.M. Camarinha-Matos et al. (Eds.): DoCEIS 2014, IFIP AICT 423, pp. 85–92, 2014.

becoming more self-x, which can stand for self-evolvable, self-reconfigurable, self-tuning, self-diagnosing and so on.

The aim of this paper is to introduce a method to utilize assembly features in an assembly to determine the required assembly process and resource capabilities in the evolvable assembly system (EAS) [7], [8]. In this paper an approach is proposed to improve the adaptability of the evolvable systems by extending the EAS paradigm to the design of a product based on its assembly features.

Assembly features are defined as "features with significance for assembly processes", and are divided into mating (connection) features (such as final position, insertion path/point, tolerances), handling features, (characteristics that give the locations on an assembly component such that it can be safely handled by a gripper during assembly) [9], and form features, which are "A set of geometric entities (surfaces, edges, and vertices) together with specifications of the bounding relationship between them and which have engineering/functional implications and/or provide assembly aid, such as a center line of a hole, on an object" [10], [11]. In other words, form features are geometrical mating entities, which include mating features. Figure 1 illustrates assembly features definition.

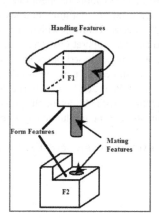

Fig. 1. Assembly features, modified from [12]

In this paper, a methodology is proposed to convert assembly requirements into required assembly processes and capabilities through analysis of assembly features. In this context assembly requirements are those assembly features or characteristics which require a set of moving and joining processes in order to transform a set of components to an assembled or semi- assembled product. An assembly-feature based model is proposed based on definition, representation and extraction of handling, mating and form features. Based on these extracted features, a new set of joining and transporting (moving) features are defined, represented and modeled in an ontology (process-feature ontology) in order to define, in details, the required assembly process and resource capabilities, which will be matched with the available processes and resources in the EAS ontology [7]. The final ultimate aim is to determine and configure the new recourses (modules) in the EAS system, which will be involved to assemble a product.

2   Relationship to Collective Awareness Systems

As EAS is a modular self-reconfigurable system, the proposed methodology tries to create a link between product's assembly features and EAS resource modules represented by their Skills (- Skill is defined as the ability of a resource to perform a process [13]). That is, product and production /assembly systems become linked. In this way product designers become aware of production implications and vice-versa. This enables several stakeholders to share vital information, create awareness and even automate complex development phases (designers, sub-contractors, system integrators, etc.). Figure 2 illustrates this collective awareness link between assembly design, process and EAS.

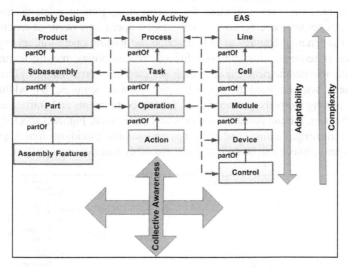

Fig. 2. Connection of design, process and EAS levels, modified from [14]

In Figure 2, the assembly design is composed of subassemblies, being stable assemblies containing two or more parts [14], and parts, which are the elementary components of an assembly [14]. An assembly activity is composed of assembly processes which collections of lower level assembling activities with a purpose of facilitating the assembling of an assembly or subassembly [15], and they are composed of tasks, operations and actions. The EAS is composed of several units; the most fundamental unit is the module unit, which is defined as "a self-composed entity with a given functionality and with well-defined interfaces, via which it interacts with other modules" [4]. In terms of collaborative design, our aim is to facilitate knowledge transfer between part, operation and modules by utilizing assembly features to define the operations and modules required for assembly.

3   Analysis and Representation of Assembly Features Information

According to [16] the basic core of assembly processes are "Moving Part x" and "Joining Parts x and y", so that handling, mating and form features have to be used to derive assembly process features, transporting and joining features, definitions to determine the required transporting and joining processes and capabilities.

Handling features is a generic form of assembly information (independent on the actual position and orientation of the component within an assembly), which can be used to store and retrieve information about feeding, fixturing and grasping [9]. Handling features are generic for some assembly components, such as the base component (the component upon which all remaining assemblies are carried out). This base component is always fixtured, fed and grasped in a predefined way. For the other assembly components, mating and form features are needed as well as handling features in order to derive grasping and feeding features to determine the required transporting processes and capabilities. Figure 4 illustrates the derivation of transporting features.

Mating features are very important for representing joining relationships between assembly components/ parts, because actual joining operations occur at mating features [17]. In order for mating features to fully describe the joining operation, an expansion of these features is required; this can be achieved by combining mating features for an assembly with the geometrical entities being selected for form features. The extended mating features will be used to determine the joining features. By adding joining information (joining methods, groove shapes, joining components and entities, and joining constraints) by user, joining processes features will be derived. Figure 3 specifies different transporting (grasping, feeding, and fixturing) and joining (welding, fitting, gluing, and fastening) features, which can be derived from assembly features.

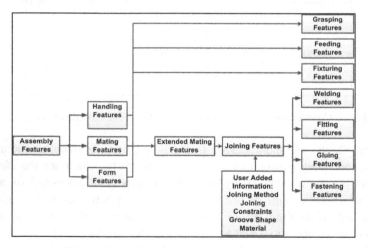

Fig. 3. Derivation of assembly processes features

The required assembly processes can be determined depending on the derived assembly process features. The required resource capabilities can be determined based on those assembly processes. Technical and functional constraints of production system's resources can be determined based on geometrical, non- geometrical, functional and constraints information of an assembly.

Derived assembly features need to be represented in a product model in order to share these features between experts in different domains in the EAS paradigm (product, process and resources). Figure 4 shows a proposed feature--based model which is based on product topology.

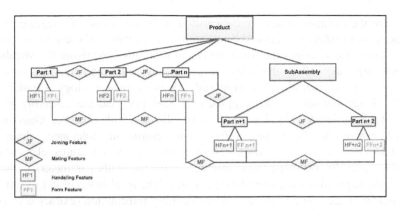

Fig. 4. Product feature-based model which is based on product topology

A product model contains a combination of different single parts. Some of these parts are assembled first to subassemblies and then to the main product. In our proposed model, handling and form features are defined for each part, while mating features are defined among form features of different parts. Joining features are defined among different parts. Figure 5 illustrates the concept of the proposed feature-based model with mapping of three parts constrained in an aligned mating relationship and screw joining relationships.

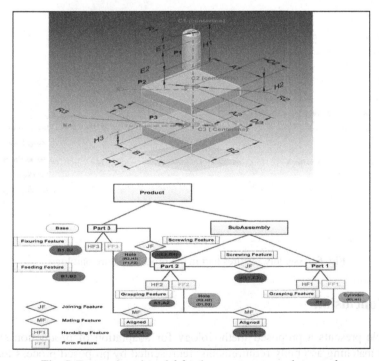

Fig. 5. Feature-based model for three parts case study example

To share the extracted assembly information in figure 5, and to integrate product and process in to EAS, a process-feature ontology is proposed (Figure 6). According to [18] "an ontology defines the basic terms and relations comprising the vocabulary of a topic area, as well as the rules for combining terms and relations to define extensions to the vocabulary". The proposed process-feature ontology is based on an assembly design ontology (AsD) proposed by Kim [19] and the assembly process class in EAS ontology [15]. Figure 6 shows the assembly features class in this ontology, where all the joining information (constraint, tolerance, shape and configuration) is classified as instances of joining process features classes (ex. riveting feature). By matching those instances with the skills parameters of joining modules, the required modules will be determined. The level of detail in this ontology can be extended according to the available assembly information (extracted or added by user), or according to the parameters of the joining resource capabilities in the production system.

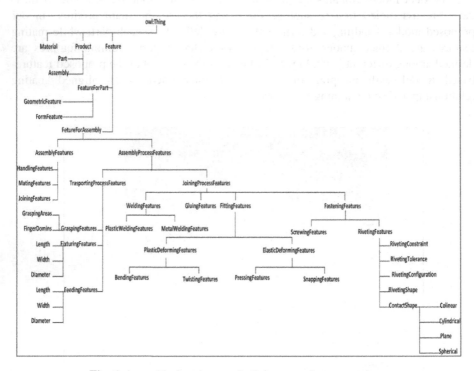

Fig. 6. Assembly features part in the process-feature ontology

4   Conclusion

The article presents a proposed methodology for adaptation of production systems based on matching assembly requirements, represented by proposed process features, to the processes and resources, represented by skills, in the EAS ontology.

A proposed feature-based model is introduced based on extracted handling and form features for each part, a mating feature between parts in a product assembly. These three basic assembly features will be used to define joining and transporting features in the next stage. The complexity of the extracted features will increase gradually from the basic assembly features to the process features.

The actual extraction of assembly features from CAD files will be covered in later work. Since assembly features includes geometrical (form, handling features), non-geometrical (mating features) and functional information (constraints, configurations), a collection of extraction methods have to be used. Extraction methods for CAD information include automatic, semi-automatic and manual methods. For assembly features a semi-automatic method will be compatible with our proposed approach in this paper; since the user-added information plays an important role in defining process features.

References

1. ElMaraghy, H.A.: Flexible and reconfigurable manufacturing systems paradigms. J. International Journal of Flexible Manufacturing Systems 17(4), 261–276 (2006)
2. Koren, Y.: General RMS Characteristics, Comparison with Dedicated and Flexible Systems. In: Reconfigurable Manufacturing Systems and Transformable Factories, pp. 27–46. Springer, Dashchenko (2006)
3. Nylund, H., Salminen, K., Andersson, P.: Digital Virtual Holons – An Approach to Digital Manufacturing Systems. In: Manufacturing Systems and Technologies for the New Frontier, pp. 103–106. Springer, Mitsuishi (2008)
4. Onori, M., Semere, D.T., Lindberg, B.: Evolvable systems: An approach to self-X production. In: Huang, G.Q., Mak, K.L., Maropoulos, P.G. (eds.) DET2009 Proceedings. AISC, vol. 66, pp. 789–802. Springer, Heidelberg (2010)
5. Onori, M.: A re-engineering perspective to assembly system development. J. Industrial Robot 32(5) (2005)
6. Onori, M., Neves, P., Akillioglu, H., Maffei, A.: Dealing with the unpredictable: An Evolvable Robotic Assembly Cell. In: Int. Conf. on Changeable, Agile, Reconfigurable and Virtual Production (CARV 2011), Montreal, Canada, pp. 160–165 (2011)
7. Semere, D., Onori, M., Maffei, A., Adamietz, R.: Evolvable assembly systems: coping with variations through evolution. J. Assembly Automation 28(2), 126–133 (2008)
8. Akillioglu, H., Neves, P., Onori, M.: Evolvable assembly systems: Mechatronic Architecture Implications and Future Research. In: 3rd CIRP Conference on Assembly Technologies and Systems (2010)
9. Van Holland, W.: Assembly Features in Modelling and Planning. Ph.D. Thesis, Delft University of Technology, Holland (1997)
10. Liu, H., Nnaji, B.: Design with spatial relationships. J. Manufacture Systems 10(6) (1991)
11. Shah, J.J., Rogers, M.T.: Feature Based Modeling Shell: Design and Implementation. In: Proceedings of the ASME Conference on Computers in Engineering, Montreal, Quebec, Canada, pp. 343–354 (1988)
12. Sung, R.: Automatic Assembly Feature Recognition and Disassembly Sequence Generation. Ph.D. Thesis, Heriot-Watt University, Edinburgh, UK (2001)

13. Pfrommer, J., Schleipen, M., Beyerer, J.: PPRS: Production skills and their relation to product, process, and resource. In: 2013 18th IEEE Conference on Emerging Technologies & Factory Automation (ETFA), pp. 1–4 (2013)
14. Lanz, M.: Logical and Semantic Foundations of Knowledge Representation for Assembly and Manufacturing Processes. Ph.D. Thesis, Tampere Univ of Technology, Finland (2010)
15. Evolvable Ultra- Precision Assembly Systems (EUPASS): Assembling Process Ontology Specification. Technical report, EUPASS std0007 (2008)
16. Smale, D., Ratchev, S.: A Capability Model and Taxonomy for Multiple Assembly System Reconfigurations. In: 13th IFAC Symposium on Information Control Problems in Manufacturing, Moscow, pp. 1923–1928 (2009)
17. Kim, K.: Assembly operation tools for e product design and realization. Ph.D. Thesis, University of Pittsburgh, USA (2003)
18. Neches, R., Fikes, R., Finin, T., Gruber, T., Senator, T., Swartout, W.: Enabling technology for knowledge sharing. J. Al Magazine 12, 36–56 (1991)
19. Kim, K.-Y., Manley, D.G., Yang, H.: Ontology-based assembly design and information sharing for collaborative product development. J. Computer-Aided Design 38(12), 1233–1250 (2006)

A Multi Agent Architecture to Support Self-organizing Material Handling

Andre Rocha, Luis Ribeiro, and José Barata

CTS, Uninova, Dep. de Eng. Electrotécnica, Faculdade de Ciências e Tecnologia,
Universidade Nova de Lisboa, 2829-516 Caparica, Portugal
ad.rocha@campus.fct.unl.pt
{ldr,jab}@uninova.pt

Abstract. Emerging market conditions press current shop floors hard. Mass customization implies that manufacturing system have to be extremely dynamic when handling variety and batch size. Hence, the ability to quickly reconfigure the system is paramount. This involves both the stations that carry out the production processes and the transport system. Traditionally system reconfiguration issues have been approached from a optimization point of view. This means allocating a certain batch of work to specific machines/stations in an optimal schedule. Although in a an abstract way these solutions are elegant and sound sometimes the number and nature of their base assumptions are unrealistic. Approaching the problem from a self-organizing perspective offers the advantage of attaining a fair solution in a concrete environment and as a reaction of the current operational conditions. Even if optimality cannot be ensured the solutions attained and the online re-adjustments render the system generally robust. This works extends the IDEAS Agent Development Environment (IADE) developed in the FP7 Instantly Deployable Evolvable Assembly Systems (IDEAS) project which has demonstrated the basic concepts of the proposed approach. The main architectural changes are presented and justified and the prospects for the analysis and self-organizing control are presented.

Keywords: Multi Agent, Transport System, Material Handling, Self-Organization, Load Balancing, Architecture.

1   Introduction

The transport system is usually composed by a network of conveyor belts, AGVs or both. These shop floor components have the responsibility to route material across the system to the different resources.

The load balancing optimization problem, as approached traditionally, normally disregards the role of these elements and perceives the transport system as a more or less passive entity. From an Assembly Line Balancing (ALB) perspective [1] several approaches have been traditionally considered. The Simple Assembly Line Balancing Problem was defined to cover the different manufacturers' goals, with four different variants. The SALBP-E aims to maximize the efficiency of the production line,

L.M. Camarinha-Matos et al. (Eds.): DoCEIS 2014, IFIP AICT 423, pp. 93–100, 2014.

SALBP-1 has as main goal the minimization of the number of stations needed in the system. If the goal is to minimize the execution time, knowing the number of stations in the system, the SALBP-2 formulation is considered. The SALBP-F problem formulation attempts to balance the optimization between number of stations and the desired time of execution[1].

The formulation of these problems only produce results, directly applicable to real systems, in very specific cases.

The ALB approach is therefore very dependent on the system characteristics and on the goals of a specific manufacturer and assembly installation. There is a gap between the academic models and simplifications, and the real scenario.

To cope with the pressing requirements, several approaches have been proposed. These new approaches have resulted in new production paradigms namely: Bionic Manufacturing Systems [2], Holonic Manufacturing Systems [3, 4], Reconfigurable Manufacturing Systems [5] and Evolvable Assembly Systems [6]. The present approach is proposed under the scope of Evolvable Assembly Systems (EAS) and is an improvement of the IADE architecture [7-9] that integrates control and reconfiguration with material handling aspects. Like IADE, the proposed architecture supports a transport system constituted by conveyors and implements the necessary mechanisms that allow, at each moment, the transport system elements to calculate optimal routes and knowing status of the entire transport network. The transport elements restricts their interaction and information exchange to a local scope and rely in self-organization to support the dynamics of the transport system. The main improvements over the IADE architecture is related to the interactions between the agents that have been streamlined to improved performance, plug-ability and, as importantly, the ability to formalize and guide the structure and collective behavior of the transport system.

2   Relation with Collective Awareness Systems

Manufacturing is, worldwide, a strategic sector from a socio-economic perspective. Hence there is a growing body of technologies that serve the purpose of integrating production plants with higher level logistic and management tools.

Although already significantly automated the existing technologies are not yet able to promote to the desired extent multilevel collective awareness in manufacturing.

Improving the self-configuration abilities of current shop-floors is decisive for the production sector and enables new ways of enterprise organization. By making systems more intelligent and aware, making their users aware on why the system is evolving, or proposing to evolve, in a certain direction and enabling the users to understand the impact of designing and managing a system under certain conditions contributes to the creation of more sustainable production practices while open up the door to tackled emerging business opportunities.

Awareness implies instant access to information but not only any assess. It implies an intelligent infrastructure that is able to understand and triggers different actions from its components and users so that the overall system converges to some target functioning points.

The proposed work focus on creating collective awareness at factory level, in particular in the transport elements which are envisioned as a fundamental pillar of the overall logistics, and promoting the generation of knowledge that can be used at other levels to promote collective awareness in manufacturing.

In particular this knowledge can be used to upgrade/update the system increasing its performance and efficiency and hence its sustainability.

3   Related Work

Several approaches and concepts have emerged to circumvent the limitations of traditional ALB. The Automated Material Handling Systems (AMHS) [10, 11] are very used in industry. The AMHS allow an automated routing of materials between stations and along different routes. However, these systems are difficult to modify. Their reduced flexibility, with respect to dynamic route management makes their usage restricted to a few systems and application scenarios. In order to develop approaches capable to tackling these problems, many researchers are focused in the use of multi agent-based architectures [12]. Multi agent-based design although not solving complex problems by default may, in certain cases, simplify some modeling aspects of the problems and enable a self-organizing approach that, not providing optimal solutions, still offers an efficient solution in a suitable time frame. In [12, 13] a multi agent approach is proposed to eliminate the combinatorial explosion associated with traditional scheduling. The test discussed in [12] suggests that a centralized Holonic approach attains scheduling solutions close to the optimal solution in the tested scenarios. A similar approach is found in [14, 15] in the context of FMS. More bio-inspired approaches based on ant foraging and stigmergy to explore in a distributed way alterative paths in an assembly system. are reported in [16] were some conceptual principles that can drive such a system are presented and justified while in [17] an architecture is proposed to use faster than real time simulation to provide continuous adjustments to the physical system. These approaches are not designed to explore and learn from the structure an organization of the transport system. In this context, the long term purpose of the research detailed in this paper is to investigate:

- What is the impact of network topology and distinct self-organization metric in the overall performance of the system?

The work hereby detailed is the first step towards this ambitious goal and relates with the definition and justification of the reference agent based transport architecture.

4   Agent Architecture

4.1   Notion of Skill

The agents in the architecture later detailed expresses their abilities as skills [18]. In this context, each skill has an interface that contains all necessary information for this execution. This information contains among other things the skill type. The type is important for the execution process and can assume one of two values: Atomic Skill or Complex Skill. Atomic Skills are responsible for the low level execution (hardware level i.e. I/O management) associating the skill data to a specific controller. The Complex Skill is responsible for executing processes of high complexity, these skills are constituted by a work-flow. This workflow is constituted by Atomic Skills or Complex Skills.

4.2   Generic Agents

The proposed architecture similarly to the IADE architecture is composed by five type of agents, although two of the agents are different as detailed in Table1. Table 1 describes the functional differences between the agents from the two architectures.

Table 1. Differences between IADE architecture and the new architecture

Agent	IADE architecture	New architecture
Handover (HA)	This entity is responsible to route the products. When a product arrives, this agent consults his routing table and checks what is the destination of the product. It checks, with this information, the next hop (Handover) and which Conveyor Agent or Transport Entity Agent can transport the product to the next hop.	
Source (SoA)	Abstracts the entry point of the system. This agent is used to enable products to enter the system.	It is responsible by the product entry, like in IADE architecture. The SoA is now perceived by the system, as a node (HA) and has the ability to route any product.
Sink (SiA)	Abstracts the products' exit points in the system.	Similarly to the SoA besides taking products from the system also behaves as an HA.
Transport Entity (TEA) / Conveyor (CA)	(TEA) This agent is able to abstract a conveyor belt defined between two HAs. This entity controls the execution of all the stations plugged and associated with a conveyor and all queues between these stations. This agent is also responsible for managing the plug and unplug of stations to and from the docking points inside the conveyor	(CA) It is responsible to control the product transition between two other entities in the transportation system (i.e. HA, SoA, SiA, DPA). This new agent doesn't contain any docking points, so it doesn't manage these operations as in the IADE stack.
Docking Point (DPA)		This entity abstracts the points in the system where it is possible plug stations. These points in the previous architecture were controlled by the TEA. During execution, this entity controls the product execution in this point of the system. In other situations has the same behavior as HA.
Yellow Pages (YPA)	It provides an improved yellow pages service. All agents in the platform can use the service to find other agents or skills, without a complete specification.	

4.3   Conceptual Assessment of the Architectural Changes

The main architectural differences are related to the entities that abstracts conveyor belts. The Conveyor Agent (CA), which is a new entity, is less complex than the previous case. The original specification, detailed in [8] was strongly inspired by the typical mechanical design of conveyors and stations' docking points. In this context, each conveyor contains several points where a station can be plugged. Each one of these docking points has a pre-stopper that prevents a pallet from entering the station when another pallet is already there and an additional stopper that fixes a pallet in place when is operated by the station. This creates several queues that precede each station inside each conveyor. In the IADE stack the user is allowed to add a docking point to an existing conveyor and the TEA autonomously resizes the queues while re-indexing the station's positions. Furthermore the TEA has to manage all the traffic inside all these queues. This also renders the interactions between the routing devices and the TEAs more complex since the routing devices have to index all the stations associated to a conveyor replicating already existing information.

There is an improvement in information management and performance. It may appear that this change prevents the dynamic addition of docking points however, from a mechanical point of view the introduction of a new docking point normally entails down time. In this context the tradeoff is that when a new point is introduced in the IADE architecture the user has to stop the conveyor and connect the docking point and then restart the conveyor reconfiguring it according with the new stations; in the current version the current TEA/CA is dissolved and two new CAs are created before and after the docking point. The overall reconfiguration effort is comparable with the adequate support tools. There is also a better isolation of the agents functions. The CA is mainly responsible for calculating its traversing cost. The docking point manages the stations and the skills therein and the HA is solely responsible for the computation of the best routes.

The new architecture also facilitates the modeling of the system as a network whereby transport entities, which do not have to be restricted to conveyors, behave as links and all the other entities as nodes. This new view if fundamental to study the impact of network topology in the performance of the system as it enables the extraction of configuration patterns that can improve or degrade performance. For instance is a transport system built after a small world model more efficient that the traditional line? Is it more robust? How is if affect by different metrics of transport cost?

The data model, discussed in the next section has also been improved to facilitate scalability and the specialization of new classes that can be seamlessly incorporated in the present architecture. In the IADE architecture a flat data model is followed and specialization of classes is not allowed.

5   Implementation

A simplified representation of the new data model is presented in Figure 1. All the classes are derived from the transport entity concept. A transport entity is something that is able to store products and can receive or dispatch them. A conveyor is, in this

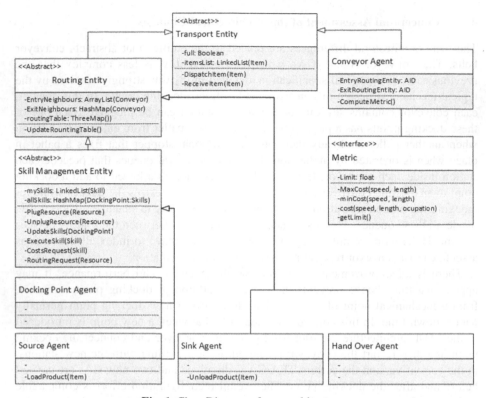

Fig. 1. Class Diagram of new architecture

context a transport entity that can compute its traversing cost using a specific metric. Each conveyor has its specific metric with a formula to calculate its traversing cost and a threshold value that defines when this cost should be updated in the neighbor HAs resend the cost update.

A conveyor is a link in the transport network as would be an AGV (not considered in this initial version of the stack).

The network's nodes are treated as routing entities. These connect several conveyors and can compute the optimal routes according to the current system status. This computation uses the dijkstra algorithm which is computed in steps inside the agent scheduler to ensure a non blocking performance by the JADE based agents.

Some of these nodes are specialized on skill management. For instance docking points are able to handle stations and their skills, This mediates the execution of skills and, plug and unplug actions. The source agent processes the state update from the docking point agents and uses this information for route the products when these enter in the system.

A sink is a specific node where it is possible remove the products when they end their execution. From a technical point of view the implementation uses the Java Agent Development Environment (JADE). Although JADE is cannot fulfill hard real time constraints its performance [19] is still acceptable to handle the dynamics of most transport systems.

6   Conclusion and Further Work

In the IADE architecture it was possible to prove that a self-organized response can be used to control a transport system in a robust manner and with acceptable performance. The architecture presented in this work improves in IADE conceptually and from a performance point of view as the role of the agents has been streamlined and the overall number of interactions was reduced. The main architectural optimization is the addition of DPA. This new agent manages the points of the system where it is possible plug and unplug stations. The CA, previously the TEA, consequently was simplified, and this simplification reduces the complexity and the computation. This reduction increases the performance of the entire system. The stabilization of the architecture and the test currently ongoing are however only the starting point of a far more ambitious work that related with the development of self-organization metrics to regulate the adaptive response of agent-based transport systems.

References

[1] Boysen, N., et al.: Assembly line balancing: Which model to use when? International Journal of Production Economics 111, 509–528 (2008)

[2] Ueda, K.: A concept for bionic manufacturing systems based on DNA-type information. In: Proc.s of the IFIP TC5/WG5. 3 Eight International PROLAMAT Conf Human Aspects in Computer Integrated Manufacturing, pp. 853–863 (1992)

[3] Van Brussel, H., et al.: Reference architecture for holonic manufacturing systems: PROSA. Computers in Industry 37, 255–274 (1998)

[4] Babiceanu, R.F., Chen, F.F.: Development and applications of holonic manufacturing systems: a survey. J. Intelligent Manufacturing 17, 111–131 (2006)

[5] Koren, Y., et al.: Reconfigurable manufacturing systems. CIRP Annals-Manufacturing Technology 48, 527–540 (1999)

[6] Onori, M., et al.: "An architecture development approach for evolvable assembly systems," in The 6th IEEE Int Symp. on Assembly and Task Planning: From Nano to Macro Assembly and Manufacturing (ISATP 2005), pp. 19–24 (2005)

[7] Ribeiro, L., et al.: Self-organization in automation-the IDEAS pre-demonstrator. In: IECON 2011-37th Annual Conference on IEEE Industrial Electronics Society, pp. 2752–2757 (2011)

[8] Rocha, A.D.B.D.S.: An agent based architecture for material handling systems (2013)

[9] Ribeiro, L., et al.: A product handling technical architecture for multiagent-based mechatronic systems. In: IECON 2012 - 38th Annual Conference on IEEE Industrial Electronics Society, pp. 4342–4347 (2012)

[10] Haneyah, S., et al.: Generic planning and control of automated material handling systems: Practical requirements versus existing theory. Computers in Industry (2012)

[11] Haneyah, S., et al.: A generic material flow control model applied in two industrial sectors. Computers in Industry (2013)

[12] Shen, W., et al.: Agent-based distributed manufacturing process planning and scheduling: a state-of-the-art survey. IEEE Transactions on Systems, Man, and Cybernetics, Part C: Applications and Reviews 36, 563–577 (2006)

[13] Guo, Q.-L., Zhang, M.: An agent-oriented approach to resolve scheduling optimization in intelligent manufacturing. Robotics and Computer-Integrated Manufacturing 26, 39–45 (2010)

[14] Turgay, S.: Agent-based FMS control. Robotics and Computer-Integrated Manufacturing 25, 470–480 (2009)

[15] Rubrico, J.I., et al.: Online rescheduling of multiple picking agents for warehouse management. Robotics & Computer-Integrated Manufacturing 27, 62–71 (2011)

[16] Peeters, P., et al.: Pheromone based emergent shop floor control system for flexible flow shops. Artificial Intelligence in Engineering 15, 343–352 (2001)

[17] Sallez, Y., et al.: A stigmergic approach for dynamic routing of active products in FMS. Computers in Industry 60, 204–216 (2009)

[18] Barata, J.: Coalition based approach for shopfloor agility (2005)

[19] Ribeiro, L., et al.: A study of JADE's messaging RTT performance using distinct message exchange patterns. Presented at the IECON 2013-39th Annual Conference on IEEE Industrial Electronics Society. IEEE (2013)

Self-organization Combining Incentives and Risk Management for a Dynamic Service-Oriented Multi-agent System

Nelson Rodrigues[1,3], Eugénio Oliveira[2,3], and Paulo Leitão[1,3]

[1] Polytechnic Institute of Bragança, Campus Sta Apolonia, Apartado 1134, 5301-857
Bragança, Portugal
[2] Faculty of Engineering - University of Porto, Rua Dr. Roberto Frias s/n, 4200-465 Porto
[3] LIACC - Artificial Intelligence and Computer Science Laboratory, R. Dr. Roberto Frias,
4200-465 Porto, Portugal
{nrodrigues,pleitao}@ipb.pt, eco@fe.up.pt

Abstract. Companies are nowadays placed in very complex and dynamic environments, making their competitiveness mandatory. This competitiveness can also be achieved through the reconfiguration of systems' network. In this paper, high level of self-organization of service-oriented multi-agent systems is explored, aiming to achieve more trustworthy and automatic reconfiguration processes, in dynamic and open environments. The correct self-organization model directly impacts the success of agents' behaviours to actively change or create new appropriate services dynamically. This paper advocates on the influence, in a distributed and cooperative way, of risk management, similarities and incentives to work together in order to speed up the self-organization agent's network. By leveraging the SOA and MAS benefits it is possible to reduce time, effort and money. Based on identified benefits and needs, several research leads for my thesis plan are here proposed towards the realization of self-organization capabilities aiming to accomplish more trustworthy and automatic reconfiguration processes, in volatile and open environments.

Keywords: Multi-agent systems, Self-organization, Service-orientation, Risk Management.

1 Introduction

Innovative solutions and technologies were created to solve the companies' problems that are playing in very dynamic and competitive environments, subject to the markets pressures demanding highly customized and quality products at reduced price. Companies, to increase their competitiveness, need to implement more flexible and responsiveness systems [1] and increase their participation in cooperative networks, which request different solutions. Currently, decentralized approaches are being pointed to address this challenge [2], based on the distribution of control functions over a network of decision-making entities. Multi-agent systems (MAS) and Service Oriented Architectures (SOA) are two examples of such approaches, based on the same principles of distribution, which can be combined to extract benefits from the two

L.M. Camarinha-Matos et al. (Eds.): DoCEIS 2014, IFIP AICT 423, pp. 101–108, 2014.
© IFIP International Federation for Information Processing 2014

worlds: the intelligence and autonomy provided by MAS solutions and the interoperability offered by SOA solutions [3]. The addition of dynamic and automatic reconfiguration processes of a distributed system [4], e.g. through adapting or creating the services offered by the several intelligent entities to the new requirements, implies a deeply intelligent orchestration of services. The aim of this paper is to describe an innovative self-organization approach based on incentives mechanisms and risk management in service-oriented multi-agent systems, aiming to achieve more trustworthy and automatic reconfiguration processes, in volatile and open environments.

This paper is organized as follows. Section 2 describes the similarity of the work with the collective awareness systems. Section 3 presents the literature review and Section 4 briefly presents the paper's contribution as well as the formalization of the proposed model. Section 5 overviews the critical analysis of this proposal and states forward the research directions. Finally, Section 6 wraps up the paper with the conclusions.

2   Relationship to Collective Awareness Systems

This contribution is related to collective awareness systems, since it uses a MAS approach to distribute the knowledge and decisions points, being necessary to perform the management of the agent's colligations that are self-organized and to select the best service. This research needs to be driven to the collective awareness system, since the agents behaves as a global brain to offer better services; local knowledge about the environment allows the agents to take local decisions, which has a huge global impact. Additionally, the management becomes critical since too many agents communicating can imply the congestion of the agent's network.

3   Literature Review

This section reviews the current state of the art in several related fields, providing the initial context around service-oriented domains, and how agents and risk management are being used. To determine the value of risk management, it is considered trust and reputation values to measure threats and predict uncertainty. In a quick perspective, agents are seen as intelligent entities that offer support for robustness, flexibility and adaptability, in case of management of services [3]. For a better autonomy process, they follow decentralized approaches [5] that require the cooperation between the agents. In some situations, this cooperation can take a form of competition, being sometimes necessary to request advices or make alliances, which require the consideration of the level of the risk management (based on trust and reputation).

Fig. 1 summarises the important works focusing individually each one of the referred domains. Once analysing Fig. 1 it is obvious the reasons that motivated this study. Some of these approaches are able to support SOA [14], [15], [16], however they do not introduce the self-organization paradigm. On other hand, the other works lack on components that covers this work's interests (self-organization phenomena achieved by using incentives and risk management metrics supported by multi-agents). Whenever a network of agents follows an overall goal, and this network does not encourage the cooperation, incentive mechanisms can improve this cooperation [6]. Additionally, is

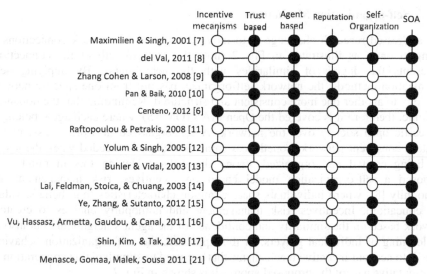

Fig. 1. Summary of the literature review

also considered works that look into the self-organization phenomena, meaning approaches that evolve by small cooperation for a greater goal.

Essentially there is lack of proposals with self-organization that have in common multi-agent systems and risk management models, which means, an opportunity to research in self-organized systems.

4   Research Contribution and Innovation

Complex systems may benefit from self-* characteristics since the agents should be able to change their own behaviours facing the overall system. The individual agents use an intelligent behaviour to actively dynamically change or create new services, without a request but instead in a response to a condition change. As previously stated, the literature review demonstrates some efforts in decentralized approaches to solve the dynamic and competitive environments. However, the existing works lack in some important points, as described previously. This work aims to propose self-organization and incentive mechanisms that can offer a dynamic and automatic self-* properties (e.g., reconfiguration, adaptation) based on agent networks, where each agent combines the learning and risk management about the interaction and cooperation to evolve and offer better services. The system, as a whole, must also find opportunities to evolve by analysing their dynamic environment, the agents should be able to self-configure and self-organize, allowing increase of the system performance on-the-fly. Additionally, this approach pretends to decompose the risk management in trust and reputation models that allow the agents to work with another level of confidence when faced by unexpected scenarios, being capable of evolving in its configuration to a state where there is a better state in order to offer new/modified services with offer highly customized and trustworthy services with better quality.

4.1   Self-organization Mechanism

The addressed research challenge is to move the system from weak connections to strongest ones, as illustrated in Fig. 2 (note that the strength of the connections represents the degree of similarities with the neighbours). By applying self-organization methods, the network self-organize, passing from one specific network topology to another one more consistent and structured. Reminding that the domain is dynamic, therefore it is covered the Open MAS (OMAS), where each agent belong to a specific open society over the network. A self-organized network is essential to create a consistent network, nonetheless some confidence is needed when the agents are being created and organized among themselves. Having this in mind, it is proposed a self-organization model based on incentives, risk management and homophily [8], where individuals share common characteristics (e.g. beliefs, values and education). Incentives, risk management and homophily enables to create a network based on the similarity and confidence of the agent's neighbours but also on the learning of individual players. In this approach, the self-organization behaviour takes into account incentive mechanisms that are used to stimulate this cooperation. A representative part of the proposed approach is shown in Fig. 2.

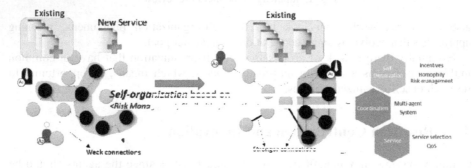

Fig. 2. Self-organization network topology (note that each edge has a preferred binding, which corresponds to a stronger connection of each entity)

After a self-organization phenomenon has occurred in a distributed and decentralized system, robustness has increased, allowing surviving without some entities or with incomplete information. Self-organization is structured (Def. 1), which represents the function to achieve or not society and the connections.

Definition 1. Being the *self-organization (SelfO)* working in an OMAS with the structure, $SelfO(Soc, Nt, Oenv, Rm, Simi, In, fSelfO) \rightarrow (Soc', Nt')$, where:

- $Soc = \{Ag_1, ..., Ag_n\}$ represents a, finite, society of agents Ag ;
- *Oenv* characterises the environment accessible by the agents, where reside different societies of agents;
- $Nt \subseteq Soc \times Soc$, is a set of connections where $(Ag_i, Ag_j) \in Oenv$ represents a direct relationship;
- *Rm* represents the risk management function of a specific *Nt* ;
- *Simi* represents the similarity degree function of the assigned *Nt* ;
- *In* means the incentives function of the assigned *Nt* ;

- *fSelfO* : *Soc*×*Oenv*×*Rm*×*Simi*×*In* →[0,1] ,   represents   the   self-organization function between a threshold ε (to be parameterized), [,..,ε,..1] which corresponds to organize or not;
- *Soc'* and *Nt'* represents the next finite society and their connections, after self-organization, respectively.

The model of agents also represents the strength of connections, which can evolve to create a proper network towards the goal, depending on the network topology. In order to increase the cooperation, incentive mechanisms will allow the formation of more stable colligations. In practice, the incentives are given from a normative environment, meaning that there is no supervision entity obligating to accept these incentives.

Definition 2. *Mechanism of incentive (In)* based on the Centeno's work [6] is represented as a tuple *In*(*St*,*Oenv*,*Soc*, *fIn*) →[0,1] , where the agent has incentives to create coalitions of the environment and can accept or ignore the incentive depending on the impact it will have on their utility.

- $St = \{St_1,...,St_m\}$ represents the internal states of an agent;
- *fIn* : *St*×*Oenv*×*Soc* →[0,1] represents the incentives mechanism function;

Taking into consideration that incentive mechanisms should be tuned for specific topologies, it is realized that to establish weak connections it is required greater incentives, and stronger connections do not require higher values. The self-organisation methodology has a direct relation with the strength of the connections, which means that stronger connections allow fastest and stronger reorganization. Working in an OMAS has a certain amount of risk, and a stronger and reliable connection structure must be created. Gathering the concepts of trust and reputation models it will be possible to create a risk management model.

Definition 3. *Risk management* is defined as *Rm*(*Nt*,*St*,*Trt*,*Rp*) →[0,1] , where:

- *Trt* represents the trust measure adapted from [18], being defined by *Trt*(*Xtr*,*Ytr*,*Cxt*,*t*, *fm*) →[0,1] where *Xtr* represents the trustor, *Yte* means the trustee, in a context *Cxt*, for a specific goal *G* at the time *t* with the function *fm* : *Xtr*,*Yte*,*Cxt*,G →[0,1] ;
- *Rp*(*Tp*,*Xtr*,*Yte*,*Cxt*,*G*,*t*, *fr*) →[0,1] , where $tp = \{tp_1,...,tp_k\}$ represents a set of recommendations, given by third-party recommender *Tp* of a trustor agent *Xtr*;

Following the principle that the level of risk depends on trust and reputation of certain agents and certain goals at given times for example, the risk shifts if there is no trust/reputation or only for certain goals. The risk level suffers alteration according to time, if the interaction occurs in the last minute or a week ago the risk level is not the same. Social structures allow to create better organizations without central decision points. One of the social structures characteristics is the homophily, based on the similarity degree of other entities [8], which allows to create stronger connections.

Definition 4. *Similarity* is defined as a tuple *Simi*(*Ro*,*Sv*,*Ont*, *fSim*) →[0,1] *where:*

- $Ro = \{Ro_1,...,Ro_\lambda\}$ represents a set of roles of each agent *Ag;*
- *Sv* is a set of service descriptions defined by another structure *Sv*(*Sn*,*Sd*,*Isv*,*Osv*,*Psv*,*Esv*) , composed by the service name *Sn*, the description *Sd*, the configuration input *Isv*, output *Osv*, preconditions *Psv* and effects *Esv*;

- *Ont* represents a set of concepts $\{c_1,...,c_\mu\}$ and properties $\{p_0,...,p_v\}$. This variable is important to evaluate the similarities;
- *fSim* : $Sv \times Sv \times Ro \rightarrow [0,1]$ represents the degree function;

It is expected that after a self-organization procedure, the system continues to have weak connections, and others became stronger (see Fig. 2). This cooperation among the entities automatically increases the probability of achieving objectives more effectively as predict future events.

4.2   Agent's Internal Model

The OMAS structure has several agents with several behaviours, outstanding to the complexity, an autonomic process should be developed for the agents behaviours. Inheriting these characteristics from the computer science, an autonomic system should remain self-configured, self-optimized, self-healed and self-protected. The agents developed should be intelligent enough to trigger the self-healing and self-protection behaviours, this means, learning when non-intentional accidents could occur, or have the intelligence to detect, diagnose and repair such accidents.

This insight fulfils very well the system architecture in two separated layers: in one layer it is represented the intelligence layer, namely the agents, and in the second layer, each agent is responsible to offer the services.

Definition 5. The agent's internal model is specified as $Ag(St, Oenv, K, ft, hUT, Sv) \rightarrow St'$ where St' represents the next state, K signifies the knowledge of agent of its set of direct neighbours inside the society, ft is the transition function $ft : Oenv \times St \rightarrow St'$, and hUt represents the utility function of each agent for a specific state and knowledge $hUt : St \times Oenv \times K \rightarrow R$ (remember that not all topologies are motivated to reach its maximum utility [19]);

The agent tends to maximize their own utility, although this local maximization might sometimes not be the best for the global system. Without the supervision, the agents need to collaborate to achieve the global maximum. At this point, and since the objective is to provide services more trustworthy, the agents intend to increase their risk management accuracy $SelfO(Rm, Simi, In) \rightarrow (Soc', Nt')$ by $\max_{i=1...l}\{Rm_i\}$, where l is the system interactions generated. Fig. 3 depicts the behaviour of decision support and composition, namely, what the agent performs after reasoning about the available

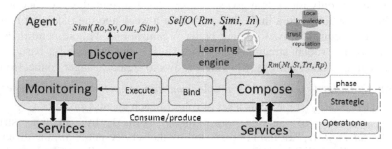

Fig. 3. Dynamic and automatic reconfiguration process engine of each agent

services as well as the current network neighbour, to cooperate and owning the information that exists in the network configuration to compose trustworthy services.

Just like in the self-organization, the agents will try to compose the best service in on-going loops. From a strategic perspective, the monitoring of services for planning, e.g. functionality and availability, is controlled. Other perspective is the focus on the operational task, which explore algorithms to compose services, to achieve the goals.

Summing up, the system tends to evolve and organize itself taking into account statistical fluctuations and internal agent's history. Self-(behavioural and structural) organization are applied, by changing the agent's behaviours and network structure.

5   Discussion of Results and Critical View

The insights of this work are to create a vision and to formalize concepts necessary to future lines of research. Aiming to verify the potentiality of this approach, two case studies are explored. The first in one business-to-business (B2B) and Virtual Enterprises (VE) domain, by exploring the benefits of using decentralized entities based on models of trust and reputation, by taking advantage of a framework that already provides a trust negotiation based on the normative environment called ANTE (Agreement Negotiation in Normative and Trust-enabled Environments) [20]. The second case study is related to the Smart Grids domain, aiming to achieve a consistent network based on some properties, such as reliability. The network can be smart enough to recognize when each entity is starting to fail or even reduce its performance and react appropriately. To validate the proposed approach more serious tests must be performed after this first attempt to identify the research opportunities. Beyond the chosen domain it is necessary to create the environment propitious to the evolution of the system and characterization of the entities feedback in order to be able to use previous equations in their overall, this mode is possible to show the advantages of this approach, which are beyond human capabilities to solve such complex.

6   Conclusions and Future Work

This work intends to propose one contribution focusing the dynamic offer, highly customized and trustworthy, of services with better configurability and responsiveness. An innovative self-organization approach for service management in service-oriented multi-agent systems is proposed, describing a reconfigurable architecture that considers risk management, incentives and homophily visions, to achieve self-* characteristics. In such self-organized systems several research questions may arise, namely how and when to evolve, what is the degree of the similarities to take into account in the trust model and finally, how all these combined pieces can offer a better configuration that modifies or creates more trustworthy services. To validate the hypothesis, two use cases focusing distributed environments will be considered, namely B2B and Smart Grids. The preliminary achievements show that the self-organization component can enhance services' selection by increasing the degree of connections, increasing the trustworthy. Future work will try to answer some remaining open questions, such as, "how to handle the correct management of the network's features", which directly impacts the success of the service governance.

References

1. Colombo, A.W., Jammes, F., Smit, H., Harrison, R., Lastra, J.L.M., Delamer, I.M.: Service-oriented architectures for collaborative automation, in Industrial Electronics Society. In: Proc. of the 31st Conf. IEEE Industrial Electronics Society, IECON 2005 (2005)
2. Leitão, P.: Agent-based distributed manufacturing control: A state-of-the-art survey. Eng. Appl. Artif. Intell. 22(7), 979–991 (2009)
3. Huhns, M.N.: Agents as Web services. Internet Computing 6(4), 93–95 (2002)
4. Maximilien, E.M., Singh, M.P.: A Framework and Ontology for Dynamic Web Services Selection. IEEE Internet Computing 8(5), 84–93 (2004)
5. Khan, S.P., Ismaeel, S., Ahmad, H.F., Suguri, H., Akbar, M., Elahi, A.: Enabling Negotiation Between Agents and Semantic Web Services. In: AWIC, vol. 43, pp. 284–291 (2007)
6. Centeno, R.: Mecanismos Incentivos para la Regulación de Sistemas MultiAgente Abiertos basados en Organizacione, Universidad Rey Juan Carlos (2012)
7. Maximilien, E.M., Singh, M.P.: Reputation and endorsement for web services. SIGecom Exch. 3(1), 24–31 (2001)
8. del Val, E.: Decentralized semantic service discovery based on homophily for self-adaptive service-oriented MAS. In: Proc. of the 10th International Conference on Autonomous Agents and Multiagent Systems, vol. 3, pp. 1347–1348 (2011)
9. Zhang, J., Cohen, R., Larson, K.: A Trust-Based Incentive Mechanism for E-Marketplaces. In: Falcone, R., Barber, S.K., Sabater-Mir, J., Singh, M.P. (eds.) Trust 2008. LNCS (LNAI), vol. 5396, pp. 135–161. Springer, Heidelberg (2008)
10. Pan, Z., Baik, J.: A QOS enhanced framework and trust model for effective web services selection. J. Web Eng. 9(2), 186–204 (2010)
11. Raftopoulou, P., Petrakis, E.G.M.: iCluster: a self-organizing overlay network for P2P information retrieval. In: Proc. of the IR Research, 30th European Conference on Advances in Information Retrieval, pp. 65–76 (2008)
12. Yolum, P., Singh, M.P.: Engineering self-organizing referral networks for trustworthy service selection. Trans. Sys. Man Cyber. Part A 35(3), 396–407 (2005)
13. Buhler, P.A., Vidal, J.M.: Adaptive Workflow = Web Services + Agents. In: Proc. of the International Conference on Web Services, pp. 131–137 (2003)
14. Lai, K., Feldman, M., Stoica, I., Chuang, J.: Incentives for Cooperation in Peer-to-Peer Networks (2003)
15. Ye, D., Zhang, M., Sutanto, D.: Self-organization in an agent network: A mechanism and a potential application, Decision Support Systems, vol. Decision Support Systems 53(3), 406–417 (2012)
16. Quang-Anh Nguyen, V., Hassasz, S., Armetta, F., Gaudou, B., Canal, R.: Combining Trust and Self-Organization for Robust Maintaining of Information Coherence in Disturbed MAS. In: Fifth IEEE International Conference on Self-Adaptive and Self-Organizing Systems (SASO 2011), pp. 178–187 (2011)
17. Shin, J., Kim, T., Tak, S.: Incentive mechanism for service differentiation in P2P networks. In: Proc. of the 2009 Int. Conf. Hybrid Information Techn., pp. 333–339 (2009)
18. Oliveira, E.: Software Agents: Can We Trust Them? In: Proceedings of IEEE 16th International Conference on Intelligent Engineering Systems (INES 2012), pp. 15–20 (2012)
19. Barabási, A.L., Albert, R.: Emergence of scaling in random networks. Science 286(5439), 509–512 (1999)
20. Cardoso, H., Urbano, J., Brandão, P., Rocha, A., Oliveira, E.: ANTE: Agreement Negotiation in Normative and Trust-Enabled Environments. PAAMS, 261–264 (2012)
21. Menasce, D., Gomaa, H., Malek, S., Sousa, J.P.: SASSY: A Framework for Self-Architecting Service-Oriented Systems. IEEE Software 28(6), 78–85 (2011)

Part V
Monitoring and Supervision Systems

A Service-Oriented and Holonic Control Architecture to the Reconfiguration of Dispersed Manufacturing Systems

Robson Marinho da Silva[1,2], Mauricio F. Blos[2], Fabrício Junqueira[2],
Diolino J. Santos Filho[2], and Paulo E. Miyagi[2]

[1] Universidade Estadual de Santa Cruz, Ilhéus, BA, Brazil,
rmsilva@uesc.br
[2] Universidade de São Paulo, SP, Brazil,
{blosmauf,fabri,diolinos,pemiyagi}@usp.br

Abstract. Manufacturing control systems must quickly react to variations of product, process specifications, fault occurrence, changes in the resources functional capabilities, and other operational demands. Besides, to gain competitive advantages, dispersed manufacturing systems must cooperate with each other. The combination of holonic control system and service-oriented architecture techniques can be effective to integrate these heterogeneous environments since the agents must also cooperate to achieve their services. Therefore, this paper introduces a service-oriented and holonic control architecture using Petri net to represent the workflow of design method itself, structure and dynamic of the entities. The modeling of an example is presented to demonstrate application at self-organization and adaptation, fault treatment, degeneration, and relatively shorter implementing time.

Keywords: reconfiguration, holonic control system, service-oriented architecture, dispersed manufacturing system.

1 Introduction

Manufacturing systems have evolved from mass production, lean manufacturing and flexible manufacturing to the reconfiguration of dispersed manufacturing systems (RDMS) which is conceived to be "adjustable" into pursuance of the businesses processes. Reconfiguration is associated with changes (i) in addition, modification and removal of some part of a workflow; (ii) in processing time, number and availability of existing resources; and (iii) in development of mechanisms for fault treatment [1].

A workflow management framework must be able to deal with the changes mentioned. The holonic control system (HCS) technique [2], [3], [4] allows the integration of heterogeneous environments. Furthermore, the service-oriented architecture (SOA) [5], [6], [7] has been used to undertake inter-enterprise collaboration.

Hence, this paper introduces a service-oriented and holonic control architecture (SOHCA) and its method to design RDMS. By exploring the potential of techniques which ensure the integration of design, the method uses the Petri net markup language (PNML) [8] and production flow schema (PFS) [9] techniques.

L.M. Camarinha-Matos et al. (Eds.): DoCEIS 2014, IFIP AICT 423, pp. 111–118, 2014.
© IFIP International Federation for Information Processing 2014

2   Relationship to Collective Awareness Systems

Collective awareness systems (CAS) allow its members to work together to improve self-organization and adaptation capacity. This is possible by integration of value-added activities, information, resources and knowledge between heterogeneous businesses workflow. RDMS is also related to the effective degree of collaboration, because its subsystems need to be aware of each other to attain the global objective [10].

We exploit HCS and SOA techniques to automatically allow the RDMS and provide a middleware for integration between the factory control and shop floor control layers in dispersed systems. SOHCA and its design method enable designers to develop collaboration and sequence diagrams, which assimilate the implementation of CAS characteristics, regarding workarounds to: (i) develop supervisory control system with interoperability and portability; (ii) propose strategies to recover the functionality of the system or maintain operations so that parts affected by the fault are disabled without affecting other parts (also called degeneration), and (iii) allow designers to create workflow for adding or modifying the productive processes, regarding the reuse of models to avoid repetition and overlapping tasks.

3   State of the Art and Related Works

There are already many suggestions applying the concept of holons or agents with different scope, focus and methodology [1], [2], [3], [4], [10]. Mendes *et al.* [5] explain how to combine the service-oriented agents in industrial automation, sharing resources in the form of services by sending requests between agents. Nagorny *et al.* [6] proposed the use of hardware and software accessed remotely, and where operational activities are described as services. Morariu *et al.* [7] present the design and implementation of the customer order management module which integrates HCS with SOA enabled shop floor devices using industry standards. Furthermore, control architecture should be based on standards ontologies, as proposed in ANSI/ISA 88/95[11] and ADACOR [10]. In previous studies [12], [13], [14], we applied the HCS with fault tolerance mechanisms for intelligent building, while in [15] our methodology was applied for manufacturing system (MS) and for the description of the control mode switching between two operational modes.

We observe that: (i) a suitable ontology can provide a semantic model for interaction in RDMS; (ii) a trading mechanism is more favorable than a request-response communication format between holons; (iii) patterns emerge without an environment that facilitates the development of new models; (iv) there are few proposals regarding strategies for the system reconfiguration; (v) another challenge is how to compose holarchies that accomplish a productive task with constraints to consider the lead time; and (vi) there are still a small number of applications combining HCS and SOA concepts, or more specifically with RDMS.

4   SOHCA and Proposed Design Method

The proposed method to design RDMS is structured in stages (Fig. 1): *analysis of requirements; modeling; analysis of models; implementation; operation*; and *integration*. The *integration* stage ensures a closed-loop for re-design and re-engineering.

In the *analysis of requirements* stage specifications are defined, such as: aim, control devices, orders, tasks, operations, interactions between parts of the system, and strategies for the cases of fault treatment and degeneration. Based on this survey, the holons are identified. Figure 2 illustrates the SOHCA holons. The *product holon* (*PrH*) manages the requests of products, searches for the corresponding holons, creates a work order (*wo*) comprising the product type and the amount. The set of *PrHs* composes the *production plan*, which represents the union, transformation and operation of the intermediated products to get the final products, i.e., the required input types and output types between *PrHs*. The *task holon* (*TH*) manages the recipe for business processes, such as, manufacturing, work order, operation and resource; and the reconfiguration strategies. The *OpH* represents the resources, such as humans or equipment (control objects). The supervisor holon (*SuH*) contains all the knowledge to coordinate the services of all holons.

At the *modeling* stage, the PFS and PN models represent the processes. To model the workflow of a holon, a **place** (terms of PN are **highlighted**) represents a state in the workflow while a **transition** represents an event or operation that conducts the flow from one state to another. The synchronization of PN models is made by **enabler** and **inhibitor arcs**, as well as by **auxiliary places**. To process an order, each holon is modeled as an atomic service that can be combined with atomic services by other holons. *OpHs* models include the control object which must regard fault states of these objects. The *PrH*, *TH* and *SuH* specify a set of operations, precedence constraints of the operations and control supervisory. PN models are used to represent the holons workflow and activities. The calculations of timing constraints are made using the **temporized transitions** in PN models.

In the *analysis of models* stage structure and dynamic behavior of the models are simulated and verified. The qualitative analysis is based on structural analysis using editors of PN models. Quantitative analysis is performed through the simulation of **temporized transitions**. At this stage, scenarios are identified with models built for each one. The models must meet the restrictions and achieve the objectives outlined in the hypothesis. It is also possible to review the control system models identifying the **places** and **transitions** that must be considered in each service.

The *implementation and operation* stages move toward the implementation and real-time monitoring system for supervision and control aspects. For low-level control applications, the code generation is made following the IEC60848 GRAFCET language. The code generation of a high-level language is made in Java using JADE and its extensions [15]. The real-time monitoring system is accomplished by synchronizing the operation of the PN models with sensor signals that represent the devices state. Communication channels are specified in the SOHCA ontology, a Java method is called and it creates the channels with the specified IP addresses and ports, while other holons associate a behavior to the specified port for each communication channel.

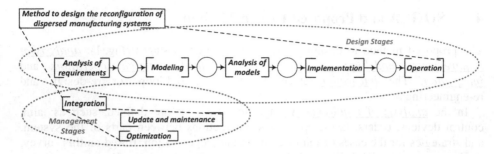

Fig. 1. Schema to design the reconfiguration of dispersed manufacturing systems

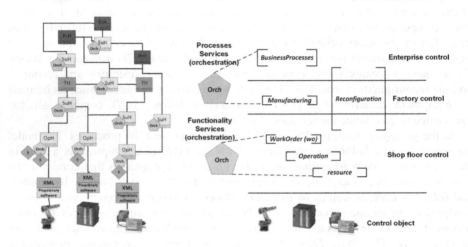

Fig. 2. SOHCA. Holarchies are formed (orchestration) to represent the control levels. The system is represented by the composition of these holarchies (choreography). Orchestration and choreography are terms related to SOA technique.

5   Application Example

Here, a benchmarking dispersed manufacturing system (BDMS), illustrated in Fig. 3, is used to describe the application of the proposed method. For each web service (WS) the devices, their control functions, commands and signals of actuation and detection are identified. The identification is made according to the specification DIN/ISO 1219-2:1996-11 and the codes recommended in specification IEC 61346-2:2000-12. For example, in the nomenclature 1S2: 1 =circuit number, S =device code, and 2 =device number.

The *temporized transitions* are used for calculating the holarchies formation. For example, let $[H_n]$ be the production sequence n formed by *PrHs* to obtain a final product. Let us consider the following sequences (the caption of Fig. 3 lists the abbreviations used): $[H_1]: [bcb] \rightarrow [bcb + ap] \rightarrow [bcb + ap + s] \rightarrow [bcb + ap + s + co]$ and $[H_2]: [rcb] \rightarrow [rcb + ap] \rightarrow [rcb + ap + s] \rightarrow [rcb + ap + s + co]$.

Suppose the lead time set by $PrH - [bcb + ap + s + co]$ is $t_e[bcb + ap + s + co]$. This imposes a timing constraint on $PrH-[bcp + ap + s]$ in $[H_1]$ sequence. Let $t_e[bcb + ap + s]$ be the latest time that $PrH-[bcb + ap + s]$ must complete all its operations. Let $t_e[t_n]$ be the time in which **transition** n must complete its operations. To meet the due date, the constraint of $t_e[bcb + ap + s] \leq t_e[bcb + ap + s + co] + t_e(t_1 + t_2 + \cdots + t_n)$ must be satisfied by $PrH-[bcb + ap + s]$ and so on for other holons, i.e., $t_e[bcb + ap + s]$ must be less than or equal to $t_e[bcb + ap + s + co]$ minus the sum of the times of **transitions** of the $PrH-[bcb + ap + s + co]$. A variable named cost Hn is associated to PrH to formation of holarchies. Let C_N be the cost of a production sequence N of a H_n or a PrH. Let c_n be the cost of a **transition** n of PN model. For composing holarchies, SOHCA compares the cost of $C_{[H1]} = C_{[bcb]} + C_{[bcb+ap]} + C_{[bcb+ap+s]} + C_{[bcb+ap+s+co]}$ and $C_{[H2]} = C_{[rcb]} + C_{[rcb+ap]} + C_{[rcb+ap+s]} + C_{[rcb+ap+s+co]}$ to decide what is the better sequence for production is at a certain time.

Figure 4 has PN and PFS models for some holons and their workflow. Figure 5 presents the production plan, and reconfiguration examples. Figure 6a shows the PFS and PN models, and the programmable logic controller (PLC) I/O addresses list of some detectors and sensors of the D-WS. The mapping between the PFS and the JAVA implementation is depicted in Fig. 6b. Scenarios were simulated to validate the advantages of the reconfiguration and degeneration mechanisms. For example, by controlling the production speed through the pneumatic pressure and by disabling the swivel arm (represented by OpH, $[sa_{D\,WS}]$) then its function was assumed by another OpH, $[R$-$WS]$.

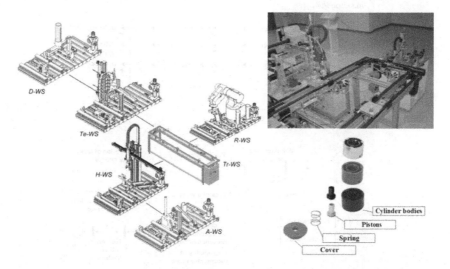

Fig. 3. BDMS composed of workstations (WSs): distributing (D-WS), testing (Te-WS), transporting (Tr-WS), handling (H-WS), assembling (A-WS) and robot (R-WS). The production plan is joining work pieces (wps): a cylinder body (black [bcb], red [rcb] or aluminum [acb]), a piston (black [bp] or aluminum [ap]), a spring [s] and a cover [co].

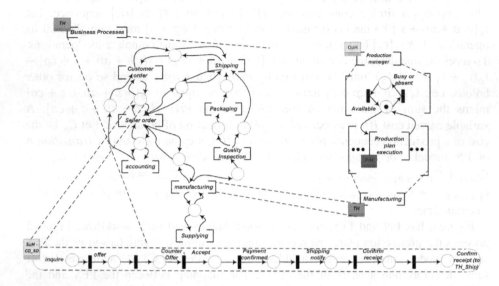

Fig. 4. Choreography of *THs-[Business Processes]*, an orchestration of *OpH* and *TH*, and the detailing of *SuH-[CO_SO]* for the message exchange between the holons

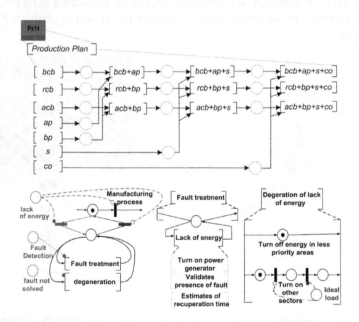

Fig. 5. PrH *[production plan]*, fault treatment and degeneration workflow

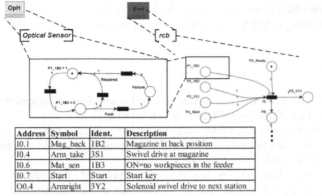

Address	Symbol	Ident.	Description
I0.1	Mag_back	1B2	Magazine in back position
I0.4	Arm_take	3S1	Swivel drive at magazine
I0.6	Mat_sen	1B3	ON=no workpieces in the feeder
I0.7	Start	Start	Start key
O0.4	Armright	3Y2	Solenoid swivel drive to next station

(a) *OpH-[optical sensor]*, OpH -[*rcb*] and the addresses list of PLC

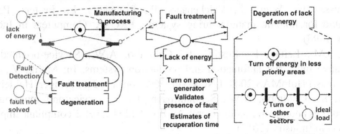

(b) PN, PFS of the *TH-[Accounting]* and the transformation into JAVA language

Fig. 6. Examples of *implementation stage*

6   Conclusions and Further Work

This paper presented service-oriented holonic control architecture (SOHCA), and its method that describes the required data and techniques to design supervisory control system to the reconfiguration of dispersed manufacturing systems (RDMS). Different scenarios were elaborated for running an application example of a benchmarking dispersed manufacturing system. The scenarios met the restrictions and achieved the objectives outlined in the hypothesis. SOHCA responded in a faster and in a collaborative manner and showed to be useful to protect the system when hardware problems occur, implemented different thresholds of production and demonstrated operational advantages, such as better and more efficient use of manufacturing resources, production speed and ability to deliver products faster. The qualitative analysis was based on structural analysis using PN editors and quantitative analysis was performed through the simulation with **temporized transitions**. It was also possible to review the control system models identifying the places and transitions that must be considered in each service. The reconfiguration was not only applied to solve fault occurrence. It also was applied to improve the system performance by increasing the production gain or the number of final products. SOHCA is generic and developed based on SOA and HCS techniques; it can be tailored for specific manufacturing applications. A larger project is being developed which involves in addition to modeling, simulation and validation of PN models, ontology description, tools for designer, and other case studies.

Acknowledgments. The authors would like to thank the partial financial support from the agencies: CNPq, CAPES, FAPESP, and from the universities: UESC and USP.

References

1. Hsieh, F.S., Lin, J.B.: A self-adaptation scheme for workflow management in multi-agent systems. Journal of Intelligent Manufacturing, 1–18 (2013)
2. Brennan, R., Fletcher, M., Norrie, D.: An agent-based approach to reconfiguration of real-time distributed control systems. IEEE Transactions on Robotics and Automation 18(4), 444–451 (2002)
3. Chirn, J., McFarlane, D.: A holonic component-based approach to reconfigurable manufacturing control architecture. In: Proceedings of 11th International Workshop on Database and Expert Systems Applications, pp. 219–223. IEEE (2000)
4. Van Brussel, H., Wyns, J., Valckenaers, P., Bongaerts, L., Peeters, P.: Reference architecture for holonic manufacturing systems: PROSA. Computers in Industry 37(3), 255–274 (1998)
5. Mendes, J.M., Leitão, P., Restivo, F., Colombo, A.W.: Service-oriented agents for collaborative industrial automation and production systems. In: Mařík, V., Strasser, T., Zoitl, A. (eds.) HoloMAS 2009. LNCS, vol. 5696, pp. 13–24. Springer, Heidelberg (2009)
6. Nagorny, K., Colombo, A.W., Schmidtmann, U.: A service-and multi-agent-oriented manufacturing automation architecture: An IEC 62264 level 2 compliant implementation. Computers in Industry (2012)
7. Morariu, C., Morariu, O., Borangiu, T.: Customer order management in service-oriented holonic manufacturing. Computers in Industry 64(8), 1061–1072 (2013)
8. Billington, J., Christensen, S., van Hee, K.M., Kindler, E., Kummer, O., Petrucci, L., Post, R., Stehno, C., Weber, M.: The Petri Net Markup Language: Concepts, Technology, and Tools. In: van der Aalst, W.M.P., Best, E. (eds.) ICATPN 2003. LNCS, vol. 2679, pp. 483–505. Springer, Heidelberg (2003)
9. Hasegawa, K., Miyagi, P.E., Santos Filho, D.J., Takahashi, K., Ma, L., Sugisawa, M.: On resource arc for Petri net modelling of complex resource sharing system. Journal of Intelligent and Robotic Systems 26(3-4), 423–437 (1999)
10. Leitão, P., Restivo, F.: Adacor: A holonic architecture for agile and adaptive manufacturing control. Computers in Industry 57(2), 121–130 (2006)
11. Instrumentation, S., Society, A.: Enterprise-control system integration: Part 1:models and terminology. 95.00.01 (2000)
12. da Silva, R.M., Miyagi, P.E., Santos Filho, D.J.: Design of active holonic fault-tolerant control systems. In: Camarinha-Matos, L.M. (ed.) Technological Innovation for Sustainability. IFIP AICT, vol. 349, pp. 367–374. Springer, Heidelberg (2011)
13. Silva, R.M., Arakaki, J., Junqueira, F., Santos Filho, D.J., Miyagi, P.E.: A procedure for modeling of holonic control systems for intelligent building (HCS-IB). Advanced Materials Research 383, 2318–2326 (2012)
14. Silva, R.M., Arakaki, J., Junqueira, F., Santos Filho, D.J., Miyagi, P.E.: Modeling of active holonic control systems for intelligent buildings. Automation in Construction 25, 20–33 (2012)
15. Silva, R.M., Junqueira, F., dos Santos Filho, D.J., Miyagi, P.E.: Design of reconfigurable and collaborative control system for productive systems, 1st edn. ABCM Symposium Series in Mechatronics, pp. 813–822. Rio de Janeiro, RJ (2012)

Mitigation Control of Critical Faults in Production Systems

Jeferson A.L. de Souza, Diolino J. Santos Fo, Reinaldo Squillante Jr.,
Fabricio Junqueira, and Paulo E. Miyagi

University of São Paulo, São Paulo, Brazil
{jeferson.souza,diolinos,reinaldo.squillante,
fabri,pemiyagi}@usp.br

Abstract. The inherent complexity of critical production systems, coupled with policies to preserve people´s safety and health, environmental management, and the facilities themselves, and stricter laws regarding the occurrence of accidents, are the motivation to the design of Safety Control Systems that leads the mitigation functionality. According to experts, the concept of Safety Instrumented Systems (SIS) is a solution to these types of issues. They strongly recommend layers of risk reduction based on hierarchical control systems in order to manage risks, preventing or mitigating faults, and to lead the process to a safe state. Additionally some of the safety standards such as IEC 61508, IEC 61511, among others, guide different activities related Safety Life Cycle design of SIS. The IEC 61508 suggests layers of critical fault prevention and critical fault mitigation. In the context of mitigation control system, the standard provides a recommendation of activities to mitigate critical faults, by proposing control levels of mitigation. This paper proposes a method to implement the mitigation layer based on the risk analysis of the plant and the consequences of faults of its critical components. The control architecture, based on distributed and hierarchical control systems in a collaborative way, will make use of the techniques of risk analysis raised and mitigation actions, based on the knowledge of an expert, implemented by fuzzy logic.

Keywords: Critical Systems, Mitigation Control System, Safety Instrumented System, Fuzzy Logic.

1 Introduction

In this first decade of the century XXI many studies have indicated that automation processes are undergoing transformations that have been strongly influenced by the advance of technology and computing resources, becoming increasingly complex due to their dynamic and needed to address issues such as global market competitive production and technology used, among other factors [1], [2], [3]. Given this new scenario, industrial processes and their control are becoming more complex. Additionally, organizations have focused on policies to achieve and to demonstrate people's safety and health, environmental management system, and controlling risks.

L.M. Camarinha-Matos et al. (Eds.): DoCEIS 2014, IFIP AICT 423, pp. 119–128, 2014.

In this context, any industrial system, as modern and innovative as can be, could be considered to pose a serious risk to people's health, the environment and equipment [4]. Although many studies have been presented for diagnosis and treatment of faults, a review of fault-tolerant reconfigurable control system can be found in [5], but accidents still occur. These issues are fully justified because there is no zero risk in process industries since: (i) physical devices do not have zero risk of failure [6], (ii) human operators do not have zero risk of error and (iii) there is no computational software project developed that can predict all the possibilities [7].

According to experts, the concepts of safety instrumented systems (SIS) is a solution to these types of issues and strongly recommend layers of risk reduction based on control systems organized hierarchically in order to manage risks by either preventing or mitigating faults, and to bring the process to a safe state. In this sense, some safety standards such as IEC 61508 [8], IEC 61511 [9] among others, guide activities related with a SIS Safety Life Cycle (SLC), such as design, installation, operation, maintenance, tests and others [10], [11].

According to IEC 61508, the term "fault" is defined as an abnormal condition that can cause a reduction or loss of the ability of a functional unit, and is defined two layers of SIS: the prevention layer and the mitigation layer. Recently, [12] proposes the implementation of a SIS prevention layer.

This work is initially proposed a systematic for modeling and validating layer of mitigation control within SIS. This approach considers the cause of the fault, its severity and its consequence for the system, through the application of risk analysis techniques such as *Failure Modes and Event Analysis* - FMEA, *Fault Tree Analysis* – FTA [13], and the *What-If* technique [14], based on a database of occurrence of faults or on knowledge of an expert or operator. The effects and the consequences of the occurrence of a critical fault, listed on the risk analyses study, are monitored and treated by the SIS sensors and actuators, respectively, independently of the BPCS devices, as predicts the IEC 61508 [15], [16]. The effect of every critical fault, or safety instrumented function (SIF), results in mitigation actions, determined by the *What-If* technique yet implemented.

Fuzzy logic is utilized for the generation of the control algorithm. It has the advantage of not using differential equations or complex mathematical models for determining the dynamic behavior of the system [17], and can therefore use the proposed mitigation actions, which in turn were the result of applying the techniques of risk analysis, for the determination of fuzzy control algorithms. Another advantage of fuzzy logic is the analysis of the parameter time derivative [18], thus contributing to an anticipatory action of the proposed mitigation layer. The generation of control codes for programmable logic controllers (PLC) can be made based on IEC 61131-7 [19], which deals with the conversion of the generated fuzzy logic algorithm to conventional PLC languages, based on IEC 61131-3. [20].

2   Relationship to Collective Awareness Systems

Recent political awareness with focus on sustainability and recycling, the use of resources and raw materials from renewable resources, along with the practices of waste management and emission control of pollutants, coupled with more rigid and punitive laws to production systems that do not meet the new regulations, results in new control systems in manufacturing plants.

In this context, a mitigation control layer is needed because, in addition to the market requirements of competitiveness, such policies result in a choosing collective awareness of choice by consumers for companies committed to focus on the environment and sustainability. Besides the increase in the complexity of the control systems for these new requirements, the proposed mitigation layer, based on IEC 61508, is precisely the preservation of men and the environment, just required for productive systems that wishing to adapt to these new practices of collective awareness recently observed.

3   Proposal of Layer of Mitigating Control System

3.1   Description of the Proposed Method

The proposed method is summarized by the following five steps, described on sections 3.1.1 to 3.1.5 below:

3.1.1   Determination of the Critical Elements

To determine the critical elements of the process under study we utilize the risk analysis techniques FTA, FMEA and What-If. The FMEA, to associate a severity level to the occurrence of fault of a component indicates which components must be monitored in the mitigation layer. Faulted components that pose risks to operators, the environment and equipment, besides violating the legislation are classified to maximum severity.

Furthermore, components which fault under no danger considerable not part of our analysis.

Because the FMEA to be centered on the component, combination of faults and a possible domino effect over other components may be analyzed by the FTA in conjunction to What-If technique. It is possible, to determine the how and the why of the fault, therefore rendering a more comprehensive study.

3.1.2   Detection of Effects Caused by the Occurrence of Faults of the Critical Elements

Each effect arising from the fault of a critical component must be monitored by a specific sensor for its fault mode. According to IEC 61508, such sensors must be independent of the BPCS. To avoid spurious faults and reading errors, it is recommended to use redundant architectures [11], such as the criteria voting 2oo3 (two of three).

3.1.3   Mitigation Actions of the Effects of Faults of Critical Elements

For each effect of a critical fault, detected by the SIS mitigation sensors, a mitigation action must be implemented by SIS mitigation actuators, controlled by the SIS mitigation control layer, aiming to preserve people, environment and equipment.

To determine de mitigation actions will make use of *What-If* technique, based on human knowledge and records of occurrence of faults, its effects and the actions proposed to mitigate its effects.

Some mitigation actions can be matched to faults occur in different components, but not necessarily input signals from SIS mitigation sensors are the same. That is, for different input signals, from different sensors, mitigation actions may be the same. In this step, besides determination of the mitigation sensors can arrive at the conclusion that the sensors would be the same prevention. In this case, it is recommended doubling of signals from sensors for prevention PLC we use in our mitigation.

After this study and compilation of mitigation actions, will determine which actuators required for each mitigation action.

3.1.4   Construction of Models for Implementation of Control Algorithms

In this step will be used results of the *What-If* technique already implemented in section 3.1.3 to determine the level or percentage of the measured variable values for activation layer of prevention and / or mitigation using the absolute value of the measured variable and its temporal variation or derivative of the measured parameter.

The results of this study will form the basis of fuzzy algorithms for mitigation control layer.

3.1.5   Control Codes Generation Based on IEC 61131-7

For each mitigation action determined by the fuzzy control algorithm, the next step is to convert the generated control algorithm for a language of IEC 61131-3 to implement in the Safety PLC for mitigation.

The IEC 61131-7 deals with the implementation of fuzzy algorithms in FCL (Fuzzy Control Language), based on IEC 61131-3 [24] ST (Structured Text) for the implementation in conventional PLCs.

4   Example of Application

To illustrate the method proposed, an application example for critical faults to be mitigated by SIS Mitigation layer in a natural gas compression station is presented. Natural gas is a mixture of highly flammable hydrocarbons. To be extracted from the environment must be pressurized in compressor stations to its carriage due to consumer centers.

4.1   Process Description

The natural gas station has one or more natural gas supply lines, called suction, from a gas pipeline which transports this natural gas. At the station entrance, natural gas goes through filters equipment before being compressed by the turbo compressor machine. A portion of this gas is directed to the utility unit. The utility unit accounts for controlling the gas temperature and pressure for use in the compression station, such as fuel gas for the turbo-compressor machine, gas heaters and gas power generators. After the natural gas is compressed by turbo compressor machine, it is sent back to the gas pipeline through discharge lines, called headers. The PI&D of a turbo compressor uni tis shown in Fig. 1.

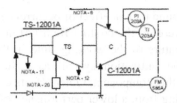

Fig. 1. P&ID of a compressor unit of the gas compression station

4.2   Application of the Proposed Method

We apply the proposed method, based on the SIS prevention layer proposed by
[11] for the case of the compressor gas compression station. A more elaborate
study should be done, considering all critical components indicated by the
application of FMEA and FTA techniques. This work presents an example for a
system component, in order to exemplify the application of the proposed method.

4.2.1 Determination of the Critical Elements

Applying the FMEA technique can be seen that the compressors are critical to
effectively our system, because they operate under high temperature, pressure and
speed, in addition to use as fuel the compression fluid itself, which is natural gas, just
explosive. A fault in this equipment certainly put under unacceptable risk operators,
the environment and the equipment itself, besides violating government standards for
safety. Hence its severity is maximum, and must be entered in our mitigation layer.
Table 1 illustrates a FMEA for compressor, and Fig. 2 the FTA for the top event
"High Temperature Lubricants Shaft".

Table 1. Proposed FMEA for temperature increase of the lubricating oil of shaft bearing
compressor

Component	Potential Fault Mode		Potential Effects of Fault	Potential Causes of Fault	Deteccion - Control	Recommended Actions	Severity Associated
Turbo Compressor	Lubrication		Increase in temperature of the lubricating oil bearing	Saturated Oil	Temperature Sensors	Shut-Down (Preventive)	10
						Dioxide Carbon (Mitigate)	
	Damage	Oil Pump					
		Shaft		Bearing Wear			
		Speed control		Hardware / Software			
	Overload	Speed		Speed Control			
		Torque		Condensate Excess			

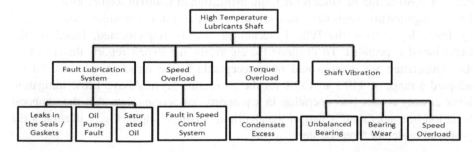

Fig. 2. Suggested FTA for the top event "High Temperature Lubricants Shaft"

Both FMEA and FTA found that an effect of occurrence of fault in the compressor is to increase the temperature on the cooling fluid turbine shaft, being able to have also an increase in temperature of the working fluid in the discharge line. We will perform our study on the mitigation system as a function of monitoring the temperature parameter for this component. Other effects can be measured as changes in discharge pressure coming from a lower performance of the compressor operating under fault.

4.2.2 Detection of Effects Caused by the Occurrence of Faults of the Critical Elements

For the effects of faults listed in the previous step, we have temperature sensors coolant axis of compressors, independently of the BPCS. Such sensors will be designated TAT 211 – Temperature Axis Turbine – for each unit present in the natural gas station. So we have the TAT 211 A, TAT 211 B, TAT 211 C and TAT 211 D as input signals our mitigation PLC. Again, a redundant architecture of these sensors as well as the implementation of algorithms for detecting spurious faults [11] should be implemented.

4.2.3 Mitigation Actions of the Effects of Faults of Critical Elements

To mitigate the effects caused by the occurrence of a fault in the compressor, beside the action of shutdown from the prevention layer, suggested action to mitigate the effects is the forced cooling of the compressor, if preventive layer is not sufficient or if the temporal variation of temperature proves too high. Will be used both carbon dioxide cylinders large that are already installed in natural gas station, and have the purpose of fire combat if an outbreak of fire. The release of carbon dioxide is currently done manually, through the action of fire brigade teams, specially trained for this purpose. The proposal would be the installation of pipelines leaving the cylinders to compressors with proportional valves connected to the outputs of mitigation´s Safety PLC. As the intensity of mitigation action, the valve would release the carbon dioxide in the same proportion.

4.2.4 Construction of Models for Implementation of Control Algorithms

From mitigation proposals have the construction of the control algorithms implemented by fuzzy logic, from the What-If technique already implemented, based on the expertise of a specialist. To illustrate the algorithm, the expert reports that 150% of the temperature set point would be unacceptable to the turbo compressor. So we adopted a range of 110% to 130% for the prevention layer. Above 120% mitigation layer already comes into operation in a proportional action. Note that the temporal variation in temperature is part of the algorithm´s control input. Fig. 3 illustrates the proposed model for temperature:

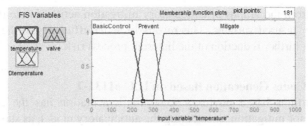

Fig. 3. Fuzzy Membership functions for temperature

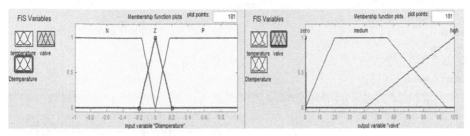

Fig. 4. and Fig. 5. Membership functions to temperature derivative and percentage of valve opening

According to the membership functions adopted in the Fig. 3 above, has three regions for temperature: Basic Control, Prevent and Mitigate. The input of time derivative of temperature was set to three values: zero, positive and negative. As for output, which is proportional to the valve opening was also set to three positions: zero or closed valve, high or 100% open and medium, open at 50%. Fig.4 and Fig.5 illustrate the above.

The rules of the fuzzy algorithm, according to What-If technique are as follows in Table 2.

Table 2. Fuzzy rules for mitigation layer

IF	TEMPERATURE	OPERATOR			
1	MITIGATE				HIGH – 100% OPEN
2	BASIC CONTROL				ZERO - CLOSED
3	PREVENT	AND		P	MEDIUM – 50%
4	MITIGATE1	AND		N	MEDIUM – 50%

The output signal, or the proportional action of mitigation, here designated by proportional valve opening, can be seen by the generated surface on Fig.6.

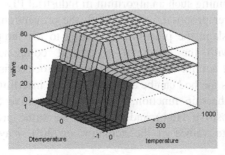

Fig. 6. Surface generated by the fuzzy algorithm by the fuzzy rules defined for the mitigation model

We can see from the graphs of anticipatory mitigation action due to the temporary increase of the measured variable. This results in better efficiency of the system, thus contributing to a further reduction of the inherent process risk.

4.2.5 Control Codes Generation Based on IEC 61131-7

From the algorithms based on fuzzy logic implementation has the control codes to Safety PLC for mitigation, considering the anticipatory model, as shown in Fig. 7.

```
FUNCTION_BLOCK FUZZYCONTROL
VAR_INPUT
        temperature : REAL;
        dtemperature : REAL;
END_VAR

VAR_OUTPUT
        valvule : REAL;
END_VAR

FUZZIFY temperature

        TERM BasicControl := (0,1) (200,1) (250,0);
        TERM prevent := (200,0) (250,1) (300,1) (350,0);
        TERM mitigate := (300,0) (350,1) (1000,1);
END_FUZZIFY


FUZZIFY dtemperature

        TERM N := (-1,1) (-0.2,1) (0,0);
        TERM Z := (-0.2,0) (0,1) (0.2,0) ;
        TERM P := (0,0) (0.2,1) (1,1);

END_FUZZIFY


DEFUZZIFY valvule

        TERM zero := 0;
        TERM medium := (0,0) (20,1) (55,1) (95,0);|
        TERM high := (40,0) (100,1);

        ACCU : MAX;
        METHOD : COG;
        DEFAULT := 0;

END_DEFUZZIFY

RULEBLOCK No1
        AND : MIN;

        RULE 1 : IF temperature is mitigate THEN valvule IS high;
        RULE 2 : IF temperature is BasicControl THEN valvule IS zero;
        RULE 3 : IF (temperature is prevent) and (dtemperature is P)
                 THEN valvule IS medium;
        RULE 4 : IF (temperature is mitigate) and (dtemperature is N)
                 THEN valvula IS medium;

END_RULEBLOCK

END_FUNCTION_BLOCK
```

Fig. 5. Control code generated, implemented in FCL (Fuzzy Control Language) according to IEC 61131-7

5   Conclusions

A method for the implementation of mitigation layer in critical industrial systems was proposed, based on the IEC 61508 and IEC 61511 standards, which recommend layers of risk reduction based on cooperative and hierarchical control prevention and mitigation of critical faults. Based on the results of applying the risk analysis techniques can be evaluated, due to the effects of their faults, what the critical elements present in the process. Based on the knowledge of an expert and making use of the What-If technique already deployed, implement corresponding mitigation actions using fuzzy logic, becoming such an algorithm in industrial PLCs languages based on IEC 61131-7. This layer proposal, coupled with the prevention layer, contributes to reduce the inherent risk in the process and adding to the temporal analysis of the variable associated with the effect of a critical component fault results in anticipatory mitigation action, resulting in a higher process risk reduction.

A refinement of this method can be accomplished by inserting a larger set of terms for de derivative membership function, such as PS (Positive Short), PM (Positive Medium) and PH (Positive High) and adopting the same procedure for negative derivative. Intermediate values of the actuator, eg 30% may be associated with these new values, which will surely determine new fuzzy rules in the algorithm. Other mitigation actions can be proposed, and this model must be implemented for the other critical elements of plant. Such elements may have other parameters that indicate the fault component and also other mitigating actions.

Acknowledgments. The authors would like to thank the Brazilian governmental agencies CNPq, FAPESP, and CAPES for their financial support to this work.

References

1. Chen, C., Dai, J.: Design and high-level synthesis of hybridcontroller. In: Proc. of IEEE Intern. Conf.of Networking, Sensing & Control (2004)
2. SantosFilho, D.J.: Aspectos do Projeto de Sistemas Produtivos. PHDThesis, Escola PolitécnicadaUniversidade deSãoPaulo, Brazil (2000)
3. Wu, B., Xi, L.-F., Zhuo, B.-H.: Service-oriented communication architecture for automated manufacturing system integration. Int. J. Computer Integrated Manufacturing 21(5), 599–615 (2008)
4. Sallak, M., Simon, C., Aubry, J.: A fuzzy probabilistic approach for determining safety integrity level. IEEE Transaction on Fuzzy Systems 16(1), 239–248 (2008)
5. Zhang, Y., Jiang, J.: Bibliographical review on reconfigurable fault-tolerant control systems. Annual Reviews in Control 32, 229–252 (2008)
6. Summers, A., Raney, G.: Common cause and commonsense, designing failure out of your safety instrumented systems (SIS). ISA Transactions 38, 291–299 (1999)
7. Miyagi, P.E.: ControleProgramável–Fundamentos do controle de sistemas a eventos discretos. Editora Edgard Blucher Ltda, SãoPaulo, SP, Brazil (2007)
8. IEC,I.E.C., Functional safety of electrical / electronic / programmable electronic safety-relatedsystems (IEC61508) (2010)
9. IEC,I.E.C., Functionalsafety-safety instrumented systems for the process industry sector-part 1(IEC 61511) (2003a)
10. Lundteigen, M.-A., Rausand, M.: Architectural constraints in IEC61508: Do they have the intended effect? In: Reliability Engineering and System Safety, pp. 520–525. Elsevier SciencePublisher Ltd. (2009)
11. Bell, R.: Introduction to IEC61508. In: Proceedings of ACS Workshop on Tools and Standards, Sydney, Australia (2005)
12. Squillante Jr., R., Santos Filho, D., Riascos, L., Junqueira, F., Miyagi, P.: Mathematical method for modeling and validating of safety instrumented system designed according to IEC61508 and IEC61511. In: InCobem 2011 (2011)
13. Modarres, M., Kaminskiy, M., Krivstov, V.: Reliability Engineering and Risk Analysis: apractical guide, 2nd edn. CRCPress (2010)
14. Souza, E.A.: O treinamento industrial e a gerência de riscos–uma proposta de instrução programada.Master Thesis, Universidade Federal de Santa Catarina, Brazil (1995)
15. Squillante Jr., R., Fo, D.J.S., de Souza, J.A.L., Junqueira, F., Miyagi, P.E.: Safety in supervisory control for critical systems. In: Camarinha-Matos, L.M., Tomic, S., Graça, P. (eds.) DoCEIS 2013. IFIP AICT, vol. 394, pp. 261–270. Springer, Heidelberg (2013)

16. Cavalheiro, A., Santos Fo, D., Andrade, A., Cardoso, J.R., Bock, E., Fonseca, J., Miyagi, P.E.: Design of supervisory control system for ventricular assist device. In: Camarinha-Matos, L.M. (ed.) Technological Innovation for Sustainability. IFIP AICT, vol. 349, pp. 375–382. Springer, Heidelberg (2011)
17. Popa, D.D., Craciunescu, A., Kreindler, L.: API-Fuzzy controller designated for industrial motor control applications. In: ISIE IEEE International Symposium on Applications, Industrial Eletronics (2008)
18. Legaspe, E.P., Dias, E.M.: Open source fuzzy controller for programmable controllers. In: 13th Mechatronics Forum Biennial International Conference (2012)
19. IEC,I.E.C., Programmable Controllers IEC 61131-7: Fuzzy Control programming (2000)
20. IEC,I.E.C., Programmable controllersIEC61131-part 3: Programming languages (2003b)

A General Distributed Architecture for Resilient Monitoring over Heterogeneous Networks

Fábio Januário[1,2], Alberto Cardoso[2], and Paulo Gil[1,2]

[1] Departamento de Engenharia Electrotécnica, Faculdade de Ciências e Tecnologia,
Universidade Nova de Lisboa, Portugal
[2] Centre for Informatics and Systems (CISUC), University of Coimbra, Portugal
`f.januario@campus.fct.unl.pt`, `alberto@dei.uc.pt`, `psg@fct.unl.pt`

Abstract. The growing developments on networked devices, with different communication structures and capabilities made possible the emergence of new architectures for monitoring systems. In the case of heterogeneous distributed environments, where knowledge, processing devices, sensors and actuators are distributed throughout the network, the design of such systems are challenging in terms of integration and, markedly, in security and state-awareness of the overall system. This work proposes a general distributed architecture, with supporting methods, for building a resilient monitoring system that can adaptively accommodate both cyber and physical anomalies. Its implementation relies on multi-agent systems within a distributed middleware.

Keywords: Resilient systems, distributed computing, embedded systems, multi-agents, wireless sensor and actuator network, heterogeneous networks.

1 Introduction

Modern computing networks that enable distributed computing consist of a wide range of heterogeneous devices, with various levels of resources, which are interconnected using differing networking technologies [1]. These networks may be implemented in industrial environments for connecting large number of subsystems, of different natures, including interactions with humans and supervision platforms.

Distributed heterogeneous environments present inherently some challenges and vulnerabilities. In the context of monitoring systems for fault and failure detection there is the need of efficient information processing and correct assessment of the systems' behaviour. Such endeavours require the development of dedicated methods to identify and recover from faults and failures, or in order to mitigate their impact on the overall system. Regarding the vulnerabilities of these networked systems, the cyber-intrusion is of major importance, as malignant actors can mask the system's degradation or provide fake data to higher management levels, with respect to the current system's status [2].

In this work, a resilient system is regarded as a system that maintains state awareness and an acceptable level of operational normalcy in response to disturbances, including threats of an unexpected and malicious nature [3][4]. In the

L.M. Camarinha-Matos et al. (Eds.): DoCEIS 2014, IFIP AICT 423, pp. 129–136, 2014.
© IFIP International Federation for Information Processing 2014

case of heterogeneous monitoring systems, this resilience provides the system with security and trust mechanism, thus fixing, mitigating or coping with the aforementioned vulnerabilities.

The implementation of resilient enforcement mechanisms can be accomplished by incorporating dedicated algorithms and heuristics on a distributed architecture based on multi-agents and "chunks" of middleware deployed remotely. One of the most prevalent alternatives in the context of distributed architecture design is the multi-agent system (MAS) paradigm. A distributed agent-based architecture provides flexibility to move functions to where actions or measures are needed, thus obtaining improved responses at execution time, autonomy, services continuity and superior levels of scalability [5].

Middleware architectures are becoming increasingly important as networks, services and applications become more complex. These architectures can deal with coordination, cooperation and interoperability of distributed components by bridging the gap between applications and their underlying low level software and hardware infrastructures. Moreover, they provide tools and methods that can help hiding the complexity and heterogeneity of hardware and network platforms. Also, middleware supports programmer's applications in several ways, such as providing appropriate system abstractions and reusability of code services, while helping in the network infrastructure management [6].

The present work proposes a general architecture for resilient monitoring over heterogeneous networks making use of multi-agents embedded on a distributed middleware framework, where each agent is tailored for executing specific and coordinated tasks, namely for detecting and recovering from physical and cyber malfunctions.

2   Relationship to Collective Awareness Systems

According to the European Commission, collective awareness systems are information and communication technologies (ICT) systems that leverage the emerging "network effect", by combining open online social media, distributed knowledge creation, and data from real environments, in order to create awareness of problems and possible solutions requesting collective efforts, while enabling new forms of innovation [7].

These systems can be adopted in various scenarios, like in social environments, enterprises collaboration, industrial processes and decision methods. In these contexts, a network infrastructure allowing communication and perception is considered necessary to assess the context aware, by making use of a collaborative approach.

The electric power production and distribution is a critical system where monitoring is of major important to prevent or accommodate systems' malfunctions. This is one striking example where resilient systems can contribute to the integrity of the overall system, by preventing cyber and physical attacks to infrastructures. Another field with high security requirements, is that of governmental communications, where intrusion threats in their communication networks are taken very seriously. Finally with the automation of many industrial processes it is necessary to develop monitoring systems in order to ensure the correctness of operation.

To provide security and dependability mechanisms to this kind of distributed systems, new set of heuristics and algorithms have to be developed, which opens a window of opportunity for new contributions to the field. In this context, the present work intends to give a contribution to the development of collective awareness systems, by proposing a general distributed architecture for enhancing the overall resilience of monitoring systems over heterogeneous networks.

3   Resilient Systems

The research area of resilient systems is a relatively new topic that involves systems design taking into account issues, such as cyber security, physical security, process efficiency and stability, process compliancy and state awareness.

In [8] a hybrid system model is used to address physical layer control design and cyber level security policy making for cyber-physical systems that are subject to cascading effects from cyber attacks and physical disturbances. In [9], the authors propose a hybrid theoretical framework to analyse and design, in a quantitative and holistic way, robust and resilient control systems that can be subject to different types of disturbances, at different layers of the system. The work in [10] presents an intelligent resilient control algorithm for a wireless network control system based on the quantification of the concept of resiliency in terms of the quality of control. The authors developed an intelligent resilient control algorithm that maintains operational normalcy in face of wireless interference incidents, such as radio frequency jamming and signal blocking, while [11] proposes a resilient condition assessment monitoring system, able to dynamically adapt and reconfigure, depending on the assessed conditions.

In literature one can find several works making use of MAS to implement intelligence mechanisms. In [12], it is presented an overview on how to achieve critical infrastructure resilience through advanced control engineering. The authors developed a hierarchical multi-agent dynamic system framework to model distributed control system dynamics and enforce resilience. In [2], a generalized design methodology was suggested for analysing and accommodating anomalies in cyber physical systems. By using computational intelligence techniques it is provided a design method to integrate cyber and physical data and, thus ensuring the appropriate response to both benign and malicious actions. In [13], a distributed multi-agent architecture is presented to implement a resilient supervision system over wireless sensor and actuator network. This system implements resilience enforced mechanisms, by incorporating dedicated algorithms and heuristics in an architecture based on agents, so as to guarantee the state awareness and an acceptable level of operational normalcy, in response to disturbances.

Taking into account previous constrains on resilient systems, this work intends to extend resilience policies to heterogeneous and distributed environments. With the modern computing networks infrastructures and distributed resources these systems become an important research area to provide security and dependability. Aiming at this purpose, the present work proposes a general resilient architecture based on the MAS paradigm.

3.1   Architecture Overview

The proposed architecture is based on a MAS embedded on a distributed middleware platform, for which issues, such as those of communication, network topologies, routing protocols and integration are assumed to be dealt with elsewhere. Fig. 1 presents a distributed environment comprising three main components, namely applications, middlewares and processes/devices or nodes.

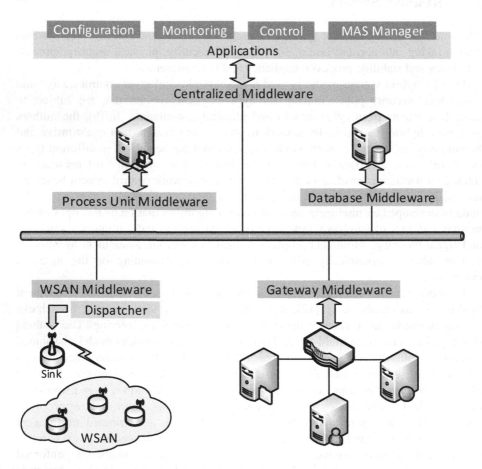

Fig. 1. General distributed architecture

The processes/devices or nodes consist of physical components and the necessary hardware and software infrastructures. These components can be represented by processing units that process data stemming from the network, or from databases that store and provide information and knowledge regarding the network, gateways interfacing with another network that uses different protocols, and sensor and actuator networks (wireless or not) to interact with the environment, by sensing and actuating on physical infrastructures. Fig. 1 shows a wireless sensor and actuator network (WSAN), where the sink node and the dispatcher are responsible for transferring the data from the WSAN into the middleware and for coordinating the underlying communications.

The application layer provides users with a number of applications allowing, in a transparent way, the interaction with networks, devices and plants in the perimeter of the networks. The monitoring application enables to follow-up the system status, while the control application manages, commands or manipulates the behaviour of the system, using sensors and actuators. The configuration application enables to setting up the system's parameters and available devices. Finally, the MAS manager configures and changes the multi-agent system's attributes.

The middleware layer allows the interaction between the application layer and the plant/devices. The middleware is in this architecture deployed, in a decentralized fashion, on several components in order to bring about the functionalities to the location where they are called out. In heterogeneous networks this topology has the advantage of each "chunk" of middleware could be implemented taking into account the associated hardware or software constrains shown up in the process/device layer.

Fig. 2 presents the main components of the proposed middleware. The integration layer facilitates data exchange between the process/device infrastructures and applications. Data from sensor and distributed devices, as well as additional diagnosis information is fed into the middleware, analysed and forwarded to the corresponding destination. Further, actuation commands and data is passed down from the middleware to actuators or devices, while the monitoring layer evaluates the performance of the middleware at runtime by applying data quality metrics. This element ensures that processing delays are within required bounds. The configuration layer enables the definition of commands to configure the data uploaded and downloaded from the devices, as well as agents in the security and dependability modules.

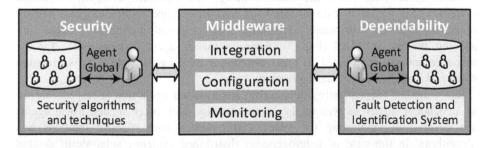

Fig. 2. Resilient middleware based in MAS

Dependability. For enabling the monitoring system to meet dependability targets it is necessary to incorporate fault tolerant and self-healing mechanisms into the resilient design. These mechanisms will ensure end-to-end performance communication in environments where growth, scalability, closed loop stability and performance, are important features.

Security. When sensitive data received from sensors or servers and data sent to actuators or other devices are transmitted through uncontrolled environments, security is inescapably compromised. If not protected, transmitted data may be accessed, corrupted, or even destroyed, by unauthorized users. Consequently, security mechanisms should be developed and implemented in order to be able to adapt to changes in the application environment, to system requirements and system resources.

3.2   Multi-agent System

Agents can be regarded as computing entities in a given environment, presenting a certain degree of autonomy, and possessing the ability to feel and act in order to accomplish a given mission. Some of their inherent features include the following [14]:

- Autonomy: agents are independent entities, able to accomplish a given task, without any programming or direct intervention;
- Reactivity: capability of perceiving their environment and respond quickly and effectively to changes;
- Pro-activity: ability to take initiative goals and behave in order to meet them;
- Cooperation: agents have the ability to interact and communicate with one another for the sake of their own teleonomy;
- Intelligence: in order to evaluate and take over a task, in an autonomous way, an agent should incorporate intelligent techniques;
- Mobility: agents have the ability to move its code from a node to another one in a system, offering mobility properties in distributed computational devices.

Multi-agent systems offer the means to design and implement complex distributed systems. This paradigm extends previous approaches and methodologies, namely object-oriented or distributed computing. The proposed architecture takes into account the use of MAS (Fig. 3), where a master agent is responsible for the management routines related to other dependent local agents, and for coordinating some tasks, such as configuration requests.

Each module (dependability or security) has a master agent that is responsible for all dependent agents, as well as the communication with other agents. Local agents are devoted to monitoring the state of the system with specific methods and algorithms. In the case of heterogeneous distributed systems, subsystems do not possess the same characteristics and vulnerabilities, so these agents have to be configured to provide the necessary functionalities to each subsystem. In the case of a detected event, the agent has the ability to act accordingly, while guaranteeing the maintenance of the system until the problem is completely solved.

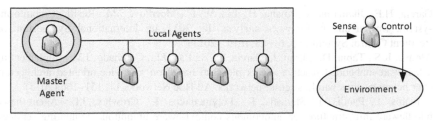

Fig. 3. Multi-agent system architecture

In this architecture, mobile agents have an important role, as their use allows reducing the network traffic. One of the applications for mobile agents deals with communications between middlewares, namely for configurations of distributed agents, with commands sent by the user. With these features MAS allows the implementation of different techniques and show to be a good way so as to provide resilience to heterogeneous networks.

4   Conclusion and Future Work

This paper addressed the problem of resilient monitoring over heterogeneous networks. A general architecture based on multi-agent systems embedded on a distributed middleware framework. The resilience is achieved by combining two main features/resources: *i*) dependability with the implementation of a fault detection and identification system, aiming at maintaining the system in safe operation state; *ii*) security envelope by developing measures to protect the system from cyber and physical attacks.

Finally, it should be mentioned that the present work is still under progress. Further developments will include the exhaustive research in security and dependability areas in order to provide metrics and methods to improve the framework responsiveness.

Acknowledgments. Januário, F. acknowledge Fundação para a Ciência e Tecnologia (FCT), Portugal for the Ph.D. Grant SFRH/BD/85586/2012. This work has been partially supported by iCIS-Intelligent Computing in the Internet of Services, Project CENTRO-07-ST24-FEDER-002003.

References

1. Tynan, R., O'Hare, G.M.P., Marsh, D., O'Kane, D.: Multi-agent System Architectures for Wireless Sensor Networks. In: Sunderam, V.S., van Albada, G.D., Sloot, P.M.A., Dongarra, J. (eds.) ICCS 2005. LNCS, vol. 3516, pp. 687–694. Springer, Heidelberg (2005)
2. Rieger, C.G., Villez, K.: Resilient control system execution agent (ReCoSEA). In: 2012 5th International Symposium on Resilient Control Systems, pp. 143–148. IEEE (2012)
3. Rieger, C.G., Gertman, D.I., McQueen, M.A.: Resilient control systems: Next generation design research. In: 2009 2nd Conference on Human System Interactions, pp. 632–636. IEEE (2009)

4. Garcia, H.E., Jhamaria, N., Kuang, H., Lin, W.-C., Meerkov, S.M.: Resilient monitoring system: Design and performance analysis. In: 2011 4th International Symposium on Resilient Control Systems, pp. 61–68. IEEE (2011)
5. Alonso, R.S., Tapia, D.I., Bajo, J., García, Ó., de Paz, J.F., Corchado, J.M.: Implementing a hardware-embedded reactive agents platform based on a service-oriented architecture over heterogeneous wireless sensor networks. Ad Hoc Networks 11, 151–166 (2013)
6. Soldatos, J., Pandis, I., Stamatis, K., Polymenakos, L., Crowley, J.L.: Agent based middleware infrastructure for autonomous context-aware ubiquitous computing services. Comput. Commun. 30, 577–591 (2007)
7. Digital Agenda for Europe: Collective Awareness Platforms for Sustainability and Social Innovation, https://ec.europa.eu/digital-agenda/en/collective-awareness-platforms
8. Zhu, Q., Başar, T.: A dynamic game-theoretic approach to resilient control system design for cascading failures. In: Proceedings of the 1st International Conference on High Confidence Networked Systems - HiCoNS 2012, p. 41. ACM Press, New York (2012)
9. Zhu, Q., Basar, T.: Robust and resilient control design for cyber-physical systems with an application to power systems. In: IEEE Conference on Decision and Control and European Control Conference, pp. 4066–4071. IEEE (2011)
10. Ji, K., Wei, D.: Resilient control for wireless networked control systems. Int. J. Control. Autom. Syst. 9, 285–293 (2011)
11. Garcia, H.E., Lin, W.-C., Meerkov, S.M.: A resilient condition assessment monitoring system. In: 2012 5th International Symposium on Resilient Control Systems, pp. 98–105. IEEE (2012)
12. Rieger, C.G., Moore, K.L., Baldwin, T.L.: Resilient control systems: A multi-agent dynamic systems perspective. In: IEEE International Conference on Electro-Information Technology, EIT 2013, pp. 1–16. IEEE (2013)
13. Januário, F., Santos, A., Lucena, C., Palma, L., Cardoso, A., Gil, P.: Resilient Supervision System over WSAN: A Distributed Multi-Agent Architecture. In: 3rd International Conference on Sensor Networks (2014)
14. Paolucci, M., Sacile, R.: Agent-Based Manufacturing and Control Systems: New Agile Manufacturing Solutions for Achieving Peak Performance (2005)

Part VI
Advances in Manufacturing

Challenges and Properties for Bio-inspiration in Manufacturing

João Dias Ferreira[1], Luis Ribeiro[2], Mauro Onori[1], and José Barata[2]

[1] EPS Group, Dep. of Production Engineering
Kungliga Tekniska Högskolan
Stockholm, Sweden
{jpdsf,onori}@kth.se
[2] CTS, UNINOVA, Dep. de Eng. Electrotcnica, F.C.T.
Universidade Nova de Lisboa
Monte da Caparica, Portugal
{ldr,jab}@uninova.pt

Abstract. The increasing market fluctuations and customized products demand have dramatically changed the focus of industry towards organizational sustainability and supply chain agility. Such critical changes inevitably have a direct impact on the shop-floor operational requirements. In this sense, a number of innovative production paradigms emerged, providing the necessary theoretical background to such systems. Due to similarities between innovative modular production floors and natural complex systems, modern paradigms theoretically rely on bio-inspired concepts to attain the characteristics of biological systems. Nevertheless, during the implementation phase, bio-inspired principles tend to be left behind in favor of more traditional approaches, resulting in simple distributed systems with considerable limitations regarding scalability, reconfigurable ability and distributed problem resolution.

This paper analyzes and presents a brief critical review on how bio-inspired concepts are currently being explored in the manufacturing environment, in an attempt to formulate a number of challenges and properties that need to be considered in order to implement manufacturing systems that closely follow the biological principles and consequently present overall characteristics of complex natural systems.

Keywords: Bio-inspiration, Self-Organization, Manufacturing Systems.

1 Introduction

Nowadays, manufacturing companies are facing a very challenging reality. Not only society is demanding more customized high quality products than ever, but also markets are becoming increasingly characterized by an erratic behavior. Hence, sustainability and responsiveness at the shop-floor level are fundamental factors to survive and thrive in such harsh and challenging conditions. The short life-cycle of products and the need to promptly respond to product demand fluctuations and short-term business opportunities is compelling companies to shift from traditional shop-floor approaches, towards more modular and reusable systems.

L.M. Camarinha-Matos et al. (Eds.): DoCEIS 2014, IFIP AICT 423, pp. 139–148, 2014.

Mainly in the last two decades, academia has been developing a number of innovative production philosophies that provide the theoretical background necessary for the implementation of such distributed and decoupled systems. As result of this effort a number of manufacturing paradigms focused on agility emerged, supported by the advances in ICT (Information and Communication Technology) and based on AI (Artificial Intelligence) and complexity science concepts. Some examples are Bionic Manufacturing Systems (BMS) [1], Holonic Manufacturing Systems (HMS) [2], [3], Reconfigurable Manufacturing System (RMS) [4], [5] and Evolvable Production Systems (EPS) [6], [7], etc. Despite some conceptual differences, most modern manufacturing paradigms theoretically rely on bio-inspired principles to attain the necessary adaptation and responsiveness to unpredictable scenarios. Nevertheless their application, so far, has been limited to specific manufacturing problems, such as: scheduling, planning, layout formation, modeling, fault detection, etc [8]. Consequently the resulting systems are typically distributed, but not necessarily highly scalable, reconfigurable and able to tackle problems in a distributed manner. The network-like architecture and modular functional complexity of modern shop-floors supported by the innovative production paradigms closely follow the structural patterns presented by complex natural systems. In this context, the following question is raised: **what are the main challenges and required properties necessary to foster the development of real production systems that closely mimic biological systems behavior and consequently present overall characteristics of complex natural systems?**

To some extent, this paper attempts to answer this question by presenting a critical analysis on how bio-inspired concepts are currently being explored and used under the scope of modern production systems. For this purpose, a brief review of some applications of bio-inspired approaches in manufacturing is presented, in order to highlight the advantages of bio-inspiration, leading to the formulation of a set of challenges that need to be tackled and to the identification of a number of required properties necessary to enact a production system as bio-inspired.

2   Relationship to Collective Awareness Systems

The origins of Life can be traced back to more than 3500 Million years ago. Since then, primitive organisms were subjected to an evolutionary process in which the occurrence of changes in the environment conditions fostered the reproduction of individuals that better matched their characteristics against the environment functional requirements. Hence, generation after generation individuals developed more adequate characteristics to handle the challenges posed by their natural habitat.

Today's natural diversity is the reflection of such evolutionary process. Biological systems are typically distributed and usually composed by numerous individuals that exhibit simple behaviors. However, through the exchange of meaningful local information the individual entities are able to foster the emergence of a collective action that drives the system towards common goals. The collective behavior itself is greater and considerably more complex than the sum of the behaviors of its parts [9].

Swarms of insects, such as ants, bees, termites, fireflies, among others, are recurrent examples of biological collective systems present in the literature.

Nonetheless, collective systems can be encountered in all sizes, forms and scales. Critical to the emergence of the collective behavior is however the self-awareness of the individual entities. Each entity, despite its simplicity and limited cognitive ability, needs to be able to assess the local surroundings and its internal state in order to make adequate decisions and consequently establish meaningful interactions in response to internal or environmental changes. In this context, collective awareness should not be perceived as an intrinsic characteristic of collective systems per se, but instead as a property that emerges in result of the awareness and social abilities of the individual entities. Even though optimal decisions can only be enacted with a global perspective, the local nature of the interactions and the efficiency of the communication mechanisms endow biological collective systems with the necessary edge to quickly respond as a whole to environmental perturbations. Collective biological systems can therefore be perceived as a unique entity that is aware of the environment and consequently act according to it.

Modern manufacturing systems, as they are idealized by modern manufacturing paradigms also consist on many 'intelligent' autonomous modular entities with social capabilities, that dynamically establish interactions with each other, in order to achieve common objectives. However, the development and implementation of innovative industrial applications which rely on bio-inspired principles tend to stumble in some important gaps between the conceptual academic models and their technological realization, as pointed out in [10]. Hence, one as to master the intricacies of biological complex systems in order to effectively translate their regulatory principles into modern modular manufacturing structures. Only in this way it is possible to truly profit from their dynamics, and achieve a collective level of awareness that allows the system to successfully face environmental changes and uncertainties.

3   On the Application of Bio-inspired Concepts in Manufacturing

Considering the before mentioned paradigms, the increasing structural modularity and autonomy of manufacturing systems means that production floors are becoming evermore complex, rendering traditional control approaches insufficient. On the other hand, ICT-based decentralized control solutions that support advanced AI, are naturally designed to efficiently support the current requirements imposed by modern manufacturing systems. However these solutions are naturally of complex nature. Bio-inspired mechanisms are currently recognized as a powerful and viable solution to tackle complex engineering problems, particularly in situations where there is a lack of available information or in cases where the search space is so large that brute-force optimization algorithms are out of the question [11]. In this context, it is the purpose of this section to discuss and analyze how bio-inspired techniques are currently being applied and explored within the manufacturing context.

Evolutionary concepts have been widely used in schedule optimization, such as flexible manufacturing scheduling problem [12] and flow shop scheduling [13]. Furthermore many evolutionary applications have also been reported in planning

[14], [15], [16], manufacturing cell formation problems [17], [18] and modeling manufacturing processes [19], [20], among others.

Biological collective systems behaviors, commonly colonies of insects, have also been applied to a number of manufacturing problems. An application based on stigmergy to regulate the self-organizing behavior is presented in [21]. Ants' behaviors are also an option to tackle layout problems [22] and vehicle routing problems [23]. Bee's food foraging behavior has also been the source of inspiration to solve job scheduling problems [24], and the printed circuit board (PCB) assembly planning problem [25]. In addition, bacterial foraging behavior provides adequate principles to tackle assembly line balancing [26]. The flashing lights of fireflies were also explored in the development of scheduling mechanisms [27], cell formation problems [28] and solving machining models [29]. Mechanisms based on the neural behavior have been widely applied in recognition problems [30, 31] and fault detection [32], among others.

Finally, the dynamics of the biological immune systems have also been a source of inspiration in manufacturing applications. Similarly to other biological sources, immune behavior has been explored to handle scheduling [33], layout problems [34] and fault diagnosis [35] and detection [36].

Bio-inspired methods, as the previous short literature review also illustrates, are generally focused on the parallel resolution of problems, not necessarily computationally distributed. Moreover, these methods are typically explored as optimization techniques. Hence, it is only natural that the majority of the applications of these techniques within the industrial context are also applied in an optimization perspective. In this sense, and according to the searching approaches generally followed by metaheuristic methods, the algorithms iterate over several possible solutions until a specific stopping criteria is satisfied. Once the stopping criterion is satisfied, the solution is assumed by the system. It is however, in the variety of solutions, distributiveness of information, type of interactions and operators used to explore and select the solutions that the biological inspiration lays. Nevertheless, bio-inspired approaches are generally not envisaged to consider the manufacturing problem from a distributed problem resolution and holistic perspective, but instead to simply tackle narrow and specific manufacturing problems. For a more extensive review on bio-inspired methods and applications please refer to [8].

4   The Relevance of Evolution and Adaptation

As previously stated, modern manufacturing paradigms clearly support the introduction of bio-inspired concepts in the so traditional field of industrial production. Some of the more desirable from a manufacturing point of view are: adaptation, evolution, self-organization and emergence. These are the key fundamental concepts behind the success of biological systems. Evolution in the natural world is a long term optimization/adaptation process. Evolutionary adaptation is a pervasive feature of biological organisms and it is the outcome of natural selection, mutation and genetic drift. The conjugation of these properties leads to an optimization process which is only constrained by the genetic code itself. In such evolutionary process, the individuals with features that better match the environmental requirements, have higher probability to survive and therefore reproduce. Consequently, those features that foster the survivability of particular individuals are passed to the next generations increasing their adequacy towards the environment. The existence of biological diversity implies however that populations do not reach adaptive peaks and that evolutionary adaptation is an open-ended optimization process that is always reacting to the environmental changing conditions, in order to provide the best functional solutions to the current state of the environment. Since manufacturing systems cannot physically adopt a similar genetic evolutionary process, evolution in this case implies the development and integration/removal of new equipment/modules in the system, to explore alternatives in the system reconfiguration, leading to new evolutionary states through which the system should be able to meet the new manufacturing requirements.

On the other hand, short term adaptation (behavioral adaptation) is the concept behind the responsiveness and agility of biological systems. It is therefore, the critical property that endows the system the ability to handle changes and uncertainties of dynamic environments. From a manufacturing perspective, adaptation is similarly a short-term process that concerns the system aptitude to revise its logical parameters and reorganize or redesign its set of processes. The extension of the adaptation process should be as big as necessary to overcome requirements or disturbances, within the constraints of the actual physical system.

Collective living systems are characteristically composed by numerous simple 'homogeneous' (possess a common interface or means to seamlessly interact with other components of the system) constructs. Typically, these systems rely on simple principles and local information, without requiring any centralization of the control and information flow. In this sense, behavioral adaptation is mainly attained by self-organization mechanisms. Interaction patterns which are locally and asynchronously established, according to the individuals internal and surrounding status, support the main regulatory principles that lead to the emergence of the necessary individuals functional adaptation and coherent global system behavior. It is however curious that the biggest advantage of these collective systems, is also their greatest weakness. As stressed before, an optimal decision can only be taken when there is a general overview of all the system information. Even though, it is possible to compose a consistent global view by the exchange of local information, it is impractical and inefficient with the increasing number of entities [10]. Nevertheless, the distribution of both the knowledge and decision nodes ensures the system responsiveness and robustness to malfunctions, reducing the effect of deviations or catastrophic failures of the individuals. In other words, the system is not dependent on specific individuals. In a certain extent, every individual unit is relatively negligible to the proper functioning of the system. Yet, every entity contributes to the whole, and the whole supports the individual entities.

This represents one of the major gaps of most modern manufacturing paradigms. The centralization of the knowledge and regulatory control on certain hierarchical structures, simply results in a computationally distributed system in which the control mechanisms are still dependent on key higher hierarchical entities to manage the execution of critical processes. Although the effects of any malfunctioning or removal in lower level entities are easily suppressed by relocating the process, if there is sufficient redundancy or simply by replacing the component. The failure or removal of a key hierarchical node however is critical to the correct functioning of the system.

Similarly to the biological systems, central to manufacturing systems' evolution and behavioral adaptation are the concepts of self-organization and emergence that need to be properly supported by the control architectures, in order to endow the system with the ability to autonomously handle both logical and physical changes.

5   Towards Bio-inspired Complex Manufacturing Systems

The currently growing interest and development of modular mechatronic production systems provides the opportunity to explore highly reconfigurable and scalable architectures with simultaneous and distributed resolution of various manufacturing problems. In this context, the integration of regulating bio-inspired principles, particularly bio-inspired self-organization, may be the way to unravel the potential of modern manufacturing paradigms. Even though these techniques have been fairly used in modern approaches [8], [10], when down to the implementation, bio-inspired methods tend to be replaced by negotiation-based procedures which tend to be efficiently poor. Consequently, current modern manufacturing systems architectures performance and scalability is typically constrained by the architecture's artificially devised self-organizing approaches and control structures. In this sense and in order to attain biological-like agility bio-inspired mechanism should be considered from a considerably more holistic perspective. This necessarily introduces a number of challenges that need to be tackled:

- Despite the technological and distributed support provided by modular mechatronic systems, the bio-inspired structural characteristics and concepts need to be properly abstracted by a robust technical architecture.
- Production requirements, typically centralized on single entities, should be distributed over the manufacturing system components in order to reduce the necessary specification and simultaneously foster the optimization and distributed execution of production workflows.
- The reconfigurable ability, scalability and distributed coherent execution of production plans must be the outcome of efficient and semantic interaction patterns that support the implementation of regulatory bio-inspired self-organizing principles. Although the feasibility of the state transitions is not so relevant for the normal execution of the bio-inspired algorithms itself (providing that the viability of the solutions is ensured), it is critical for systems that holistically follow the same principles. Every transition may have a direct impact in the physical shop-floor consequently the devised regulatory principles need to ensure the transition feasibility. In this way the system should be able to closely mimic the robustness and adaptation of biological systems.
- As opposed to natural systems, production systems evolutionary cycles have to be proportional to the very demanding production time frames [10].

Recent developments in distributed manufacturing architectures have attempted to attain agility and sustainability through the implementation of complexity abstraction layers, based on dynamic self-organizing logical hierarchical structures. Nevertheless, these efforts have only been partially successful. In this sense, through the literature review and analysis of the most common biological inspiration sources, a set of properties required by distributed systems to support biological-like structures and dynamics were identified:

1. **Use of genetic operators** - Similarly to nature, crossover and mutation operators are employed to introduce diversity in the population so that the full extent of the solution space can be properly explored. Although most bio-inspired approaches use genetic operators to perform evolutionary adaptation, from a mechatronic perspective these operators could be used to evolve particular parameters and achieve individuals' adaptation. It is important to stress that this is not an essential property to enact an approach as bio-inspired.

2. **Large populations** - Biological systems are typically characterized by large populations of individuals. Group size is typically determined by the resulting advantage, where the size of the group affects its performance. Small populations imply a magnification of individual mistakes. Too large populations may generate a fearsome resource competition leading to the destruction of population members. Within the midrange is the optimal population size. In the mechatronic context processes and components redundancy should be optimized in order to provide the desired level of robustness, in case critical failures affect the system components.

3. **Homogeneity of the individuals** - Biological systems are generally composed by identical individuals. This ensures that coherent and semantic interaction patterns are established as a response to disturbances in the environment. Mechatronic components do not necessarily need to present similar physical properties. However, compatible interfaces should be defined so that a meaningful collective behavior is attained.

4. **Decoupled nature of the entities** - Biological entities usually present high levels of autonomy, in the sense that different individuals interact by using well defined interfaces rather than tightly depending on each other. This implies that the removal or addition of components should have little or no impact in the normal system behavior. In other words, each mechatronic entity should contribute to the collective behavior, whilst every entity should be as negligible as possible to the proper functioning of the system.

5. **Stochastic behavior** - In nature, as opposed to the engineering world, systems do not have any specific goal rather than reproduce and survive. Individual decisions that influence the overall function and dynamics of the collective are the result of an interplay between deterministic and stochastic events presented by the environment. Similarly, evolutionary adaptation is also the outcome of exposing living systems to some events of stochastic nature such as mutations, genetic drift and natural selection. In this sense, it is therefore important to have, in the mechatronic context, a limited amount of randomness in order to explore all possible, available and feasible solutions (the stochastic process can be virtually simulated [37]).

6. **Asynchronous and local interactions** - Local interactions are one of the key aspects in the emergence of coherent and meaningful biological collective behaviors. Notoriously, the regulatory principles that foster the emergence of the natural self-organizing mechanisms are supported by the asynchronous establishment, between population individuals, of semantic and efficient interaction patterns. Consequently, comparable interaction patterns need to be

devised in artificial systems in order to attain similar behavior. Furthermore, living systems are usually characterized by physical limitations that heavily constrain the individuals perception of the environment. This however implies that they are extremely responsive to environmental disturbances. Thus, despite the fact that self-organization in artificial systems does not necessarily preclude the use of global information, it seriously has a negative impact with the increasing size of the system. Therefore, in order to closely match the mechanisms of biological self-organization, a trade-off between the responsiveness of the system and the optimality of the decision needs to be found.

7. **No centralized knowledge or decision nodes** - Unlike the approach that has been typically followed by current state-of-the-art architectures, biological systems do not rely on entities with centralized knowledge and decision making capabilities. Instead, each individual is an autonomous decoupled entity that acts according to its internal and surrounding information. In this way, natural systems are able to, in a distributive manner, handle the disturbances or problems posed by the environment. This is a relevant differentiation aspect between natural and artificial distributed systems. Hence, a bio-inspired self-organizing system that presents the agility and robustness of biological systems can only be enacted with the full distribution of both knowledge and decision making capabilities.

8. **Emergence of collective behavior** - Emergent properties are a hallmark of collective animal behavior. In this context, the adequate implementation of the previous properties should foster the emergence of a robust, coherent, and meaningful collective behavior. Moreover, the system should present characteristics such as responsiveness, agility, robustness, scalability, pluggability, adaptability and evolutionary properties as result of the implemented bio-inspired self-organizing mechanisms.

6   Conclusions

Organizational sustainability and supply chain agility have become critical for the success of regular manufacturing enterprises. Companies with a pluggable, scalable and highly reconfigurable shop-floor have the ability to better cope with uncertainties and therefore have a faster response to new business opportunities. In this context, modern manufacturing paradigms play a crucial role, since they provide the methodology and background necessary to implement such systems. However, although the importance and validity of modern production paradigms is generally consensual, their implementation in real industrial systems has been, until now, minimal and only partially successful.

To some extent, this problem can be attributed to the fact that modern manufacturing systems heavily rely on biologically inspired concepts that are not properly considered and supported in the following development phases. Modern production systems, as envisioned by innovative paradigms, rely on modular independent constructs that naturally make the system a complex structure. However, the biological mechanisms that critically regulate collective living systems with similar structure are still faced simply as a desirable characteristic of the systems, but not as the main control mechanism. A more holistic integration of these mechanisms in the manufacturing context could possibly not only foster systems autonomy, scalability and reconfigurable ability but also its distributed problem resolution capabilities, enabling the system to more efficiently respond to changes and uncertainties in the environment. It is the authors' belief that considering the highlighted challenges and identified properties is a small step towards that goal.

References

1. Ueda, K.: A concept for bionic manufacturing systems based on dna-type information. In: Proc. of the IFIP TC5/WG5. 3 Eight International PROLAMAT Conference on Human Aspects in Computer Integrated Manufacturing, pp. 853–863. North-Holland (1992)
2. Gou, L., Luh, P.B., Kyoya, Y.: Holonic manufacturing scheduling: architecture, cooperation mechanism, and implementation. Computers in Industry 37(3), 213–231 (1998)
3. Bussmann, S., McFarlane, D.C.: Rationales for holonic manufacturing control. In: Proc. of Second Int. Workshop on Intelligent Manufacturing Systems, pp. 177–184 (1999)
4. Koren, Y., Heisel, U., Jovane, F., Moriwaki, T., Pritschow, G., Ulsoy, G., Van Brussel, H.: Reconfigurable manufacturing systems. CIRP Annals-Manufacturing Technology 48(2), 527–540 (1999)
5. Mehrabi, M.G., Ulsoy, A.G., Koren, Y.: Reconfigurable manufacturing systems and their enabling technologies. International Journal of Manufacturing Technology and Management 1(1), 114–131 (2000)
6. Onori, M.: Evolvable assembly systems - a new paradigm? In: 33rd Int. Symposium on Robotics (ISR), pp. 617–621 (2002)
7. Onori, M., Alsterman, H., Barata, J.: An architecture development approach for evolvable assembly systems. In: 6th IEEE Int. Symposium on Assembly and Task Planning: From Nano to Macro Assembly and Manufacturing (ISATP 2005), pp. 19–24. IEEE (2005)
8. Ferreira, J.D.: Bio-inspired Self-Organisation in Evolvable Production Systems. Tekn. Lic. dissertation, Royal Institute of Technology, Sweden (2013)
9. Holland, J.H.: Emergence: From chaos to order. Oxford University Press (2000)
10. Ribeiro, L., Barata, J.: Self-organizing multiagent mechatronic systems in perspective. In: 2013 11th IEEE International Conference on Industrial Informatics, INDIN (2013)
11. Floreano, D., Mattiussi, C.: Bio-inspired artificial intelligence: theories, methods, and technologies. The MIT Press (2008)
12. Kumar, V., Murthy, A., Chandrashekara, K.: Scheduling of flexible manufacturing systems using genetic algorithm: A heuristic approach. J. Ind. Eng. Int. 7(14), 7–18 (2011)
13. Wang, L., Zheng, D.: A modified evolutionary programming for flow shop scheduling. The International J. Advanced Manufacturing Technology 22(7), 522–527 (2003)
14. Zhang, F., Zhang, Y., Nee, A.: Using genetic algorithms in process planning for job shop machining. IEEE Tran. Evolutionary Computation 1(4), 278–289 (1997)
15. Routroy, S., Kodali, R.: Differential evolution algorithm for supply chain inventory planning. Journal of Manufacturing Technology Management 16(1), 7–17 (2005)
16. Brezocnik, M., Kovacic, M., Psenicnik, M.: Prediction of steel machinability by genetic programming. Journal of Achievements in Materials and Manufacturing Engineering 16(1-2), 107–113 (2006)
17. Stawowy, A.: Evolutionary strategy for manufacturing cell design. Omega 34(1), 1–18 (2006)
18. Wu, T.-H., Chang, C.-C., Chung, S.-H.: A simulated annealing algorithm for manufacturing cell formation problems. Expert Systems with Applications 34(3), 1609–1617 (2008)

19. Chan, K., Kwong, C., Tsim, Y.: A genetic programming based fuzzy regression approach to modelling manufacturing processes. International Journal of Production Research 48(7), 1967–1982 (2010)
20. Chan, K., Kwong, C., Fogarty, T.: Modeling manufacturing processes using a genetic programming-based fuzzy regression with detection of outliers. Information Sciences 180(4), 506–518 (2010)
21. Leitão, P., Restivo, F.: Adacor: A holonic architecture for agile and adaptive manufacturing control. Computers in Industry 57(2), 121–130 (2006)
22. Solimanpur, M., Vrat, P., Shankar, R.: Ant colony optimization algorithm to the inter-cell layout problem in cellular manufacturing. European Journal of Operational Research 157(3), 592–606 (2004)
23. Yu, B., Yang, Z.-Z., Yao, B.: An improved ant colony optimization for vehicle routing problem. European Journal of Operational Research 196(1), 171–176 (2009)
24. Pham, D., Koc, E., Lee, J., Phrueksanant, J.: Using the bees algorithm to schedule jobs for a machine. In: Proceedings of Eighth International Conference on Laser Metrology, CMM and Machine Tool Performance, pp. 430–439 (2007)
25. Pham, D., Otri, S., Darwish, A.H.: Application of the bees algorithm to pcb assembly optimisation. In: 3rd International Virtual Conference on Intelligent Production Machines and Systems, IPROMS, pp. 511–516 (2007)
26. Atasagun, Y., Kara, Y.: Assembly line balancing using bacterial foraging optimization algorithm (2012)
27. Sanaei, P., Akbari, R., Zeighami, V., Shams, S.: Using firefly algorithm to solve resource constrained project scheduling problem. In: Bansal, J.C., Singh, P.K., Deep, K., Pant, M., Nagar, A.K. (eds.) BIC-TA 2012. AISC, vol. 201, pp. 417–428. Springer, Heidelberg (2013)
28. Sayadi, M.K., Hafezalkotob, A., Naini, S.G.J.: Firefly-inspired algorithm for discrete optimization problems: An application to manufacturing cell formation. Journal of Manufacturing Systems (2012)
29. Aungkulanon, P., Chai-Ead, N., Luangpaiboon, P.: Simulated manufacturing process improvement via particle swarm optimisation and firefly algorithms. In: Proceedings of the International MultiConference of Engineers and Computer Scientists, vol. 2 (2011)
30. Rangwala, S., Dornfeld, D.: Sensor integration using neural networks for intelligent tool condition monitoring. J. Engineering for Industry 112(3), 219–228 (1990)
31. Sunil, V., Pande, S.: Automatic recognition of machining features using artificial neural networks. Int. J. Advanced Manufacturing Technology 41(9), 932–947 (2009)
32. Eski, I., Erkaya, S., Savas, S., Yildirim, S.: Fault detection on robot manipulators using artificial neural networks. Robotics and Computer-Integrated Manufacturing 27(1), 115–123 (2011)
33. Ong, Z.X., Tay, J.C., Kwoh, C.K.: Applying the clonal selection principle to find flexible job-shop schedules. In: Jacob, C., Pilat, M.L., Bentley, P.J., Timmis, J.I. (eds.) ICARIS 2005. LNCS, vol. 3627, pp. 442–455. Springer, Heidelberg (2005)
34. Ulutaş, B.H., Işlier, A.A.: Parameter setting for clonal selection algorithm in facility layout problems. In: Gervasi, O., Gavrilova, M.L. (eds.) ICCSA 2007, Part I. LNCS, vol. 4705, pp. 886–899. Springer, Heidelberg (2007)
35. Hao, X., Cai-xin, S.: Artificial immune network classification algorithm for fault diagnosis of power transformer. IEEE Trans. Power Delivery 22(2), 930–935 (2007)
36. Dasgupta, D., Forrest, S.: Tool breakage detection in milling operations using a negative-selection algorithm. Technical Report CS95-5, Department of Computer Science, University of New Mexico, Tech. Rep. (1995)
37. Leitão, P., Barbosa, J., Trentesaux, D.: Bio-inspired multi-agent systems for reconfigurable manufacturing systems. Engineering Applications of Artificial Intelligence 25(5), 934–944 (2012)

The ProFlex Methodology: Agile Manufacturing in Practice

Giovanni Di Orio[1], José Barata[1], Carlos Sousa[2], and Luís Flores[2]

[1] CTS – UNINOVA, Dep. de Eng. Electrotécnica, Faculdade de Ciências e Tecnologia,
Universidade Nova de Lisboa, 2829-516 Caparica, Portugal
{gido,jab}@uninova.pt
[2] IntRoSys, SA – Global Control System Designers
2860 Moita, Portugal
{carlos.sousa,luis.flores}@introsys.eu

Abstract. Today, manufacturing production systems are required to be responsive, re-configurable, adaptable, and flexible. The fulfilment of these requirements is directly related to the method used to design and develop their control applications. The paper presents the result of an investigation study to capture current techniques adopted by logic designers during the design and development of control applications at IntRoSys. The result of this study is then used as the foundation for the development of a methodology and its related tools for achieving agility in programming manufacturing control systems.

Keywords: Automation, Automotive Industry, Industrial Standards, Controllers, IEC-61131-3, Behavioural Modelling, Code Generation.

1 Introduction

Today, European manufacturing industries are under high pressure in the global market due to a "pincer effect" as claimed in [1]: on the one side, the presence of competitors which operate in regions with low-wage and are absorbing very fast the available technologies, and on the other side, the need to keep pace with science-based innovation processes and products that are creating new markets and new business. This effect is incredible felt in the automotive industry, where the increasing demand and competition in market sharing are radically changing the way production systems are designed and products are manufactured in terms of improved flexibility/adaptability, reduced lifecycle costs, high products customization and quality, and reduced time-to-market [2].

In the recent years, the competition between manufacturing companies is going to be shifted from cost reduction to added value in products. The Original Equipment Manufacturers (OEMs) are demanding for more and more exclusive, efficient and effective production systems capable to produce as many different product variations as quickly as possible. Manufacturing companies are striving to introduce flexible and adaptive manufacturing techniques in order to better meet market needs whilst maintaining the low cost base of heavily automated mass production techniques [3]. In this context, the key to competitiveness is represented by the reduction of the

L.M. Camarinha-Matos et al. (Eds.): DoCEIS 2014, IFIP AICT 423, pp. 149–156, 2014.
© IFIP International Federation for Information Processing 2014

production costs during production system lifecycle and the capability to have systems that are able to quickly respond to markets variations and demands for new products. However as stated in [4], the capability of offering more variants per product, and introducing new products faster, is constrained by the adoption of technologies and equipment of mass production operations, implying that frequently changes in products may potentially cause changes in the plant's hardware structure and its functionality.

As existing plants cannot be rebuild due to high investment volumes, the changes in hardware components affect particularly the control software. Therefore, control software is the stage of continuous modifications during the production system lifecycle which implies an increasing in its amount and complexity [5]. Furthermore, every hardware/software change in the production environment has heavy consequences on the normal flow of operations: the production system needs to be shut-down, mechanically changed and reprogrammed in order to face the new configuration. The (re)programming task is often commissioned by the OEMs to external suppliers that are then responsible for fast and secure rump-ups. The time needed for this task is directly proportional to the dimension of changes, presence of a well-suited documentation, particular familiarity with the standard used by the OEMs, presence of an up-to-date control software version [6]. Moreover, the presence of rigid industrial proprietary standards, encompassing electrical systems and diagnostic as well as hardware types and control software structure, has two fundamental implications as exposed in [7]. From one side, standards contribute to reduce the flexibility of the programmer while discouraging any technique/technology other than the defined in the standard. From the other side, rigid and strong software structure aids the development of tools and methodologies for automatically generating the source code.

The present paper is motivated by the need to conduct an analysis of current methods for implementing monitoring and control solutions in the automotive industry in order to outline the challenges faced by the system integrators. The paper also serves as a starting point for research community for providing the baseline of current industrial software design and development practices. The baseline has then been used as the foundation of the ProFlex methodology intended to enhance the (re)programming task of control software during the shut-downs.

2   Contribution to Collective Awareness Systems

To face globalization challenges manufacturing industry needs to move towards new paradigms and methodologies aiming at enabling distributed knowledge creation and intelligent usage of manufacturing production data from the production environment. The main goal is creating awareness about relevant problems faced by industry in order to elaborate solutions by exploiting collective efforts. To remain competitive, manufacturing companies need to understand where and how to change. This paper aims at contributing to collective awareness by providing a theoretical background about the way current monitoring and control solutions are designed and developed and, most important, presenting a way to model manufacturing components/resources.

The findings of this paper have the intent to open the door to the introduction of new technologies and paradigms in industry that are expected to support environmentally awareness by framing the current practices for designing and developing control applications for the automotive sector.

3   Review on Current Design Methodologies

Today, programmable logic controllers (PLCs) represent the *de-facto* standard for implementing control and monitoring solutions in many manufacturing companies due to their features and characteristics. Since PLCs are established as the device of choice for automation. Then a good starting point for gaining background information and knowledge on how to develop PLC programs is the introduction of the International Electrotechnical Commission (IEC) 61131-3 standard.

3.1   IEC 61131-3 Standard

As stated in [8], after the introduction of the PLCs in 1969, various companies developed its own platform with different run-time environments, operating systems, and programming languages. To reduce the complexity for the PLCs users and their fragmentation, the IEC 61131-3 standard has been published.

The IEC 61131-3 contains five programming languages for developing programs to be executed inside a PLC, which are supposed to be supported by any compliant vendor. The main goals of the IEC 61131-3 standard is to unify all the programming concepts for developing industrial control applications while abstracting from proprietary peculiarities and creating a common user experience when programming and configuring industrial controllers. The supported programming languages are:

1. Instruction List (IL) – A textual programming language similar to assembly and consisting of simple operation codes.
2. Structured Text (ST) – A programming language with syntax similar to Pascal, Basic or FORTRAN suited for programming more complex algorithms.
3. Functions Blocks Diagram (FBD) – A graphical language providing a mechanism for encapsulating functionality into a module with a common external interface, and designed to promote the code reusability.
4. Sequential Functional Chart (SFC) – A graphical language derived from the GRAFCET. It is made up of graphical elements called steps and transitions; a step represents a specific state in process and/or machine sequence while a transition represents the set of conditions that allows the evolution from one step to another.
5. Ladder Logic Diagram (LLD) – A graphical language derived from the electrical relay diagrams and created to be easily understood by booth engineers and technician without any software skill.

Most of the industrial applications are designed using one of these languages. However as exposed in [9], and recently confirmed a poll realized by the Control

Engineering U.S. and Control Engineering Poland magazine [10], 96% of industrial developers uses LLDs. FBDs are the next most popular at 67%, followed by IL (37%). The large predominance of the LLD language for developing industrial applications is due to several factors that are well exposed in [11]. However, the main goal in this scope is not the evaluation of current practices for developing monitoring and control solution, but only their analysis in order to enable the identification of the challenges related to the development of logic control programs. With this in mind, this analysis represents the foundation upon which new methodologies can be designed for improving the control logic development process.

3.2   Control Logic Development Process

As stated in [7], if from one side there is an extensive published material outlining the code structures available to the practitioners of programmable controllers, from the other side, there is little reported on how these techniques are really applied for developing ready-to-use and above all standard compliant industrial control applications.

A practical perspective on the design and development process for control logic has been firstly given by Lucas and Tilbury in their work [12]. During their investigation, conducted from September to December 2001 into logic design techniques for automotive industry, Lucas and Tilbury have noticed that the control logic, needed to monitor and control a manufacturing process/machine, is obtained by gathering and combining information from three sources, namely: project specific specifications, unspecified requirements and standard specifications (see Fig. 1).

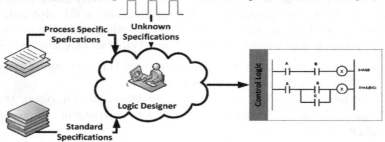

Fig. 1. Control Logic Development Process [12]

The project specific specifications comprise several documents that are primary used during the control logic development activity, namely: project specifications (functional requirements), mechanical drawings, electrical drawings and schematics, printout of previous projects, PLC memory map and process/machine timing bar (used to specify the desired behaviour of the process/machine along time during production activities). At this point, logic designers are responsible to integrate fragmented piece of information, extracted from these documents, and combine it with standard specifications (usually in the form of old project) to create the control logic. Finally, the control logic is refined taking into account the unspecified requirements that, in turn, can include late changes and/or unexpected constraints in the process/machine.

4   ProFlex Methodology

4.1   Research Activity

In order to gain a solid knowledge about current control logic development process for the automotive industry, a set of observations were conducted at IntRoSys SA in Portugal. Since 2004 the company has established itself in the market as one of the reference Controls Houses for the European Manufacturing Industry.

The observations were targeted at capturing current practices and activities followed by system integrators when creating control logic. The set of observations will be, then, used as inputs for developing an innovative methodology to improve the control logic development process by pushing software engineering techniques in the world of automation.

The investigation revealed that the control logic development process is today still the same as the one depicted by Lucas and Tilbury. The entire set of documents analysed and used by the logic designer during the development process is the same for all the car manufacturers meaning that all the information needed to create the control and monitoring logic can be extracted from them. Moreover, the particular software structure can be derived from standard template projects provided by the car manufacturers. Finally, new control and monitoring logic could be implemented on-site if late changes in process and/or machine are realized. In this context, the main activities carried out by logic designers are four, namely: Documents analysis, Modification of standard template functions/projects, Human Machine Interface (HMI) development and new code Development.

The fulfilment of the first three activities leads to 90% of control and monitoring logic ready to installation and test. The other 10% includes modifications on-site necessary to fit the logic to the real state of things. The conducted investigation has been used as the basis for the formation of the ProFlex methodology by providing necessary information on how to enhance the control logic development process.

4.2   Proposed Methodology

The methodology proposed in this paper address the automatic generation of standard compliant and ready-to-use control and monitoring logic starting from the behavioural model of the resources composing the considered manufacturing process. More extensive explanation of the methodology can be found in [13].

The proposed methodology is depicted in Fig. 2. The main activities of the methodology are five, namely:

- A0: Monitoring and Control system requirements analysis.
- A1: Identification of the resources of the considered manufacturing process.
- A2: Construct a universal behavioural model of each resource.
- A3: Software Generation.
- A4: Deployment and Tests.

The activities are organized in two macro activities the **Model Development Cycle** and the **Control Development Cycle** respectively. The activities A0, A1 and A2 are executed during the Model Development Cycle while the activities A3 and A4 are executed during the Control Development Cycle.

Furthermore, inside the Model Development Cycle, the **Agentification Cycle** is executed. During the Agentification Cycle real manufacturing resources are considered as autonomous components which behaviour is modelled using a unique and formal representation. The result of this activity is a self-contained piece of software (agent) to be executed inside the PLC and, thus, implemented using one of the five languages of the IEC-61131 standard.

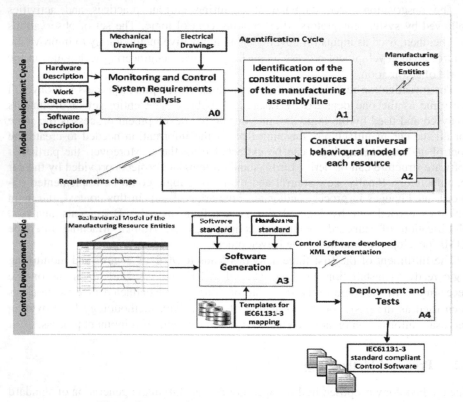

Fig. 2. Conceptual model of the Proposed Methodology

The methodology is focused on a kind of "lingua franca", represented by the behavioural model of the system resources, as a way to enable fast coding, improve industrial acceptance of developed code and, most important, reduce the error during the programming activity.

4.3   A Tool for Applying the Model

The following tool has been implemented to assist the logic designers in applying and implementing the ProFlex methodology (see Fig. 3). The tool is in the form of an Excel spreadsheet providing a table which will be used to model the behaviour of the manufacturing resources inside the considered manufacturing process.

The behaviour of the resources is modelled using impulse diagrams that enable the description of the temporal course of states of system resources together with signals among them to activate state changes. To describe the value sequences and relations of signals over time, a global timescale is considered.

4.4   Case Study

In order to validate and assess the proposed methodology, two robotized manufacturing cells located at IntRoSys have been used. The cells include a loading station used to load car components, a robot manipulator and a welding station. The cells are identical in terms of functionalities, i.e. once a part is loaded the robots perform the same actions on it according to the sequence: pick-weld-place. However, they have been designed and developed using two different proprietary standards, from two distinct car manufacturers. The usage of different standards directly implies the way the control software is organized, structured and built, even if the process for developing the control logic is the same. As a matter of fact, the project specific specifications are analysed by the logic designer and all the acquired information is directly parsed into LLDs. The time and the effort needed to develop the control logic strictly depend on the logic designer experience, familiarity with the particular used standard, and the way the control logic is organized. In this context, the ProFlex methodology has been applied to develop the control logic for both the presented cells. In a first moment the **Model Development Cycle** activity is applied aiming to identify the system requirements and according to them to design the behavioural model of the resources of the robotic cell (during this activity the Excel spreadsheet is used Fig. 3). Once the behavioural model is built then the **Control Development Cycle** activity can be executed. This activity is completely automatic and aims to create a standard compliant ladder logic code based on the designed behavioural model.

According to a poll realized at IntRoSys, the methodology enables potentially the reduction of the overall time required to create the control logic and the possible errors related to this activity. Moreover, the initial modelling activity permits to abstract the development process from the particular OEMs standard used, while equalizing the effort needed during the development activity.

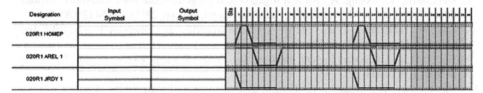

Fig. 3. Excel Tool for applying the ProFlex methodology

5   Conclusions

The paper present a software engineering approach based on a behavioural model of manufacturing resources for developing PLC-based OEMs standard compliant control logic. The way manufacturing resources are modelled, is the result of the research activity on current control logic development process. The key element and innovative aspect of the ProFlex methodology is the usage of a unique model to generate different structured and organized control logics. Further, a case study has been presented with the goal to show the feasibility and repeatability of the proposed approach. Finally, the approach, denoted with the acronym ProFlex, makes the control systems satisfy requirements for flexibility and reconfiguration while offering competitive advantage to system integrators by enabling faster and securer ramp-ups. As a matter of fact, any change/modification in the hardware configuration of the manufacturing process/machine simply triggers a new modelling activity.

Acknowledgments. This work was funded by the Regional Operational Programme of Lisbon (POR Lisboa), in the scope of the National Strategic Reference Framework of Portugal (QREN), part of the European Regional Development Fund (FEDER). This work is also supported by FCT Fundação para a Ciência e Tecnologia under project grant Pest-OE/EEI/UI0066/2011.

References

1. Jovane, F., Westkämper, E., Williams, D.J.: The ManuFuture Road: Towards Competitive and Sustainable High-adding-value Manufacturing. Springer (2009)
2. Rosiná, F., Temperini, S.: Advanced Maintenance Strategies for a Sustainable Manufacturing. In: 10th IFAC Workshop on Intelligent Manufacturing Systems, Lisbon (2010)
3. Hajarnavis, V., Young, K.: An Assessment of PLC Software Structure Suitability for the Support of Flexible Manufacturing Processes. IEEE Trans. on Autom. Sci. Eng. 5(4), 641–650 (2008)
4. Michalos, G., Makris, S., Papakostas, N., Mourtzis, D., Chryssolouris, G.: Automotive assembly technologies review: challenges and outlook for a flexible and adaptive approach. CIRP J. Manuf. Sci. Technol. 2(2), 81–91 (2010)
5. Wagner, T.: Applying Agents for Engineering of Industrial Automation Systems, pp. 1097–1097. Springer (2003)
6. Barata, J.: Coalition Based Approach For ShopFloor Agility. Orion, Amadora (2005)
7. Hajarnavis, V., Young, K.: An investigation into programmable logic controller software design techniques in the automotive industry. Assem. Autom. 28(1), 43–54 (2008)
8. Otto, A., Hellmann, K.: IEC 61131: A general overview and emerging trends. IEEE Ind. Electron. Mag. 3(4), 27–31 (2009)
9. Lucas, M.R.: Understanding and assessing logic control design methodologies. University of Michigan (2003)
10. Pietrusewicz, K., Urbanski, L.: Control system software programming. Control Eng. 59, 32–37 (2012)
11. Walter, T.: Ladder logic: Strengths, weaknesses. Control Eng. (March 2007)
12. Lucas, M.R., Tilbury, D.M.: The practice of industrial logic design. In: Proceedings of the American Control Conference, vol. 2, pp. 1350–1355 (2004)
13. Di Orio, G., Barata, J., Sousa, C., Flores, L.: Control System Software Design Methodology for Automotive Industry. Presented at the IEEE International Conference on Systems, Man, and Cybernetics - IEEE SMC 2013, Manchester, UK (2013)

Enterprise Competency Modeling - A Case Study

Reza Vatankhah Barenji and Majid Hashemipour

Department of Mechanical Engineering, Eastern Mediterranean University,
Famagusta, TRNC, Via Mersin 10, Turkey
Reza.vatankhah@emu.edu.tr

Abstract. The purpose of this paper is analyzing the crucial play in carrying out enterprise competency within the enterprise. This study uses the method of case study and has directed the survey on Chika Food Industry (ChFI). It is anticipated through the case study of this company that it will be possible to implement the finding of enablers concluded by other papers, thus showing the inter-relationship between organizational management and information technology management perspectives.

Keywords: Enterprise knowledge modelling, enterprise competency, modelling, case study.

1   Introduction

In recent years, the artificial intelligence and enterprise modelling communities have developed important enterprise models and/or ontologies, including: the Toronto Virtual Enterprise (TOVE), the Open Information Model (OIM), Computer Integrated Manufacturing Open System Architecture (CIMOSA), IDEON, Business Process Modelling Language (BPML), and Collaborative Network Organisation (CNO) [1]. Furthermore, there exist a number of frameworks to model Collaborative Networked Organizations (CNO) in general [2] including the Zachman's Frameworks, GERAM – Generalized Enterprise Reference Architecture, and ARCON - A Reference Model for Collaborative Networks. In particular, ARCON consolidates existing frameworks through generic abstractions to model basic elements of CNO [2].

In addition to the enterprise model, it is important to capture and manage the knowledge and skills of enterprises' internal competencies [4]. Enterprise Competency is a crucial factor in business scenarios, in that it provides a more nuanced description of an enterprise's [5] or individual's [6,7] profile. Such a profile demonstrates the knowledge, skills, experience, and attributes necessary to effectively implement a defined function [1,8].That Competency is an essential component of enterprise engineering, acting as a new means to consider knowledge capitalisation [9], associated with new vision of performance [7,10], as well as new forms of ontology [11,12]. First, the understanding and auditing of competencies acquired, required, and desired by a company and second, representing them in a structured manner, are beneficial steps for enhancing the company's performance. These issues motivate the co-author of this paper to propose a PhD thesis to the author.

L.M. Camarinha-Matos et al. (Eds.): DoCEIS 2014, IFIP AICT 423, pp. 157–164, 2014.
© IFIP International Federation for Information Processing 2014

The main aims of this research thesis are: (a) understand capability and competency concepts (b) introduce an approach to store, manage and maintenance capability and competency of an organization in different levels of abstraction (c) suggest some criteria for using competency as ontology for organization integration. The central research questions which are addressed in this research thesis;

RQ1) how to model an enterprise with its existing competencies?

RQ2) what are the templates, procedures and methods to store, maintain and manage competency of an organization?

This study uses a case-study of one organization to examine the dynamics of successful competency modelling practices, and to consider the extent to which such practices can be generalized and adapted by others. Therefore, the overall effect of this theoretical approach is to bridge a gap between the abstract concepts that we employ to understand enterprise competency and the practical, context-dependent realities facing business organizations.

2   Contribution of the Paper to "Collective Awareness Systems"

Competency modeling is an approach for configuring an organizations knowledgebase scheme. In contrast to the other organizational knowledge modeling approaches that covers only the information and knowledge of the organization from only one perspective, competency modeling has the capability of providing an organizational knowledge model from different viewpoints. Similar to the other organizational data models, it can be used on BtoB context in order to provide a platform for sharing information among the stockholders of a network. Furthermore, the competency-based knowledge model can provide an appropriate context for doing better collaboration. The authors believe is that, competency modeling approach, can be employed as an appropriate infrastructure for developing a robust "Collective Awareness System" on organizational management context.

3   Sector's Capability - Concepts and Model

The formalization of sector capability is as follows. Let's consider for subsequent modelling a set of sectors at an enterprise E= {S1, S2, S3, …}.

Definition 1 (Sector capability) – Capability can be understood as sector's ability to perform activities, tasks, acts or processes possible through corresponding resources and knowledge, aimed at achieving a specified number of outcomes.

For modelling the remaining concept, let's consider the set of capabilities at sector α: $C_\alpha = \{ C_{\alpha 1}, C_{\alpha 2}, …, C_{\alpha n} \}$ in which each element $C_{\alpha i}$ stands for a capability. The following definition introduces the concept of capability, which is built upon three building aspects. It can be specified as a set

$C_{\alpha i} = \{X_{\alpha i}, R_{\alpha i}, K_{\alpha i}\}$, i=1…n; Such that: $X_{\alpha i} = \{x_{\alpha 1}, x_{\alpha 2}, x_{3\alpha}, .., x_{\alpha j}\} = \{x_{\alpha j} | x_{\alpha j}$ is a activity for i^{th} capability at sector $\alpha\}$, i=1,…,n, j=1,…,m $R_{\alpha i} = \{r_{\alpha 1}, r_{\alpha 2}, r_{3\alpha}, .., r_{\alpha j}\} =$

$\{r_{\alpha j}|r_{\alpha j}$ is a resource for i^{th}capability at sector $\alpha\}$, $i=1\ldots n$, $j=1,\ldots,m$; $K_{\alpha i} = \{k_{\alpha 1}, k_{\alpha 2}, k_{3\alpha}, \ldots, k_{\alpha j}\}\{k_{\alpha j}|k_{\alpha j}$ is a knowledge for l^{th}capability at $\alpha\}, i=1\ldots n, j=1,\ldots, m$

Definition 2 (Sector's task-oriented capability) – is a sub-set of a sector capability set, this sub-set represents capabilities which are needed to run a specific outcome or specific goal.

For sector α it can be shown as C_{α}^{*} where: $C_{\alpha}^{*} \subseteq C_i$; $C_{\alpha}^{*} = \{C_{\alpha 1}, C_{\alpha 2}, C_{\alpha 3}, \ldots, C_{\alpha n}\} = \{C_{\alpha k} \mid C_{\alpha k}$ is a selected capability at sector α for a specific task $\}$; $k=1,\ldots, n$

4   Cross-Functional Coordination and Integration Processes

Cross-functional co-ordination of capabilities of a sector has been identified as a key operation for enterprise competency creation process [13]. The successful achievement of the enterprise's global goals depends not only on the appropriate co-ordination of sectors' capabilities, but the proper integration of the capabilities at enterprise level is also vital. Additionally, a potential defect in one node (sector capabilities) may jeopardise the enterprise competency model [10,14]. The interdependencies (sequence/parallelism, synchronisation, data flow, precedence conditions) among capabilities, at the various sectors, must be properly integrated in order to achieve the enterprise global goals. 'Cross-functional co-ordination' and 'Cross-functional integration' of capabilities is defined as:

Definition 3 (Cross Functional Co-ordination (CFC) of capabilities) – is a link among capabilities within a sector, this link seeks to fund relations between the activities of the capabilities using sector's 'product/service workflow diagram.' CFC is act as union for the other component of the capability (i.e. resource$\{R_{\alpha 1} \cup R_{\alpha 2} \cup R_{\alpha 3} \cup \ldots \cup R_{\alpha n}\}$, knowledge$\{K_{\alpha 1} \cup K_{\alpha 2} \cup K_{\alpha 3} \cup \ldots \cup K_{\alpha m}\}$,). CFC is the set of ordered pairs$(x, \mathfrak{x})$; where x is the independent activity and the $\mathfrak{x}$ is dependent on x. CFC$(C) = \{(x, \mathfrak{x}) \mid x \in C$ and CFC$(x) = \mathfrak{x}\}$;

$$FC(x) \begin{cases} = 0 \text{ ; if a is not sector to the other activities} \\ = x \text{ ; is reachable from product/service workflow diagram} \end{cases};$$

where: C- is a capability set; $x, \mathfrak{x}, x$- is a activity, task, act or process

Definition 4 (Cross Functional Integration (CFI) of capabilities) - CFI is a link among capabilities of sectors within an enterprise. This link seeks to fund relations among the activities of the capabilities at the enterprise using enterprise's 'product or service structural model'. CFI acts as union for the other component of the capability between sectors (i.e. resource$\{R_{\alpha 1} \cup R_{\alpha 2} \cup R_{\alpha 3} \cup \ldots \cup R_{\alpha n}\}$, knowledge$\{K_{\alpha 1} \cup K_{\alpha 2} \cup K_{\alpha 3} \cup \ldots \cup K_{\alpha m}\}$,).
$CFI_{\alpha\beta}(C_{\alpha}, C_{\beta}) = \{(x_{\alpha}, \mathfrak{x}_{\beta}) \mid x \in C_{\alpha}$ and $CFI(x_{\alpha}) = \mathfrak{x}_{\beta}\}$

$$CFI(x_{\alpha}) \begin{cases} = 0 \text{ ; if } x_{\alpha} \text{ is not sector on the other activities at sector } \beta \\ = x_{\beta} \text{ ; is reachable based on product/service structural model} \end{cases}$$

Definition 5 (Enterprise's competency) –Is defined as cross functional co-ordination and integration of task-oriented capabilities aimed at achieving a global outcome or goal.

Enterprise's competency definition can be formulated as:

$$\text{Competency} \mid_G^{1,2} = C_1^* \otimes C_2^* = CFI_{12} \left[CFC\left(\bigcup_{i=1}^{n} \{C_{1i}^*\} \right), CFC\left(\bigcup_{i=1}^{m} \{C_{2i}^*\} \right) \right]$$

$$\text{Competency} \mid_G^{1,2,3} = C_1^* \otimes C_2^* \otimes C_3^*$$

$$= CFI_{12} \left[CFC\left(\bigcup_{i=1}^{n} \{C_{1i}^*\} \right), CFC\left(\bigcup_{i=1}^{m} \{C_{2i}^*\} \right) \right], CFC \text{Competency} \mid_G^{1,2,\cdots,n} = C_1^* \otimes C_2^* \otimes \ldots \otimes C_n^*$$

$$= CFI_{12\ldots(n-1)} \left[CFI_{12\ldots(n-2)} \left[\cdots \left[CFI_{12} \left[CFC\left(\bigcup_{i=1}^{n} \{C_{1i}^*\} \right), CFC\left(\bigcup_{i=1}^{m} \{C_{2i}^*\} \right) \right], \ldots \right], CFC\left(\bigcup_{i=1}^{k} \{C_{ni}^*\} \right) \right]$$

Where: G- Represents a specific outcome or goal; 1, 2, 3,…, n- Is an index for representing sectors; C_m^*- Task-oriented capability for Sector m as defined previously; $\bigcup_{i=1}^{n} \{C_{\alpha i}^*\} = \{C_{\alpha 1} \cup C_{\alpha 2} \cup C_{\alpha 3} \cup \ldots \cup C_{\alpha n}\}$; $\otimes$ cross functional integration and co-ordination; CFI_{nm}- Cross Function Integration between sector n and sector m; CFC- Cross Function Co-ordination.

5   Case Study

The competency modeling implementation is a part of knowledge management strategy for a corporation and with knowledge as an intangible asset, the usefulness of it usually cannot be seen in the short run. Therefore, this research uses the method of a case study and has directed our survey on Chika Food Industry (MFI). The reason that we have chosen this company is that it has already carried out knowledge management strategy so a part of knowledge for competency modeling proposes is accessible for the enterprise knowledge base system. (A)Identify and list required capabilities of sector: In this example the goal of the sectors is producing a conserve base 'Macaroni &Sauce' food. After identification process, the listed capabilities are then sequenced so that they follow the order in which they will be performed. Successful completion of these attempts often requires a good knowledge of process planning, manufacturing features and manufacturing resources. 'production' and 'Laboratory' of ChFI have sets of capabilities: $C_{Pro}=$ {'Cooking','Mixing','Filling','Weighing','Freezing','Sealing','Palletizing'}; $C_{Lab}=$ { 'Microbiologic test' }Since the goal is producing 'Macaroni & Sauce'', 'Production' and 'Laboratory' sectors' task oriented capability set is as: $C_{Pro(M\&S)}^*=$ {Cooking ','Mixing', 'Weighing', 'Sealing', 'Palletizing'}; $C_{Lab(M\&S)}^*=$ { 'Microbiologic testing'}. (B)Assign resources, activities and knowledge to the sequenced capabilities: For the resources, activity and knowledge assign processes of acquired capabilities, interviews of personal appreciation, samples, references is used. For instance, the 'Cooking' and 'Mixing' capabilities at 'Production' sector also 'Microbiological test' at 'Laboratory' sector has the following sub elements (raw ingredients Code are:

$1,2,3,4,5):$ 'Cooking$_{@Pro.}$' $=$

$$\left\{ \begin{array}{c} \{\text{Combine\&}Cook(1,2), \text{Combine\&}Cook\ (3,4,5)\} \\ \{\text{Cooking kettleA, Cooking kettleB, Technician1}\} \\ \{\text{Manuals/KettleA, Manuals/KettleB, Recipe1, Recipe2, Bill of Material1}\} \end{array} \right\} =$$

$$\left\{ \begin{array}{c} \{x_{C1,}x_{C2,}\} \\ \{r_{C1,}r_{C2,}, r_{C3,}\} \\ \{k_{C1,}k_{C2,}k_{C3,}k_{C4,}k_{C5,}\} \end{array} \right\};$$

'Mixing$_{@Pro.}$'$=\left\{ \begin{array}{c} \{\text{Draining, Mixing}\} \\ \{\text{Mixinging kettle, Draining kettle, Technician2}\} \\ \{\text{MixingKettleManuals, Recipe3, Bill of Material2}\} \end{array} \right\}=\left\{ \begin{array}{c} \{x_{M1,,}x_{M2}\} \\ \{r_{M1,}r_{M2,}r_{M3}\} \\ \{k_{M1,}k_{M2,}k_{M3}\} \end{array} \right\}$

'Microbiological test$_{@Lab}$' $=$

$$\left\{ \begin{array}{c} \{\text{Microbiologic testing}\} \\ \{\text{Sample Precreation Machine, Oven, Testing Machine}\} \\ \{\text{Manual1, Manual2, Manual2, Worksheet}\} \end{array} \right\}=\left\{ \begin{array}{c} \{x_{MT1}\} \\ \{r_{MT1,}r_{MT2,}r_{MT3}\} \\ \{k_{MT1,}k_{MT2,}k_{MT3,}k_{MT3}\} \end{array} \right\}$$

(C)Interactions of capabilities within sectors and between the sectors: The sequence diagram in Fig.1 demonstrates the interaction of capabilities among these sectors.

(D)Capability modeling: A capability knowledgebase is developed to assure that the knowledge of capabilities at the sectors is capitalized. At present, the knowledgebase is developed under ACCESS and is operational. The relational model of the capability knowledgebase is represented by Fig.3. The use of a standard incoming application adds knowledge gathering process to the capability knowledgebase system.

(E)'Cross-functional co-ordination' and 'Cross-functional integration' of capabilities :The 'cross-functional co-ordination' and 'cross-functional integration' sub-categories concerns the linking of enterprise competency aspects. The 'Cross-functional co-ordination' process (definition 3) was adapted to all the identified capabilities at the sectors. For do this, the sector's capabilities sequence diagram (figure 1) is used. As an example: Cross Functional Co-ordination (CFC) Cooking$\rightarrow$ Mixing:

$$\left\{ \begin{array}{c} \{(x_{C1,}x_{M1,}), (x_{C1,}x_{M2}), (x_{C2,}x_{M2})\} \\ \{r_{C1,}r_{C2,}, r_{C3,}, r_{M1,}r_{M2,}r_{M3}\} \\ \{k_{C1,}k_{C2,}k_{C3,}k_{C4,}k_{C5,}, k_{M1,}k_{M2,}k_{M3}\} \end{array} \right\}$$ Using the capabilities sequence diagram

among the sectors (figure2), the 'Cross-functional integration' process (definition4) was adapted to the identified capabilities at the enterprise. As an example: Cross Functional Integration (CFC) Cooking$\rightarrow$ Microbiological test: $=$

$$\left\{ \begin{array}{c} \{(x_{C1,}0), (x_{C1,}0), (x_{C2,}x_{MT1})\} \\ \{r_{C1,}r_{C2,}, r_{C3,}, r_{MT1,}r_{MT2,}r_{MT3}\} \\ \{k_{C1,}k_{C2,}k_{C3,}k_{C4,}k_{C5,}, k_{MT1,}k_{MT2,}k_{MT3,}k_{MT3}\} \end{array} \right\}$$

(F)Enterprise competency representation: At this stage all the competency aspects were stored, and all the competency associated sub-categories were linked as well; the next step is to represent enterprise competency. Using enterprise competency definition (definition 5) the example blow depicts competency creation process at the enterprise. For simplification in this example only three capabilities ('Cooking' and

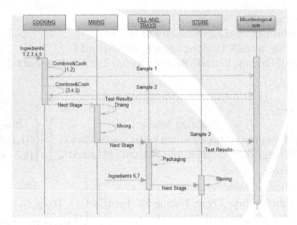

Fig. 1. Interactions of capabilities among the design and manufacturing sectors

'Mixing' from 'Production' sector and 'Microbiological test' from laboratory department) are taken in to consideration. Competency (Cooking, Mixing) →(Microbiological test):

$$\left\{ \begin{array}{c} \{(x_{C1}, x_{M1}, 0), (x_{C1}, x_{M2}, 0), (x_{C2}, x_{M2}, x_{MT1})\} \\ \{r_{C1}, r_{C2}, r_{C3}, r_{M1}, r_{M2}, r_{M3}, r_{MT1}, r_{MT2}, r_{MT3}\} \\ \{k_{C1}, k_{C2}, k_{C3}, k_{C4}, k_{C5}, k_{M1}, k_{M2}, k_{M3}, k_{MT1}, k_{MT2}, k_{MT3}, k_{MT3}\} \end{array} \right\} ;$$

Fig.2 depicts the dialog boxes in which the competency are shown. The dialog boxes also show the features of the competency stored in the knowledgebase. The experimental software developed can show capability attributes by clicking on the particular sign beside each row.

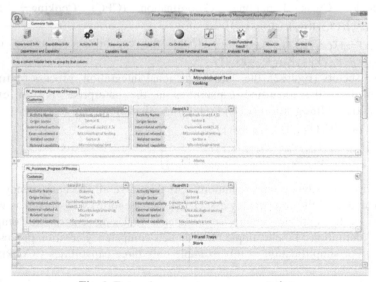

Fig. 2. Enterprise competency representation

6   Conclusions

Under the influence of the enterprise engineering paradigm with enterprise's productability, companies need to start to actively implement competency management with the goal of obtaining important information for their future decision making processes. This research first concluded that 'cross-functional co-ordination', 'cross-functional integration', and 'sector capability' are three of the sub-categories in enterprise competency modelling. Furthermore, based on past published papers, resource, activity, and strategy (knowledge related resource and activity) are three of the aspects for sector capability modelling. A generic sector capability model is proposed; also cross-functional co-ordination and cross-functional integration of the capabilities are defied as major advancements for intra-enterprise competency modelling. Through the case study of ChFI, we implement the academic 'enterprise associated' concepts with real practice in the industry. The developed experimental system for the case study of ChFI offers four benefits, in that they a) enhance the organizations willingness to collaborate, b) boost the organization's competitiveness, c) facilitate appropriate decision-making, and d) finally help to integrate the entire organization.

References

1. Ljungquist, U.: Core competency beyond identification: presentation of a model. Management Decision 45(3), 393–402 (2007)
2. Camarinha-Matos, L.M., Afsarmanesh, H. (eds.): Collaborative networks: Reference modeling. Springer (2008)
3. Javidan, M.: Core competence: what does it mean in practice? Long Range Planning 31(1), 60–71 (1998)
4. Harzallah, M., Vernadat, F.: IT-based competency modeling and management: from theory to practice in enterprise engineering and operations. Computers in Industry 48(2), 157–179 (2002)
5. Huat Lim, S., Juster, N., de Pennington, A.: Enterprise modelling and integration: a taxonomy of seven key aspects. Computers in Industry 34(3), 339–359 (1997)
6. Trejo, D., Patil, S., Anderson, S., Cervantes, E.: Framework for competency and capability assessment for resource allocation. Journal of Management in Engineering 18(1), 44–49 (2002)
7. Capece, G., Bazzica, P.: A practical proposal for a "competence plan fulfillment" key performance indicator. Knowledge and Process Management 20(1), 40–49 (2013)
8. Joshi, K.A.I.L.A.S.H.: Cross-functional integration: the role of information systems. Journal of Information Technology Management 9, 21–30 (1998)
9. Barenji, R., Hashemipour, M., Barenji, A., Guerra-Zubiaga, D.: Toward a framework for intra-enterprise competency modeling (2012)
10. Barenji, R.V., Hashemipour, M., Guerra-Zubiaga, D.A.: Toward a Modeling Framework for Organizational Competency. In: Camarinha-Matos, L.M., Tomic, S., Graça, P. (eds.) DoCEIS 2013. IFIP AICT, vol. 394, pp. 142–151. Springer, Heidelberg (2013)

11. Barenji, R.V.: Towards a Capability-Based Decision Support System for a Manufacturing Shop. In: Camarinha-Matos, L.M., Scherer, R.J. (eds.) PRO-VE 2013. IFIP AICT, vol. 408, pp. 220–227. Springer, Heidelberg (2013)
12. Barenji, A.V., Barenji, R.V., Sefidgari, B.L.: An RFID-enabled distributed control and monitoring system for a manufacturing system. In: 2013 Third International Conference on Innovative Computing Technology (INTECH), pp. 498–503. IEEE (2013)
13. Barenji, R.V., Barenji, A.V., Hashemipour, M.: A multi-agent RFID-enabled distributed control system for a flexible manufacturing shop. The International Journal of Advanced Manufacturing Technology, 1–19 (2014)
14. Pépiot, G., Cheikhrouhou, N., Furbringer, J.M., Glardon, R.: UECML: Unified enterprise competence modelling language. Computers in Industry 58(2) (2007)

Part VII

Human-Computer Interfaces

Comparison of Hand Feature Points Detection Methods

Tomasz Grzejszczak, Adam Gałuszka, Michał Niezabitowski, and Krystian Radlak

Silesian University of Technology, Faculty of Automatic Control,
Electronics and Computer Science, 16 Akademicka Street, 44-100 Gliwice, Poland
{Tomasz.Grzejszczak,Adam.Galuszka,Michal.Niezabitowski,
Krystian.Radlak}@polsl.pl

Abstract. This paper presents the research and comparison of four methods of hand characteristic points detection. Each method was implemented and modified in order to test their capabilities on database for hand gesture recognition. All methods are explained, tested and compared to others with other leading to final remarks. The main purpose of the research is to choose the best algorithm giving the most information about human hand that would lead to create a human – computer interaction program.

Keywords: Hand gestures, HCI, Characteristic points detection, Sign language.

1   Introduction

From the time when first machine was created, there was a need of controlling it. From the levers, through buttons to the present touchscreens, the way of controlling our devices is tending to be more and more intuitive. On the other hand the most intuitive way of communication between people were body language and gestures. Thus in order to create the future, there is a need of gesture detection algorithms. This lead to the question: is it possible to detect the human hand characteristic points accurate enough to classify the shown gestures? What methods of computer vision approach are the best for this problem?

This paper contains the first author's method of hand characteristic points detection [1]. Moreover, the authors of the paper compare the above mentioned method with 3 other algorithms [2 - 4]. All of presented methods were modified in order to fit them into the common testing environment that was performed on created by author gestures base [1, 5 - 7].

2   Relationship to Collective Awareness Systems

The presented solutions on hand feature points detection can be the core of human computer interaction program that can have many applications, such as sign language gestures classification or intelligent home control, and it means that the main contributions of our work coincide with the main theme of the conference "Collective Awareness Systems".

L.M. Camarinha-Matos et al. (Eds.): DoCEIS 2014, IFIP AICT 423, pp. 167–174, 2014.
© IFIP International Federation for Information Processing 2014

Presented research analyzes the hand detection algorithms which are the core of human computer interaction. Having the good algorithm of hand detection can lead to accurate computer control program, which can lead to various applications, from simple advertisement or information systems to control of a robotic arm manipulator in a sterile environment. All of those applications are the part of extended user communities, linking humans and smart objects.

3   Data Base Tests

The presented solutions on hand feature detection were tested on database [7] for hand gesture recognition (HGR). The data are organized into three series acquired in different conditions containing over 1500 images with hands of over 30 individuals presenting about 30 gestures from Polish Sign Language, American Sign Language and some special gestures, such as pointing or grabbing.

Each single gesture that is referred as a one image from database is in fact a set of few things:

- original RGB images (jpg files);
- ground truth binary skin presence masks (bmp files);
- hand feature points location (xml files).

The original RGB image is a photo taken during the base creation. Basing on this image, an expert created a ground truth binary image, where value 0 is referred to background, and 1 is referred to skin region. Finally, expert marked up to 25 characteristic points, such as:

- wrist points;
- 5 finger tips;
- 14 knuckle points;
- 4 points representing space between fingers.

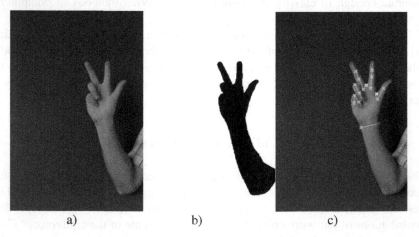

Fig. 1. a) Original hand image, b) Binary mask, c) Hand feature points visualization

```
<IMAGE WIDTH="400" HEIGHT="604" COLOR="TRUE">
 <hand>
  <FeaturePoint x="312" y="333" category="Wrist" type="WristThumb"/>
  <FeaturePoint x="258" y="345" category="Wrist" type="WristPinky"/>
  <FeaturePoint x="309" y="260" category="Thumb" type="Knuckle1"/>
  <FeaturePoint x="328" y="228" category="Thumb" type="Knuckle2"/>
  <FeaturePoint x="347" y="206" category="Thumb" type="Fingertip"/>
     ......
```

File 1. Part of XML File with ground troth feature points localization

All tests of the presented methods were performed on the newest HGR1 database, containing 899 images of 24 gestures from Polish Sign Language.

It is also important to point out that all of the presented methods assume a binary mask as the initial image. In order to obtain a mask, one should perform a segmentation algorithm such as skin detection [6, 8- 15]. All of presented results are based on analyze of ground truth mask, because analyzing output of skin detection algorithm would carry the error of those algorithms.

4   Hand Feature Points

Human hand has 27 bones, where 14 of them are phalanges of fingers. In a simple approximation, hand position or gesture can be described by the orientation of hand, position of wrist and mutual relations between knuckles and fingertips. Human anatomy does not allow bone bending, the actual bending is performed in a joints. This fact can lead to some simplifications in hand modeling. The palm region is the biggest, inflexible object in hand, bounded by wrist, finger first knuckles and concave points between fingers. Fingers consist of 3 phalanges (2 in case of thumb), which can be descried as a constant length line segment bounded by two knuckle points or fingertip. In conclusion, those are the reasons why exactly 25 characteristic points have been chosen in database.

However, not all points can be detected. Some of them cannot be seen in particular hand orientation, other are limited by algorithm capabilities. Most of the presented algorithms are able to detect only the wrist points and finger tips.

4.1   Method 1. Analyze of Contour Points Distance from Wrist Point

The method was inspired by Nobuhiko Tanibata [3]. In this article author described the whole laboratory stand for Japanese Sign Language gesture recognition. The article contains the multiple solutions on whole body gesture detection. One of them is a method of counting raised fingers by analyzing the distance of contour points from wrist point. The method, however assumed that the user would wear a high sleeve, covering the hand, which gives easy palm segmentation. In order to have the method work in any conditions, a set of improvements needed to be added.

In summary, this whole method consists of two curtail steps. First step, developed by the first author and described briefly in paper [1] provides the hand orientation and wrist detection. Second step, inspired by Nobuhiko Tanibata [3], was expanded in order to detect finger tips. The algorithm of the whole method is presented below:

1. Obtain mask and its contour;
2. Find the longest diameter of contour;
3. Calculate the image profile. Each column of the profile image is a sum of pixels perpendicular to longest diameter in a given point;
4. Find the local minimum of profile and obtain the wrist points in original image;
5. Calculate the distance of each pixel of hand contour from wrist middle point;
6. Analyze the extrema of distance function and obtain the fingertip points.

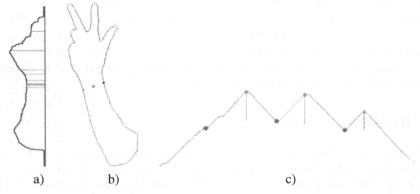

a)   b)   c)

Fig. 2. a) Hand profile along the longest diameter, b) Hand contour with selected wrist region and detected fingers with use of extrema localization on c) function of distance of contour points from wrist point

4.2   Method 2. Distance Transform

Method developed by Maciej Czupryna [2] is based on distance transform. New image is created with the same size as a source image, and each pixel from this new image has a value equal to distance to closest white pixel on source image.

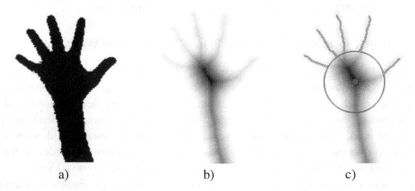

a)   b)   c)

Fig. 3. a) Source image, b) distance transform of source image, c) analyzed image

As it can be observed on Fig. 3, a white pixel representing a background on source image is also represented as value 0 (white pixel). The pixels from palm region are located far from background. In conclusion, the farther from background, the higher value of output image pixel.

Basing on the distance transform image, one can easily obtain the palm region with center in the maximal value of distance transform, and radius equal to 1.3 times value of maximum. The fingers are detected as a constant line of non-rapidly changing local maxima of distance transform. The beginning of these lines are considered as a knuckle and the end as a fingertip.

4.3   Method 3. Template Matching with Use of Cross Correlation

In article [15] author describes a method of template matching. Observing that a fingertip is round ended, the algorithm is based on matching the template image presenting black circle into the source image with use of cross correlation.

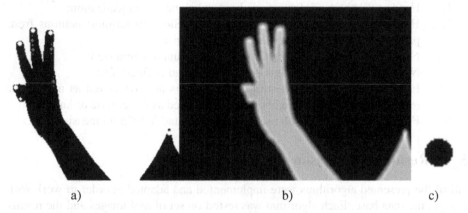

a)   b)   c)

Fig. 4. a) Source image with annotated results (white – correctly detected points, grey – rejected points), b) Cross correlation of source image with c) template

The finger tips should be located in the local maxima of cross correlation output image. However it is important to filter the false positives. First method is based on closeness of detected points. Each detected point restricts an area in which no other points can be detected. Second method is based on the observation that a fingertip is surrounded by huge amount of background. While detecting a new local maximum, the test is performed if for a black pixel located on a border of template (about 20 pixels in radius from detected point) there exists a white pixel on an opposite side of template border. This rule can be observed on Figure5, where gray points are these which did not applied to the following rule.

4.4   Method 4. Analyze of Contour Points Distance from Wrist Point

The rather complicated method of characteristic points detection was presented in [4] where the authors applied the Self Organizing and Self Growing Neural Gas

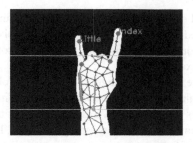

Fig. 5. Example of hand image with SGONG, hand slope detection, palm region detection and finger recognition

(SGONG) on the image of hand. The neural gas was spread on hand image and in stable state, it uniformly covered the hand region.

The neural gas was just a first step of characteristic points detection. The set of rules were applied in order to gather as much information as possible:

1. Hand profile brings information about palm and wrist localization;
2. Hand slope is calculated using 2 methods. Slope of leftmost neurons from palm region and a slope between palm and wrist;
3. Neuron connections passing through background are removed;
4. Neurons with only one neighbor are considered as fingertips;
5. From finger tips, neurons with two neighbors are considered as mid-finger region, and with more neighbors are considered as finger base or knuckle;
6. Fingers are identified with use of hand slope and 3 different measurements.

5   Results Comparison

All of the presented algorithms were implemented and adapted in order to work with our gesture data base. Each algorithm was tested on set of 899 images and the results are presented in Figures 6 and 7. On each image, few characteristic points were found, and these points positions were compared to those from database. The output data is presented in form of cumulative distribution function of detected points displacement. For example, method 3 with total of 3789 detected characteristic points has total of 3294 points with displacement error less than 14 pixels. This gives 89% of all points.

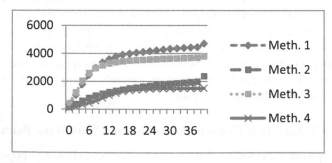

Fig. 6. Distribution of error of characteristic points detection for all presented methods

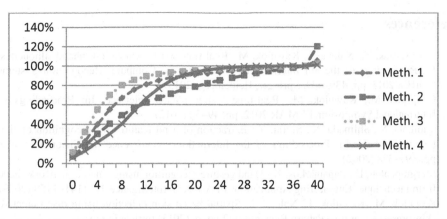

Fig. 7. Distribution of error of characteristic points detection for all presented methods. The values above 100% represents the rest of points with too high error.

6   Discussion of Results and Critical View

In conclusion to chapters 3 and 4, the following remarks have been done.

First method of finger detection with use of distance between contour points and wrist point brings the highest amount of points detection. However, there are images, where the algorithm identify the hand orientation in opposite way, leads to detection in region of elbow. These points are located in last column, with error values above 40.

Second method based on distance transform did not well in images with joined fingers. Bringing the small amount of characteristic points with the lowest accuracy, this method is considered as fast bus inaccurate.

Third method with template matching used to match a finger in sharp mask edges, as it can be shown on example on Fig. 4 where a template is matched on shoulder. This caused the error to be similar to method 1, however with much less characteristic points detected.

Fourth method, that used SGONG, brings the best characteristic points matching, with nearly flat line, while expanding of distribution function. However this method has many limitations with proper hand orientation, causing for example horizontally oriented hand not to be considered in tests. This brings the smallest detection rate. Moreover this algorithm is much slower in comparison to other.

In further work the first algorithm would be used and improved as it brings the most information and detects the highest amount of characteristic points. The improvement should deal with proper hand orientation leading to less false points detection. Moreover an reasoning algorithm would be developed, able to classify the viewed gesture into the proper gesture type.

Acknowledgements. The research presented here were done by authors as parts of the research projects BKM/514/RAU1/2013/t.2, BK/214/RAU1/2013 and BKM/514/RAU1/2013/t.1, respectively.

References

1. Grzejszczak, T., Nalepa, J., Kawulok, M.: Real-time wrist localization inhand silhouettes. In: Proceedings of the 8th International Conference on Computer Recognition Systems CORES 2013, pp. 439–449. Springer, Heidelberg (2013)
2. Czupryna, M., Kawulok, M.: Real-time vision pointer interface. In: Proceedings of International Symposium ELMAR 2012, pp. 49–52 (2012)
3. Tanibata, N., Shimada, N., Shirai, Y.: Extraction of hand features for recognition of sign language words. In: Proceedings of the International Conference on Vision Interface, pp. 391–398 (2002)
4. Stergiopoulou, E., Papamarkos, N.: Hand gesture recognition using a neural network shape fitting technique. Engineering Applications of Artificial Intelligence 22, 1141–1158 (2009)
5. Kawulok, M., Kawulok, J., Nalepa, J.: Spatial-based skin detection using discriminative skin-presence features. Pattern Recognition Letters (2013) (article in press)
6. Kawulok, M.: Fast propagation-based skin regions segmentation in color images. In: Proceedings of 10th IEEE International Conference and Workshops on Automatic Face and Gesture Recognition, pp. 1–7 (2013)
7. Kawulok, M., Grzejszczak, T., Nalepa, J., Knyc, M.: Database for hand gesture recognition (2012), http://sun.aei.polsl.pl/~mkawulok/gestures/
8. Kawulok, M., Kawulok, J., Nalepa, J., Papiez, M.: Skin detection using spatial analysis with adaptive seed. In: Proceedings ofIEEE International Conference on Image Processing (ICIP 2013), pp. 3720–3724 (2013)
9. Kovac, J., Peer, P., Solina, F.: Human skin color clustering for face detection. IEEE REGION 8 EUROCON 2003, Computer as a Tool 2, 144–148 (2003)
10. Jones, M., Rehg, J.: Statistical color models with application to skin detection. International Journal of Computer Vision 46, 81–96 (2002)
11. Kakumanu, P., Makrogiannis, S., Bourbakis, N.G.: A survey of skin-color modeling and detection methods. Pattern Recogn. 40, 1106–1122 (2007)
12. Phung, S.L., Chai, D., Bouzerdoum, A.: Adaptive skin segmentation in color images. In: IEEE International Conference on Acoustics, Speech, and Signal Processing, vol. 3, pp. 353–356 (2003)
13. Soriano, M., Martinkauppi, B., Huovinen, S., Laaksonen, M.: Skin detection in video under changing illumination conditions. In: 15th International Conference on Pattern Recognition (ICPR 2000), vol. 1, pp. 839–842 (2000)
14. Solar, J.R., Verschae, R.: Skin detection using neighborhood information. In: 6th IEEE International Conference on Automatic Face and Gesture Recognition, pp. 463–468 (2004)
15. Sato, Y., Kabayashi, Y., Koike, H.: Fast Tracking of Hands and Fingertips in Infrared Images for Augmented Desk Interface. In: Proceedings of 4th IEEE International Conference on Automatic Face and Gesture Recognition, pp. 462–467 (2000)

Adaptive Human-Machine-Interface of Automation Systems

Farzan Yazdi and Peter Göhner

Institute of Industrial Automation and Software Engineering, Pfaffenwaldring
47,70550 Stuttgart, Germany
{farzan.yazdi,peter.goehner}@ias.uni-stuttgart.de

Abstract. Automation systems are inseparable part of everyday life; heating systems, ticket vending machines or blood glucose meters are few examples of such systems, showing the diversity of their application domain. This diversity implies the variety of different user groups with assorted capabilities interacting with such systems in different contexts. Hence, the requirements of the human-machine interfaces of such systems are strongly varying, depending on the context of use. Attempts in developing high interactive systems, such as user centered development or universal design have failed; either they are costly or system specific. Furthermore, many context-relevant aspects are only known at run-time. In this paper, we propose a generic concept, which adapts the human-machine interfaces of automation systems at run-time, according to the context of use. It addresses not only the representational aspects but also the semantics and the connection to the underlying technical system. The concept is implemented as an evaluating prototype.

Keywords: human-machine-interaction, usability, adaptive user interface, context of use, context sensitive user interface, automation systems.

1 Introduction

The application growth of automation systems has brought a wider range of users among those who interact with such systems. Hence, future-oriented automation systems must satisfy a broad range of expectations. Furthermore, the demographic changes bring new requirements on automation systems to be used by users with limited capabilities. The new trends in the domain of automation systems are "Have-It-Your-Way" solutions for better usability [1] and cost reduction in maintenance, training, support, etc. [2]. **Usability** is "the extent to which a product can be used by specified users to achieve specified goals with effectiveness, efficiency, and satisfaction in a specified context of use" [3]. This definition abstractly frames out the guidelines of ergonomics and usability of human-machine-interaction. Similar definitions used in the literature point out the same aspects as effectiveness, efficiency and satisfaction. While usability has been long discussed in the domain of web and smartphone applications, it is quiet new in automation systems domain. In some cases, the provided guidelines for the web applications can be adopted to automation

L.M. Camarinha-Matos et al. (Eds.): DoCEIS 2014, IFIP AICT 423, pp. 175–182, 2014.

system [4], however, automation systems have different requirements, imposed by their software-hardware architecture.

Automation systems are systems consisting of a technical process running in a technical system that is automated by components necessary for automation. These components can be sensors, actors and directly wired components to interact with the technical system. The automation functionality is realized on automation computers that are interconnected by a communication infrastructure. Finally, there are components to display information and to input user interventions to interact with the users of the system [5]. We refer to the last component as User Interface (UI). It includes not only the graphical UI but also the modalities, hardware and the interconnections with the technical system. The UI is exposed to the environmental conditions, HW/SW- circumstances and the users, referred to as Context of Use (CoU).

Addressing the usability issues, the researchers and engineers – both in academic and industrial areas – investigate methods, concepts and technologies to increase the usability of automation systems and to provide more intuitive and flexible interaction to the users. Principally, two categories of solutions can be determined:

- Development process-oriented for correct integration of usability requirements;
- Run-time-oriented to provide suitable interaction methods on demand.

Consideration of the development process has the advantage of providing the necessary system architecture, required for integrating the usability requirements. However, it can be very costly and cannot fully allow sensibility to all aspects that affect human-machine interaction at run-time. Furthermore, predefinition of different UIs suitable for different user groups and environmental conditions is impossible. Therefore, the run-time solutions is the reasonable alternative and the new trend in this field [6]. It has the advantage of being able to react to demands that are only known at run-time without the necessity to undergo costly analysis procedure of anticipating and forecasting different run-time situations.

2   Relationship to Collective Awareness Systems

This research project addresses the domain of automation systems and focuses on the enhancement of the usability of such systems. It refers to both industrial plant automation systems and product automation systems. The proposed concept adapts the UI of such systems according to the CoU. The *development of such systems and the new trends and visions* affect the proposed concept. The concept contributes to the development of *computational and perceptional systems* of collective awareness platforms. It provides the automation system with the *intelligence to perceive the CoU* and to react upon different circumstances. Furthermore, it proposes methods and technologies of *industrial informatics* for the necessary infrastructure.

3   Survey on Usability Issues and Existing Solutions in the Domain of Automation Systems

Automation system are in their operation very dependent on run-time parameters. Depending on the user's individual properties the quality of interaction can differ; e.g., user's experience or his/her disabilities. Additionally, environmental effects, such as background light or noise, can influence the interaction. These factors are examples of the CoU. CoU includes the user's profile, the tasks, means of work (hardware, Software and material) and the physical and social environment [7].

In order to increase the usability of automation systems, it is necessary to react upon different states of CoU [8]. One challenge is that the CoU can change dynamically at run-time; the computing platform might change, network bandwidth may alter and user role or capabilities can differ. Hence, a remedy at run-time is required. This has initiated a trend towards context-sensitive UIs. However, the categorization and capturing of the CoU, in a way that is readable for the machine is not trivial; the details of the affecting factors are domain and system specific. For instance, the affecting parameters in the domain of e-learning systems are different from those of automation systems. Many solutions regarding the capturing and representing the CoU have been proposed [9, 10]. However, there are no uniform and generic solution for reflecting the dynamic changes of CoU onto the UI [8]. Moreover, this reflection leads to unstable representation of UI. Therefore, the maintenance of usability is another important task to carry out [8]. Realization of context-sensitive UIs is another effort, which is often limited to the used technologies in the underlying automation system and the interaction tasks. Finally, the semantic clarity of the interaction tasks should be sustained in the adapted system.

The hardware-software architecture of the automation system is, besides CoU, relevant to the usability of the system, since it provides the interaction means. While at run-time the system architecture is hard to manipulate, often reconfigurations and deployment of system modalities can account for improvement of usability.

The existing run-time solutions can be divided into model-based, migratory or optimization approaches. Model-based approaches attempt to reduce the complexity of the CoU by abstracting it into models [11, 12, 13]. The models are structured with meta rules, defining the expected parameters. The models are then specified at run-time and are used by an automated component to generate the suitable UI. Some solutions predefine different configurations of UI and decide at run-time which configuration to use. This has the advantage that the implementation of such systems is simple and less computations are required. However, it requires thorough analysis of the CoU and is not flexible to new situations. Another method for generating the UI is by models that describe different aspects of UI [14]. The advantage here is the fact that depending on the amount of models used, the details of a UI can be defined to the granularity level needed. The problem here is that there is no unified approach to model UI. Specifically, in the domain of automation systems where real-time requirements are often the case, modeling of temporal behavior of the UI is non-trivial [15]. Moreover, generation of such models at run-time is a tremendous effort.

Migratory approaches deal with allocation and configuration of different system modalities and available interaction components [16]. With the popularity of smartphone, it is nowadays very common to perform tasks using multi-devices. Hence, an infrastructure to support cross-device and cross-platform interaction is necessary. Often web or cloud technologies are employed for exchanging data [17]. The migration can be done partially or fully, according to the requirements of the specific system [18]. Usage of multimodal concepts and embedding devices, which come along with the user (e.g. smartphones), are the advantages of these solutions. Users already know these devices; hence, the interaction is much simpler. However, the existing solutions are limited to software applications. Our tests on Programmable Logic Controller (PLC)-based and Controller Area Network (CAN)-based interaction systems, which are very common in the field of automation systems, showed that manipulation of the technical system from uncoupled devices are very hard to achieve. The platform dependencies and security issues make this even harder.

The optimization solutions are popular among computer science researchers. They mainly focus on the representation of the UI and according to CoU calculate all possible configurations/compositions of the UI. An optimization function traverses the possibilities and find the best solution to be exposed. The UI can be provided either using modelling techniques or using run-time interpreters, e.g. [19] or the smart toolbar grouping in Microsoft-Office 2007 or later. This category of approaches is scalable and very precise in its results. However, since they are very computation intensive, they are less suitable for the automation domain. Moreover, design of the optimization function is not trivial for so many parameters of the CoU.

4   Context of Use in Adaptive User Interfaces

The CoU in its initial definition from the ISO-Norm [3] is not specific enough. Regarding the goals of the concept and after thorough survey of the state of the art, we have summarized the following aspects of CoU, relevant for automation systems:

User Tasks: User tasks characterize the semantics of the interaction. They stand for the elements used on a UI and construct the sequence of activities needed to be done in order to complete a task [20]. There are four types of tasks [21]: User tasks, system tasks, Interaction tasks and Abstract tasks (A set of tasks, which can be grouped together in accordance to their semantic goals). This categorization is important, when reallocation of tasks by the user and by the automation system is necessary [22].

User Role: An automation system specifies a set of tasks (and consequently system functions) to a certain user role. Different user roles affect directly the type of tasks a user might probably intend to perform. For instance, maintenance staff, operator and system administrator are three user roles, which can indicate what a user belonging to a certain user role group would perform and what experience he/she has.

Users' Individual Properties (User Profile): [23] classifies user properties, regarding the context of use, in three categories: User role, user experience, knowledge and skills and personal attributes. [24] contemplates the human factor in its actions to perform everyday tasks. Here the user activities are divided into three abstract steps: receiving information, processing information and realizing the information. For receiving information, regarding the current technological possibilities, three human senses sight, hearing and touch are used. The cognitive capabilities of the user (fluid and crystallized intelligence) contribute to processing information and individual power, endurance and coordination capabilities correspond to realizing information. [25] investigates in a similar way as above the human factors in Human Machine Interaction (HMI). It sets categories of human factors and attempts to provide value ranges that define the normal values and their extents, using statistics and anthropometrics.

Environment: It is often in literature distinguished between physical environment and organizational environment. The environment (place) in which the interaction occurs is referred to as workspace. The physical environment characterizes the workspace conditions (i.e. atmospheric, auditory, thermal and visual conditions and the environmental instability), the workspace design (i.e. space and furniture, user posture and location) and the workspace safety (i.e. health hazards and required protective clothing and equipment) [7][23]. The organizational environment defines aspects such as the organizational structure or attitude and culture. These aspects can affect the HMI from a social and psychological point of view, hence, it will not be delved into any deeper here. [26] references five typical examples of environment influences, which can strongly affect the usability: Glare caused by bright light, environmental noise, physically limited workplaces, distractions and the tasks with special concentration demands.

5   The Concept of Adaptive User Interfaces

Based on the previous discussions and identified deficiencies of existing solutions, the concept has to fulfill the following requirements: 1) It must adapt the UI to the CoU. 2) The adaptation should occur at run-time and aims at existing systems (released systems). 3) It must consider the characteristics of automation systems. 4) It must be generic. The derived requirements stipulate abstractly the boundaries of the approach required to fulfill them. The concept has the goal to consider the existing automation system, i.e. automation systems on the market and to adapt and alter the user interface according to the CoU. Based on the definition of CoU given above, the adaptation should be *Task-oriented*, *User specific* and *Environment specific*. As mentioned earlier, the new trend to use mobile devices for the purpose of interaction is essential for the concept to be useful for current and future systems. We refer to the interaction devices – coupled or uncoupled to/from automation system – as Interaction component, hence, the adaptation should also be *interaction component specific*.

The task-oriented adaptation necessitates the computer-based analysis of the tasks at run-time. The challenges here are that the semantic description of tasks are difficult to capture at run-time [8] and that a description methods covering the requirements mentioned previously, including all necessary details and generic for the domain of automation systems does not exist. For the description method, we have proposed an XML-based description. Each task corresponds to an XML tag. The composite tasks are described as interlaced tags. Each tag have an ID and a task type as properties. For each task type – user tasks, system tasks, interaction tasks and abstract tasks – a keyword is dedicated. The temporal behavior of the tasks is adopted from [20], since it thoroughly describes all possible compositions of the tasks. The applicability of this proposed method has been evaluated using the prototype of a ticket vending machine. In addition to the tasks, the description of the original UI (UI before being adapted) is necessary for the adaptation; e.g., it is necessary to know which tasks belong to which UI-dialogue. UIML [27] is suitable for this purpose, since it not only describes the representation of the UI but also the connection to the underlying system functions. This is specifically interesting for event-based realizations of UI, which is often the case in automation system; e.g. refreshing a sensor value upon an interrupt. In addition, the discussed parameters of the CoU in previous chapter are the basis for the proposed concept.

System's performance can affect the usability when the interaction flow is disturbed. Therefore, the adaptation process has been divided into two stages: static and dynamic adaptation. This separation allows a pre-adaptation process to take place, before the high computational tasks start. The static adaptation refers to those parameters of CoU, which are static during an interaction session; e.g. user individual properties or user role. The dynamic adaptation takes care of parameters changing during an interaction session; e.g. light intensity or the interaction component. Figure 1 illustrates an overview on the concept, its components and the CoU.

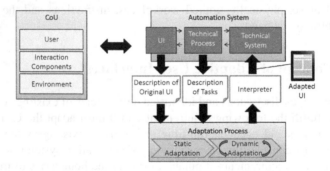

Fig. 1. Illustration of the concept its components and the CoU

A rule-based logic projects the situation constraints onto the UI arrangement. The rules are designed in three levels: 1) Modalities and their configuration; e.g. when user is visually impaired then visual interaction is not applicable. 2) Tasks and semantics; e.g. when user role x then load task set y or if user experience x then separate task y into sub-tasks. 3) Representation and layout; e.g. if display resolution less than x then group widgets. In the design of the rules, we have used the existing standards and guidelines on usability, e.g. [28]. The representation decisions rely on the method used to describe the original UI. It traverses the original UI and adapts each element of the description to the CoU. The result of the adaptation process is an adapted description of UI, which should be interpreted at run-time for different goal platforms. This makes the concept generic to different automation systems. To provide a multiplatform solution, we have proposed an interpreter, which from one side receives a unified description of UI, specified by the concept. From the other side it must be adapted to each goal platform, once in an initialization phase. In an HTML-based platform, a standard web-browser could act as an interpreter. The details of the interpreter is outside the scope of this paper. Hence, it will not be further explained.

6   Summary and Outlook

In this paper, the basics of usability of human machine interaction and CoU in the domain of automation systems were discussed. The deficiencies and shortcomings of the existing solutions has been used to motivate the investigations for a new solution. A view on relevant aspects of CoU provide the basis of the concept. The proposed concept of adaptive user interfaces is described. Various decisions on used methods in the concept has been made, based on experiments or implemented scenarios. The plausibility and conformity of the concept has been assessed using the prototype ticket-vending machine. It realizes multimodal/multiplatform interaction in two scenarios (operation and maintenance). The supported modalities are touch-LCD, speech input/output and gesture/mimic control. It provides evaluation of a broad range of interaction concepts. The platforms used in the system are Java, C#.NET and HTML5 with Jscript. In addition to the evaluation of the concept, regarding its plausibility and applicability, the prototype confirmed the necessity of a component to maintain the usability at run-time, upon changes of UI to the CoU.

The proposed concept assumes that the information on CoU are available. However, it is interesting, if the concept could work with incomplete information on CoU, which is often the case in reality. One idea to compensate the missing data would be to employ usage history or statistical information. This aspect is one of the milestones of our future works. Furthermore, it is interesting to compare the evaluation results with another system, with different characteristics, e.g. minimal performance systems or systems with complex internal communication system.

References

1. Nivethika, M., Vithiya, I., Anntharshika, S., Deegalla, S.: Personalized and adaptive user interface framework for mobile application. In: International Conference on Advances in Computing, Communications and Informatics (ICACCI), pp. 1913–1918 (2013)
2. Cooke, L., Mings, S.: Connecting usability education and research with industry needs and practices. IEEE Transactions on Professional Communication 48, 296–312 (2005)
3. ISO 29249 - Ergonomics of Human System Interaction (2006)
4. Rabin, J., McCathieNevile, C.: Mobile Web Best Practices 1.0 – Basic Guidelines. W3C Recommendation (July 29, 2008), http://www.w3.org/TR/mobile-bp/
5. Maga, C., Jazdi, N., Göhner, P.: Requirements on Engineering Tools for Increasing Reuse in Industrial Automation. In: 16th IEEE International Conference on Emerging Technologies and Factory Automation (ETFA 2011), pp. 1–7 (2011)
6. Blumendorf, M., Lehmann, G., Albayrak, S.: Bridging models and systems at runtime to build adaptive user interfaces. In: Proceedings of the 2nd ACM SIGCHI Symposium on Engineering Interactive Computing Systems, pp. 9–18. ACM, New York (2010)
7. ISO 9241-11:1998, Ergonomic requirements for office work with visual display terminals (VDTs) – Part 11: Guidance on usability
8. Vanderdonckt, J., Grolaux, D., Van Roy, P., Limbourg, Q., Macq, B.M., Michel, B.: A Design Space for Context-Sensitive User Interfaces. In: Proceedings of the ISCA 14th International Conference on Intelligent and Adaptive Systems and Software Engineering, Novotel Toronto Centre, Toronto, Canada, IASSE 2005, pp. 207–214 (2005)
9. Calvary, G., Coutaz, J., Thévenin, D.: Embedding Plasticity in the Development Process of Interactive Systems. In: Proceedings of Workshop on User Interfaces for All (2000)

10. Crease, M., Brewster, S., Gray, P.: Caring, Shar-ing Widgets: A Toolkit of Sensitive Widgets. In: Proceedings of BCS Conference on Human Computer HCI 2000, pp. 257–270. Springer, Berlin (2000)
11. Vellis, G., Kotsalis, D., Akoumianakis, D., Vanderdonckt, J.: Model-Based Engineering of Multi-platform, Synchronous and Collaborative UIs - Extending UsiXML for Polymorphic User Interface Specification. In: 16th Panhellenic Conf. on Informatics, pp. 339–344 (2012)
12. da Silva, P.P.: Object modelling of interactive systems: the UMLi approach. Thesis sumbited to the University of Manchester for the fegree of Doctor of Philosophy in the faculty of science and engineering (2002)
13. Trætteberg, H.: A hybrid tool for user interface modelling and prototyping. In: Proceedings of the Sixth International Conference on Computer-Aided Design of User Interfaces CADUI 2006, Bucharest, Romania, June 6-8, ch. 18. Springer, Berlin (2007)
14. Meixner, G., Seissler, M., Breiner, K.: Model-Driven Useware Engineering. In: Hussmann, H., Meixner, G., Zuehlke, D. (eds.) Model-Driven Development of Advanced User Interfaces. SCI, vol. 340, pp. 1–26. Springer, Heidelberg (2011)
15. Nóbrega, L., Jardim Nunes, N., Coelho, H.: Mapping concurTaskTrees into UML 2.0. In: Gilroy, S.W., Harrison, M.D. (eds.) DSV-IS 2005. LNCS, vol. 3941, pp. 237–248. Springer, Heidelberg (2006)
16. Bandelloni, R., Paterno, F., Salvador, Z.: Dynamic discovery and monitoring in migratory interactive services. In: Fourth Annual IEEE International Conference on Pervasive Computing and Communications Workshops, pp. 603–607 (2006)
17. Raising the Floor Community, Global Public Inclusive Infrastructure (GPII), http://gpii.net/
18. Ghiani, G., Paternò, F., Santoro, C.: User Interface Migration Based on the Use of Logical Descriptions. In: Migratory Interactive Applications for Ubiquitous Environments. Human-Computer Interaction Series, vol. 2011, pp. 45–59 (2011)
19. Gajos, K.Z.: Automatically Generating Personalized User Interfaces. PhD thesis, University of Washington, Seattle, WA, USA (2008)
20. Paternò, F., Mancini, C., Meniconi, S.: ConcurTaskTrees: A Diagrammatic Notation for Specifying Task Models. In: IEEE Proceeding INTERACT 1997 Proceedings of the IFIP TC13 Interantional Conference on Human-Computer Interaction, pp. 362–369 (1997)
21. Paternò, F., Meixner, G., Vanderdonckt, J.: Past, present, and future of model-based user interface development. i-com Journal for Interactive und Cooperative Media 10(3), 2–11 (2011)
22. Schweitzer, G.: Mechatronics for the Design of Human-Oriented Machines. Transactions on IEEE/ASME Mechatronics 1, 120–126 (1996)
23. Maguire, M.: Context of Use within usability activities. J. Human-Computer Studies 55, 453–483 (2001)
24. Hentschel, C., Wagner, A., Spanner-Ulmer, B.: Analysis of the application of the assembly-specific evaluation method EAWS for the ergonomic evaluation of logistic processes. In: Annual International Conference of the Institute of Ergonomics and Human Factors, pp. 221–226. CRC Press, Taylor & Francis, London (2012)
25. Federal Aviation Administration, Human Factors Division: FAA Human Factors Awareness Web Course, https://www.hf.faa.gov/Webtraining/index.htm
26. ISO 9241-20:2008, Ergonomics of human-system interaction – Part 20: Accessibility guidelines for information/communication technology (ICT) equipment and services
27. Abrams, M., Phanouriou, C., Batongbacal, A.L., Williams, S., Shuster, J.: UIML: An Appliance-Independent XML User Interface Language. In: Proceedings of 8th International World-Wide Web Conference WWW'8. Elsevier Science Publishers (1999)
28. World Wide Web Consortium W3C, Web Content Accessibility Guidelines (WCAG) 2.0 (2009), http://www.w3.org/Translations/WCAG20-de/

CARL: A Language for Modelling Contextual Augmented Reality Environments

Dariusz Rumiński and Krzysztof Walczak

Poznań University of Economics,
Niepodległości 10, 61-875 Poznań, Poland
{ruminski,walczak}@kti.ue.poznan.pl

Abstract. The paper describes a novel, declarative language that enables modelling ubiquitous, contextual and interactive augmented reality environments. The language, called CARL – Contextual Augmented Reality Language, is highly componentised with regards to both the structure of AR scenes as well as the presented AR content. This enables dynamic composition of CARL presentations based on various data sources and depending on the context. CARL separates specification of three categories of entities constituting an AR environment – trackable markers, content objects and interfaces, which makes the language more flexible and particularly well suited to building collective awareness systems based on ubiquitous AR-based information visualisation.

Keywords: Augmented reality, AR, AR services, contextual services, CARL.

1 Introduction

Augmented reality (AR) technology enables superimposing rich computer generated content, such as interactive 3D animated objects and multimedia objects, in the real time on a view of a real environment. Augmented reality enables building advanced localised information visualisation systems and therefore forms a solid basis for the development of collective awareness systems.

Widespread use of the AR technology has been enabled in the recent years by remarkable progress in consumer-level hardware performance, in particular, in the computational and graphical performance of mobile devices and quickly growing mobile network bandwidth. Education, entertainment, e-commerce and tourism are notable examples of application domains in which AR-based systems become increasingly used.

There are a variety of tools available for authoring AR content and applications. These tools range from general purpose computer vision and graphics libraries, requiring advanced programming skills to develop applications, to easy-to-use point-and-click packages for mobile devices, enabling creation of simple AR presentations. However, the existing tools are designed for manual authoring of AR presentations – either through programming or visual design. To enable more widespread use of AR

L.M. Camarinha-Matos et al. (Eds.): DoCEIS 2014, IFIP AICT 423, pp. 183–190, 2014.

technology for visualisation of different kinds of up-to-date data, needed in collective awareness systems, a different paradigm is required. The interactive presentations must be created automatically based on the available data sources, through selection of data and automatic composition of AR scenes. A key element enabling selection of information for visualisation is context, which incorporates aspects such as user location, time, privileges, preferences, device capabilities, environmental conditions and others. To enable meaningful contextual selection and automatic composition of non-trivial interactive AR interfaces, semantic web technologies may be used.

In this paper, we present foundations of a novel high-level programming language, called *CARL – Contextual Augmented Reality Language*, which enables building ubiquitous, contextual and interactive AR presentations. In the presented approach, the AR content that is presented to users is created by selecting and merging content and rules originating from different service providers, without the necessity to explicitly switch from one to another. Moreover, in CARL, the rules for tracked markers (where the information will appear), content objects (what will be presented) and interfaces (how the information will be accessible) are separated, enabling flexible composition of presentations meeting real business requirements.

The reminder of this paper is organised as follows. Section 2 presents related works in the context of building AR environments, in particular focusing on the use of AR for building collective awareness systems. Later, in Section 3, the concept of building ubiquitous contextual AR presentations is explained. In Section 4, the CARL language is presented. In Section 5, the current implementation of CARL is outlined. Finally, in the last section, conclusions and directions for future research are presented.

2   AR Environments and Collective Awareness Systems

A number of research works have been devoted to the problem of modelling and building AR environments combining the view of real-world objects with multimedia content. One of well-known toolkits that supports rapid design and implementation of AR applications is DART – Designer's Augmented Reality Toolkit [1]. Many AR systems are based on the ARToolKit library, which however requires advanced programming skills and technical knowledge to create AR applications [2].

GUI-based authoring applications that enable mixing virtual scenes with the real world have been presented in [3-8]. These systems provide user-friendly functionality for placing virtual objects in mixed reality scenes without coding and can be used for building AR environments, but they target only desktop computer environments and do not provide functionality to build open, interoperable and dynamic AR systems.

Since mobile devices have gained popularity, a number of mobile AR toolkits and applications have been developed. For example, ComposAR-Mobile is a cross-platform content authoring tool for mobile phones and PCs based on an XML scene description [9]. Lee and Seo have presented the U-CAFÉ framework, which can be used to build ubiquitous AR environments [10]. It supports contextual presentation and collaboration between users on various devices. However, it does not provide means for combining content from different services and modelling user interactions within the AR presentations.

Stiktu is a mobile AR authoring application that enables attaching ("sticking") simple content, such as text messages and images, to particular views of real places [11]. It enables users to express and exchange opinions with other users who have created virtual objects. The main limitation of this application is the lack of a possibility of creating complex objects, such as interactive 3D models.

Aurasma is an application that enables augmenting arbitrary real-world views using a mobile device's camera and then sharing the created AR content with other users [12]. Layar, Junaio and Wikitude are other examples of AR applications that enable users to share AR content [13-15]. These platforms are good examples of collective awareness systems, where users contribute to "augmenting" the real world with virtual objects. Growing popularity of such applications indicates that there is public interest in new forms of information visualisation, even if the functionality of these systems is still highly limited.

Several studies have been performed in the domain of declarative languages for building virtual environments. InTml is an XML-based language for describing 3D interaction techniques and hardware platforms in VR applications [16]. MPML-VR enables describing multimodal presentations in 3D virtual spaces [17]. VR-BML is a high-level XML-based behaviour programming language developed as a part of the Beh-VR approach [18]. These languages, however, have been designed for 3D/VR applications without taking into account AR-specific functionality.

A number of high-level languages have been also designed specifically for building AR applications. APRIL is an XML-based scripting language for authoring AR presentations within the Studierstube framework [19]. SSIML/AR is a language for structured specification and development of AR applications [20]. AREL is a JavaScript binding of the Metaio/Junaio SDK API allowing development of AR applications based on web technologies, such as HTML5, XML and JavaScript [21]. MR-ISL is a high-level language for defining mixed reality interaction scenarios [22].

A common motivation for developing novel platforms and declarative AR languages is to simplify the process of programming AR applications. However, the existing languages are not intended for building dynamic contextual AR information visualisation systems. In such systems, different types of data must be retrieved in the real time from distributed sources and automatically composed into interactive AR scenes.

3   Contextual Augmented Reality Environments

Existing AR platforms support two main forms of augmentation. In the first form of augmentation – *directional augmentation* – information is associated with points of interest in physical locations in the real world and is presented on mobile devices as directions on how to navigate to these locations (e.g., hotels and restaurants in Wikitude or Junaio). Coordinates of a mobile device are monitored through GPS or other sensors. This form of augmentation is usually functionally and graphically very simple – it rarely goes beyond displaying simple graphical objects with textual descriptions and the content is not geometrically registered in the real environment.

The second form of augmentation – *natural augmentation* – is more powerful. Position and orientation of content to show is identified directly through a device's camera by tracking known markers (patterns) in the captured images. This kind of augmentation can be based on complex interactive 3D content, which can additionally incorporate other multimedia objects, such as images, sounds and videos. The presented content can be geometrically integrated with particular objects in the real environment extending them with presentation of additional virtual objects or information. However, a common problem with the development of systems based on natural augmentation is their limited scope with regards to the number of markers tracked and – consequently – content objects presented. Systems of this type can be successfully used as interactive extensions to printed advertisements or AR additions to video games, but their wider applicability, in particular for building information visualisation and collective awareness systems, is largely limited.

As a solution to this problem, we propose the notion of *CARE – Contextual Augmented Reality Environment* (Fig. 1). CARE is a new approach to building AR applications, which combines the advantages of both directional and natural augmentation systems. In CARE, the AR scenes that are presented to users are based on natural augmentation, but they are dynamically composed in the real time based on a number of data sources and the current context, which includes such elements as location, time, user privileges and preferences, device capabilities and environmental conditions. Contextual approach to AR application development is a necessity to enable access to a variety of data sources, guarantee scalability and provide for seamless operation. In CARE, the same AR application can be used to display elements coming from different content- and service-providers. Moreover, due to the use of semantic linking, content coming from different sources can be geometrically and functionally combined into integrated AR scenes.

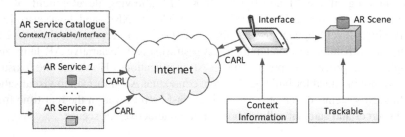

Fig. 1. Architecture of the CARE environment

To enable implementation of CARE environments, a language is required that would enable passing augmentation information from distributed service providers to the AR browser. Such a language must be designed to support dynamic composition of complex interactive AR scenes, and must enable passing information about different entities: *trackable markers*, *content objects* and *interfaces*. In practical applications, these aspects may be separated. For example, trackable markers may come from municipal services, content objects from AR service providers, while interfaces from application developers. Only in such heterogeneous environment, practical ubiquitous AR environments may be realised.

In the next section, we present a language, called CARL – Contextual Augmented Reality Language, designed to meet these criteria.

4 The CARL Language

CARL enables specification of three types of entities that form an AR presentation: *Trackables*, *Content Objects* and *Interfaces*. The independence of trackables, content objects and interfaces in connection with the dynamism of each of these elements requires the use of loose coupling between descriptions of these entities. The problem can be solved by the use of semantic modelling and semantic linking, which is however out of the scope of this paper. Below, the main elements of the CARL language are presented.

4.1 Trackables

The *Trackables* element, shown in Listing 1, groups *Trackable* elements that are used for describing all real-world objects that an application can detect and track in 3D space. Every *Trackable* element has its unique *URI*, which identifies associated binary resource data used in the tracking process. The *Trackable* element has three child elements: *Begin*, *Active* and *End*.

```
<Trackables>
  <Trackable uri="…">
    <Begin>
      <ContentObjectBegin id="1"/>
      <ContentObjectBegin id="2"/>
      <InterfaceMessage value="…"/>
    </Begin>
    <Active>
      <DistanceChanged change="0.1" distanceMin="2" distanceMax="5"
angleMin="30" …>
        <ContentObjectAction id="1" action="show"/>
        <ContentObjectAction id="1" action="playAnimation"/>
      </DistanceChanged>
    </Active>
    <End>
      <ContentObjectEnd id="1"/> <ContentObjectEnd id="2"/>
    </End>
  </Trackable>
<Trackable uri="…" />…
```

Listing 1. Specification of trackables in CARL

In the runtime, when an application detects a trackable object, the *Begin* section is executed. In the presented example, it initializes content objects using the *ContentObjectBegin* command and displays a message using the *InterfaceMessage* command. The *Active* section contains rules of user interaction when the trackable is visible. In the example, the *DistanceChanged* event is detected and appropriate action is performed. The *End* section is executed when the trackable object disappears from the camera view.

4.2   Content Objects

The second element of AR environment specification in CARL are *ContentObjects*, containing multimedia content to be shown to end-users while augmenting real-world objects (Listing 2a). Each *ContentObject* has an *id*, which is used while calling object actions. The *id* can have the form of a fixed literal or a semantic expression to enable loose coupling with other elements of the dynamically created AR scenes.

Content objects support states. The initial object state is specified by the *initialState* attribute. The state can be changed by the use of the *ContentObjectState* command in actions. The *Resources* element contains information about particular *Components* and *Locations* of component resources. Several locations can provide alternative URIs for components meeting particular criteria, e.g., regarding the model complexity or language.

The *Actions* element specifies actions that can be called on this content object. An action is declared within the *Action* element, which has a mandatory *name* attribute and an optional *state* attribute. If the *state* is specified, the action (if called) will be executed only when the object is in the given state. Within the *Action* element, commands based on the VR-BML [18] language are used. These commands can be used to load, manipulate and animate object components, change object state, execute interface operations and control other objects.

```
<ContentObjects>
 <ContentObject id="1" initialState="hidden">
  <Resources>
   <Component id="c1">
    <Location details="low" uri="..." />
    <Location details="high" uri="..." />
   </Component>
  </Resources>
  <Begin/>
  <Actions>
   <Action name="show" state="hidden">
    <SetPosition component="c1" value="..."/>
    <Activate component="c1" active="true">
    <ContentObjectState value="shown"/>
   </Action>
  </Actions>
  <End/>
 </ContentObject>
...
```

```
<Interface>
 <TouchSingle target="…">
  <ContentObjectAction
action="highlight">
 </TouchSingle>
 <TouchSingle target="…">
  <ContentObjectAction
id="1" action="hide">
  <ContentObjectAction
id="2" action="show">
 </TouchSingle>
 <Pan target="…">
  <ContentObjectRotate>
 </Pan>
 <Pinch target="…">
  <ContentObjectScale>
 </Pinch>
</Interface>
```

a)   b)

Listing 2. Specification of content objects (a) and interfaces (b) in CARL

4.3   Interfaces

The third main part of a CARL environment specification are interfaces (Listing 2b). An *Interface* element contains specification of possible interactions on an end-user device. Listing 2b shows an example of an interface specification, in which *touch*, *pan* and *pinch* gestures can trigger actions. An interface can recognize interactions with objects and interactions with trackables. Interaction with objects enables activation and manipulation of objects. Interaction with trackables can be used, e.g., to display objects associated with the tracked markers (e.g., information about a known product).

5   Implementation

CARL has been implemented within the *Mobile Augmented Reality Authoring Tool (MARAT)* for virtual museum exhibitions [23]. Currently the application enables only playback of CARL presentations, but the intention is to implement also content authoring directly on a mobile device within MARAT.

The following figures show examples of CARL usage in the MARAT application. Figure 2a presents an example when an end-user moves his/her finger on a device screen. A *Pan* event triggers an action responsible for object rotation (*ContentObjectRotate* – c.f. Listing 2b). In turn, Figure 2b shows example of using a *Pinch* gesture, which triggers an action responsible for object scaling (*ContentObjectScale*).

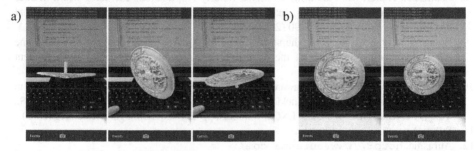

Fig. 2. Triggering the ContentObjectRotate action on a Pan event (a) and the ContentObjectScale action on a Pinch event (b)

6   Conclusions

In this paper, the overall concept of Contextual Augmented Reality Environments and fundamentals of the CARL language, designed to support creation of such environments, have been presented. CARL enables modelling ubiquitous, contextual and interactive AR environments. The language is highly componentised with regards to the structure of the AR scenes as well as the presented content, enabling dynamic composition of complex AR environments. The CARL language can support building ubiquitous AR-based data visualisation and collective awareness systems.

Future research will incorporate several facets. First, the method of semantic coupling of elements of dynamic AR environments will be elaborated. Second, techniques for combining CARL specifications from independent sources will be investigated, with particular focus on security. Finally, methods of automatic creation of CARL specifications through WYSIWYG actions on mobile devices, to enable direct on-site authoring of AR environments, will be designed.

References

1. MacIntyre, B., Gandy, M., Dow, S.: DART: A Toolkit for Rapid Design Exploration of Augmented Reality Experiences. In: Proc. of ACM Symposium on User Interface Software and Technology (UIST), pp. 197–206 (2004)

2. ARToolKit, http://www.hitl.washington.edu/artoolkit
3. Anagnostou, K., Vlamos, P.: Square AR: Using Augmented Reality for urban planning. In: Proc. of the 3rd IC on Games and Virtual Worlds for Serious Appl., pp. 128–131 (2011)
4. Lee, G.A., Kim, G.J., Billinghurst, M.: Immersive authoring: What You eXperience Is What You Get (WYXIWYG). Communications of the ACM 48(7), 76–81 (2005)
5. Wang, M.-J., Tseng, C.-H., Shen, C.-Y.: An easy to use augmented reality authoring tool for use in examination purpose. In: Forbrig, P., Paternó, F., Mark Pejtersen, A. (eds.) HCIS 2010. IFIP AICT, vol. 332, pp. 285–288. Springer, Heidelberg (2010)
6. Seichter, H., Looser, J., Billinghurst, M.: ComposAR: An Intuitive Tool for Authoring AR Applications. In: Proc. of the 7th IEEE/ACM Int. Symp. on Mixed and Augmented Reality (ISMAR), pp. 147–148 (2008)
7. Grimm, P., Haller, M., Paelke, V., Reinhold, S., Reimann, C., Zauner, R.: AMIRE - authoring mixed reality. In: Proc. of the First IEEE International Augmented Reality Toolkit Workshop (2002)
8. Zauner, J., Haller, M.: Authoring of Mixed Reality Applications Including Multi-Marker Calibration for Mobile Devices. In: Proc. of the 10th Eurographics Symposium Virtual Environments (EGVE), pp. 87–90 (2004)
9. Wang, Y., Langlotz, T., Billinghurst, M., Bell, T.: An Authoring Tool for Mobile Phone AR Environments. In: Proc. of the New Zealand Computer Science Research Student Conference, pp. 1–4 (2009)
10. Lee, J.Y., Seo, D.W.: A context-aware and augmented reality-supported service framework in ubiquitous environments. In: Enokido, T., Yan, L., Xiao, B., Kim, D.Y., Dai, Y.-S., Yang, L.T. (eds.) EUC-WS 2005. LNCS, vol. 3823, pp. 258–267. Springer, Heidelberg (2005)
11. Stiktu, http://stiktu.com/
12. Aurasma, http://www.aurasma.com/
13. Layar, https://www.layar.com/
14. Junaio, http://dev.metaio.com/junaio/
15. Wikitude Developer – Devzone, http://developer.wikitude.com/
16. Figueroa, P., Green, M., Hoover, H.J.: InTml: A Description Language for VR Applications. In: Proc. of Web3D 2002 Symposium, pp. 53–58. ACM, New York (2002)
17. Okazaki, N., Aya, S., Saeyor, S., Ishizuka, M.: A Multimodal Presentation Markup Language MPML-VR for a 3D Virtual Space. In: Workshop Proc. on Virtual Conversational Characters: Applications, Methods, and Research Challenges (in conj. with HF 2002 and OZCHI 2002), Melbourne, Australia (2002)
18. Walczak, K.: Beh-VR: Modeling Behavior of Dynamic Virtual Reality Contents. In: Zha, H., Pan, Z., Thwaites, H., Addison, A.C., Forte, M. (eds.) VSMM 2006. LNCS, vol. 4270, pp. 40–51. Springer, Heidelberg (2006)
19. Ledermann, F., Schmalstieg, D.: APRIL: a high-level framework for creating augmented reality presentations. In: Proceedings of IEEE Virtual Reality 2005, pp. 187–194 (2005)
20. Vitzthum, A., Hussmann, H.: Modeling augmented reality user interfaces with SSIML/AR. J. Multim. 1(3), 13–22 (2006)
21. AREL - Augmented Reality Experience Language, http://dev.metaio.com/arel/overview/
22. Walczak, K., Wojciechowski, R.: Dynamic Creation of Interactive Mixed Reality Presentations. In: Proc. of the ACM Symposium on Virtual Reality Software and Technology, pp. 167–176. ACM (2005)
23. Rumiński, D., Walczak, K.: Creation of Interactive AR Content on Mobile Devices. In: Abramowicz, W. (ed.) BIS Workshops 2013. LNBIP, vol. 160, pp. 258–269. Springer, Heidelberg (2013)

Part VIII
Robotics and Mechatronics

On the Design of a Robotic System Composed of an Unmanned Surface Vehicle and a Piggybacked VTOL

Eduardo Pinto[1,2], Pedro Santana[3,4], Francisco Marques[1,2], Ricardo Mendonça[1,2], André Lourenço[1,2], and José Barata[1,2]

[1] Universidade Nova de Lisboa, Faculdade de Ciências e Tecnologia, Departamento de Engenharia Electrotécnica, 2829-516 Quinta da Torre, Portugal
[2] Uninova – Instituto de Desenvolvimento de Novas Tecnologias, Faculdade de Ciências e Tecnologia, Edifício Uninova, 2829-516 Quinta da Torre, Portugal
[3] Instituto Universitário de Lisboa (ISCTE-IUL), Departamento de Ciências e Tecnologias da Informação, Avenida das Forças Armadas, 1649-026 Lisboa, Portugal
[4] Instituto de Telecomunicações, Lisboa, Portugal

Abstract. This paper presents the core ideas of the RIVERWATCH experiment and describes its hardware architecture. The RIVERWATCH experiment considers the use of autonomous surface vehicles piggybacking multi-rotor unmanned aerial vehicles for the automatic monitoring of riverine environments. While the surface vehicle benefits from the aerial vehicle to extend its field of view, the aerial vehicle benefits from the surface vehicle to ensure long-range mobility. This symbiotic relation between both robots is expected to enhance the robustness and long lasting of the ensemble. The hardware architecture includes a considerable set of state-of-the-art sensory modalities and it is abstracted from the perception and navigation algorithms by using the Robotics Operating System (ROS). A set of field trials shows the ability of the prototype to scan a closed water body. The datasets obtained from the field trials are freely available to the robotics community.

Keywords: cooperative robots, unmanned aerial vehicles, UAV, autonomous surface vehicles, ASV, environmental monitoring, riverine environments.

1 Introduction

The monitoring of riverine and maritime environments has been shown in the last years as one of the most important activities to reveal the negative impact of human activities in Nature [1], [2], [3], and increase the awareness of people regarding climatic changes. The unattractiveness and difficulty of monitoring remote aquatic environments by humans render the automation of the process highly valuable.

A fine spatiotemporal mapping of environmental variables across extensive water bodies hampers the application of typical fixed sensor networks. Furthermore, such a solution is unable to return samples for laboratorial analysis, which is a requirement for several monitoring procedures. Several projects had dealt with these limitations by

L.M. Camarinha-Matos et al. (Eds.): DoCEIS 2014, IFIP AICT 423, pp. 193–200, 2014.

introducing autonomous surface vehicles [4], [5], [6]. These projects, however, are still unable to cope with inherent limitations of a surface-level perspective of the environment. The major limitation concerns the ability to properly perceive the far field, which ultimately limits the autonomy of the vehicle when facing cluttered and shallow water bodies. Previous work capitalized on the benefits of multi-robot systems to overcome this problem by relying on a helicopter to provide the human operator with improved situation awareness [6]. The RIVERWATCH[1] experiment, a part of the EU funded FP7 project ECHORD[2], takes this idea up to the next level, i.e., without demanding the presence of a human operator. Concretely, RIVERWATCH considers an autonomous surface vehicle (ASV) with a multi-rotor unmanned aerial vehicle (UAV) piggybacked, which is deployed when a higher vantage point is required. Hence, the surface vehicle benefits from the aerial vehicle to extend its field of view. Conversely, the aerial vehicle benefits from the surface vehicle to ensure long-range mobility. Although the use of aerial images to help surface-level navigation has been explored in a parallel work, the actual acquisition process has not been considered [8].

This paper provides both an overview of the RIVERWATCH experiment and the details of the practical aspects related to the hardware and software architectures. Details regarding high-level software components, such as those enabling perception and safe navigation, will be published elsewhere.

2 Relationship to Collective Awareness Systems

We consider that the perception of the environment shared by both surface and aerial vehicles implements a form of collective awareness. The environment monitoring products generated by the system contribute to an ecological global awareness.

3 The RIVERWATCH Architecture

Fig. 1 depicts the major computational and communications hardware components selected for the RIVERWATCH system. These have been selected in order to ensure large computational and communications capacity and, thus, openness to future demands. From the software perspective, this openness is ensured by the use of the Robotics Operating System (ROS) as backbone. ROS[3] offers a cross-language inter-process communication system and several state-of-the-art software components freely available.

---

[1] RIVERWATCH homepage: `http://riverwatchws.cloudapp.net`
[2] ECHORD homepage: `http://www.echord.info/`
[3] ROS homepage: `http://wiki.ros.org`

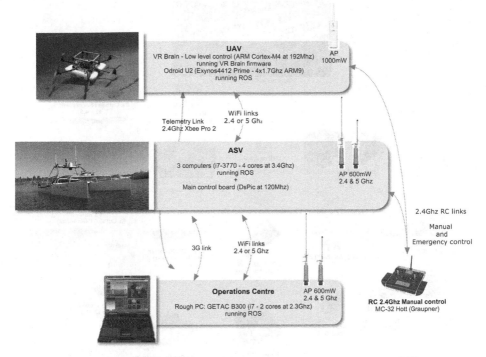

Fig. 1. Computational and communications hardware in RIVERWATCH

3.1   The Multi-rotor UAV

Several commercial multi-robot aerial platforms had been evaluated, namely, AirRObot, Asctec, Microdones, Draganfly, and Cyberquad. These products are patented and closed, which limits considerably their interest for research activities. This issue can be avoided by recurring to open source solutions, such as MikroKopter, Mikro's Aeroquad, NG-UAVP, UAVX, UAVP, OpenPilot, Arducopter, Multipilot 32, and VBrain. From these, the VBrain commercialized by the Italian company Virtualrobotix[4], was selected as the base for the RIVERWATCH's aerial platform. This solution is fully open sourced, including the mission control software and hardware, and it is provided with the most powerful microcontroller from all the evaluated possibilities.

Regarding the mechanical design, a six-rotor configuration was preferred to the common four-rotor solution. First, a six-rotor solution, a *hexacopter*, provides more lifting capability. Second, it endows the system with graceful degradation, as it is able to land with one motor off, without yaw control though. Moreover, it may still be able to fly with two motors off, provided that they stand on the neutral torque bar. Fig. 2, illustrates the designed hexacopter and its supporting hardware architecture.

---

[4] http://www.virtualrobotix.com/

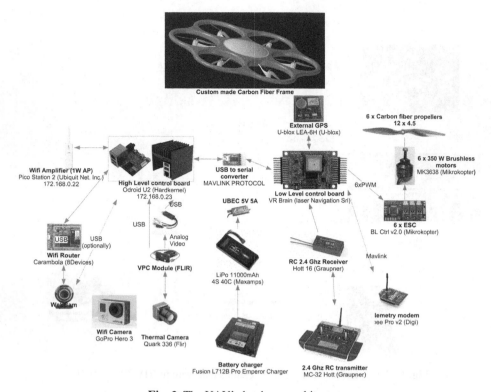

Fig. 2. The UAV's hardware architecture

4   The ASV

The ASV is based on a 4.5 m *Nacra* catamaran, which has received special carbon fibre reinforcements for the roll bars and motor supports. The hulls have been filled with special PVC closed cells foam, making it virtually unsinkable. The trial tests here presented have been done with the propulsion in a differential locomotion configuration using a Haswing Protuar 2 Hp motor in each hull. The motors can be driven manually for safety reasons and if required for debugging purposes. A docking station with an H-marked for facilitated detection and tracking from aerial images taken by the UAV was fitted to the ASV's deck. A net was set around the docking station as lateral protections to the UAV when docked. Fig. 3 depicts the ASV's hardware architecture. Aiming at autonomous behaviour, the ASV is equipped with a set of state-of-the-art navigation-related sensors. Concretely, it is equipped with a GPS-RTK (Proflex 800 from Ashtec SAS), an IMU (PhidgetSpacial 3 axis from Phidgest Inc), a long range tilting laser scanner (LD-LRS2100 from Sick), a fixed underwater sonar (DeltaT 837B from Imagenex), and a multi-camera vision system (Ladybug3 from Pointgrey).

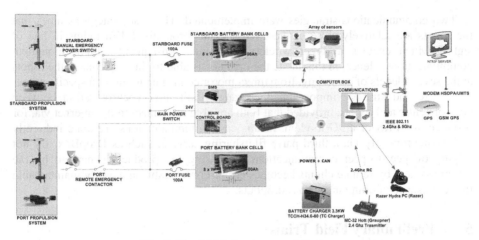

Fig. 3. The ASV's hardware architecture

4.1   The Operations Control Centre

The operations control centre is a web-based application that allows the remote operator to perform offline missions planning and online missions execution monitoring. This section describes briefly the hardware and software aspects related to the maintenance of the communication channels between control centre and robots (see Fig. 4).

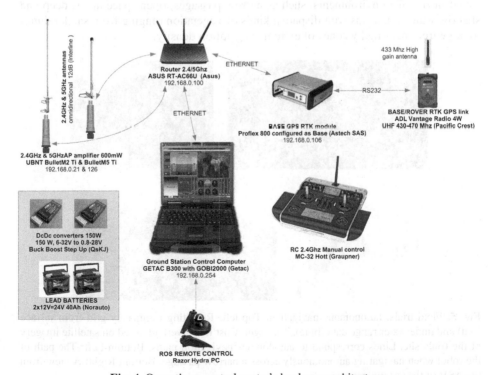

Fig. 4. Operations control centre's hardware architecture

Two communications strategies were implemented. The first strategy assumes that the robots are relatively close to the control centre (roughly 1 Km radius in line of sight), which enables a direct wireless link and, consequently, high throughput communications. Hence, in this mode the operator is able to directly control the robots at the several levels of autonomy, from direct motor control up to mission specification.

The second strategy imposes neither range limit nor synchronisation between control centre and robots, provided that both are able to connect to the internet via, for instance, a GPRS or a 3G modem. In this case communications are done indirectly and asynchronously via a third-party file sharing service, such as DropBox. By not relying on peer-to-peer communications, accessing the products generated by the robotic system by remote clients becomes easily configurable and easily maintainable in the presence of communications dropouts.

5   Preliminary Field Trials

To validate the proposed system, a large set of field trials was carried out with the robots in autonomous and tele-operated modes. The autonomous mode is based on a set of navigation and coordination algorithms. The goal of the tests, in what regards this paper, is to show that the hardware architecture and the robots mechanical design fits the purpose of robust navigation in riverine environments. The tests were carried out in a private lake in the Sesimbra region in Portugal, with an area of roughly 1.5 Km^2. This site offered in a single place most of the environmental traits that can be found in riverine environments, such as narrow passages, open space areas, deep and shallow waters, margins with disparate kinds of vegetation ranging from sander dunes to large trees passing by zones of extreme vegetation density.

Fig. 5. Field trials: autonomous navigation. Top-left: Resulting occupancy grid from surface (red) and underwater range data (blue). Top-right: Cost map overlaid in red on satellite imagery of the trials site. Lines correspond to autonomous navigation paths. Bottom-Left: The path of the robot when navigating autonomously across a narrow passage. Bottom-Right: A view from the ASV of the narrow passage.

Fig. 5 depicts several results obtained throughout the field trials. It highlights the ability of the described system to provide the reliability necessary for the perceptual and navigation algorithms to autonomously scan the environment that will be published in future articles. Fig. 6 also depicts a situation in which the UAV is taking off the ASV. All sensory data produced throughout the field trials are publicity available as ROS-enabled log files.

Fig. 6. Field trials: UAV taking off the ASV. Left: moment right after the taking off takes place. Right: A few moments after the take off onset.

6   Conclusions and Further Work

Although the use of heterogeneous robots working as a collective is an old idea, only in RIVERWATCH it has been realised in the context of aquatic-aerial robotic teams for environmental monitoring. This approach serves the purpose of enabling long-lasting robust operation. Long lasting from using the ASV as energy supplier and robust from using the UAV to provide far field navigation cost information. This paper focused on the hardware and technological aspects of the RIVERTWATCH experiment with the expectation of fostering the development of new prototypes for the execution of related experiments. Our future work will be centred in making RIVERWATCH a full autonomous system, and so is necessary to solve problems as: Energy harvesting by the ASV and charge of the UAV; Full tests and evaluation of the landing algorithms; Full characterization of the innovative dual propulsion system; Consider seriously the development of a smart "catch" mechanism that is vital for safe landings in adverse meteorological conditions.

Acknowledgments. This work was co-funded by EU FP7 ECHORD project (grant number 231143) and by the CTS multi-annual funding, through the PIDDAC Program funds.

References

1. Hawbecker, P., Box, J.E., Balog, J.D., Ahn, Y., Benson, R.J.: Greenland outlet glacier dynamics from Extreme Ice Survey (EIS) photogrammery, American Geophysical Union, San Francisco (2010)
2. Church, J.A., White, N.J.: A 20th century acceleration in global sea level rise. Geographic Research Letters 33(1) (2006)

3. Allison, I., et al: The Copenhagen Diagnosis - Updating the World on the Latest Climate Science. The University of New South Wales Climate Change Research Centre (CCRC), Elsevier, Sydney (2011)
4. Sukhatme, G.S., Dhariwal, A., Zhang, B., Oberg, C., Stauffer, B., Caron, D.A.: Design and development of a wireless robotic networked aquatic microbial observing system. Environmental Engineering Science 24(2), 205–215 (2007)
5. Bhadauria, D.I. (s.d.): A Robotic Sensor Network for monitoring carp in Minnesota lakes. In: IEEE International Conference on Robotics and Automation (ICRA), pp. 3837–3842. IEEE Press, Anchorage (2010)
6. Tokekar, P., Bhadauria, D., Studenski, A., Isler, V.: A robotic system for monitoring carp in Minnesota lakes. Journal of Field Robotics 27(6), 779–789 (2010)
7. Murphy, R.R., Steimle, E., Griffin, C., Cullins, C., Hall, M., Pratt, K.: Cooperative use of unmanned sea surface and micro aerial vehicles at Hurricane Wilma. Journal of Field Robotics 25(3), 164–180 (2008)
8. Heidarsson, H.K., Sukhatme, G.: Obstacle detection from overhead imagery using self-supervised learning for autonomous surface vehicles. In: 2011 IEEE/RSJ International Conference on Intelligent Robots and Systems (IROS), pp. 3160–3165. IEEE (2011)

Tracking a Mobile Robot Position Using Vision and Inertial Sensor

Francisco Coito, António Eleutério, Stanimir Valtchev, and Fernando Coito

Faculdade de Ciências e Tecnologia – Universidade Nova de Lisboa,
2829-516 Caparica, Portugal
francisco.coito@sapo.pt, a.eleuterio@campus.fct.unl.pt,
{ssv,fjvc}@fct.unl.pt

Abstract. Wheeled mobile robots are still the first choice when it comes to industrial or domotic applications. The robot's navigation system aims to reliably determine the robot's position, velocity and orientation and provide it to control and trajectory guidance modules. The most frequently used sensors are inertial measurement units (IMU) combined with an absolute position sensing mechanism. The dead reckoning approach using IMU suffers from integration drift due to noise and bias. To overcome this limitation we propose the use of the inertial system in combination with mechanical odometers and a vision based system. These two sensor complement each other as the vision sensor is accurate at low-velocities but requires long computation time, while the inertial sensor is able to track fast movements but suffers from drift. The information from the sensors is integrated through a multi-rate fusion scheme. Each of the sensor systems is assumed to have it's own independent sampling rate, which may be time-varying. Data fusion is performed by a multi-rate Kalman filter. The paper describes the inertial and vision navigation systems, and the data fusion algorithm. Simulation and experimental results are presented.

Keywords: Mobile robotics, multi-rate sampling, sensor fusion, vision, inertial sensor.

1   Introduction

The number of industrial mobile robots applications has increased significantly in the last decades and its use in other indoor environments also shows a growing interest. Due to its simple mechanical structure and agility wheeled robots are still the first choice when it comes to industrial or domotic applications. Knowing exactly where a mobile entity is and monitoring its trajectory in real-time is an essential feature. Most applications, however, still rely on a more or less complex infrastructure to fulfill this task and that reduces the robot's autonomy [1-3].

The guidance system for a mobile robot is usually composed by three modules: trajectory generation and supervision system, control system and navigation system. The navigation system aims to determine reliable information on the robot's position, velocity and orientation and provide it to the other modules. Knowing the location of

L.M. Camarinha-Matos et al. (Eds.): DoCEIS 2014, IFIP AICT 423, pp. 201–208, 2014.

an autonomous mobile equipment and monitoring its trajectory in real-time is still a challenging problem that requires the combined use of a number of different sensors. The most frequently used sensors are inertial measurement units (IMU) [1-8]. With inertial sensors the robot's position and velocity can be computed by integrating the acceleration measurements given by the sensors, but the position and the orientation errors grow over time as a result from the accumulation of the measurements' noise and bias. As the computation position from acceleration requires two integration steps, the position error grows much faster than the error of orientation, which requires only one integration step. Thus, when an IMU is used for potion measurement it is usually combined with another position measurement system with lesser drifting error.

To tackle with this task a number of solutions are proposed in the literature. Some authors use multiple IMUs or combine them with electromechanical odometers [2,7]. As all the sensors are prone to drift errors the problem is not eliminated but the precision is improved. The use of GPS as an absolute positioning system is also proposed [5,8], however this is not an option for indoor operation. A number of solutions for indoor applications use active beacons as a means to determine the absolute position of the robot [1-3]. This type of approach requires the existence of a fixed infrastructure that involves additional cost and reduces the robot's autonomy.

1.1   The Proposed System

In this work we propose the use a vision based system in combination with the inertial system to determine the robots position and velocity. The use of vision together with inertial sensors has been used by other authors [4,6]. Most systems consider the use of one or more cameras with known positions on the robot's environment. An IMU attached to a moving camera is proposed in [6] to determine the grip position of an industrial robotic manipulator. Experimental results also make use of measurements from mechanical odometers included on the robot traction system.

1.2   Multi-rate Sensor Fusion

When fusing the information from multiple sensor systems each senor presents a different time delay. While inertial systems are fast, allowing sample rates in the order of several kHz at least, when vision algorithms are involved the computation time associated with feature evaluation and coordinate computation causes a significantly larger time delay. As a result the vision system position estimates are delayed, both to the robot real position, as well as to the IMU's information.

In summary, while inertial measurements are frequent and accurate on the short term, but present drift, vision information has long (variable) time delays, which renders them inappropriate for driving control purposes. Thus, the two systems are complementary to each other and the fusion of their information can improve the system overall performance.

The integration of the IMU and vision systems is done by meas of a Kalman filter (KF). As the KF is not computationally intensive, it is easily applied in real-time applications such as inertial navigation systems.

2   Contribution to Collective Awareness Systems

The impact information technology on every day's life is enormous and has jumped from plants to offices and the household. Most equipment embeds one or more computers and the ability to access and use web based information is increasing daily. However, the equipment ability to integrate the human environment is still limited by their lack of mobility. Households are unstructured environments and, in spite the progress in the mobile robotics area, this is still a challenge for the autonomy of mobile equipment and the ability to move around freely without supervision.

3   Position Estimation

In this work a wheeled robot is considered, moving on an almost plane horizontal surface, located indoors.

3.1   The Inertial Measurement System

The mobile robot dynamics is represented in two coordinate frames: the robot frame and the world frame. The robot frame is fixed onto the moving robot, usually with the origin at the center of mass, which is assumed to lie between the wheels. The world frame is fixed on the field. Fig. 1 illustrates how the two frames are related.

The IMU sensor is placed at the center of the robot. The gyroscope is aligned with the vertical axis. The position of the robot is given by:

$$
\begin{bmatrix} \dot{x} \\ \dot{y} \end{bmatrix} = \begin{bmatrix} \cos(\theta(t)) & \sin(\theta(t)) \\ -\sin(\theta(t)) & \cos(\theta(t)) \end{bmatrix} \begin{bmatrix} v_L \\ v_T \end{bmatrix},
\tag{1}
$$

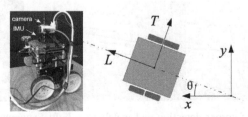

Fig. 1. Left: Mobile robot. Right: Coordinate frames: $<x, y>$ world frame; $<L, T>$ robot frame; θ orientation.

Where <x, y> is the robot position on the world frame; <v_L, v_T> are the longitudinal and transversal robot speed on it's own frame and θ is the orientation.

The values for θ v_L and v_T result from the integration of the measured rotation rate and translational accelerations. In order to minimize the error these integral values are computed through a KF.

3.2   The Vision System

The vision based odometry system uses the FAST corners method [9] to extract features from each image. Two consecutive images are compared and their features are matched. From this comparison the camera displacement is computed through a kinematic transformation of a frame onto the next.

The FAST corner features are referred to as corners. Each corner corresponds to a central pixel location that, on its surrounding circle of pixels (with a certain diameter), contains an arc of N or more contiguous pixels that are all much brighter than the central pixel. A corner is also found if the all the pixels on the arc are much darker than the center. Each corner on a frame is characterized by a set of features such as coordinates, pixel arc size, the arc average color and corner color.

In order to find a match between two corners from different images, all the corners in both images are compared. The comparison takes into account all the above features, as well as each corner signature. The signature of a corner in an image is the set of distances between that corner and every other corner in the same image. This feature is computationally burdensome, but the probability of two corners having a similar signature is small.

In the comparison each feature is weighted by the reverse of the frequency of occurrence. Hence, the features that are less probable to occur in different corners are given a higher relevance for corner comparison. When the same corner is found in two different images it is said to be a matching. Two matching (two points) allows the definition of corresponding coordinate frames in each of the images.

Experimental results are presented in Fig. 2. The displacement of the coordinate frame between the two images allows computing the camera movement. To increase the precision the frame displacement is computed as an average from several matching. By calculating the motion of the robot between frames, a track of the motion can be drawn.

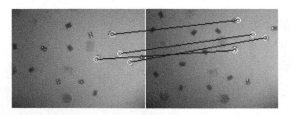 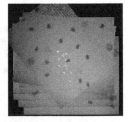

Fig. 2. An example of corner detection and matching – Left: Sequence of two images taken from a ceiling. All FAST corners feature points are marked as white points. Those in the middle of white circumferences are chosen for matching between frames, and black lines point out corner matching pairs. Right: Mapping of the robot motion.

Fig. 3 represents schematically the vision system geometry. As proposed by [10] the camera is located on top of the robot directed to the ceiling. It is aligned with the middle of the wheel axis line (where the IMU is also located). A known limitation stems from the fact that the pixel displacement depends not only on the robot movement, but also on the distance to the ceiling (*D*). As seen in the figure, there is a linear relation between the distance D (from the robot to the ceiling) and the camera generic field of view A. Hence, it is possible to calculate A by knowing D.

At present it is assumed that this value is known in advance. In the future the camera will be integrated with a distance sensor. Precision is also affected if the camera angle is not precisely perpendicular to the ceiling. This is not a limitation as the robot is aimed to work on a horizontal surface. The use of a 3-axis accelerometer-gyroscope IMU allows the detection of strong disturbances and the correction of constant displacement angles.

In order to minimize error accumulation, the first image of the set used to measure the robot's movement is kept as long as the number of matchings is large enough to ensure precision. Only when that fails to happen a new initial image is selected.

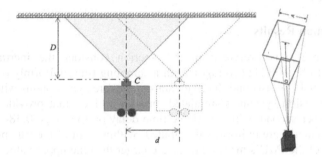

Fig. 3. Vision system image acquisition geometry. C: camera; D: distance from the target (ceiling); d: robot displacement.

3.3   Multi-rate Sensor Fusion

The two sensor systems present very different sampling rates. The vision system's sampling period is long and changes randomly. Due to its computation time the position measurements are time delayed. The proposed fusion scheme uses a discrete multi-rate KF [4] designed to tackle with these characteristics.

The state estimation is based on a discrete-time state-space description of the system dynamics:

$$\begin{aligned} \mathbf{Z}_{k+1} &= \mathbf{A}_k\,\mathbf{Z}_k + \mathbf{G}_k\,n_k, \\ \mathbf{Y}_k &= \mathbf{C}_k\,\mathbf{Z}_k + w_k \end{aligned} \qquad (2)$$

where $\mathbf{Z}=[v_x \ x \ v_y \ y \ \theta]^\mathrm{T}$ is the system state; Y_k is the vector of measured values at time k; n_k and w_k are considered to be noise. The time varying matrices are:

$$A_k = \begin{vmatrix} 1 & 0 & 0 & 0 & 0 \\ \Delta t_k & 1 & 0 & 0 & 0 \\ 0 & 0 & 1 & 0 & 0 \\ 0 & 0 & \Delta t_k & 1 & 0 \\ 0 & 0 & 0 & 0 & 0 \end{vmatrix} \quad G_k = \begin{bmatrix} 1 & 0 & 0 \\ 0 & 0 & 0 \\ 0 & 1 & 0 \\ 0 & 0 & 0 \\ 0 & 0 & 1 \end{bmatrix}, \tag{3}$$

The sampling period Δt_k is the interval between consecutive data fusion moments, and may not be constant. As in [4], at each fusion moment the set of available measured signals depends on their own sampling rates, therefore the output equations may vary because only available measurements are considered. To use the position data provided by the vision system it is necessary to take into account the computation delay. As proposed by [11], model (2) is used to embed the time delay of the vision measurements in the multi-rate KF.

4   Results

4.1   Simulation Results

The simulation results presented in this section consider the inertial system's sampling rate to be 100 Hz (average), with a sampling time uniformly distributed in the range from 9-11 milisecond. Inertial data samples are synchronous with the fusion moments. The vision system is simulated through a block that provides 2 position measurements per second, with a variable time delay on the range 0.38-0.42 second. Vision measurements are uniformly distributed within 5 cm of the true position. The noise and bias of the IMU's measurements are larger than the upper values taken from the manufacturer datasheet. For simulation purposes image position information is assumed to lie within ±3cm from the real position. The robots initial position and orientation are assumed as the origin of the real world coordinate frame. The trajectory is performed at a constant linear speed of 0.04 m/s. Different rotation rates are used along the test. Fig. 4 presents the simulated results. The robot's trajectory is tracked without significant bias. If there is any bias it does not increase with time. The position error is under 3 cm over most of the test (see Fig. 5). The deviation of the position measurement increases only when the robot performs a fast turn, with a small radius. The error is strongly affected by the vision system sample rate and delay. A faster vision system leads to smaller errors.

4.2   Experimental Results

The mobile robot from Fig. 1 is used to develop the position measurement system.

The measurement unit couples an inertial sensor (IMU) and upward directed camera positioned on top of the IMU. Both sensors are vertically aligned. The robot has two mechanically connected wheels on each side. The measurement of the wheel speed for both sides of the robot is available. The robot is coupled with a device to graphically mark the performed path.

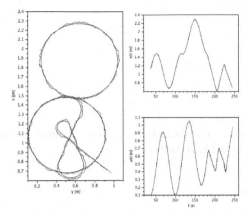

Fig. 4. Simulation results: Robot position vs. measured position. Left: Robot movement and measured position. Right: x and y measurements and true values.

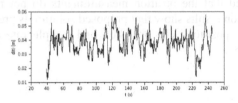

Fig. 5. Simulation results: Position error (in meters)

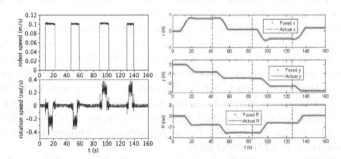

Fig. 6. Experimental results - Left: Raw measurement data. Right: x, y and θ estimation and actual values; vertical lines indicate when absolute position measurement are obtained.

At the present stage the robot's absolute position is determined only when it stops at a still position. In Fig. 6 are presented experimental results obtained with the position measurement system. The robot performs a trajectory with four sections. Within each section the robot moves with a linear speed of 10 cm/s and turns with an angular speed of 0.2 rad/s (approx.). At the end of a section the robot stops and absolute position is determined. A Kalman filter is used to estimate the robot's position (x, y) and heading (θ) from the IMU and wheel speed measurements. The estimates are computed every 25 ms. Absolute position measurements are also included in the estimation process when they are available. The time gap between these measures is of 40 s (approx.).

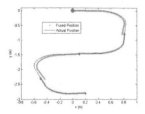

Fig. 7. Experimental results - Line: Robot position. Dot: position estimate. Cross: Absolute position measurement. Circle: Initial position.

5   Conclusions

The paper addresses the problem of measuring the position and speed of an autonomous mobile robot for indoor applications. In order to enhance the robot autonomy it is assumed that the position measurements do not require an external infrastructure. Simulation results show that proposed solution is presents a viable way to determine the robot position. The precision and sample rate adequate for most control applications.

References

1. Kim, S.J., Kim, B.K.: Dynamic Ultrasonic Hybrid Localization System for Indoor Mobile Robots. IEEE Trans. Ind. Elect. 60(10), 4562–4573 (2013)
2. Lee, T., Shirr, J., Cho, D.: Position Estimation for Mobile Robot Using In-plane 3-Axis IMU and Active Beacon. In: IEEE Int. Symp. Ind Electronics, pp. 1956–1961. IEEE Press (2009)
3. Tennina, S., Valletta, M., Santucci, F., Di Renzo, M., Graziosi, F.: Minutolo.: Entity Localization and Tracking: A Sensor Fusion-based Mechanism in WSNs. In: IEEE 13th Int Conf. on Digital Object Identifier, pp. 983–988. IEEE Press, New York (2011)
4. Armesto, L., Chroust, S., Vincze, M., Tornero, J.: Multi-rate fusion with vision and inertial sensors. In: IEEE Int. Conf. on Robotics and Automation, pp. 193–199. IEEE Press (2004)
5. Bancroft, J.B., Lachapelle, G.: Data Fusion Algorithms for Multiple Inertial Measurement Units. Sensors 11(7), 6771–6798 (2011)
6. Hol, J., Schön, T., Luinge, H., Slycke, P., Gustafsson, F.: Robust real-time tracking by fusing measurements from inertial and vision sensors. J. Real-Time Image Proc. 2(3), 149–160 (2007)
7. Cho, B.-S., Moon, W.-S., Seo, O.-J., Baek, K.-R.: A dead reckoning localization system for mobile robots using inertial sensors and wheel revolution encoding. Sensors 11(7), 6771–6798 (2011), J. Mech. Sc. Tech. 25-11, pp. 2907-2917 (2011)
8. Marín, L., Vallés, M., Soriano, Á., Valera, Á., Albertos, P.: Multi Sensor Fusion Framework for Indoor-Outdoor Localization of Limited Resource Mobile Robots. Sensors 13(10), 14133–14160 (2013)
9. Rosten, E., Porter, R., Drummond, T.: Faster and Better: A Machine Learning Approach to Corner Detection. IEEE Tans. Pattern Anal. Mach. Int. 32(1), 105–1019 (2010)
10. Lucas, A., Christo, C., Silva, M.P., Cardeira, C.: Mosaic based flexible navigation for AGVs. In: IEEE Int Symp. Industrial Electronics, pp. 3545–3550. IEEE Press (2010)
11. Hu, Y., Duan, Z., Zhou, D.: Estimation Fusion with General Asynchronous Multi-Rate Sensors. IEEE Trans. Aerosp. El. Sys. 46(4), 2090–2102 (2010)

A Shell-Like Induction Electrical Machine

João F.P. Fernandes and P.J. Costa Branco

LAETA/IDMEC, Instituto Superior Técnico,
Universidade de Lisboa, Lisbon, Portugal
{joao.f.p.fernandes,pbranco}@ist.utl.pt

Abstract. This paper proposes to recover the concept of spherical induction electrical machines to conceive a *shell-like* actuator with multi-DOF (Degrees-Of-Freedom). The actuator is formed by a *shell* stator and a spherical rotor. This work contains the feasibility study of that solution when applied as an active joint actuator in assistive devices. Its electromechanical characteristics are first analyzed using an analytic model that includes: the distribution of the magnetic potential vector and thus the components of the magnetic flux density in the airgap due to a sinusoidal current distribution imposed in the stator; the model also shows the induced electromotive forces and associated current density distribution in the rotor; and at last the radial and tangential components of the force density in the rotor. The *shell-like* actuator is concretized as an active joint for assisting movement of the lower leg of a typical 70kg person. Based on its requirements, the joint actuator electromechanical characteristics are analyzed according to its sensitivity to a set of electrical and mechanical variables.

Keywords: Electromechanic energy conversion, Electric machines, spherical induction motors, assistive devices.

1 Introduction

Nowadays, multi-DOF motion devices have been mostly formed by a composition of classic electric motors. While today multi-DOF manipulators provide enough accurate motion, the type and number of joints used means a heavy, bulky and expensive structure. Particularly in the area of assistive devices, as active joints [1], and in some industrial applications [2, 3, 4], these manipulators have to be compact, light, with a simple structure and also competitive cost.

This research recovers the concept of classic induction machines to conceive an innovative *shell-like* spherical actuator which allows multi-DOF motion while reducing the complexity of the solution. This work model and analyses its electromechanical characteristics and feasibility as a multi-DOF actuator to operate as an active joint assisting the movement of the lower leg of a typical 70kg person. The key characteristics to be analyzed due to demanded requirements are the amplitude and range of the density forces induced in the rotor and the range of its angular speed.

The paper is divided in the following three steps: *the concept* where the design is explained; *the analytical model* with the statement of its assumptions and computation of its electromechanical characteristics; and in the end the *assistive device* application feasibility and comparison of the analytical and its finite element simulation.

L.M. Camarinha-Matos et al. (Eds.): DoCEIS 2014, IFIP AICT 423, pp. 209–216, 2014.
© IFIP International Federation for Information Processing 2014

Figures 1 (a) and 1 (b) show the origin of the *shell-like* induction machine concept. The shoulder joint in Fig. 1(a) is multi-axial and possesses the greatest range of motion of any human joint. Meanwhile, the hip joint of the anterior femur in Fig. 1 (b), it presents a shallower shell providing a lower range of motion. A geometrical representation of these joints is shown in Fig. 2. Figure 3 illustrates the conversion of the biomechanical nature to a spherical and *shell-like* induction machine.

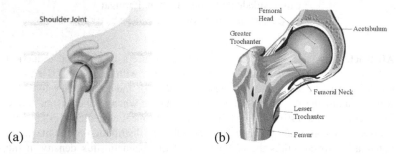

(a)   (b)

Fig. 1. Main joints of human body. (a) Shoulder joint, (b) hip joint.

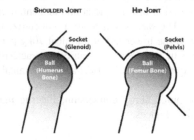

Fig. 2. Large and short range motion due to variable joint socket areas

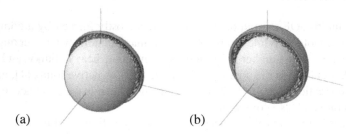

(a)   (b)

Fig. 3. (a) *Shoulder joint* to a shoulder-like spherical actuator. (b) *Hip joint* to a hip-like spherical actuator.

2   Contribution to Collective Awareness Systems

Although the main focus application is in assistive devices, i.e., in *biomedical engineering* systems, its applications can be easily extended to *robotics and integrated manufacturing* systems [2, 3, 4] due to its 3DOF and compact solution, and also used to electric motion devices, as an active wheel [5].

In industrial and motion applications where 3DOF are required, this solution presents a simple control due to its reduced complexity, contributing to the simplification of the *control and decision, electronics* and *signal processing systems*.

3   The Concept of a *Shell-Like* Induction Electrical Machine

The *shell-like* design is inspired on the main joints of the human body as shown in Section 1. The stator acts like a socket where the spherical rotor fits, with its periphery defining the range of motion. Figures 2 and 3 illustrated the concept behind the *shell-like* actuator as result of a high range and medium range joints.

The machine uses the same concept of classic induction motors to produce an electromagnetic torque between rotor and stator parts. Figure 4 indicates that the stator is formed by an outer ferromagnetic material layer and an inner layer composed by a distribution set of slotless coils. These will create a travelling magnetic flux density wave that will be commanded to generate an electromagnetic torque in any 3D direction.

The rotor has two layers: the outer one using an electric conductive material and an inner layer of a high magnetic permeability material.

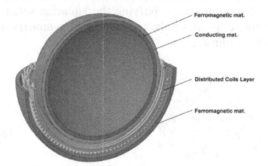

Fig. 4. Shell-like induction machine design

4   Analytical Model

Figure 5 shows a schematic for the analytical model, which was formulated under the following assumptions: one single homogeneous zone (airgap); a thin current density layer defined as the inner stator surface; a thin electric conducting material layer (σ – electrical conductivity) defined as the outer surface of the rotor; a high permeability ferromagnetic material on the stator and also in the rotor; and negligence of border effects due to the small size of the airgap in relation to the stator length.

To produce an electromagnetic torque it is needed the presence of a travelling wave of electromotive force (EMF) in the rotor. This will be originated by the airgap travelling magnetic wave due to the current density circulating in the stator. The EMF wave is an image of the stator current density being expressed by (1), where k is the wavelength of the stator's EMF (equivalent to the number of poles of the stator):

$$\vec{J}_s = Re\{J_m e^{j(\omega t - k\varphi)}\vec{u}_\theta\} \tag{1}$$

As result the airgap potential vector will be similar to the current density.

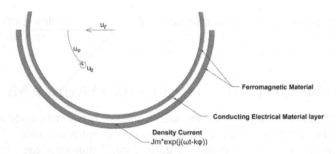

Fig. 5. Cross section of *shell-like* induction machine

The following generic equation defines the potential vector as the composition of two components: a radial dependent component $A(r)$ and a time/space dependent component $e^{j(\omega t - k\varphi)}$:

$$\vec{A}_z = A(r)e^{j(\omega t - k\varphi)}\,\vec{u}_\theta \tag{2}$$

This problem simplifies into the laplacian solution of the radial component $A(r)$. Although initially using spherical coordinates, this *shell-like* design will allow us to change to cylindrical coordinates, simplifying the laplacian solution computation of (2). The following figure illustrates the uniqueness of symmetry around axis -y that allows the change of coordinates.

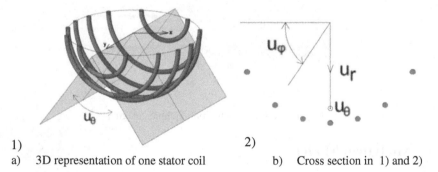

1)   2)

a)   3D representation of one stator coil   b)   Cross section in 1) and 2)

Fig. 6. Stator's windings geometry in cylindrical coordinates

Therefore the laplacian of potential vector in cylindrical coordinates is defined by:

$$\nabla^2 \vec{A}_\theta = \left(\frac{\partial^2 A(r)}{\partial r^2} + \frac{1}{r^2}\frac{\partial^2 A(r)}{\partial \varphi^2} + \frac{\partial^2 A(r)}{\partial \theta^2} + \frac{1}{r}\frac{\partial A(r)}{\partial r}\right)\vec{u}_\theta = 0 \tag{3}$$

Solution of (3) is the typical Cauchy-Euler solution as in (4):

$$\vec{A}_\theta = (C_1 r^k + C_2 r^{-k})e^{j(\omega t - k\varphi)}\,\overrightarrow{u_\theta} \tag{4}$$

Constants C_1 and C_2 in (4) must be defined by two boundary problem conditions. Using the integral form of Ampere's Law, it is possible to establish the relation between magnetic flux density components and density currents on the stator and rotor. These boundary conditions are defined by equations a and b in (5), and illustrated in Fig. 7(a) and 7(b), respectively.

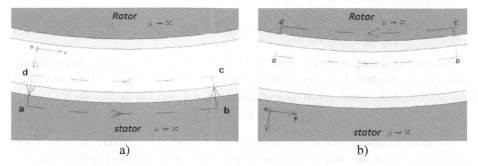

Fig. 7. Boundary condition – Ampere's law in the a) stator and b) rotor

$$a) \int_c^d H\, dl = \int_a^b J_S\, \vec{n}\, d\varphi \;, \qquad b) \int_a^b H_\varphi\, d\varphi = \int_a^b J_i\, \vec{n}\, d\varphi \tag{5}$$

The stator current density $\vec{J}_s$ is the source for the rotor EMF, resulting in the rotor induced density current $\vec{J}_i$, given by (6), taking in account the electric field $\vec{E}$ and the linear tangential velocity of the rotor, $\vec{V}$.

$$\vec{J}_i = \sigma(\vec{E} + \vec{V} \times \vec{B}) = \sigma\left(-\frac{\partial \vec{A}_\theta}{\partial t} - \omega_r \frac{\partial \vec{A}_\theta}{\partial \varphi}\right) \tag{6}$$

Defining the slip parameter $S = (\omega - k\omega_r)$, constants C_1 and C_2 become given by (7) and (8). Term r_r is the radius of the rotor, r_s is the radius of the stator.

$$C_1 = \frac{J_{m\,eq}\mu_0 r_s^{\,k+1}}{k} \frac{\left(1 + j\frac{\mu_0 \sigma S}{k} r_r\right)}{\left(-r_r^{\,2k} + r_s^{\,2k} + j\frac{\mu_0 \sigma S}{k}(r_r^{\,2k} + r_s^{\,2k})\right)} \tag{7}$$

$$C_2 = \frac{J_{m\,eq}\mu_0 r_s^{\,k+1} r_r^{\,2k}}{k} \frac{\left(1 - j\frac{\mu_0 \sigma S}{k} r_r\right)}{\left(-r_r^{\,2k} + r_s^{\,2k} + j\frac{\mu_0 \sigma S}{k}(r_r^{\,2k} + r_s^{\,2k})\right)} \tag{8}$$

4.1 Electromechanical Characteristics

The magnetic flux density is given by the rotational of the potential vector eq. (9). As result, the magnetic flux density presents two components, $\vec{B}_r$ and $\vec{B}_\varphi$, and the induced current only one component, $\vec{J}_\theta$.

$$\vec{B} = \nabla \times \vec{A}_\theta, \qquad \vec{B} = \left(\frac{1}{r}\frac{\partial \vec{A}_\theta}{\partial \varphi}\right)\vec{u}_r + \left(-\frac{\partial \vec{A}_\theta}{\partial r}\right)\vec{u}_\varphi \tag{9}$$

$$\vec{B}_r = -j\frac{k}{r}(C_1 r^k + C_2 r^{-k})e^{j(\omega t - k\varphi)}\vec{u}_r \tag{10}$$

$$\vec{B}_\varphi = -\frac{k}{r}(C_1 r^k - C_2 r^{-k})e^{j(\omega t - k\varphi)}\vec{u}_\varphi \tag{11}$$

$$\vec{J_i} = \sigma\left(-\frac{\partial \vec{A}_\theta}{\partial t} - \omega_r \frac{\partial \vec{A}_\theta}{\partial \varphi}\right) = -j\sigma S(C_1 r_r^{\ k} + C_2 r_r^{\ -k})e^{j(\omega t - kd\varphi)} \qquad (12)$$

With the magnetic flux density given by (10) and (11) and the induced current density from (12), the magnetic force density in the rotor is computed as in (13). This shows the resultant force density having two components, a radial $\vec{F}_r$ and a tangential one $\vec{F}_\varphi$.

$$\vec{F} = \vec{J_i} \times \vec{B} = \vec{F}_r + \vec{F}_\varphi \qquad (13)$$

For motion applications, it is important to analyze the resultant average force density in the rotor and thus the average torque. This can be obtained as shown in (14) taking in account the active surface area of the rotor, its radius and the average force density in (15).

$$\langle T_\varphi \rangle = r_r \iint \langle F_\varphi \rangle \, d\theta \, d\varphi, \qquad \langle T_r \rangle = r_r \iint \langle F_r \rangle \, d\theta \, d\varphi \qquad (14)$$

$$\langle F_\varphi \rangle = \frac{1}{2} Re\{\hat{J}_i \hat{B}_r^{\ *}\}, \qquad \langle F_r \rangle = \frac{1}{2} Re\{\hat{J}_i \hat{B}_\varphi^{\ *}\} \qquad (15)$$

4.2   Parameters' Sensibility

Figures 8 (a) and (b) show the dependence of the tangential and radial force density in the rotor in function of parameter k, which is the wavelength of stator's EMF. These results were obtained for a 5cm rotor radius with a 2mm thick aluminum material, a 2mm airgap and a stator current density of $6 \times 10^6 A/m^2$ (typical value of current density in electrical machines).

Results show that when increasing the wavelength of EMF (parameter k), both density forces magnitude are reduced, moving the maximum tangential density force to higher values of slip S'. The radial component presents a repulsing force between stator and rotor.

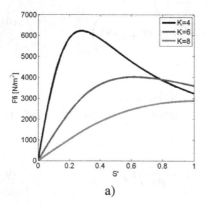
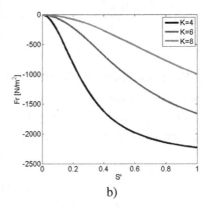

a)   b)

Fig. 8. Influence of the EMF's wavelength in the density force components on rotor: (a) tangential component; (b) radial component. $S' = (\omega - k\omega_r)/\omega$

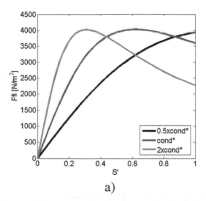
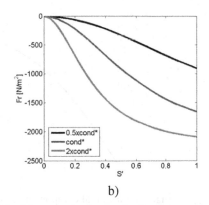

a)   b)

Fig. 9. Influence of electric conductivity's material of the rotor in the density force components: (a) tangential component; (b) radial component. $S' = (\omega - k\omega_r)/\omega$.

Figures 9(a) and 9(b) indicate that the change of the rotor's electric conductivity has a similar effect as in classic induction machines. The maximum point of the force will be shifted to higher values of synchronism when decreasing the conductivity.

5   Assistive Device Application

In this section it is analyzed the application of the *shell-like* actuator as an active joint for assistive devices. Application is a device for assistive movement of the lower leg for a typical 70kg person, usually used in orthopedics. Table 1 resumes the specifications of the active joint. For this purpose, the *shell-like* induction motor will have a simple design with a stator shell covering half of the rotor. This design warranties the range of motion in both Sagittal and Coronal planes.

Table 1. Specifications for the orthopedic active joint

Total weight to lift with a 0.25m distance to the center of mass	2 kg (average lower leg for 70 kg person)
Average linear velocity	3 Km/h [0,83m/s]
Range of Motion for Sagittal plane	[-10° 90°]
Range of Motion for Coronal plane	[-25° 25°]
Rated torque	5 Nm
Range of radius	[2,5 5] cm

To study its feasibility, it was considered a 5cm radius rotor with a 2mm thick aluminum conducting material, a 2mm thick airgap, a 50 Hz stator current density of $6 \times 10^6 A/m^2$ and $k=4$. For comparison, our design was simulated in 2D finite element method (FEM) program. FEM results are compared with the analytical solutions. For example, Figure 10 plots the resultant average torque obtained from the analytical model (continuous line) and that one determined using the FEM model (discrete marks).

The maximum torque obtained is about 4 Nm, therefore near to the specified one. The maximum angular velocity corresponds to 3.9m/s (78.5 rad/s), including the average linear velocity of 0.83m/s (16.6 rad/s). This preliminary analysis indicates that the *shell-like* actuator design is capable of be the joint of the orthopedics assistive device.

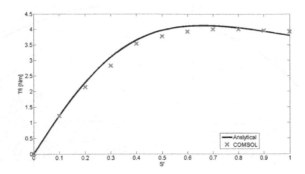

Fig. 10. Tangential component of the torque: analytical and FEM results

6   Conclusions

The purpose of this paper is to study the feasibility of the shell-like induction machine, especially in assisting devices applications. The combination of the stator shell geometry and distribution of its coils allow a wide range of motion with multi-DOF. By the analysis of the potential vector with a simple model of the machine it was possible to extrapolate its behavior and the amplitude of forces possible for small sizes. Despite the simple model, it was possible to verify the parameter's influence on the main electromechanical characteristics. In addiction the FEM results obtained for a more realistic model, i.e., non-homogeneous geometry with non-simplified structure, were close to the analytical results.

The feasibility study concluded that this machine is capable of originating amplitude of torque needed in an orthopedic assistive device as the proposed leg joint.

Acknowledgements. This work was supported by FCT, through IDMEC, under LAETA Pest-OE/EME/LA0022.

References

1. van Ninhuijs, B.: Feasibility Study of a Spherical Actuator in an Arm Support. Master graduation paper, Electromechanics and Power Electronics Group, Eindhoven University of Technology, Netherlands (August 2011)
2. Suzuki, T., Yano, T., Takatuji, T.: Wafer inspection equipment and system, Japanese Patent No.3641688 (1995)
3. Lee, K.M., Son, H., Joni, J.: Concept Development and Design of a Spherical Wheel Motor (SWM). In: Proc. of 2005 IEEE International Conference on Robotics and Automation, pp. 3663–3668 (April 2005)
4. Kawano, H., Hirahara, T.: Pre-loading Mechanism and Angular Positioning Control Algorithm for Multi-DOF Ultrasonic Motor with a Weight Load. In: Proc. on 18th International Congress on Acoustic, Mo5.I.2, pp.I-569–I-572 (2004)
5. Kagawa, A., Ono, E., Kusakabe, T., Sakamoto, Y.: Absorption of hydrogen by Vanadium-rich V-Ti-Based Alloys, J. Less-Common Met. 172(174), 64–70 (1991)

Part IX
Petri Nets

Elementary Events for Modeling of Human-System Interactions with Petri Net Models

Rogério Campos-Rebelo[1,2], Anikó Costa[1,2], and Luís Gomes[1,2]

[1] Universidade Nova de Lisboa, Faculdade de Ciências e Tecnologia, Portugal
[2] UNINOVA – Centro de Tecnologias e Sistemas, Portugal
{rcr,akc,lugo}@uninova.pt

Abstract. This paper presents a proposal for structuring events for system models expressed using IOPT nets (Input-Output Place-Transition Petri nets). Currently, a non-autonomous event within an IOPT model is defined based on change of input signals with respect to a specific threshold, when two consecutive execution steps are considered. New types of events are proposed, allowing the definition of an event activated not only by crossing a fixed threshold, but also by considering a change in associated signal values on a specific amount (belonging to an interval of values). The concept is further extended allowing the definition of an event based on signal values presented on previous execution steps. The proposal results on a classification of several types of events, namely threshold events, momentum events, impetus events, as well as delayed events and logical events. Usage of these types of events allows improvements in terms of expressiveness and compactness of the resulted model.

Keywords: Petri nets, embedded systems, human-system interaction.

1   Introduction

System complexity has grown considerably over the last decades. Some of the reasons for this growth are related with the increase in systems' processing capability and memory, as well as associated with the development of more robust technology. This complexity is due, mainly, to the need to produce more efficient systems, and more user friendly.

The growing development of the Service Oriented Architecture (SOA) and the Event-Driven Architecture (EDA) combined with the reduction of cost in sensor technology have led, in the last years, to a big development in the use of events in the analysis of signals and system' behavior [1].

Use of event modeling allows the development of more complex systems expressed through a much more compact model, supporting the development of much more user friendly interfaces.

In order to cope with problems related with the system's complexity, several approaches were proposed to decompose the system development using multiple steps. Model Based Development approach, using specific languages for system specification, is one of these examples.

L.M. Camarinha-Matos et al. (Eds.): DoCEIS 2014, IFIP AICT 423, pp. 219–226, 2014.
© IFIP International Federation for Information Processing 2014

Graphical languages are normally recommended to support this approach. Some examples of graphical languages are statecharts [2], state-machines [3] or Petri Nets, which are used as reference modeling formalisms on this work.

The many developments that Petri Nets have suffered since its first presentation in 1962 [4] have led to the growth of their use in various areas. One of those areas is engineering, taking advantage of some Petri net intrinsic features, such as the ability to model synchronization, concurrency, and conflicts [5], as well as the simultaneous visualization of the structure and behavior of the system [6]. These features allow Petri nets to capture the dynamics of the system, making them also useful in simulation [7].

One of the issues that have taken a significant contribution to the growth of its use was the inclusion of non-autonomous characteristics. This has led to increasing the impact of their use, when compared to other modeling formalisms used for embedded systems design [8]. These non-autonomous characteristics allow the connection of the net with the environment, making them well suitable for modeling the system's controller behavior.

Some examples of non-autonomous Petri net classes are the Interpreted Petri nets [9], synchronized Petri nets [10] or IOPT nets (Input-Output Place-Transition nets) [11].

Whenever modeling human systems interactions are addressed, the number of available modeling primitives is limited.

Following the line of the work presented in [12], this paper presents a more complete and elaborated set of elementary events, benefiting from a scalable definition, addressing the effectiveness of Petri nets modeling for human-system interaction systems, also keeping applications within the controllers' modeling.

Augmenting the traditional way to define an event (as a signal trespassing a threshold level), new ways to define an event associated with signal changes were proposed in [12] related to amplitude and changing speed of variation on signal values.

In this work, this concept is extended allowing a generalized definition supporting the use of the specified signal variations.

On the other hand, for each variation level two different types of behaviors were previously defined. In this work, this definition is also extended.

2   Relationship to Collective Awareness Systems

Collective awareness systems are systems that allow interaction and knowledge development in collaboration by several people.

The work proposed in this paper presents a set of elementary events that are defined to be used in the future behavior definition based on events from one or more variables from one or more systems. These characteristics and its exact definition make them suitable for defining behaviors that can be shared or even set together.

On the other hand, the fact that the IOPT nets can be used to implement these events and these tools implemented online and interconnected in the same system ensures a greater capacity for development of collective awareness systems.

3   Signal Analysis

It is considered a signal as the value of a certain time variable. This variable can be physical or from the output of a system. As presented in [11], two ways can be considered to analyze a signal:

- Analysis of the signal value at a given moment in order to verify that this value meets a certain condition.

- Definition of events associated to that signal to analyze if the behavior of the signal in time obeys to the behavior that we want to find.

These two types of analysis are mainly different in terms of duration of the analysis. Whereas a condition is inspecting the value of the signal at a given instant of time. In an event, the behavior of the signal in time is analyzed.

From another point of view, a signal analysis can also be made taking into account different degrees of differences between consecutive values. The value of the signal, its derivative (considering two instants of analysis), its second derivative or the following derivatives can be analyzed. Given this analysis one can analyze not only a point or a variation between two points but also behaviors taking into account three or more instants on the signal.

In Table 1 these two types of dependencies on signal evolution is presented, showing the various possibilities for the analysis of a signal.

Table 1. Events and Conditions Definition

	X	ΔX	$\Delta^2 X$	$\Delta^3 X$	
Events	$X_{n-1} \leq K$ $\wedge$ $X_n > K$	$\Delta X_{n-1} \leq K$ $\wedge$ $\Delta X_n > K$	$\Delta^2 X_{n-1} \leq K$ $\wedge$ $\Delta^2 X_n > K$	$\Delta^3 X_{n-1} \leq K$ $\wedge$ $\Delta^3 X_n > K$	...
Conditions	$X_n = K$	$\Delta X_n = K$	$\Delta^2 X_n = K$	$\Delta^3 X_n = K$	...

In terms of levels of variation (Δ), defined as a difference of signal value in two consecutive instants, it is possible to define higher order of differences.

It can be seen that for all variations of X, the conditions only analyze a given value of signal X of the index n, while an event is analyzed based on the behavior between two index values n and n-1.

In the table, it can also check the possibility of analysis of the signal behaviors at various instants of the signal. Given that $\Delta X_n = X_{n-1} - X_n$ and $\Delta X_{n-1} = X_{n-2} - X_n 1$, an event in ΔX analyzes signal values in three instants (X_{n-2}, X_{n-1} e X_n). Likewise analyzing the event condition $\Delta^2 X$ it has that $\Delta^2 X_n = \Delta X_{n-1} - \Delta X_n = (X_{n-2} - X_{n-1}) - (X_{n-1} - X_n)$ and $\Delta^2 X_{n-1} = \Delta X_{n-2} - \Delta X_{n-1} = (X_{n-3} - X_{n-2}) - (X_{n-2} - X_{n-1})$. Thus in $\Delta^2 X$ four values of the signal are analyzed (X_{n-3}, X_{n-2}, X_{n-1} e X_n).

4   Elementary Events

An elementary event is defined as an event that cannot be obtained from the composition of other events of the same type.

In this work, two types of elementary events are determined that are distinguished by their behavior: elementary open events and elementary closed events.

4.1   Elementary Closed Events

Elementary Closed Events are events that define the analysis for the entire spectrum of signal values. In other words, any value that the signal can have satisfies one of the conditions of the event, either the Pre- or the Post-Condition.

An Elementary Closed Event is in such a way "Closed" that the event can be detected just by looking at one of the conditions. For instance, knowing that the signal satisfies a pre-condition at instant (n-1) and fails to satisfy this condition at instant n, it is guaranteed that he satisfies the post-conditions, which allows the detection of the signal without the need to analyze pre- and post- conditions.

In Table 2 the events defined with these characteristics are shown.

Table 2. Elementary Closed Events definitions

Pre-Condition	Post-Condition
$X_{n-1} > K$	$X_n \leq K$
$X_{n-1} \geq K$	$X_n < K$
$X_{n-1} < K$	$X_n \geq K$
$X_{n-1} \leq K$	$X_n > K$
$X_{n-1} = K$	$X_n \neq K$
$X_{n-1} \neq K$	$X_n = K$

All events presented in the table are associated to value of signal X to simplify the presentation, but all these types of events can be defined in any $\Delta^n X$.

4.2   Elementary Open Events

Elementary Open Events are events that do not define the analysis for the entire spectrum of signal values. In other words, there are values that the signal may have that are not covered by the equations that define the signal.

The Table 3 presents the defined set of types of elementary open events.

Table 3. Elementary Open Events definitions

Pre-Condition	Post-Condition
$X_{n-1} > K$	$X_n < K$
$X_{n-1} < K$	$X_n > K$
$X_{n-1} < K$	$X_n = K$
$X_{n-1} > K$	$X_n = K$
$X_{n-1} = K$	$X_n > K$
$X_{n-1} = K$	$X_n < K$

An elementary closed event can be analyzed taking into account only one of the conditions (pre or post-condition) in the two consecutive steps of analysis. For this kind of events, if the pre-condition was satisfied in the step n-1 and the same pre-condition is not satisfied in the step n, the event occurs because the post-condition is satisfied. For elementary open events, both conditions (pre and post) are needed to be analyzed.

To better understand, this type of events, it will be analyzed the event of the fourth row of the table, with the conditions $X_{n-1} > K \wedge X_n = K$. In this case, if the signal satisfies the pre-condition in instant (n-1), being greater than K and fails to satisfy this condition in instant n becoming less than K, while the pre-condition to be no longer satisfied the event does not occur. Thus, it is not possible to define this type of events using only one equation as proved possible for elementary closed events.

It is not possible to define an elementary open event by composition of other events of the same type. This is the reason that makes it elementary. Despite this, being of another type, it is possible to define elementary closed events by composition of elementary open events.

4.3   Complete List of Elementary Events

With the previous definitions, it is possible to define a large number of types of events that define the behaviors of interest in the associated signal.

Putting together the elementary open events and the elementary closed events a set of twelve types of events is obtained. A set of identifiers is used to represent events in a compact form.

The twelve events resulting from the merger mentioned above are defined for each of the various levels of variation (X, ΔX, $\Delta^2 X$, $\Delta^3 X$, ..., $\Delta^n X$).

The Table 4 presents this set of events for X and ΔX, showing the conditions of the event and a graphical representation of the behavior of the signal that makes this event to occur.

For each level of variation, twelve types of events were defined, each one representing a specific behavior of the signal. These types of events are represented with a specific operator.

In the variation level 0, the value of (X) is analyzed, detecting crossing of a specific value. For example, it is possible to define an event to detect if   the temperature of the car is higher than 90 degrees. In variation level 1, the first difference (ΔX) is analyzed, associated with the velocity of variation of the signal. For example, an event that occurs whenever the temperature of the car is increasing more than 5 degrees per minute can be defined. Finally, the other levels represent the next variations, starting with level 2, the second difference ($\Delta^2 X$) (representing the acceleration of variation) up to level n represented by ($\Delta^n X$).

Table 4. Complete Elementary Events list

	X		ΔX	
$<>$	$X_{n-1} < k$ $\wedge$ $X_n > k$		$\Delta X_{n-1} < k$ $\wedge$ $\Delta X_n > k$	$\Delta > k$ / $\Delta < k$
$><$	$X_{n-1} > k$ $\wedge$ $X_n < k$		$\Delta X_{n-1} > k$ $\wedge$ $\Delta X_n < k$	$\Delta < k$ / $\Delta > k$
$>$	$X_{n-1} \le k$ $\wedge$ $X_n > k$		$\Delta X_{n-1} \le k$ $\wedge$ $\Delta X_n > k$	$\Delta > k$ / $\Delta \le k$
$\ge$	$X_{n-1} < k$ $\wedge$ $X_n \ge k$		$\Delta X_{n-1} < k$ $\wedge$ $\Delta X_n \ge k$	$\Delta \ge k$ / $\Delta < k$
$<$	$X_{n-1} \ge k$ $\wedge$ $X_n < k$		$\Delta X_{n-1} \ge k$ $\wedge$ $\Delta X_n < k$	$\Delta < k$ / $\Delta \ge k$
$\le$	$X_{n-1} > k$ $\wedge$ $X_n \le k$		$\Delta X_{n-1} > k$ $\wedge$ $\Delta X_n \le k$	$\Delta \le k$ / $\Delta > k$
$=$	$X_{n-1} \ne k$ $\wedge$ $X_n = k$		$\Delta X_{n-1} \ne k$ $\wedge$ $\Delta X_n = k$	$\Delta = k$ / $\Delta \ne k$
$=+$	$X_{n-1} \ne k^+$ $\wedge$ $X_n = k$		$\Delta X_{n-1} \ne k^+$ $\wedge$ $\Delta X_n = k$	$\Delta = k$ / $\Delta > k$
$=-$	$X_{n-1} \ne k^-$ $\wedge$ $X_n = k$		$\Delta X_{n-1} \ne k^-$ $\wedge$ $\Delta X_n = k$	$\Delta = k$ / $\Delta < k$
$\ne$	$X_{n-1} = k$ $\wedge$ $X_n \ne k$		$\Delta X_{n-1} = k$ $\wedge$ $\Delta X_n \ne k$	$\Delta \ne k$ / $\Delta = k$
$\ne+$	$X_{n-1} = k$ $\wedge$ $X_n \ne k^+$		$\Delta X_{n-1} = k$ $\wedge$ $\Delta X_n \ne k^+$	$\Delta > k$ / $\Delta = k$
$\ne-$	$X_{n-1} = k$ $\wedge$ $X_n \ne k^-$		$\Delta X_{n-1} = k$ $\wedge$ $\Delta X_n \ne k^-$	$\Delta < k$ / $\Delta = k$

4.4   Delayed Analysis of Events

Until now, all events have been presented as analysis of variations in the signal between two consecutive steps of analysis (X_n and X_{n-1}). Nevertheless, it is possible to set all these events and an analysis based on two distinct instants (not consecutive) separated by p steps of analysis (instead of one).

In this sense, delayed events are not a new type of events. They are a characteristic that can be changed in any type of elementary events. With the addition of this characteristic, the analysis of the event is done taking into account the values (X_n e X_{n-p}). The analysis accomplished in the previous section is done considering two consecutive steps as a particular case where the value p = 1.

In Fig. 1 the way to analyze an event with delay is represented.

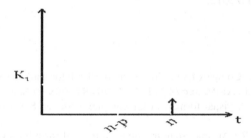

Fig. 1. Analysis of a delayed Event

To analyze this event a p-sized window is created and the signal inside this window is not used to the analysis of the event.

Taking into account the large number of types of events is not possible to use one single example to explain the full potential of these types of events. As a matter of fact, due to lack of space, it is not possible even to present a simple system's model taking advantage of any of these new modeling capabilities.

5   Conclusions and Further Work

The use of these new types of events provides effective modeling capabilities when using IOPT nets models. It allows the creation of smaller models for the same system, making them easier to read and easy to interpret.

The definition of different variations of the signal allows the analysis of a set of behaviors that allow a better knowledge of the signal, allowing not only changes on the value of the signal, but how these changes occurs, like its velocity or acceleration or even the acceleration of the velocity of the signal.

Taking into account the following objective to define events that represent more complex behavior of associated signs by composition of simple events, the definition of these elementary events becomes a robust starting point for this composition.

Those robust definitions allow the composition of events from different signals and even from different variations of the signal.

As future work, it is foreseen to define a set of transformations, as well as composition between elementary events in order to create a composed event. To do this, a standard compact representation for all elementary events and to the interaction between them needs to be defined.

Acknowledgments. The first author was supported by a Grant from project PTDC/EEI-AUT/2641/2012, financed by Portuguese Agency" FCT - Fundação para a Ciência e a Tecnologia".

This work was partially financed by Portuguese Agency" FCT - Fundação para a Ciência e a Tecnologia" in the framework of projects PEst-OE/EEI/UI0066/2011 and PTDC/EEI-AUT/2641/2012.

References

1. Eckert, M., Bry, F.: Complex Event Processing (CEP). Informatik-Spektrum 32(2), 163–167 (2009), http://dx.doi.org/10.1007/s00287-009-0329-6
2. Statecharts, H.D.: A visual formalism for complex systems. Sci. Comput. Program 8(3), 231–274 (1987)
3. Börger, E., Stärk, R.: Abstract state machines: a method for high-level system design and analysis. Springer, Heidelberg (2003)
4. Petri CA. Kommunikation mit Automaten (Communication with Automata) Hamburg (1962), http://edoc.sub.uni-hamburg.de/informatik/volltexte/2011/160
5. Peterson, J.: Petri Nets. ACM Comput. Surv. 9(3), 223–252 (1977)
6. Zurawski, R., MengChu, Z.: Petri nets and industrial applications: A tutorial. Ind. Electron IEEE Trans. 41(6), 567–583 (1994)
7. Murata, T.: Petri nets: Properties, analysis and applications. Proc. IEEE 77(4), 541–580 (1989)
8. Gomes, L., Barros, J.P., Costa, A.: Modeling Formalisms for Embedded Systems Design. In: Zurawski, R. (ed.) Embed. Syst. Handb. (2005)
9. Moussa, F., Riahi, M., Kolski, C., Moalla, M.: Interpreted Petri Nets used for Human-Machine Dialogue Specification in Process Control: principles and application to the Ergo-Conceptor+ tool. Integr. Comput. Aided Eng., 87–98 (2002)
10. Moalla, M., Pulou, J., Sifakis, J.: Synchronized petri nets: A model for the description of non-autonomous sytems. Math. Found. Comput. Sci., 374–384 (1978)
11. Gomes, L., Barros, J.P., Costa, A., Nunes, R.: The Input-Output Place-Transition Petri Net Class and Associated Tools. In: 5th IEEE Int. Conf. on Ind. Informatics, pp. 509–514 (2007)
12. Campos-Rebelo, R., Costa, A., Gomes, L.: On structuring events for IOPT net models. In: Camarinha-Matos, L.M., Tomic, S., Graça, P. (eds.) DoCEIS 2013. IFIP AICT, vol. 394, pp. 229–238. Springer, Heidelberg (2013)

From SysML State Machines to Petri Nets Using ATL Transformations

Rui Pais[1,2,3], João Paulo Barros[2,3], and Luís Gomes[1,2]

[1] Universidade Nova de Lisboa, Faculty of Sciences and Technology, Portugal
ruipais@uninova.pt
[2] UNINOVA, Center of Technologies and Systems, Portugal
lugo@fct.unl.pt
[3] Instituto Politécnico de Beja, Escola de Superior Tecnologia e Gestão, Portugal
jpb@uninova.pt

Abstract. The ATLAS Transformation Language (ATL) is a well-known hybrid model transformation language that allows both declarative and imperative constructs to be used in the definition of model transformations. In this paper, we present ATL transformations providing an integrated structural description of the source and target metamodels and the transformation between them. More specifically, the paper presents translation rules of Systems Modeling Language (SysML) state machines models into a class of non-autonomous Petri net models using ATL. The target formalism for the translation is the class of Input-Output Place Transition Nets (IOPT), which extends the well-known low-level Petri net class of Place/Transition Petri nets with input and output signals and events dependencies. Based on this Petri net class, a set of tools have been developed and integrated on a framework for the project of embedded systems using co-design techniques. The main goal is to benefit from the model-based attitude while allowing the integration of development flows based on SysML state machines with the ones based on Petri nets.

Keywords: ATL, Transformation Models, SysML, UML, State Machines, Petri Nets, PNML, MDE, MDA, IOPT.

1 Introduction

The increasing complexity of new-generation systems raises major concerns in various critical application domains, in particular with respect to the validation and analysis of performance, timing, and dependability-related requirements. Model-driven engineering (MDE) [1] approaches aimed at mastering this complexity during the development process have emerged and are being increasingly used in industry. They address the problem of complexity by promoting reuse and partial or total automation of specific phases of the development process.

By taking advantage of Petri nets visual representation and precise semantics, this paper contributes to their use as an intermediate formalism between SysML behavior models and code generation. More specifically, it presents a tool that transforms a

L.M. Camarinha-Matos et al. (Eds.): DoCEIS 2014, IFIP AICT 423, pp. 227–236, 2014.
© IFIP International Federation for Information Processing 2014

Systems Modeling Language (SysML) state machine, to an Input-Output Place Transition Net (IOPT) [2] target model. The process of transformation — the production of an XMI [3] file — is realized using the Atlas Transformation Language (ATL) [4]. The XMI file will be subsequently used as an entry to a Model Driven Architecture (MDA) [5, 20] process.

Present work benefits from the results of previous projects where a set of tools was created. These tools allow creation of IOPT net models, as well as many others operations. Ultimately, they also permit code generation for different platforms. Those tools rely on an Ecore metamodel for the IOPT class, already presented in [2].

Translation techniques to convert state machines elements into correspondent Input Output Place Transition Nets items were already presented elsewhere [7]. Later, these translations were extended to include other state machines elements and several strategies to translate state machines with pseudostate history attribute were also presented [8].

This research extends previous works [7, 8] with the implementation of translation techniques, using ATL, to transform SysML state machine models to IOPT net models. The integration of these translations on a Petri net-based framework for the development of embedded systems using co-design techniques permits the use of SysML state machines as an additional modeling language.

The paper is structured as follows: Section 2 presents motivation, innovations and its relation to Collective Awareness Systems, while Section 3 describes some important concepts of Model to Model transformations. The implemented transformations rules using ATL are presented in Section 4. Finally, Section 5 presents topics for discussion and Section 6 concludes.

2   Relation to Collective Awareness Systems

An awareness system can be defined as a system intended to help people construct and maintain awareness of each other activities, context or status, even when the participants are not co-located [9].

As testified in [10], "Internet has changed the way we develop, perform and understand business", hence it is urgent to innovate and exploit the full potential of the Future Internet. To achieve this goal, a set of recommendations is provided and an analysis of important areas to consider is presented. More specifically, the analysis of Collective Awareness Platforms for Sustainability and Social Innovation (CAPS), points out that its basic layer includes (smart) objects that capture the environment reality. Many of these (smart) objects are embedded systems, which are frequently specified using Petri Nets. These are used for modeling and analyzing complex systems that exhibit characteristics of concurrency, synchronization, simultaneous, distributed, resource sharing, etc., taking advantage of tools that permit modeling visualization, simulation, property verification (e.g. deadlock, starvation, bottlenecks, and execution time), and code generation.

With this research, we contribute to the development of embedded systems allowing system specification using a well-known modeling formalism: SysML state machines. With better specification we are promoting better embedded systems, the core of smart objects.

3   Model to Model Transformations

Software development is becoming more and more complex with the increasing complexity of system requirements, necessity to integrate frameworks, libraries, communication platforms, etc.

Maintenance, changing requirements, production cost, specification reusability, and a lot of other factors contribute to a demand to improve software specification. One way to reduce technical complexity is the use of the Model-Driven Software Development paradigm, as it facilitates a more abstract specification of software based on modeling languages [11].

In the context of model-driven engineering, models are the main development artifacts and model transformations are among the most important operations applied to models. Model-to-model transformations constitute an important ingredient in model-driven engineering.

The following sections present the main concepts used in this context.

3.1   Ecore

The model used to represent models in the Eclipse Modeling Framework (EMF) is called Ecore. Ecore is itself an EMF model, and thus is its own metamodel [12].

Ecore is a modeling language (in fact a meta-meta-modeling language) to describe domain specific meta-models. It is used to define all entities that will exist on the domain specific models of interest, define the characteristics of these entities, as well as their relationships.

3.2   ATL Language

The ATLAS Transformation Language is a model to model transformation language originally developed as a response to the Object Management Group (OMG) Request for Proposals for the Query/View/Transformation (QVT) standard [13], implemented in the Eclipse modeling tools.

ATL is a well-known hybrid model transformation language that allows fully declarative, hybrid, or fully imperative constructs to be used in transformation definitions.

An ATL transformation is unidirectional and operates on read-only source models to produce write only target models. This way, source model can be navigated during transformation, but it is neither possible to write on the source model, nor to navigate target model. Yet, ATL provides automatic traceability links between target and source elements.

ATL model transformation process is illustrated on Figure 1. This provides an overview of the transformation process used for the generation of an IOPT model (IOPTnet.xmi) from a SysML state machine model (StateMachine.xmi) through a transformation model (SM2IOPT.atl). In this diagram, the dotted line specifies model to model transformation; the normal arrows specify conformance between source state machine model (StateMachine.xmi) and its metamodel (StateMachine.ecore), as well as between this metamodel and the Eclipse Modeling Framework (EMF) metametamodel. On the same way, the generated model (IOPTnet.xmi) conforms to its metamodel (IOPT.ecore), and the latter conforms to the EMF metametamodel.

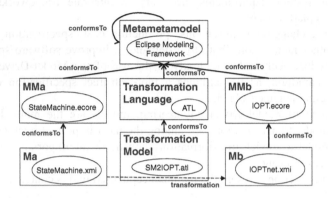

Fig. 1. ATL Model Transformation (adapted from [21])

3.3   SysML State Machines

The Systems Modeling Language (SysML) is a general-purpose modeling language that supports the specification, analysis, design, verification, and validation of a broad range of complex systems, which may include hardware, software, information, processes, personnel, and facilities [14]. The respective State Machine package defines a set of concepts that can be used for modeling discrete behavior through finite state transition systems [14].

3.4   Input-Output Place Transition Nets

Petri nets have been successfully used for concurrent systems specification [15] for the analysis of behavioral properties and performance evaluation, as well as for systematic construction of discrete event simulators and controllers [16].

The Input-Output Place Transition class (IOPT) extends place-transition nets with non-autonomous constructs, allowing explicit modeling of input and output events and signals [2]. The Ecore metamodel for the IOPT net class was already presented in [2, 17].

4   Transformations Rules Using ATL

In this section, we present a set of transformation rules and helpers that allow the generation of IOPT models from SysML state machines models. The result of each transformation is an XMI file on PNML format.

Some translation techniques to convert state machines elements into correspondent Input Output Place Transition Nets items were already presented elsewhere [7, 8].

The respective implementations are here presented. Yet, as the transformation application is extensive, we only show some code blocks, representative of interesting transformation aspects.

Transformation rules are the core of a model transformation. A transformation rule can be implemented with declarative, fully imperative, or hybrid constructs.

The declarative style of specifying transformations are encouraged as best practice [18], letting imperative features of the language to the most complex transformations when declarative constructs are not sufficient.

Declarative transformations rules use pattern matching to match the source model elements by type. A rule guard permits an additional filtering on pattern matching. For each matched element one or more target model elements are created.

Transformations traceability is an important language feature that offers trace links between source and target models, and between rule applications. This is important, as it allows one rule to reference elements created in other rules.

The matching rule 'package2PetriNetDoc' is one of simplest transformation rule.

```
rule package2PetriNetDoc {
    from
        inn : SysML!Package(inn.isPackage())
    using {
        ruleName:String= 'package2PetriNetDoc'.debug('!Starting);
    }
    to
        petriNetDoc :   PNML!PetriNetDoc(
            nets <- thisModule.allStateMachines->collect (p |
                        thisModule.resolveTemp (p, 'net'))
        )
    do {
        ruleName.debug('!Ending rule');
    }
}
```

For each source model element of type 'Package' it collects the traceability links generated by target pattern 'net'; it uses the imperative parts 'using' and 'do' for debug proposes. For each source model element of type 'Package', this rule creates a target 'PetriNetDoc'. This rule uses the 'allStateMachines' helper that is presented later in this paper.

To illustrate and exemplify the implemented translations, we present the translation of state machine **choice pseudo-state**.

A choice pseudo-state has one incoming and several outgoing edges, and it is used as a switch of control flow based on zero or more guard condition. Depending on the guards' conditions, the control flow is redirect from one incoming edge to exactly one outgoing edge. Figure 2 presents a simple state machine with a choice pseudo-state, and Figure 3 presents its translation to an IOPT net.

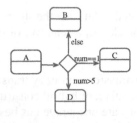

Fig. 2. State machine with choice pseudo-state

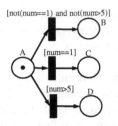

Fig. 3. Translated **IOPT net**

This translation was implemented by three main rules:

1. One rule that translates state machine states to IOPT places;
2. One rule to translate state machine choice guards;
3. One rule to translate state machine choice state to a set of IOPT transitions and arcs. This last transformation rule is presented next.

Matching rule 'choicePseudostate' transforms a source 'choice pseudo state' into a set of target model transitions, a set of input arcs, and a set of output arcs. This rule calls various 'unique lazy rules' to generate arcs on target models. The 'using' declarative part allows the declaration of variables that increase code legibility by giving names to different elements used in transformations.

```
--- join vertices serve to merge several transitions emanating
--- from source vertices in different orthogonal regions.
rule choicePseudostate {
    from
        inn: SysML!Pseudostate(inn.isChoicePseudostate())
    using
    {
        sourceTransition : SysML!Transition = inn.incoming.first();
        targetTransitions : Sequence(SysML!Transition) = inn.outgoing;
        sourceState : SysML!State = sourceTransition.target;
        targetStates : Sequence(SysML!State) = targetTransitions->
                                         collect(tr | tr.target);
    }
    to
        trans:distinct PNML!Transition foreach(tr in targetTransitions)(
            name <- thisModule.createTransitionName(sourceTransition,
                sourceVertex, inn),
            priority <- thisModule.iopt_createPriority(1)
        ),
        inArc: distinct PNML!Arc foreach(tr in targetTransitions)(
            id <- thisModule.createInArcName(tr, sourceState),
            source <- thisModule.createIoptInRefPlace(sourceState),
            target <- thisModule.createIoptInRefTransition(tr)
        ),
        outArc: distinct PNML!Arc foreach(tr in targetTransitions)(
            id <- thisModule.createOutArcName(tr, tr.target),
            source <- thisModule.createIoptOutRefTransition(tr),
            target <- thisModule.createIoptOutRefPlace(sourceState)
        )
}
```

Helpers can be used to define attributes, constants, and methods like functions. They can be reused throughout several transformations and in different contexts as utility functions, rule guards, and model properties analysis, as well as to navigate the source model. A task that is common to many transformation examples is the definition of **helpers** to get all elements of a specific type. Next, we present some of these helpers.

(1) Helper 'allStatesMachines' gets all state machines from a source model.

```
--- helper to get all statemachines
helper def: allStateMachines: Set(SysML!StateMachine) =
      SysML!StateMachine -> allInstances();
```

(2) The matching between source models elements and transformation rules is defined by conditions guards defined on rules level. As these conditions are normally extensive, we have created **guard helpers** to increase overall legibility. Helper 'isJoinPseudostate' is a simple guard helper that verifies if vertex element is a pseudostate of the 'join pseudo state' kind. In a complete state machine, a join vertex must have at least two incoming transitions and exactly one outgoing transition [19].

```
--- Verify if pseudostate is of kind 'join pseudo state'
helper context SysML!Vertex def: isJoinPseudostate(): Boolean =
      self.oclIsTypeOf(SysML!Pseudostate)
      and self.kind = #join
      and self.incoming->size() >= 2
      and self.outgoing->size() = 1;
```

(3) Transforming elements from source model to target model, normally, doesn't depend only on the kind of source item, but also on the context where it is inserted and/or on the context of interconnected elements. It was necessary to create a vast number of **functional helpers** to navigate interconnected elements and to obtain the necessary collections or properties. Functional helper 'getOwners' gets all owners of context element, which can be of any type, and returns a sequence of all owners from context element to root item.

```
--- Get all owners of a given element
helper context OclAny def: getOwners(): Sequence(SysML!Element)=
   let itemOwner : SysML!Element =  self.owner in
      if self.oclIsUndefined() then
            Sequence {}
      else
            if(itemOwner.oclIsKindOf(SysML!Element)) then
               Sequence {self}->union(itemOwner.getOwners())
            else
               Sequence {}
            endif
      endif;
```

Previous helper 'getOwners' is used by various helpers to determine its relation of inclusion with other types of items. Helper 'isInsideCompositeState()' is one of these examples: it verifies if a transition is inside a composite state, which means that it must be inside of a composite state with or without a region inside it. This helper calls helper 'isCompositeState()' to determine if SysML element is of the composite state kind.

```
--- Verify if a transition is inside a composite state
helper context SysML!Transition def : isInsideCompositeState():
Boolean =
    let elements : Sequence(SysML!Element) = self.getOwners() in
        if(elements.size() >= 2) then
            elements.at(1).isCompositeState() or
            elements.at(2).isCompositeState()
        else
            elements.at(1).isCompositeState()
        endif;
```

5   Discussion

When implementing a translation from a source model to a target model, it is necessary to create rules that match source elements and generate target elements. When there is no clear relation between them, it is not easy to identify which elements from source should be used, and which matched element should be generated.

Currently ATL has four compilers: 2004, 2006, 2010, and EMF Transformation Virtual Machine (EMFTVM). Present research was implemented with compiler version 2010, which has important limitations on traceability (tracing mechanism), not permitting the use of all types of rules as they lack the tracing mechanism. Moving to EMFTVM, which has advanced language features, will allow some code improvements. ATL documentation is spread along Eclipse ATL site, wikis, papers, and forums. A book or some other documentation explaining "How to do transformation" and not "How to use methods/functions" is missing. Even so, using ATL as a model to model transformation language simplified all the translation process, allowing an emphasis on translation technics in opposition to programming languages issues. Preliminary results, allow us to conclude that the strategy to convert Behavior Models to Petri Nets is possible and of great interest to strengthening the assertion of Petri Nets as a formalism suitable for integration of models.

6   Conclusions and Future Work

The implementation of all transformations between SysML state machines and IOPT nets is a significant contribution to be added to the framework for the project of embedded systems using co-design techniques. It is now possible to go from SysML state machines to IOPT nets, verify their properties, optimize the model, generate code, visualize, and execute. As an overall view of the implementation, it is interesting to note that 80% of all transformation rules were implemented using the declarative language style, recommended as best practice [18]. The use of variables on the imperative 'using' rules part, rule guard helpers, and called rules, allowed an overall good code readability. As near future work, it is necessary to enhance understandability of transformations and improve their correctness, which will lead to the identification and analysis of the used modeling patterns, as well as refactoring and abstraction mechanisms on model to model transformations. Finally, formal techniques for checking semantics equivalence of MDA transformations [20] can be used to validate implemented transformations.

References

1. Gasevic, D., Djuric, D.: Devedzic, V., Selic, B.V., Bézivin, J.: Model Driven Engineering and Ontology Development. Springer (2009) ISBN-13: 978-3642101342
2. Moutinho, F., Gomes, L., Ramalho, F., Figueiredo, J., Barros, J.P., Barbosa, P., Pais, R., Costa, A.: Ecore Representation for Extending PNML for Input-Output Place-Transition Nets. In: IECON 2010 - 36th Annual Conference of the IEEE Industrial Electronics Society, Phoenix, AZ, USA, pp. 2010–2036 (2010)
3. OMG: OMG MOF 2 XMI Mapping Specification. v2.4.1. (2013), http://www.omg.org/spec/XMI/
4. ATL - A Model Transformation Technology, http://projects.eclipse.org/projects/modeling.mmt.atl (accessed on December 30, 2013)
5. OMG: MDA - The Architecture of Choice For A Changing World (2013) , http://www.omg.org/mda/
6. Gomes, L., Barros, J.P., Costa, A., Nunes, R.: The Input-Output Place-Transition Petri Net Class and Associated Tools. In: INDIN 2007 - 5th IEEE International Conference on Industrial Informatics, Vienna, Austria, Julho 23-26 (2007)
7. Pais, R., Gomes, L., Barros, J.P.: Towards Statecharts to Input-Output Place Transition Nets Transformations. In: Camarinha-Matos, L.M. (ed.) Technological Innovation for Sustainability. IFIP AICT, vol. 349, pp. 227–236. Springer, Heidelberg (2011)
8. Pais, R., Gomes, L., Barros, J.P.: From UML State Machines to Petri nets – History Attribute Translation Strategies. In: 37th Annual Conference on IECON 2011. IEEE Computer Society (2011)
9. Markopoulos, P., Mackay, W.: Awareness Systems - Advances in Theory, Methodology, and Design. Springer (2009) ISBN 978-1-84882-476-8
10. Future Internet Enterprise Systems (FInES): Embarking on New Orientations Towards Horizon 2020 (2013)
11. Lochmann, H.: HybridMDSD: Multi Domain Engineering with Model Driven Software Development Using Ontological Foundations, PhD Dissertation (2010)
12. Steinberg, D., Budinsky, F., Paternostro, M., Merks, E.: EMF: Eclipse Modeling Framework, 2nd edn. Addison-Wesley (2008) ISBN-13: 978-0-321-33188-5
13. OMG: Meta Object Facility (MOF) 2.0 Query/View/Transformation Specification. Version 1.1 (2011)
14. OMG: OMG Systems Modeling Language, v 1.3 (2012), http://www.omg.org/spec/SysML/

15. Choppy, C., Petrucci, L., Reggio, G.: A Modelling Approach with Coloured Petri Nets. In: Kordon, F., Vardanega, T. (eds.) Ada-Europe 2008. LNCS, vol. 5026, pp. 73–86. Springer, Heidelberg (2008)
16. Zurawski, R., Zhou, M.C.: Petri Nets and Industrial Applications: A Tutorial. IEEE Transactions on Industrial Electronics 41(6) (1994)
17. Ribeiro, J., Moutinho, F., Pereira, F., Barros, J.P., Gomes, L.: An Ecorebased Petri Net Type Definition for PNML IOPT Models. In: INDIN 2011 - 9th IEEE International Conference on Industrial Informatics, Caparica, Lisbon, Portugal, pp. 777–782 (2011), doi:10.1109/INDIN.2011.6034992 ISBN 978-1-4577-0434-5
18. Allilaire, F., Bézivin, J., Jouault, F., Kurtev, I.: ATL - Eclipse Support for Model Transformation. In: Proc. of the Eclipse Technology Exchange Eorkshop (ETX) at ECOOP (2006)
19. OMG: Unified Modeling Language™ (OMG UML), Superstructure. v2.4.1,
    http://www.omg.org/spec/UML/
20. Barbosa, P., Ramalho, F., Figueiredo, J., Junior, A., Costa, A., Gomes, L.: Checking Semantics Equivalence of MDA Transformations in Concurrent Systems. Journal of Universal Computer Science 15(11), 2196–2224 (2009), doi:10.3217/jucs-015-11-2196
21. Jiang, M., Ding, Z.: From Textual Use Cases to Message Sequence Charts. In: Information Engineering and Applications. Lecture Notes in Electrical Engineering, vol. 154, pp. 732–739. Springer (2012)

Strategies to Improve Synchronous Dataflows Analysis Using Mappings between Petri Nets and Dataflows

José-Inácio Rocha[1,2,3], Octávio Páscoa Dias, and Luís Gomes[1,3]

[1] Universidade Nova de Lisboa
Faculdade de Ciências e Tecnologias, Portugal
[2] Escola Superior de Tecnologia de Setúbal, Setúbal , Portugal
[3] UNINOVA – Centro de Tecnologia e Sistemas, Portugal
`jose.rocha@estsetubal.ips.pt`, `lugo@uninova.pt`
`octavio.pdias@gmail.com`

Abstract. Over the last decades a large variety of dataflow solutions emerged along with the proposed models of computation (MoC), namely the Synchronous Dataflows (SDF). These MoCs are widely used in streaming based systems such as data and video dominated systems. The scope of our work will be on consistent dataflow properties that can be easily demystified and efficiently determined with the outlined mapping approach between Dataflows and Petri nets. Along with this strategy, it is also highlighted that it's of a major relevance knowing in advance the proper initial conditions to start up any SDF avoiding buffer space over dimensioning. The methodology discussed in this paper improves the outcomes produced so far (in Petri net domain) at design stage aiming at knowing the amount of storage resource required, as well as has a substantial impact in the foreseen allocated memory resources by any signal processing system at the starting point and also points out new directions to minimize the buffer requirements at design stage.

Keywords: Dataflows, Petri nets, Synchronous Dataflows, Place and Transition Invariants, Models of Computation.

1 Introduction

Over the last decades a multitude of dataflow solutions emerged along with requirement always supported by the needs of each new Model of Computation subtleties and peculiarities. Researchers have developed a few number of dataflow variants, trying to model the different constraints presented in the firing rules. These constraints about the firings rules are what differentiate several developed MoCs. For example in Cycle-static Dataflow [1], tokens rates can be scheduled in a periodic way, while in Dynamic Dataflow [2] token rates can be changed after each schedule. Another variant called Parameterized Synchronous Dataflow [3] allows the tokens rates to be defined by a parameter, which can be changed only between schedules.

Synchronous Dataflow (SDF) is a Model of Computation where the number of tokens consumed or produced by an actor is known in advance at compile time,

L.M. Camarinha-Matos et al. (Eds.): DoCEIS 2014, IFIP AICT 423, pp. 237–248, 2014.

i.e. without executing the model. This way, a SDF can be viewed as a statically schedulable model [4]. Every SDF model is composed by nodes and arcs (also usually called edges or channels). The nodes (graphically correspond to circles or ellipses) represent the functional or activity elements, while the edges (symbolized by arrows) represent communication channels between them. These channels are conceptually first-in-first-out queues. Edges are tagged with weights appended on both ends by non-negative integers. Due to their activity nodes consume and produce an establish amount of tokens on each firing based on tagged weights in the arcs. Tokens (represent by black dots) can also be seen as delays; therefore edges may contain initial markings which correspond to an initial number of delays.

SDF model have demonstrate their adequacy in many signal processing and multimedia domains, in which there is no data-dependency in each schedule, i.e. the tokens rates remains constant after each actor firing, therefore the rates of tokens are fixed by nature. But this weakness (lack of expressiveness) can be viewed as its strength, since it's possible to analyze all models for deadlock or bounded buffers.

Several advantages can be outlined to SDFs models, namely the easiness in modeling signal processing multi-rate systems, the ability to generate static schedule at compile time (allowing to know the required resources in advance), and the sequential static schedules can be optimized for buffer memory minimization, as well as allowing efficient simulation and efficient code generation. In a nutshell, SDF are well-matched to signal processing and communications systems. In [5] and [6] the researchers aimed at unveiling the effective and potential maximum number of tokens for each arc in every Synchronous Dataflow to foresee at compile time the necessary amount of storage resources, i.e. at design stage, as well as the study of dataflow models under a cyclic and continuous flow of data, whereas in this paper new developments are highlighted in the scheduling analysis and on reducing the allocated buffer memory by buffer sharing.

The paper is organized in seven topics. After a short introduction and motivation about dataflows, a brief discussion about the issue under investigation in the present paper to the relationship to collective awareness systems is outlined. In a nutshell, the state of the art provides an overview upon the matter under research. Section 4. ("The proposed methodology") presents and depicts the basic ideas behind the mappings outlined between dataflow and Petri net domains. Experimental work and results are shown in section 5. In section 6 it is presented in a summarized way the tools and computational environments used to work out the conclusions. The paper ends up with some guidelines that will rule the further research work on the key area of interest.

2   Relationship to the Collective Awareness Systems

Collective awareness is an issue that is nowadays being addressed in existing literature by several researchers due to the need to improve the knowledge of the impact of the different scenarios that are growing mainly on the side of the technological "footprint". The need to seek for understanding at what surround us, which make us above all a social creature, is demanding a worldwide coordination and awareness to gather in a more fruitful way all these new collaborative emerging realities.

The proposed approach in this document maintains a close regarding to collective awareness and aims its future research to a subject which is a topic being conveyed as a worldwide awareness (a wide collective awareness) in terms of earth's sustainability, the optimization energy consumption in order to fulfill very tight power requirements demanded by a target application or an embedded system. It is a vital subject since our society is founded on energy and due to increasing demanding on portable electronic devices and massive utilization, optimization of power (energy) consumption has become a critical design point. The industry of low power consumer electronics is currently submitted to a huge growth, while the power dissipation of digital systems experiences an increase in device density, complexity and speed, and the last but not the least the growing interest in lower power chips and systems are driven by business and demanding technical needs. Researchers and system developers are aware of the tradeoff between power and performance [7]. Moreover, tools and strategies can be developed or improved to go on even further to pursue the goal of devising optimized applications and embedded systems on different platforms. Memory resources are scarce in every signal processing system, therefore the contribution of tools to ameliorate this topic are of great significance nowadays when used together with the lowest sleep and active power draws hardware techniques. Our approach intends to be a contribution aiming at knowing the minimal allocated resource memory even at and during a system's startup period.

3   Summarized Overview of the Supporting Formalisms

Actually Embedded Systems have a multitude of constraints and non-functional requirements, as a result to perform the synthesis or code generation efficiently of models the practitioner must deal with several issues, ranging from energy constraints, buffer requirements, memory efficiency and performance. To deal with these matters, two modeling formalisms are used in this work: Dataflows and Petri nets. In order to present Dataflows and Petri nets a summarized overview is provided, for a more complete and formal descriptions of Petri nets see for example [8, 9 and 10] and for Dataflows see [11, 12 and 13].

3.1   Dataflows

The basic idea of the principles of a language for parallel processing outlined by Gilles Kahn [11] in his seminal paper was to represent programs using nodes and arcs. Later on, Dennis [12] established the dataflow principles aiming at exploitation of parallelism in programs. In a dataflow, nodes are associated to computational program activity, and arcs connecting nodes convey the flow of information. Arcs are First-In-First-Out queue channels with blocking read operations and non-blocking write operations. Lee et al. [13] presented in 1987 a special kind of dataflow known as Synchronous Dataflow (SDF) where the token rates (incoming and outgoing) are fixed and established by nodes on each arc, thus improving the runtime overhead of the modeled system.   Nowadays SDFs are a mature approach and are well disseminate among stream based system, namely parallel and distributed signal processing systems, since one can detect deadlocks at compile time, as well as have demonstrate their adequacy in specifying multirate systems that can be statically scheduled. Figure 1 (a) illustrate a SDF model where some arcs contain black dots representing initial markings (which can also be known as tokens or delays).

3.2   Petri Nets

A Place-Transition Petri net is a special bipartite directed graph (consisting of two kind of nodes, places and transitions), and can be viewed as a graphical and mathematical modeling tool. As a graphical tool, Petri nets are a valuable tool since they can be used as a visual aid to simulate the concurrent and dynamic activities of the systems. As a mathematical tool, it is possible to address algebraic equations, state equations and other mathematical formalisms that govern the system's behavior and structure. The origin theory of Petri nets is based on the work of Carl Adam Petri dissertation's published in 1962. Graphically, places are drawn as circles or ellipses, and transitions as bars or rectangles. Arcs are tagged with positive integers representing weights, and are either from a place (/transition) to a transition (/place). Places may hold tokens defining the state of the system. In the graphical environment tokens assigned to places correspond to black dots or positive integer numbers. The activity of Petri nets is ruled by enabling and firing transitions. In a word, a transition is enabled if each input place is marked with a minimum number of tokens fixed by the arc weight, and whenever enabled the transition will eventually fire. In figure 1. (b) one can see an example of a Petri net model.

4   The Proposed Methodology

The fundamental idea of our methodology is to make use of an inter-relation between a Synchronous Dataflow (SDF) and a transformed equivalent Petri net (PN). The idea was first presented in [14]. This way one can encompasses and take benefit from the well known properties verification capabilities of Petri nets, namely invariants and reachability analysis. Based on this inter-relation between SDF and PN, valuable information is added to the scheduler allowing one to gain insight knowledge about the storage resource in each arc, as well as the resource allocation of the whole SDF by realizing an inverse mapping into dataflow domain.

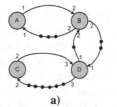   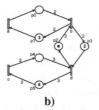

a)   b)

Fig. 1. An example of a Dataflow model, where some arcs contain initial tokens (a) and Petri net model (b)

The complete mapping strategy is depicted in figure 2. In a summarized way, one tries to find a relationship between every element in each domain, keeping in mind that each of these environments due to its nature has passive and active elements.

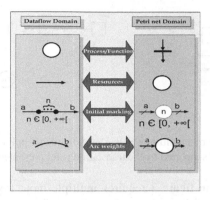

Fig. 2. Proposed mapping between dataflow and Petri net domain

This way, a process/function (represented by a circle) in a dataflow corresponds to a transition in Petri net domain (depicted by a bar). Since arcs in dataflows represent paths that conveys the oriented stream of information, arrows are used between the processes. The matching element on the Petri net side for the oriented path is a circle, embodying normally conditions, states, resources or objects. Arcs in a dataflow also include weights and initial markings denoted by non negative integer tagged at both ends of the arrow and black dots, respectively. The previous two characteristics find their matching elements in the lower part of figure 1, on the side of Petri net domain.

To illustrate our approach, a dataflow is presented in figure 1 (a) and the corresponding equivalent Petri net model is shown in figure 1 (b).

A Petri net with **m** places and **n** transitions is represented by a matrix *(mxn)* called the incidence matrix C. The entries of the incidence matrix are integers representing the token balance associated to the firing of every transition in the Petri net. If there is token gain the numbers are positive, for a loss and no change on the associated transitions, negative and zero entries in C are used respectively. This matrix is the key element to establish the transition or fundamental equation which governs the new markings on the Petri net given an initial marking set. Based on a qualitative analysis of PN two properties support our analysis: (1) Place Invariant (P-Invariant); and (2) Transition Invariant (T-Invariant). Based on the incidence matrix one creates a set of homogeneous linear equations which allows finding the so called P and T-Invariants. The structural domain analysis is aimed at designing systems without any kind of inconsistencies coping with properties which stay constant during model's execution. Maximum potential number of tokens is computed with P-Invariants, as presented in [15], whereas the maximum effective number of tokens on each arc in a dataflow is achieved summing all values presented in the reachability tree. More over with T-Invariants one can foresee the potential initial markings (M_0) for each place. Using the reachability tree one can keep track of the maximum number of tokens present on each arc during a schedule. This way the buffer size for each individual arc is foreseen on the dataflow. Gathering all these values the total buffer requirements for the arcs present in a dataflow is estimated.

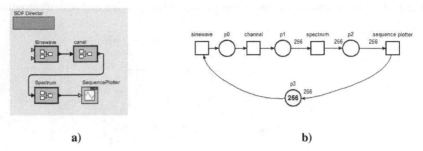

a)   b)

Fig. 3. An implementation of a multi-rate SDF Dataflow model in Ptolemy environment (a) and Petri net equivalent model supported by the translation rules presented in Fig. 2

The spectrum actor in figure 3 requires multiple input tokens to fire. This input multiplicity is established by a parameter called order, which is always a number dependent of power 2. The icon "SDF Director" governs the proposed dataflow model by defining the number of iteration to be performed and additionally determines the number of times each actor fires in one iteration using the balance equations. The multi-rate SDF model presented in figure 3. (a) is called well ordered chain structured SDF, and for these cases it is necessary to add an extra place (p_3) to make the whole Petri net reversible and live (for further reading see [9]), because at the end of each periodic schedule the digital signal processing system exhibits cyclic behavior since there is always an imbalance in the number of tokens. Besides this condition is central to evolve our methodology which is based on the P/T-Invariants in Petri net domain (see [6]).

5   Experimental Work and Results

One may argue that the reachability tree is a bottleneck due to the state space explosion problem; however, it is possible to distribute the state space in a set of computing machines [9] limiting this drawback. The experimental work was carried on two kinds of (well-known) problems, the dining philosophers and distributed database, besides the process is still dependent on the choice of having a good hash function to guarantee also a good state's distribution among the processors.

5.1   Scheduling Analysis of Synchronous Dataflow

Analysis of ill-specified SDF at the design time or even the misbehaved ones is an issue of major relevance. In a SDF our concern is related to attain a static schedule (if one exists), where a finite periodic list of actors is schedule in a sequential way. A SDF can evidence either rate-inconsistency (motivated by unbounded memory problems), or inability to run, or even deadlock caused by lack of initial buffering. As stated in Lee et al. [4] "a periodic admissible sequential schedule (PASS) is a periodic and infinite admissible sequential schedule", one concludes that after a period has been executed, the size of each buffer returns to its initial state, i.e. or to its home state (as it is known in Petri net domain).

To illustrate our idea considers the following flowchart depicted in figure 4. Taking the equivalent Petri net model one can identify easily if either a SDF model is consistent or inconsistent, or has deadlocks, using the P-T Invariants (structural analysis) and the reachability tree (behavioral analysis). Therefore, it is possible to draw a methodology (on the side of Petri net domain) for each and every SDF to find out their runnability feature. Figure 5 presents all those cases of runnability in SDFs models.

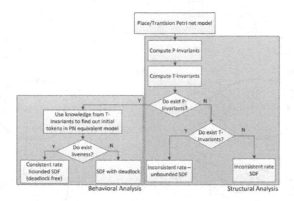

Fig. 4. Flowchart to identify the runnability (or rate consistency) of SDFs

As another example consider the simplified spectrum analyzer dataflow presented in figure 6 (a) and based on the translation rules outlined in section 4. one can get the Petri net model shown in figure 6 (b). Using the Petri net structural analysis, it is possible to know the maximum potential and effective amount of storage space for the static schedule identified, as stated before. The amount of storage space for each arc ($M(p_i)$ stands for the number of tokens in place p_i) are shown in figure 7 (a) and the associated static schedule in figure 7 (b). The total buffer requirement for the spectrum analyzer is 17 units, where each arc in dataflow domain maps to a place in Petri net domain. In the reachability tree achieved under Petri net domain one must find the home state in order to know the execution time of the periodic static schedule and at the same time can observe the execution pattern of each process in a single schedule as shown in figure 7 (b). If every process (transition) takes 2 time units, the entire schedule execution time will take place in 18 time units. From figure 7 (b), one can observe that in every periodic static schedule the Adaptive LPF (ALPF) process executes 4 times, FFT Zoom (FFT Z), Peak Detector (PD), Interpolator (INT) and Zoom Control (ZC) executes one time and that Decision (DECI) executes 2 times.

The use of the reachability tree can be seen as a bottleneck by some researchers due to the state space explosion problem; however it is possible to distribute the state space in a set of computing machines [16] limiting this drawback. Additionally to know the real impact of the initial conditions in the spectrum analyzer multiple simulations were performed using the computational environments to estimate the number of states generated, as well as the number of fired transitions in the foreseen static schedule.

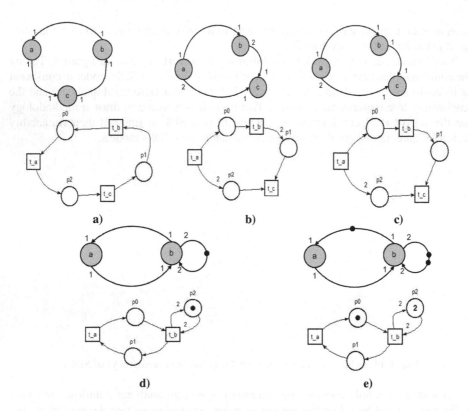

Fig. 5. Synchronous Dataflows and their equivalent Petri net models for different runnability cases; (a) rate consistent with bounded memory; (b) rate-inconsistent with unbounded memory; (c) rate-inconsistent – cannot run; (d) rate consistent with deadlock; (e) rate-consistent without deadlock

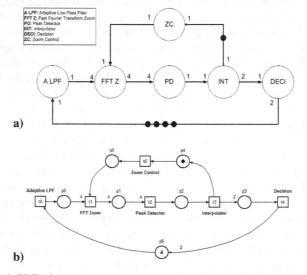

Fig. 6. SDF of spectrum analyzer (a) and Petri net equivalent model (b)

Observing *Table 1* among the various cases identified one (**case 1**) conducts the system to a minimum in the number of states and fired transitions. Therefore from the other cases in *Table 1* one can conclude that initial conditions (in place p_1 and p_4) will make the modeled system progress to some extra states that are not relevant for the periodic static schedule, thus augmenting the state space and at the same time unnecessarily over dimensioning the system. Thus it is important to develop an algorithm to find out the initial conditions that conduct the modeled system to a minimal state space and also reduces the required storage space.

5.2   Reducing Buffer Memory by Buffer Sharing

In order to reduce even further the allocated buffer space among the arcs it is important to unveil those that don't overlap their live ranges stressing that each arc is committed to a First-In-First-Out (FIFO) buffer. The strategy to follow is to observe the way each place in each state behaves traversing the state space of the schedule and then, decide what are the places that don't overlap at each state by grouping those that are disjoint in state space schedule.

Table 1. State space and fired transitions for several cases for different initial marking on places p_1 and p_4 ($M_0(p_1)$ and $M_0(p_4)$), for figure 6 (b)

Case	M0(p1)	M0(p4)	# states	# fired transitions
1	4	1	20	32
2	4	2	31	56
3	8	1	68	144
4	8	2	114	276
5	5	1	28	50
6	7	1	52	106

Amount of storage space per arc							Total
M(p0)	M(p1)	M(p2)	M(p3)	M(p4)	M(p5)	M(p6)	
4	4	1	2	4	1	1	17

a)

Schedule	Time stamp									
	1	2	3	4	5	6	7	8	9	10
1	ALPF, ZC	ALPF	ALPF	ALPF	FFTZ	PD	INT	DECI	DECI	
2										ALPF ,ZC

b)

Fig. 7. Buffer requirement for a periodic static schedule of spectrum analyzer dataflow (a), and the corresponding static schedule (b)

After this identification one is able to state which buffers can be fully shared. To illustrate our methodology, an example is used, the spectrum analyzer. The corresponding SDF is shown in figure 6. (a). The equivalent Petri net model of the spectrum analyzer (in figure 6. (b)) is used as an example to illustrate the issue of the buffer sharing. This way the following buffer sharing is possible: places p_0, p_1, p_2 and p_3 can share the same buffer and requires a buffer size of 4 units, which is the maximum amount of storage in this subset of places; places p_4 and p_5 can also share the same buffer and require a buffer with unitary dimension; and finally place p_6 needs a buffer of 4 units. Summing the contributions of each group of shared places one attains the total amount of buffer space of 9 units, representing a space memory saving of 47 percent in comparison to the initial solution.

6   Tools and Computational Environments

The development of the equivalent transformed Petri nets was performed by two non commercial tools that are freely available in the web: (1) Integrated Net Analyzer INA [17], a non graphical Petri net editor, that requires a previous knowledge of the INA syntax language to perform a net description; (2) Still TINA - TIme Petri Net Analyzer [18], a graphical editor with a stepper simulator to perform a step by step animation in a Petri net. The invariant analysis (used in section 5) was performed in both environments.

7   Conclusions and Further Work

The proposed methodology allows one start from a dataflow model, a Synchronous Dataflow, and based on a set of mapping rules achieve an equivalent Place – Transition Petri net model exploring the well-known verification properties: Behavioral (Structural) properties which are dependend (independent) on the initial marking. In this paper the focus is on the behavioral properties: reachability, boundedness, deadlocks, liveness, reversibility and home state. Using reachability tree one can identify if from a initial marking it is always possible to reach that initial state (referred in this case as home state), or know the firing sequence that brings the equivalent Petri net to the home state. Boundedness was useful to get the required buffer space for each Synchronous Dataflow arc and reveal if no design errors were committed at design stage. Liveness is a major property for the stream based SDF models since data flows sequentially, continuously and iteratively since ones tries to find a schedule, i.e., guarantee the absence of deadlocks.

On the side of structural properties the analysis is focused on the invariants (P and T) to achieve the buffer (token) capacity concerning each arc in SDF models. Initial conditions in dataflow models play an important role since with the proper initial conditions the system will not be over dimensioned, which are not desirable and should be avoided. Moreover, choosing different initial conditions one can state and conclude that the modeled system will evolve to some states that are not relevant for the periodic static schedule, thus augmenting the state space will jeopardize (more time consuming task) the search for specific states, such as for example the home state. In resume, the goal is to look for initial markings that make system runnable to complete periodic static schedules. As future work, one should find an algorithm to identify amongst all firing conditions those that lead the system to a minimal state space encompassing a finite periodic static schedule using the modeling formalism of PN. Also outline an algorithm (or an optimization algorithm) to find regions (or sub-regions) in the state space where a buffer sharing may be possible. This strategy will allow reducing to a great extent the buffer requirements.

Acknowledgments. This work was partially financed by Portuguese Agency "FCT - Fundação para a Ciência e a Tecnologia" in the Framework of projects PEst-OE/EEI/UI0066/2011 and PTDC/EEI-AUT/2641/2012 and by Instituto Politécnico de Setúbal (IPS) in the framework of projects of "3º Concurso de Projectos do IPS", reference 3CP-IPS-8-2009.

References

1. Bilsen, G., Engels, M., Lauwereins, R., Peperstraete, J.: Cycle-static dataflow. IEEE Transactions on Signal Processing 44(2), 397–408 (1996)
2. Buck, J.T., Lee, E.A.: Scheduling dynamic dataflow graphs with bounded memory using the token flow model. In: IEEE International Conference on Acoustics, Speech, and Signal Processing, ICASSP 1993, vol. 1, pp. 429–432 (April 1993)
3. Bhattacharya, B., Bhattacharyya, S.S.: Parameterized Dataflow Modeling for DSP Systems. IEEE Transactions on Signal Processing 49, 2408–2421 (2001)
4. Lee, E., Messerschmitt, D.G.: Static Scheduling of Synchronous Data Flow Programs for Digital Signal Processing. IEEE Transactions on Computers C-36(1), 24–35 (1987)
5. Rocha, J.-I., Gomes, L., Dias, O.: Petri net verification techniques on synchronous dataflow models. In: IECON 2011 37th Annual Conference on IEEE Industrial Electronics Society, pp. 3792–3797 (2011)
6. Rocha, J.-I., Páscoa Dias, O., Gomes, L.: Exploiting Dataflows and Petri Nets Mappings. In: 2013 11th IEEE International Conference on Industrial Informatics (INDIN), pp. 590–595 (2013)
7. Yeap, G.K.: Practical Low Power Digital VLSI Design. Springer (1997)
8. Giraud, C., Valk, R.: Petri Nets for Systems Engineering. A Guide to Modeling, Verification, and Applications. Springer, Heidelberg (2003)
9. Murata, T.: Petri Nets: Properties, Analysis and Applications. Proceedings of the IEEE 77(4), 541–580 (1989)
10. Peterson, J.L.: Petri Net Theory and the Modeling of Systems. Prentice Hall PTR, Upper Saddle River (1981)
11. Kahn, G.: The Semantic of a Simple Language for Parallel Programming. In: Proc. of the IFfP Congress 74, vol. 74, North-Holland Publishing Co. (1974)
12. Dennis, J.: First version of a data flow procedure language. In: Robinet, B. (ed.) Programming Symposium. LNCS, vol. 19, pp. 362–376. Springer, Heidelberg (1974)
13. Lee, E.A., Messerschmitt, D.G.: Synchronous data flow. Proceedings of the IEEE 75(9), 1235–1245 (1987)

14. Rocha, J.-I., Gomes, L., Páscoa Dias, O.: Dataflow Model Property Verification Using Petri net Translation Techniques. In: INDIN'2011 - 9th IEEE International Conference on Industrial Informatics, pp. 783–788 (2011), doi:10.1109/INDIN.2011.6034993, ISBN 978-1-4577-0434-5

15. Rocha, J.-I., Gomes, L., Páscoa Dias, O.: Analysing Storage Resources on Synchronous Dataflows using Petri Net Verification Techniques. In: IECON'2012 – The 38th Annual Conference of the IEEE Industrial Electronics Society (2012)

16. Boukala, M.C., Petrucci, L.: Towards distributed verification of petri nets properties. In: Barkaoui, K., Ioualalen, M. (eds.) Proceedings of the First International Conference on Verification and Evaluation of Computer and Communication Systems (VECoS 2007), pp. 13–24. British Computer Society, Swinton (2007)

17. INA - Integrated Net Analyser, Humboldt – Universita zu Berlin,
    http://www2.informatik.hu-berlin.de/starke/ina.html

18. TINA-TIme Petri Net Analyzer, Laboratoire d'Analyse et d'Architecture des Systémes,
    http://homepages.laas.fr/bernard/tina/

Application of Hypergraphs to SMCs Selection

Łukasz Stefanowicz, Marian Adamski, Remigiusz Wiśniewski,
and Jakub Lipiński

University of Zielona Góra, Institute of Computer Engineering and Electronics,
ul. Licealna 9, 65-417 Zielona Góra, Poland
{L.Stefanowicz,J.Lipinski}@weit.uz.zgora.pl,
{M.Adamski,R.Wisniewski}@iie.uz.zgora.pl

Abstract. The paper deals with selection of State Machine Components (SMCs) based on Hypergraphs theory. The entire selection process use Petri nets as benchmarks. As it is known, Petri nets are used for modeling of concurrency processes. The SMCs selection problem is classified as NP-Hard which means there does not exist polynomial algorithm which provides an exact solution. In the article we show three SMCs selection methods, advantages and disadvantages of each, results of comparison between traditional methods (exponential backtracking, polynomial greedy) and an exact transversal method based on hypergraphs theory, their efficiency and propriety. An exact transversal method allows to obtain exact solution in polynomial time if selection hypergraph belongs to xt-hypergraph class.

Keywords: Petri net, State Machine Component (SMC), hypergraph, exact transversal, concurrency hypergraph, sequential hypergraph, backtracking, greedy, algorithm of exact transversals.

1 Introduction

Petri nets are mainly applied in the modelling of concurrent processes since they enable describing both sequential and parallel relations [1], [2], [3]. Since the description mechanism of places and transitions is relatively simple, Petri nets are commonly applied. Unfortunately, the analysis may be impossible for the considerable sizes of the net and the exponential relation between its size and the number of states [1], [2], [4]. The selection of SM-components of a Petri net is applied in order to make the analysis possible, which helps to obtain a less complex problem [5].

The selection of SM-components may be performed with the use of classical methods, as well as with the use of the hypergraph model. It must be stated here that the application of hypergraphs allows the analysis in the polynomial time in the majority of cases [6], [7], [8].

In the article, three possible algorithms of covering have been presented, including the one which applies the hypergraph model: the backtracking algorithm, the greedy algorithm and the methods applying exact transversals of concurrency hypergraph [7],

L.M. Camarinha-Matos et al. (Eds.): DoCEIS 2014, IFIP AICT 423, pp. 249–256, 2014.

[9], [10]. The course of conduct consists in the determination of a concurrency hypergraph on the basis of the net. The hypergraph presented in a form of a matrix contains vertices (columns) which relate to places of a Petri net and also hyperedges (rows) reflecting relations between the places. Then a selection hypergraph is determined and subsequently an exact transversal, constituting the solution, is found [7], [8], [11]. The described process will be presented in details in the article.

One of goals of current PhD work is to develop algorithms using hypergraphs theory to analyze discrete states of concurrent automata. That goal implies following hypothesis: discrete states of concurrent automata can be efficiently and effectively analyzed using hypergraphs theory. Current article is a continuation of the work presented in [8].

2   Contribution to Collective Awareness Systems

The Collective Awareness Systems refer to algorithms that are considered as global systems. The most fruitful advantages are: easy diagnosis and usability. Industrial structures are constantly transformed. Infrastructure, technology resources, services and human potential permit the development of civilization and economy. Innovation is the main force of any economy, the essence is the constant search for new ideas and solutions. New technical solution should be based on Collective Awareness Systems because in this way it will be possible efficient use of facilities.

The algorithm and research methodology presented in the article are useful in case of global systems. Presented algorithms can be seen as a piece of a major system, which leads to the conclusion about the usefulness. Moreover, all methods shown in the article introduce new quality in the process of SM-Component selection. Reduction of the execution time of algorithms influences on financial savings. Proposed algorithms may be used as a part of the decomposition process of concurrent automata. Separate modules can be implemented on different FPGA devices, parts of the Collective Awareness Systems.

3   Main Definitions and Preliminary Notation

Current chapter contains the necessary definitions required to understand the topic.

Definition 1. Hypergraph H [12] is defined by a couple:

$$H=(V,E),\tag{1}$$

where:
$V=\{v_1,\ldots,v_n\}$ is a finite, non-empty set of vertices;
$E=\{E_1,\ldots,E_n\}$ is set of hyperedges, i.e., of set P(V).

Definition 2. Graph G [12] is a particular case of hypergraph H. Formally a graph is defined by a couple:

$$G=(V,E), \tag{2}$$

where:
V={$v_1,...,v_n$} is a finite, non-empty set of vertices;
E={$E_1,...,E_n$} is a finite set of unordered pairs of vertices, called edges .

Definition 3. A Petri net [1], [2], [13] is a 4-tuple:

$$PN=(P,T,F,M_0), \tag{3}$$

where:
P is a finite set of places, T is a finite set of transitions,
$F \subseteq (P \times T) \cup (T \times P)$ is a finite set of arcs, M_0 is an initial marking.

Definition 4. A SM-Component PN' of a Petri net PN is such its consistent subnet:

$$PN'=(P,T,F), \tag{4}$$

that:

$$\forall t \in T' : |\cdot t| = |t \cdot| = 1 \tag{5}$$

$$P' = \cdot T' \cup T' \cdot \tag{6}$$

$$F' = (P \times T') \cup (T \times P') \cap F \tag{7}$$

Definition 5. A transversal T (hitting set, vertex cover) of hypergraph H is set:

$$T \subseteq V, \tag{8}$$

containing vertices incident to each of the hypergraph edges.

Definition 6. Exact transversal [10] D of a hypergraph H is a set:

$$D \subseteq V, \tag{9}$$

of vertices of hypergraph H, which is incident to all the edges of hypergraph H, while each edge is incident to exactly one vertex of set D which forms an exact transversal.

Definition 7. Exact transversal hypergraph H_{XT} (xt-hypergraph) is such a hypergraph of which all the minimum transversals are simultaneously exact ones [9].

4   Problem Formulation

The selection of SM-components is essential for the decomposition process [14]. It provides the base for the target SM-components which enable covering the whole Petri net. Approximate algorithms enable obtaining the result which is mainly an redundant one, and hence the result is not optimal [15], [16]. Exact algorithms (colouring, backtracking) enable obtaining results which do not contain redundant SM-components because a complete set of solutions is searched for, thus the obtained result is optimal and the number of SM-Components minimal. The analysis applying the hypergraph model [7] enables obtaining the exact result in a polynomial time for the vast majority of cases [8]. It should be stated here that the selection is a problem of the NP-hard class therefore the algorithm allowing obtaining an exact result in a polynomial time is not known. The selection process may be divided into three basic stages [7]:

- determination of the concurrency hypergraph on the basis of the elementary net,
- determination of SM-Components (SMCs),
- proper selection of SMCs and determination of the solution of the cover.

5   The State of the Art

The selection of SM-Components belongs to the class of NP-hard problems. It means that there is no universal polynomial algorithm allowing the optimal solution to be obtained in a polynomial time. The decomposition of a Petri net in the form of SM-Components selection may be carried out with the use of classical methods which apply graphs, as well as the theory of hypergraphs. The selection basing on the theory of graphs involves graph colouring [7], whereas the theory of hypergraphs applies the method of the determination of transversals of the exact hypergraph, proposed in [7]. The method has been expanded by the application of the exact transversals hypergraph, proposed in [8]. The exact methods demonstrate the exponential computational complexity [17], hence they require application methods. They, in turn, allow obtaining a solution comprising generally excessive/redundant subnets, in a acceptable time, though [7]. The method applying the exact hypergraph is not always effective since there is no polynomial algorithm allowing stating if a given hypergraph belongs to the class of exact hypergraphs. The method proposed in [8] involves the application of the hypergraph of exact transversals, and since there is a polynomial algorithm allowing testing if a given hypergraph belongs to the xt-hypergraph class in a polynomial time [10], it is possible to obtain an optimal solution on the condition that the selection hypergraph belongs to the xt-hypergraph class. In other words, it is possible to obtain exact cover, and hence an optimal solution provided that the selection hypergraph belongs to the class of exact transversals hypergraphs.

5.1   Backtracking Algorithm

The backtracking method enables obtaining the exact solution, and thus an optimal one, since it checks all the possible combinations. As a result, the smallest cover is found. The most considerable drawback of this solution is its computational complexity, which is exponential. It means that the method will work only with small nets. The algorithm's procedure is very simple. In the first step, a subnet is selected and then it is removed. If the remaining set of SM-Components still constitutes a net cover, the result is understood to be better than the previous one. In the subsequent stages, the algorithm searches for an optimal solution for all SM-Components of the elementary net. The backtracking method will always return the best solution, which is undoubtedly its immense advantage. Its biggest drawback is its exponential computational complexity, which in practice means that the given problem may not be solved, or the solution may not be determined in an acceptable time.

5.2   Greedy Algorithm

The greedy algorithm searches for the solution on the basis of a decision which is locally optimal [18], [19]. Each step of the method relies on the selection of a currently optimal solution. One of its advantage is its computational complexity. Its major drawback is the fact that the solution seen as locally optimal not always is globally optimal. Due to the fact, that the selection of SM-Components is equivalent to the hypergraph of selections, the greedy algorithm has been presented below in a vertex version. Its procedure may be presented as follows:

- At the beginning the cover set is empty.
- The selection of an essential vertex (the only one which belongs to a given edge). If such does not exist, a vertex of the largest degree is selected. Here is the moment when the selection of the locally optimal solution takes place. If the vertex is impossible to be determined univocally, the first possible one is selected.
- The selected, locally optimal, vertex is added to the cover set and removed from the selection hypergraph.
- The set is verified in terms of constituting a cover. If not, step 2 is repeated.

5.3   Algorithm of Exact Transversals

The algorithm of exact transversals has been described in [7]. The idea of the method has been presented in figure 1. The first step is the transposition of the sequential hypergraph matrix and the cyclical reduction called Cyclic-Core [4]. It involves the reduction of dominated columns and dominating rows. Then, for such an obtained hypergraph, the first exact transversal is determined. It is worth noticing that the whole process may be performed in a polynomial time provided that the hypergraph subjected to the cyclical reduction belongs to the class of exact hypergraphs. Moreover, the hypergraph should be verifiable in a polynomial time in terms of its affiliation to the class of xt-hypergraphs (an exact hypergraph is a generalisation of the xt-hypergraph), and hence yet at the stage of the hypergraph obtaining, after the reduction, it is possible to evaluate the probability of achieving the exact solution. As presented in [8], 80% of the tested nets contained a selection hypergraph belonging to xt-class. It means that for 80% of tested cases, it is possible to achieve an exact solution in a polynomial time, which undoubtedly is a great advantage of the method.

The application of the theory of hypergraphs enables the reduction of the execution time from exponential to polynomial in the vast majority of cases.

6   Selection Process of SM-Components

The whole process of the selection of SM-Components may be presented in the following steps:

- An input/elementary net is given.
- On the basis of the input/elementary net, a concurrency hypergraph is determined in which vertices respond to the places of the Petri net, whereas hyperedges describe relations between the places.
- On the basis of the concurrency hypergraph, a sequential hypergraph is determined in which vertices respond to the places whereas hyperedges determine SM-Components [7].
- The proper selection of SM-Components with the use of the backtracking method, the greedy algorithm or exact transversals.
- Final solution is determined.

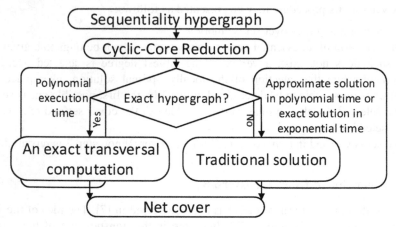

Fig. 1. Idea of exact transversal method

7   Experiments

The essence of the experiments is to use benchmarks to reveal the efficiency and effectiveness of the presented methods. The execution times for particular algorithms were compared with taking into account their sensitive moments and the results they return. Altogether, 30 benchmarks were used, nets of which belong to various classes: MG (Marked Graph), FC (Free-Choice), EFC (Extended Free-Choice) as well as SN (Simple Net). The column "a minimal number of SMCs" included a division in terms of the applied algorithms: Backtracking (B), Greedy (G), Exact transversals (T).

For the ampleness of the data, only representative benchmarks were selected and are presented in table 1. As can be seen, the algorithm using exact transversals dominates over the backtracking and greedy ones in terms of the execution times, providing similarly good results. The greedy algorithm does not provide optimal value for MG_NP net, whereas the backtracking one takes a considerable time to be executed, unlike the exact transversal algorithm which in the shortest time returns the optimal solution. Both the exponential and the exact backtracking algorithm provide the optimal solution. The exact transversal method is as good as Backtracking which can obtain results in more than 200% shorter execution time. Presented results of research were obtained on the following workstation: AMD FX4100 CPU, 8GB RAM.

Table 1. Partial results of research

Benchmark name, number of transitions, places	Cover determi- -nation time B [ms]	Selection determi- -nation time B [ms]	Cover determi- -nation time G [ms]	Selection determi- -nation Time G [ms]	Cover determi- -nation Time T [ms]	Selection determi- -nation Time T [ms]	Minimal number of SMCs B, G, T
bridge, 6, 8	119	338	128	254	2	162	2, 2, 2
cncrr002, 7, 11	281	504	288	449	2	162	5, 5, 5
Frame, 10, 13	204	748	201	387	2	188	3, 3, 3
P2N2, 7, 13	702	2901	711	1684	9	841	4, 4, 4
MG_NP, 5, 10	439	16521	438	644	5	188	3, 4, 3
Philoso2, 10, 14	382	651	312	577	2	194	6, 6, 6

8   Summary and Conclusions

The paper presents the selection of SM-Components of a Petri net applying the hypergraph model. It should be noted that the selection is an NP-hard problem. The main objective of the article was to present the possibilities of solving the selection problem with the use of hypergraphs. Thus, three algorithms were presented: the backtracking one, the greedy one and the one applying exact transversals. The last mentioned was proposed and described in [7], whereas its modified version was presented and subjected to research in [8].

The essence of the research was to demonstrate the effectiveness and efficiency of the particular algorithms. Due to the considerable sizes of research results, a group of representative benchmarks was selected and on their basis partial results were presented. It may be noted that the backtracking algorithm, which in fact provided optimal solution, takes frequently much more execution time. This is connected with the exponential computational complexity. The greedy algorithm provides the solution which is locally optimal, which not always is equivalent to a globally optimal solution. The execution time is frequently shorter than for the backtracking algorithm, but the solution may not be the best. The algorithm applying exact transversals provides exact solution for the vast majority of cases in a polynomial time [8], which is also confirmed by the presented research. It should be noted that the given solution is just as good as the one obtained with the use of the backtracking algorithm, which is characterised by an exponential computational complexity. The execution time for the algorithm of exact transversals is by over 30% lower than the execution time for the greedy algorithm and by over 200% lower than for the backtracking algorithm, the exact transversal method is as good as Backtracking.

References

1. Murata, T.: Petri nets: properties, analysis and applications. Proceedings of the IEEE 77, 541–580 (1989)
2. Karatkevich, A.: Dynamic analysis of Petri net-based discrete systems. Springer (2007)
3. Bukowiec, A., Mróz, P.: An FPGA synthesis of the distributed control systems designed with Petri nets. In: Proc. IEEE 3rd Int. Conf. on Networked Embedded Systems for Every Application, Liverpool, UK (2012) [6]
4. Karatkevich, A., Wiśniewski, R.: Computation of Petri nets covering by SM-components based on the graph theory. Electrical Review, 141–144 (August 2012)
5. Rudell, R.L.: Logic Synthesis for VLSI Design. PhD thesis, University of California (1989)
6. Wiśniewski, R., Wiśniewska, M., Adamski, M.: A polynomial algorithm to compute the concurrency hypergraph in Petri nets. Measurement Automation and Monitoring 58(7), 650–652 (2012) (in Polish)
7. Wiśniewska, M.: Application of Hypergraphs in Decomposition of Discrete Systems. LNCCS, vol. 23. University of Zielona Góra Press, Zielona Góra (2012)
8. Stefanowicz, Ł., Adamski, M., Wisniewski, R.: Application of an exact transversal hypergraph in selection of SM-components. In: Camarinha-Matos, L.M., Tomic, S., Graça, P. (eds.) DoCEIS 2013. IFIP AICT, vol. 394, pp. 250–257. Springer, Heidelberg (2013)
9. Knuth, D.: Dancing links. Millennial Perspectives in Computer Science (2000)
10. Eiter, T.: Exact transversal hypergraphs and application to boolean u-functions. Journal of Symbolic Computation 17(3), 215–225 (1994)
11. Desel, J., Juhás, G., Lorenz, R.: Concurrency Relations and the Safety Problem for Petri Nets. In: Jensen, K. (ed.) ICATPN 1992. LNCS, vol. 616, pp. 299–309. Springer, Heidelberg (1992)
12. Berge, C.: Hypergraphs: Combinatorics of Finite Sets. North-Holland (1989)
13. Bukowiec, A., Doligalski, M.: Petri net dynamic partial reconfiguration in FPGA. In: Moreno-Díaz, R., Pichler, F., Quesada-Arencibia, A. (eds.) EUROCAST. LNCS, vol. 8111, pp. 436–443. Springer, Heidelberg (2013)
14. Karatkevich, A.: SM-Components problem reductions of Petri nets. Telecommunication Review (2008) (in Polish)
15. Barkalov, A., Titarenko, L., Bieganowski, J., Miroshkin, A.: Synthesis of Compositional Microprogram Control Unit with Dedicated Area of Inputs. In: Lecture Notes in Electrical Engineering vol. 79, pp. 193–214 (2011)
16. Barkalov, A., Kołopieńczyk, M., Titarenko, L.: Design of CMCU with EOLC and encoding of collections of microoperations. (79), 262–265 (2007)
17. Blanchard, M.: Comprendre, maitriser et appliquer le Grafcet. Automatisation Production (1979)
18. DeMicheli, G.: Synthesis and Optimization of Digital Circuits. PhD thesis, McGraw-Hill Higher Education (1994)
19. Bukowiec, A., Barkalov, A.: Automata Implementation in FPGA devices with Multiple Encoding States. Electrical Review, 185–188 (2009)

Part X
Multi-energy Systems

Modeling Energy Demand Dependency in Smart Multi-Energy Systems

N. Neyestani[1], Maziar Yazdani Damavandi[1], Miadreza Shafie-khah[1], and João P.S Catalão[1,2,3]

[1] University of Beira Interior, Covilhã, Portugal
catalao@ubi.pt
[2] INESC-ID, Lisbon, Portugal
[3] IST, Univ. Lisbon, Portugal

Abstract. Smart local energy networks provide an opportunity for more penetration of distributed energy resources. However, these resources cause an extra dependency in both time and carrier domains that should be considered through a comprehensive model. Hence, this paper introduces a new concept for internal and external dependencies in Smart Multi-Energy Systems (SMES). Internal dependencies are caused by converters and storages existing in operation centers and modeled by coupling matrix. On the other hand, external dependencies are defined as the behavior of multi-energy demand in shifting among carriers or time periods. In this paper, system dependency is modeled based on energy hub approach through adding virtual ports and making new coupling matrix. Being achieved by SMESs, the dependencies release demand-side flexibility and subsequently enhance system efficiency. Moreover, a test SMES that includes several elements and multi-energy demand in output is applied to show the effectiveness of the model.

Keywords: Dependency modeling, internal and external dependency, smart multi-energy system.

1 Introduction

Environmental issues as well as other critical issues such as global warming, emission of greenhouse gases and declining fossil resources have initiated trends towards sustainable development of energy and manipulation of various energy resources capacity. Sustainable energy development and preparation of future energy portfolio have to be in a manner that besides saving the environment does not interfere with the provision of future generation energy needs. In this regard, Smart Multi-Energy Systems (SMES) and consequently, Distributed Energy Resources (DERs) are two main tools for achieving these goals [1]. SMES accelerates the integration of various renewable-based resources and facilitates the participation of DERs in system operation. Development of DERs in SMES will provide bi-directional relation between distribution level and upstream grid which brings the advantages of network loss reduction and increases the reliability on demand side [2].

L.M. Camarinha-Matos et al. (Eds.): DoCEIS 2014, IFIP AICT 423, pp. 259–268, 2014.

Moreover, DERs introduce dependency in both carrier and time domain to SMES through converters and storages. Distributed energy converters transform input energy carrier to output energy carrier and make mutually dependent carriers. In SMES with high penetration of DERs, these converters are spread all over the grid. As a result, network in various points has dependency between energy carriers [3]. Studies in such networks have to be simultaneously integrated, take into account all carriers' effects.

Two modeling approaches entitled *"energy hub system"* and *"matrix modeling"* were developed to model these dependencies ([4] and [5]). These models are based on some operation centers (energy hubs which mostly consist of cogeneration or tri-generation) and their interconnectors. The operation centers convert input energy carriers to the output required services and the interconnectors transmit energy carriers between operation centers. The energy flow in interconnectors has been investigated in [6] and [7] and new models have been proposed for integrated load flow of gas and electric networks. In [8], the energy hub system operation considering energy carriers price and operation objectives is surveyed. DERs impacts on energy hub system model are investigated in [9] and [10]. In these models the energy service vectors in the output (demand side) of operation centers are independent from each other.

Literature review reveals that researches could not successfully fulfill all the requirements of modeling dependency in demand side. Hence, this paper introduces a new concept of internal and external dependencies in SMES. Internal dependencies are assumed to be caused by converters and storages existing in operation centers and modeled by coupling matrix. External dependencies are defined as the behavior of multi-energy demand in shifting among carriers or time periods. This means that, in SMES, operation centers deliver energy to multi-energy demands which also consist of various end-use converters and storages [11]. It is presumed that demand consumes various energy carriers and one end-service may be achieved by multi input energy vectors (e.g. cooking in a household that can take place with either electric or gas oven). According to the intimate relation between multi-energy demands and SMES, it would be very complicated to model the multi-energy demands' elements inside an operation center coupling matrix. On the other hand, ignoring these external dependencies in decision making scheme of SMES may also reduce the accuracy of studies.

Two main approaches have been adopted for confronting with dependency of multi-energy demand in SMES: first group of researchers have neglected external dependencies and have modeled the network just before end-use (e.g. [4]); second group (e.g. [8] and [12]) have modeled the network with high resolution but in a very limited area such as a household and have modeled these dependencies in ultimate service level. In this method, although high resolution brings effects of multi-energy demand dependency, model is very complicated and the level of modeling should be limited.

This paper proposes a comprehensive framework for modeling both internal and external dependencies, which will help in investigating the role of multi-energy demand dependency. Multi-energy demands will bring higher levels of flexibility in the network. Modeling and manipulation of this flexibility will provide SMES operators with more freedom degrees and will reduce operation costs [13]. In this paper, energy hub approach is used for modeling internal and external dependencies in SMES. The comparison between the numerical results obtained from the proposed model and the ones from previous models demonstrates the usefulness of the proposed model.

In next section, a contribution to collective awareness systems is discussed. In Section 3, a comprehensive model is proposed for energy networks with multi-energy carriers' dependency. In Section 4, modeling the external dependencies of the network is described. Section 5 explains the results, and the conclusion is presented in Section 6.

2   Contribution to Collective Awareness Systems

Development in Information and Communication Technology (ICT) infrastructure unlocks new opportunities for SMES managers to merge independent administrative systems and to involve more players in system decision making procedure.

Considering emerging energy resources and increase of energy carriers' dependency, the needs for collaboration of independent energy system operators are inevitable. In this context, collective awareness systems can openly link independent energy entities in SMES and design the architecture of a new collaboration environment for the benefit of enterprise.

New models should be developed to implement collective awareness systems goals and to utilize huge amount of collected information for managing high interdependent SMES. These models consider energy infrastructure as integrated system which facilitates interaction between energy carriers. From modeling point of view, *Energy Hub system* and *Matrix Modeling,* as explained in Section 1, are good examples for integrated models. Involving new smart elements and more participation of demand side are key factors for enhancing the level of smartness and awareness of these models.

Emerging multi-energy demands which can convert and store energy carriers, highlighted the role of demand side players in SMES decision making process. Modeling multi-energy demand behavior will involve new layer of players to the system which improves socio-technical aspect of integrated models.

3   Comprehensive Smart Multi-Energy System Model

Considering the presence of DERs in future SMES, energy carriers surely will be dependent to each other. Dependency in these forthcoming energy systems will be in a manner that studying some part of the network will not be possible nor correct without considering the dependencies of energy carriers in input or output of the model. Therefore, modeling the effects of DERs is essential in studies.

References [4] and [5] are main studies that have proposed a model for interdependent energy networks. The SMES in this paper has also been motivated by these references and is represented by a coupling matrix that converts the input energy carriers to output carriers. Each element of this matrix denotes the conversion of one carrier into another. Each element is composed of two parameter categories: first category of constant coefficients (η_α), which is dependent to physical characteristics of system and converters; second category includes decision making parameters ($v_{\alpha,t}$) that indicate energy distribution between converters.

3.1 Energy Converter Model

In this paper, external dependencies are modeled through modules in output of the model. Output module indicates dependency of multi-energy demand that is not considered in the modeling process. This module represents the dependency between energy converters in the output with a function of carriers. This model shows that, multi-energy demand has the ability to receive definite amount of energy regardless of its carrier and convert it to its own required service. Dependency between outputs is added to demand vector through some lines, which itself increases the rows of coupling matrix. It is noteworthy that, the added lines do not represent actual outputs but virtually illustrates the dependency in output.

Hence, the output vector (λ) in the proposed model can be divided into two sections: rows indicating independent output carriers (λ_I) and rows introducing dependency in the output (λ_D). With this regard, conversion function should have new rows rather than previous models, as shown by (1):

$$[\lambda] = \begin{bmatrix} \lambda_I \\ \lambda_D \end{bmatrix} = \begin{bmatrix} C_I \\ C_D \end{bmatrix} * [p] \tag{1}$$

where C_I is the old coupling matrix that states the conversion of independent inputs to independent output and C_D is the matrix showing share of independent inputs in providing dependent demand.

3.2 Energy Storage Model

As [6] has thoroughly explained, the role of energy storages can be modeled through some changes in the matrix. In the modified model, $\dot{E}$ is the change in stored energy and can be computed from (2) and (3). S is the coupling matrix of the storage and shows how changes in energy amount of storage will affect the system output. Equation (4) shows coupling matrix considering converters and storages model.

$$\dot{E}_{\alpha,t} = E_{\alpha,t} - E_{\alpha,t-1} \approx e_\alpha Q_\alpha \tag{2}$$

$$e_\alpha = \begin{cases} \eta_{\alpha,ch}^{Storage}, & if \ Q_\alpha \geq 0 \quad (Charge/Standby) \\ 1/\eta_{\alpha,dis}^{Storage}, & if \ Q_\alpha < 0 \quad (Discharge) \end{cases} \tag{3}$$

$$[\lambda] = [C \quad -S] * \begin{bmatrix} p \\ \dot{e} \end{bmatrix} \tag{4}$$

Considering external dependencies, new storage coupling matrix will have rows stating amount of changes in independent output based on changes in energy of storage:

$$[\lambda] = [\mathbf{C} \quad -\mathbf{S}] * \begin{bmatrix} \mathbf{p}_{new} \\ \dot{\mathbf{e}} \end{bmatrix} \tag{5}$$

$$\begin{bmatrix} \lambda_I \\ \lambda_D \end{bmatrix} = \begin{bmatrix} \mathbf{C}_I & \mathbf{S}_I \\ \mathbf{C}_D & \mathbf{S}_D \end{bmatrix} * \begin{bmatrix} \mathbf{p}_{new} \\ \dot{\mathbf{e}} \end{bmatrix} \tag{6}$$

Where S_I and S_D are matrices showing changes of independent and dependent outputs versus changes in the stored energy.

4   Smart Multi-Energy System Operational Model

Modeling external dependencies can considerably reduce system operation cost. Moreover, it brings more choices for system operators and planners while making it possible to utilize facilities of multi-energy demand in operation and planning studies. Fig. 1 illustrates a typical SMES that consists of CHP unit, auxiliary boiler, and heat storage and feeds a multi-energy demand. Input carriers of the system are electricity and gas while output carriers will be electricity, gas, and heat. External dependencies between gas and electricity carrier in this network is considered through a module in output. The mathematical model of this network is shown in (7):

$$\begin{bmatrix} v_{e,t}^{out} & v_{g,t}^{CHP} \eta_e^{CHP} y_{e,t}^{out} & 0 \\ 0 & v_{g,t}^{CHP} \eta_h^{CHP} + v_{g,t}^{boiler} \eta_h^{boiler} & 1/e_h \\ 0 & v_{g,t}^{out} & 0 \\ v_{e,t}^{dd} & v_{g,t}^{dd} + v_{e,t}^{dd} y_{g,t}^{CHP} \eta_e^{CHP} & 0 \end{bmatrix} * \begin{bmatrix} p_{e,t} \\ p_{g,t} \\ \dot{h}_t^{storage} \end{bmatrix} = \begin{bmatrix} L_{e,t} \\ L_{h,t} \\ L_{g,t} \\ L_{eg,t} \end{bmatrix} \tag{7}$$

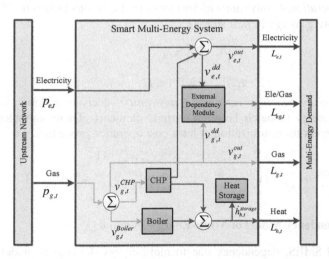

Fig. 1. A typical SMES model considering external dependency module

4.1   Objective Function and Operational Constraints

The objective function is to minimize the cost of gas and electricity used as input. The nomenclature of the following equations is shown in Table 2 of Appendix.

$$Min \quad \sum_t (p_{e,t}.\pi_{e,t} + p_{g,t}.\pi_{g,t}) \tag{8}$$

The objective is minimized considering the system operational constraints as below:

1) Input carriers constraints: each energy carrier has a supply limit which may be due to energy amount in upstream or energy transmission limits.

$$0 \leq p_{\alpha,t} \leq \overline{P}_\alpha \quad , \alpha \in e, g, eg \tag{9}$$

2) Operational constraints of CHP unit: regarding manufacturing characteristics of CHP unit, they face limits in the amount of electrical power or heat output. Furthermore, based on (12), CHP unit should be operated in certain heat to power ratio.

$$\underline{P}^{CHP} \leq p_t^{CHP} \leq \overline{P}^{CHP} \tag{10}$$

$$\underline{H}^{CHP} \leq h_t^{CHP} \leq \overline{H}^{CHP} \tag{11}$$

$$\gamma_t^{CHP} = h_t^{CHP} / p_t^{CHP} \tag{12}$$

3) Operational constraints of auxiliary boiler: heat energy output of auxiliary boiler has some limits in providing energy.

$$\underline{H}^{Boiler} \leq h_t^{Boiler} \leq \overline{H}^{Boiler} \tag{13}$$

4) Operational constraints of heat storage: Equations (14) and (15) indicate the limit for storing energy in heat storage.

$$\left| \dot{h}_t^{storage} \right| \leq r_h^{storage} \tag{14}$$

$$\underline{H}^{Storage} \leq h_t^{storage} \leq \overline{H}^{Storage} \tag{15}$$

5) Decision-making variables constraints: decision-making variables will determine energy dispatch between various elements. These variables are system freedom degrees for optimization of least-cost operation procedure.

$$v_{e,t}^{dd} + v_{e,t}^{out} = 1 \quad , 0 \leq v_{e,t}^{dd}, v_{e,t}^{out} \leq 1 \tag{16}$$

$$v_{g,t}^{CHP} + v_{g,t}^{boiler} + v_{g,t}^{dd} + v_{g,t}^{out} = 1 \quad , 0 \leq v_{g,t}^{CHP}, v_{g,t}^{boiler}, v_{g,t}^{dd}, v_{g,t}^{out} \leq 1 \tag{17}$$

4.2   Dependency Model of Multi-Energy Demand

In a typical SMES, dependency due to multi-energy demand is modeled through a block connected to the output. The external dependency indicates that demand can utilize either electricity or gas carriers. The dependency is modeled through a block in the output, which is the function of dependencies between gas and electricity carriers. In order to deal with the dependency among carriers, two decision variables are used as:

v_e^{dd}, v_g^{dd} : Decision-making variables stating the share of dependent energy demand in output of each carrier (gas and electricity):

$$f\left(v_{e,t}^{dd}, v_{g,t}^{dd}\right) = L_{eg,t} \tag{18}$$

Dependency variables for external dependency will illustrate demand share; thus, it is necessary to balance them with some coefficients and then exploit them in the model. New decision variables in output show the share of each carrier in demand provision:

$$v_{e,t}^{dd-new} = v_{e,t}^{dd} \cdot \left(p_{e,t} + p_{g,t} v_{g,t}^{CHP} \eta_e^{CHP}\right) / L_{eg,t} \tag{19}$$

$$v_{g,t}^{dd-new} = v_{g,t}^{dd} \cdot p_{g,t,s} / L_{eg,t} \tag{20}$$

5   Numerical Results

An SMES under study in this paper consists of CHP unit, auxiliary boiler, and heat storage. Inputs of this system are gas and electricity carriers while the outputs are electricity, gas, and heat. Detailed information of these elements is explained in Appendix Table 1.Moreover, information about multi-energy demand in base case and input energy carrier prices are depicted in Fig. 2 and Fig. 3. The external dependency in this study is hot water consumption of multi-energy demand, which can be supplied by both gas-fired and electrical heaters. Therefore, in a simple case heat consumption has been divided into two parts hot water and other consumption considering hot water as dependent output. The relation between electricity and gas carrier decision variable in the dependent input of multi-energy demand is shown with Eq. (21).

$$v_{e,t}^{dd-new} + \eta_g^{dd} * v_{g,t}^{dd-new} = 1 \tag{21}$$

In addition, in this study it is assumed that the system operator can control the gas and electricity dependent consumption through sending signals to load in smart environment and can benefit from the achieved flexibility. Four different levels of hot water consumption in the multi-energy demand and four different values for η_g^{dd} (indicator of efficiency of gas-fired water heater) are assumed. Therefore, total 16 case studies are produced. In these cases, the amount of output energy consumption from SMES is the same whether ignoring the dependencies. For this reason, the dependency between electricity and gas is equally reduced from both output carriers' demand; as for gas share, it is adjusted based on the efficiency of gas-fired water heater.

As it is shown in Fig. 4, by increase in hot water consumption and utilization of this situation in equal levels of gas-fired water heater efficiency, total cost will decrease. On the other hand, by increasing controllable dependent consumption for the operator, operational freedom degree will increase, which results in lower system operation cost.

Fig. 5 shows prices versus output dependency amount. The re-decrease of costs after reaching a maximum amount should be highlighted. Due to even output energy amount of local energy network, by reducing in gas-fired water heater efficiency, the system will provide all dependencies through electricity carrier when their efficiency gets lower than electrical efficiency of CHP units in the system.

Therefore, with lower efficiency of multi-energy demand converters, demand requirements in output can be achieved by taking the benefits of less electricity with higher efficiency than the gas carrier and in a total view reducing the system operation cost. Fig. 6 and Fig. 7 depict the amount of input gas and electricity carriers when $\eta_g^{dd} = 0.5$ for various levels of hot water consumption.

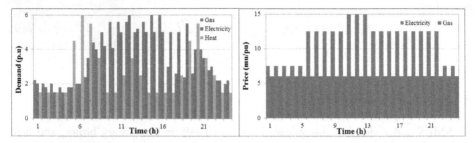

Fig. 2. Multi-energy demand data   **Fig. 3.** Input energy carriers price data

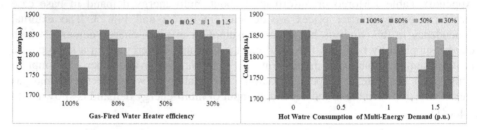

Fig. 4. System operation cost based on gas-fired water heater efficiency   **Fig. 5.** System operation cost based on hot water consumption of multi-energy demand

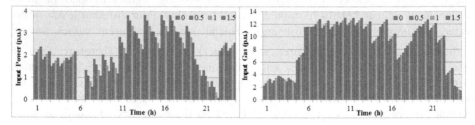

Fig. 6. Input power variation for various hot water consumption with $\eta_g^{dd} = 0.5$ **Fig. 7.** Input gas variation for various hot water consumption with $\eta_g^{dd} = 0.5$

As it can be observed, during hours 1-5, 23 and 24, when electricity price is lower, by increasing the level of flexibility tendency for electricity carrier consumption will increase while gas consumption shall decrease. However, during hours 6-22 the average electricity price is high, so the system operator prefers to provide hot water consumption through gas carrier and as a result reduces the total operation cost.

6   Conclusion

The concept of internal and external dependencies in SMES was introduced. The external dependencies deal with the ability of multi-energy demand to implement its own converters. A new model based on energy hub approach was developed to represent internal and external dependency simultaneously. For assessing the efficiency of the model, an SMES with multi-energy demand in its output was considered. The numerical results confirmed that the external dependencies have a significant impact on the operation condition of the system. In addition, in the case of manageable external dependencies by increasing the dependency the operation cost will be decreased.

Acknowledgment. This work was supported by FEDER funds (European Union) through COMPETE and by Portuguese funds through FCT, under Projects FCOMP-01-0124-FEDER-020282 (Ref. PTDC/EEA-EEL/118519/2010) and PEst-OE/EEI/LA0021/2013. Also, the research leading to these results has received funding from the EU Seventh Framework Programme FP7/2007-2013 under grant agreement no. 309048.

References

1. Alanne, K., Saari, A.: Distributed energy generation and sustainable development. Renewable and Sustainable Energy Reviews 10, 539–558 (2006)
2. Akorede, M.F., Hizam, H., Pouresmaeil, E.: Distributed energy resources and benefits to the environment. Renewable and Sustainable Energy Reviews 14, 724–734 (2010)
3. Chicco, G., Mancarella, P.: Distributed multi-generation: a comprehensive view. Renewable and Sustainable Energy Reviews 13, 535–551 (2009)
4. Krause, T., Andersson, G., Frohlich, K., Vaccaro, A.: Multiple-energy carriers: modeling of production, delivery, and consumption. Proc. IEEE 99, 15–27 (2011)
5. Chicco, G., Mancarella, P.: Matrix modelling of small-scale trigeneration systems and application to operational optimization. Energy 34, 261–273 (2009)
6. Arnold, M., et al.: Distributed predictive control for energy hub coordination in coupled electricity and gas networks. In: Intelligent Infrastructures, pp. 235–273. Springer (2010)
7. Damavandi, M.Y., Kiaei, I., Sheikh-El-Eslami, M.-K., Seifi, H.: New approach to gas network modeling in unit commitment. Energy 36, 6243–6250 (2011)
8. Bozchalui, M.C., et al.: Optimal operation of residential energy hubs in smart grids. IEEE Trans. Smart Grid 3, 1755–1766 (2012)
9. Kienzle, F., Ahcin, P., Andersson, G.: Valuing investments in multi-energy conversion, storage, and demand-side management systems under uncertainty. IEEE Trans. Sustainable Energy 2, 194–202 (2011)
10. Schulze, M., Friedrich, L., Gautschi, M.: Modeling and optimization of renewables: applying the energy hub approach. In: Proc. IEEE ICSET, pp. 83–88 (2008)
11. Mancarella, P., Chicco, G.: Real-time demand response from energy shifting in distributed multi-generation. IEEE Trans. Smart Grid (2013) (in press)
12. Houwing, M., Negenborn, R.R., De Schutter, B.: Demand response with micro-CHP systems. Proc. IEEE 99, 200–213 (2011)
13. Mancarella, P.: Smart multi-energy grids: concepts, benefits and challenges. IEEE PES General Meeting, 22–26 (2012)

Appendix. SMES Elements' Characteristics and Nomenclature

Table 1. Data of SMES elements' Characteristics

CHP Unit			Auxiliary Boiler			Heat Storage		
Output Energy (Min/Max)	η_e^{CHP}	η_h^{CHP}	Output Energy (Min/Max)	η_h^{Boiler}	$Tr_h^{Storage}$	Stored Energy (Min/Max)	$\eta_{\alpha,ch}^{Storage}$	$\eta_{\alpha,dis}^{Storage}$
0/5 pu	0.35	0.45	0/10 pu	0.9	3 pu	0.5/3 pu	0.9	

Table 2. Definition of Indices, Parameters, and Variables

Subscripts					
e	Electricity	g	Gas	t	Time
eg	Dependent demand	h	Heat	α, β	Generic energy carriers
Parameters and Variables					
E	Energy stored	η	Efficiency		
H	Heat output	P	Electrical power	v	Decision variable
L	Energy demand	γ	Heat to power ratio	π	Input energy carrier price
r	Maximum charge and discharge rate of heat storage	$\dot{e}$	(column vector) changes in stored energy	$\dot{h}$	Heat storage level difference in two consecutive time intervals
An underlined (overlined) variable is used to represent the minimum (maximum) value of that variable.					

The Distributed Generation as an Important Contribution to Energy Development in Angola and Other Developing African Countries

Joaquim Moreira Lima[1,2], José Barata[1], Miguel Fernandez[3], and Angel Montiel[2]

[1] Universidade Nova de Lisboa / UNINOVA, Quinta da Torre, 2829-516 Caparica, Portugal
jab@uninova.pt
[2] INTEL Lda, Luanda, Angola
jomoreli@netcabo.pt
[3] CIPEL, Havana, Cuba
mcastro@electrica.cujae.edu.cu

Abstract. Distributed generation (DG) is related to the use of small or medium size generating units close to the load centers. DG has great benefits for application in places distant from the major centers of production where the population needs a reliable and cheap supply of electric power. This is the case of Angola and the majority of other African countries. This paper discusses the potential that distributed generation may have in these countries highlighting four crucial aspects: the utilization of a realistic and simple optimal allocation method, the consideration of intentional islanding mode of operation, the utilization of all the existing sources of energy in the region, specially the re- newable ones, and the demand side management of energy through the application of smart metering and communication techniques.

Keywords: Distributed Generation, Intentional Islanding, Energy Efficiency, Demand Side Management, Smart Metering.

1 Introduction

Although there is a lot of definitions of distributed generation in the technical litera- ture [1-5], we are going to present our own definition, considering that, on one hand, nowadays there are a large number of technologies to be incorporated into this con- cept and on the other hand, the increasing incorporation of artificial intelligence and the new technologies of computing and communications. Facing these facts, it is proposed the following definition of distributed generation:

Distributed generation consists of the supply and storage sources of electrical and thermal energy located in the distribution network, close to the points of consumption, together with the possibility of forming intelligent and highly automated networks, able to control and integrate the operation of the different supply and storage sources with the active participa- tion of consumers in both the purchase and sale of energy, as well as in their more econom- ical and environmental use.

L.M. Camarinha-Matos et al. (Eds.): DoCEIS 2014, IFIP AICT 423, pp. 269–276, 2014.
© IFIP International Federation for Information Processing 2014

The benefits of employing distribution generation in existing distributed networks have economical, technical and environmental implications and they are interrelated. In spite of the fact that in the literature many advantages for the distributed generation are exposed, it is possible to reduce them to four main aspects that comprise in one way or another, the rest of the advantages that can be argued [6-8]. These four aspects are: are increased reliability, line loss reduction, voltage profile improvement and environmental impact reduction.

This paper seeks to explain the most important aspects that must be taken into account to justify the introduction in Angola a mix of this concept and the challenges that the application of this generation paradigm might imply.

2   Relationship to Collective Awareness Systems

The current challenges faced by the Energy sector involve complex interactions between humans and machines with intelligent decision making or behavior. More and more intelligent energy based devices are being applied to create intelligent and adaptive systems in the production, transport, and distribution of electrical energy. The grid is becoming more intelligent (SMART GRID) both at device level and at system level. It is important to understand that the SMART GRID is above all a distributed system composed of intelligent nodes (machines and humans) that regulates and control de demand and supply in a sustainable way.

The SMART GRID and distributed energy in particular are completely dependent on cyberphysical systems that are at its kernel. Considering the different actors involved and interacting with the grid what is available is a complex system in which different services are being used and created in a dynamic way to accommodate the different actors. In particular it is important to stress the importance of customers that more and more interact with producers and suppliers. The internet of things is therefore a concept well in line with this new world of energy.

It is clear that the emerging energy systems (SMART GRID) are socio-technical systems capable of harnessing collective intelligence for promoting innovation and taking better, informed and sustainability-aware decisions, Therefore, the development of Collective Awareness Systems will have a strong and beneficial impact on the creation of effective and sustainable energy systems.

3   Motivation

Angola is a country where almost all energy resources known all over the world can be found. The energy potential from national rivers is enormous and it is one of the major oil producers in Africa. This has resulted on a energy generation planning based on the intensive exploitation of these energy resources and in the construction of big and expensive infrastructures to transport the electric power to very distant places [9].

It was only in the last few years that a strategic plan was developed to include in the country's energy mix the varied and abundant sources of renewable energy.

However, it was not carried out large-scale plans to introduce the energy efficiency and power management on the demand side.

Accordingly, our proposal to improve the reliability of the Angolan electric system is based on the hypothesis that it is possible to consider the comprehensive use of distributed generation in Angola taking into account four important factors: the optimal allocation of the units of distributed generation, the possibility of operating on intentional island mode, the use of all available energy resources and the active participation of the demand side utilizing smart grids and smart metering

4   Optimal Allocation of Distributed Generation

From the very beginning of the introduction of the generation in the distribution networks this has been one of the most important problems faced by specialists and researchers due to the fact that an incorrect planning of these units can lead to an increase in the cost of both investment and exploitation thus losing one of the main advantages of this type of generation.

In the early years of this century there were several papers about the location and sizing of the units taking into account only the decrease of losses in the network [10-12]. These studies consider also that there is a source of generation in each node and limited load scenarios. Also at this time there were several important researchers that attempted to quantify the benefits from the introduction of distributed generation with the goal of providing tools for the economic feasibility studies of this type of generation. Among these works is particularly important the paper submitted by Chiradeja Ramakumar [6].

A large number of methods for the optimal location and size of the GD based on the use of genetic algorithms or fuzzy logic were proposed [13-16]. It also introduced new variants of evolutionary algorithms such as the so-called Particle Swarm Optimization (PSO), Honey Bee Mating (HBM), Tabu Search (TS), Simulated Annealing (SA), Differential Evolution (DE) and Cokoo Search (CS). All of them are heuristic or meta-heuristic methods based on iterative procedures and considering different objective functions [17-24]. Among the major aspects to be considered in the proposed objective functions are the losses in the network, the voltage profile, the greenhouse gas emissions, the savings in the costs of expansion of the transmission system and the reliability.

There has been a great contrast between the level of development achieved by the theoretical and academic studies of optimal location of distributed generation units and, in some cases, a lack of will and rejection of the utilities to its introduction due to the fact that these studies have not consider the uncertainties and risk factors present at all times. From the point of view of real-world application, geographical, climatic and social factors could change completely the results obtained by means of sophisticated methods purely mathematical since they have not been taken into account. It is also true that the advantages of one method over another are mainly based on an easier consideration of the objective functions or in the saving of computing time.

Concerning the optimal localization studies, the fact of variability and the forecasting difficult to deliver energy and the influence of these factors both the size and the location of the units of distributed generation have been considered by some authors only [25, 26].

According to these facts, the proposed method of optimal location will take some aspects into consideration: A mathematical method that could simplifies the development and implementation of the proposed software, the reliability of the whole system, the transmission losses, the voltage profiles at the more frequent load scenarios, also as a priority, the geographic, climatic and social more important factors as well as the presence of renewable energy sources and their variability as part of the possible scenarios.

This well in line with the research question of this PhD work is which computational methods and tools should be used for the optimal location of distributed generation units in the context of under developing countries and considering environmental factors.

5   Islanding as a Normal Mode of Operation of Distributed Generation Networks

A set of distributed generation units having different technologies are usually connected to the Utility Network through a common point of connection (CPC) where there is a coupling transformer.

It can happen that, mainly due to faults in the utility or distributed generation network, the CPC is disconnected and form an island or isolated system. The rules of connection of the distributed generation to the network disconnect immediately all the distributed generation units by interrupting the service users. But more recently has been questioned this policy due to the economic losses that brings to the users the energy not supplied caused by these blackouts.

It has been proposed to continue operating the distribution network in isolation mode, diminishing the quality of the energy parameters but guaranteeing the supply to users. In this way, an increased service quality is achieved. This policy has almost always been rejected by the large utilities but has been seen in recent years a greater flexibility in this regards.

The most important challenges in adopting a policy of intentional islanding operation are the sharing of the power load, which could include load shedding, the frequency and voltage control and the subsequent synchronization of the island with the network [27].

The solution of these problems, even without reducing the quality of service, is nowadays a reality due to the application of the most modern techniques of computing and communications and the introduction of new concepts and techniques of operation of networks such as the smart grids and metering and the virtual power plants [28-30].

We propose that, as a part of the incorporation of distributed generation to the Angolan mix, the operation on intentional islanding mode should be considered.

6   Sustainability of the Projects of Distributed Generation

Sustainability means an equitable distribution of limited resources and opportunities in the context of the economy, the society, and the environment seeking the well-being of everyone, now and in the future [34, 35]. A sustainable energy solution generates enough energy for all at a reasonable price in a clean, safe and reliable way.

In regard to the technologies used, machines such as internal combustion engines or Diesel, gas turbines, microturbines, fuel cells, biomass and renewable energies extracting the energy of the sun and wind as the photovoltaic solar energy and wind power, are recognized as sources of energy utilized in distribution generation [36].

Within these technologies, are the renewable energy, solar and wind, which have the biggest sustainability and greater benefits from an environmental point of view but have the disadvantage that depend on climatic factors that vary according to the time of the day and the season. On account of that, they are difficult to predict and, therefore, are not "dispatchable". In spite of that, its use and expanding potential are increasing [37, 38].

A few years ago, while recognizing the great potential of renewable energy sources as part of distributed generation in the more rational use of electric energy, it was a great concern that precisely this dispersal and not "dispatchable" characteristic of this type of sources, could give rise to a liability diluted among the large utilities, the costumers and local producers [35].

In the last years this situation has been changing mainly motivated by the development of new organization concepts of the electrical networks such as the microgrid and the virtual power plants, already mentioned, and also by the increase in international pressure for searching energy productions solutions more efficient, clean and reliable [26, 37].

Consequently we propose that clean and renewable energies available in Angola must be effectively used.

7   Demand Side Management and Smart Metering

The participation of the consumer and the local producer of energy in the implementation of energy efficiency policy is so-called Demand Side Management. Due to the lack of awareness of the Angolan population and some authorities concerning the necessity of saving electrical energy and use it more rationally this DSM concept is necessary to evaluate at the proposal to solve the Angolan electric system develop.

This participation is proposed through several methods such as financial incentives and education. The goal of demand side management is to encourage the consumer to use less energy during peak hours or to translate the energy utilization to off peak hours such as nighttime and weekends. Peak demand management does not necessarily decrease total energy consumption but could be expected to reduce the need for investment in networks components and the size of distributed generation energy sources.

There are three models for implantation the Demand Side Management:

1. *Energy efficiency.* Using less power to perform the same tasks.
2. *Demand Response.* Any preventive method to reduce, flatten or shift peak demand that includes all intentional modifications to consumptions patterns of electricity. It is related to a wide range of actions which can be taken at the customer side of the electricity meter in response to particular condition of the electricity meter (peak period, network congestion, high prices) [39].
3. *Dynamic Demand.* Advance or delay appliance operating cycle by a few seconds to increase the Diversity Factor of the whole load. The concept is that by monitoring the power factor of the power grid, as well as their own control parameters, individual, intermittent loads would switch on or off at optimal moments to balance the overall system load with generation, reducing critical power mismatches [40].

The successful implementation of the demand side management would be impossible without the application of the so-called Smart Metering based on the smart meter.

A smart meter is usually an electrical meter that records consumption of electric energy in intervals of an hour or less and communicates that information at least daily back to the utility for monitoring and billing purposes. Smart meters enable two-way communication between the meter and the central system. The possibility of this two-way communication of smart metering makes its utilization mandatory in our proposal.

8   Conclusions

We conclude that, to solve the lack of reliability of the electrical system in Angola it is mandatory to apply a new concept of distributed generation who bears in mind the needs of the different regions of the country, the available resources and the application of the last advances in the information and communication technologies, emphasizing the following aspects:

- The optimal allocation of the units of distributed generation taking into account the decrease of power losses, the improvement of the voltage profile, the characteristics of the demand and the possible variability of the sustainable available sources of energy as well as utilizing an adequate and simple software. This the main focus of this PhD research.
- To consider, as a part of the proposal, the possibility and necessity to create intentional islands that increase the availability of the electric power to the consumers.
- To utilize all the energy resources available in the region in a sustainable way and considering the possible environmental aspects.
- To apply the demand side management of the energy consumed and produced utilizing the smart metering and communication possibilities.

Acknowledgments. This work was partially supported by CTS multi-annual funding, through the PIDDAC Program funds.

References

1. Ackermann, T., Andersson, G., Soder, L.: Distributed generation: a definition. Electric Power Systems Research 57(3), 195–204 (2001)
2. Pepermans, G., et al.: Distributed generation: definition, benefits and issues. Energy Policy 33(6), 787–798 (2005)
3. Bessmertnykh, A., Zaichenko, V.: Development of distributed power generation. Herald of the Russian Academy of Sciences 82(5), 398–402 (2012)
4. Lopes, J., et al.: Integrating distributed generation into electric power systems: A review of drivers, challenges and opportunities. Electric Power Systems Research 77(9), 1189–1203 (2007)
5. González-Longatt, F., Fortoul, C.: Review of the Distributed Generation Concept: Attempt of Unification. In: ICREPQ 2005 - International Conference on Renewable Energies and Power Quality, Zaragoza - Spain (2005)
6. Chiradeja, P., Ramakumar, R.: An approach to quantify the technical benefits of distributed generation. IEEE Transactions on Energy Conversion 19(4), 764–773 (2004)
7. Poullikkas, A.: Implementation of distributed generation technologies in isolated power systems. Renewable & Sustainable Energy Reviews 11(1), 30–56 (2007)
8. Castro, M.: Distributed Generation and Renewable Energy: The Cuban Experience. In: ACEEW 2013 - Angolan Conference on Energy Engineering and Water, Luanda - Angola (2013)
9. MINEA, Ministry of Energy and Water of Angola - Plano de Desenvolvimento Energético (Energy development Plan). MINEA: Luanda - Angola (2012)
10. Hadjsaid, N., Canard, J., Dumas, F.: Dispersed generation impact on distribution networks. IEEE Computer Applications in Power 12(2), 22–28 (1999)
11. Wang, C., Nehrir, M.: Analytical approaches for optimal placement of distributed generation sources in power systems. IEEE Transactions on Power Systems 19(4), 2068–2076 (2004)
12. Griffin, T., et al.: Placement of Dispersed Generations Systems for Reduced Losses. In: 33rd Hawaii International Conference on System Sciences. IEEE Xplore (2000)
13. Celli, G., et al.: A multiobjective evolutionary algorithm for the sizing and siting of distributed generation. IEEE Transactions on Power Systems 20(2), 750–757 (2005)
14. Borges, C., Falcao, D.: Optimal distributed generation allocation for reliability, losses, and voltage improvement. International Journal of Electrical Power & Energy Systems 28(6), 413–420 (2006)
15. Cano, E.B.: Utilizing Fuzzy Optimization for Distributed Generation Allocation. In: TENCON 2007 - 2007 IEEE Region 10 Conference. IEEE Xplore (2007)
16. Varikuti, R., Reddy, M.D.: Optimal Placement of DG Units Using Fuzzy and Real Coded Genetic Algortihm. Journal of Theoretical and Applied Information Technology 7(2), 145–151 (2009)
17. Abdi, S., Afshar, K.: Application of IPSO-Monte Carlo for optimal distributed generation allocation and sizing. International Journal of Electrical Power & Energy Systems 44(1), 786–797 (2013)
18. Doagou-Mojarrad, H., et al.: Optimal placement and sizing of DG (distributed generation) units in distribution networks by novel hybrid evolutionary algorithm. Energy 54, 129–138 (2013)
19. Nekooei, K., et al.: An Improved Multi-Objective Harmony Search for Optimal Placement of DGs in Distribution Systems. IEEE Transactions on Smart Grid 4(1), 557–567 (2013)

20. Viral, R., Khatod, D.: Optimal planning of distributed generation systems in distribution system: A review. Renewable & Sustainable Energy Reviews 16(7), 5146–5165 (2012)
21. Porkar, S., et al.: Optimal allocation of distributed generation using a two-stage multi-objective mixed-integer-nonlinear programming. European Transactions on Electrical Power 21(1), 1072–1087 (2011)
22. Moradi, M., Abedini, M.: A combination of genetic algorithm and particle swarm optimization for optimal DG location and sizing in distribution systems. International Journal of Electrical Power & Energy Systems 34(1), 66–74 (2012)
23. Moravej, Z., Akhlaghi, A.: A novel approach based on cuckoo search for DG allocation in distribution network. International Journal of Electrical Power & Energy Systems 44(1), 672–679 (2013)
24. El-Zonkoly, A.: Optimal placement of multi-distributed generation units including different load models using particle swarm optimisation. IET Generation Transmission & Distribution 5(7), 760–771 (2011)
25. Niknam, T., et al.: A modified honey bee mating optimization algorithm for multiobjective placement of renewable energy resources. Applied Energy 88(12), 4817–4830 (2011)
26. Atwa, Y., et al.: Optimal Renewable Resources Mix for Distribution System Energy Loss Minimization. IEEE Transactions on Power Systems 25(1), 360–370 (2010)
27. Gomez, J., Morcos, M.: Letter to the Editor: On Islanding Operation in Systems with Distributed Generation. Electric Power Components and Systems 37(2), 234–237 (2009)
28. Dash, P., Padhee, M., Barik, S.: Estimation of power quality indices in distributed generation systems during power islanding conditions. International Journal of Electrical Power & Energy Systems 36(1), 18–30 (2012)
29. Best, R., et al.: Synchronous islanded operation of a diesel generator. IEEE Transactions on Power Systems 22(4), 2170–2176 (2007)
30. Wissner, M.: The Smart Grid - A saucerful of secrets? Applied Energy 88(7), 2509–2518 (2011)
31. Kundar, P., et al.: Definition and classification of power system stability (vol. 18, pg 1387, 2004). IEEE Transactions on Power Systems 19(4), 2124 (2004)
32. Kundur, P., et al.: Definition and classification of power system stability. IEEE Transactions on Power Systems 19(3), 1387–1401 (2004)
33. Xyngi, I., et al.: Transient Stability Analysis of a Distribution Network With Distributed Generators. IEEE Transactions on Power Systems 24(2), 1102–1104 (2009)
34. Alanne, K., Saari, A.: Distributed energy generation and sustainable development. Renewable & Sustainable Energy Reviews 10(6), 539–558 (2006)
35. Bugaje, I.: Renewable energy for sustainable development in Africa: a review. Renewable & Sustainable Energy Reviews 10(6), 603–612 (2006)
36. Huang, J., Jiang, C., Xu, R.: A review on distributed energy resources and MicroGrid. Renewable & Sustainable Energy Reviews 12(9), 2472–2483 (2008)
37. Tan, W., et al.: Optimal distributed renewable generation planning: A review of different approaches. Renewable & Sustainable Energy Reviews 18, 626–645 (2013)
38. Loughran, D., Kulick, J.: Demand-side management and energy efficiency in the United States. Energy Journal 25(1), 19–43 (2004)
39. Albadi, M., El-Saadany, E.: A summary of demand response in electricity markets. Electric Power Systems Research 78(11), 1989–1996 (2008)
40. Torriti, J., Hassan, M., Leach, M.: Demand response experience in Europe: Policies, programmes and implementation. Energy 35(4), 1575–1583 (2010)

Optimal Operation Planning of Wind-Hydro Power Systems Using a MILP Approach

Paulo Cruz[1], Hugo M.I. Pousinho[1,2], Rui Melício[1,2],
Victor M.F. Mendes[1,3], and Manuel Collares-Pereira[1]

[1] University of Évora, Évora, Portugal
{paulo.cruz1964,hpousinho}@gmail.com,
{ruimelicio,collarespereira}@uevora.pt
[2] IDMEC/LAETA, Instituto Superior Técnico, Universidade de Lisboa, Lisbon, Portugal
[3] Instituto Superior of Engenharia de Lisboa, Lisbon, Portugal
vfmendes@deea.isel.pt

Abstract. This paper addresses an approach for a day-ahead operation of a wind-hydro power system in an electricity market. A wind-hydro system is able to mitigate the intermittence and the variations on wind power, mitigating the economic penalty due to unbalance in the satisfaction of the compromises. The approach consists in a model given by a mixed-integer linear programming. This model maximizes the profit in the day-ahead market, taking into consideration the operating constraints of both the wind the farm and the pumping-hydro system. Finally, numerical case studies illustrate the interest and effectiveness of the proposed approach.

Keywords: Collective awareness, day-ahead market, wind-hydro power system, optimal operation.

1 Introduction

New operational challenges have come into view in power system due to large-scale integration of renewable energy. In particular, high levels of wind integration have strongly conditioned the power system operating security and stability due to the inherent intermittence and variability on wind power [1], implying that wind power dispatchability is a challenging task, namely when the control and management of the active power output is required [2]. Management of the system requires the use of wind power forecasting methods in order to mitigate the impact of wind power variability. But forecasting methods are not able to provide enough accuracy results for wind power. Hence, in order to accommodate the absence of the enough accuracy, uncertainty on the wind power forecast should be included on the approach for optimal operational planning. Otherwise, for a wind power producer that acts in the day-ahead market the disregarding of uncertainty can lead to monetary penalties due to a significant deviation from offers [3]. An effective way to minimize the deviation losses is to combine a wind farm with conventional generating units. Examples of such combinations can be found in [2], [3] and [4]. In [3], a comprehensive study about the economic benefits of a wind farm with pumped-hydro system is analyzed in an electricity market environment. In [5], the effect of wind power with varying degrees of integration is examined in order to determine the costs and carbon emissions associated with a hydro-dominated with gas unit's power system.

In this paper, the main contribution is to provide an effective approach based on mixed-integer linear programming (MILP) to find the optimal planning for the operation of a single entity having to manage a wind farm and a pumped-hydro system, so as to maximize the profit in the day-ahead market.

2   Relationship to Collective Awareness Systems

The technological evolution on electric power system encouraged by the expansion of distributed generation has been crucial to create collective awareness systems useful to define new energy consumption and production patterns. A collective awareness system can result from the development of powerful optimization approaches for the management of power systems, helping to make decisions. The collective awareness system not only promotes the sustainable use of energy resources in favor of an effective low-carbon economy [6], but also processes the optimal decision. In order to achieve this optimal decision, besides real-time information on data, such as, the electricity prices and wind power, collective tools are essential to allow maximizing the profit. Hence, research on technological innovation for collective tools based on approaches for solving the day-ahead operation planning of a power producer is crucial to achieve guidelines for the best bidding in an electricity market.

3   State of the Art

The increased wind power integration in power system has drawn attention to largescale energy storage techniques. In the technical literature, multiple storage technologies and optimization approaches have been analyzed from several perspectives. In [5] and [7], the impact of the intermittence and variability of wind power generation from an operation and economic perspective has been addressed respectively by a discrete dynamic quadratic program with linear constraints and by using the EnergyPLAN computer model. In order to accommodate excessive wind power generation, batteries energy storage [8] and flywheels [9] have been proposed. However, the aforementioned propose is unattractive for large-scale power systems [10]. A viable alternative to overcome the unattractiveness is the wind-hydro power system, combining a wind farm system with a pumped-hydro system, storing the excess wind energy and otherwise providing electric energy in favorable economic conditions. This alternative as shown attractiveness not only to balance the intermittence and variability of wind power in combined optimization settings [2] and [3], but also to enable the provision of firm capacity [11]. In [12], the operation of a wind-hydro power system is analyzed for the Greek islands, showing that the electricity cost is significantly reduced when conventional thermal units are changed by the wind-hydro power system. In [13], the sizing of a wind-hydro power system and the operational performance is examined for the Canary Islands, to optimize exploitation of available hydraulic and wind potential. In [14] and [15], a methodology is proposed for defining the best size of the several devices of a windhydro power system. In [2], daily planning for a wind-hydro power system is modeled by a linear programming. In [16], a dynamic programming for the operation planning of energy storage for wind farms in the electricity market is presented. In [17], a wind-hydro system is studied in order to find an energy balance analysis and economic viability. In [3], two approaches for minimizing the imbalance costs of the wind farm power output are addressed. The first approach considers only bidding for the wind farm in the day-ahead market, trying to minimize the risk of the bidding; the second one considers a wind-hydro power system to minimize the imbalance costs incurred by the wind farm system.

Unless other technology for storing energy can into play, for instance, the promising liquid storage battery [18], the state of the art has as a point in case the wind-hydro power system. This system has been shown in the literature as convenient for a wind farm in a day-ahead market, avowing imbalances due to wind power intermittence and variability.

4   Problem Formulation

The proposed MILP approach for the day-ahead operation planning of a wind-hydro power producer in an electricity market is developed in order to access an economic viability of this type of system. In the proposed approach, the operation planning is considered with hourly discretization. For each hour, the wind-hydro power producer offers in the day-ahead market based on assumption that the electricity prices are derived from a forecast. Also, the wind power is considered with an hourly discretization. The operation planning comprises at each hour the amount of: power output that should be offer; wind power; hydro power; and pumping power.

The wind-hydro power system considered for the optimal operation planning of a single entity managing a wind farm and a pumped-hydro system is shown in Fig. 1.

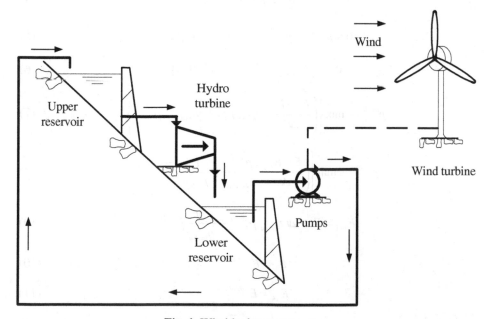

Fig. 1. Wind-hydro power system

4.1   Objective Function

The problem formulation for the wind-hydro power system aims to maximize the power producer profit in a day-ahead market. The objective function to be maximized is stated as:

$$\sum_{t\in T}(\lambda_t^{da} p_t^{da} - c_t^{pump} p_t^{pump}).\tag{1}$$

In (1), t stands for the hour index and T stands for the time horizon, λ_t^{da} is the forecast electricity price in the day-ahead market, p_t^{da} is the power output injected into the grid, c_t^{pump} is the cost of the pumping operation; and p_t^{pump} is the power consumed by the pump operation. In (1), the objective function is the difference between the revenue of selling the energy and the cost of pumping.

4.2   Constraints

The objective function is subject to a set of technical and operational constraints. The constraints are stated as:

$$p_t^{da} = p_t^W + p_t^{hydro} - p_t^{pump}\tag{2}$$

$$\underline{p}^W + \underline{p}^{hydro} \le p_t^{da} \le \overline{p}^W + \overline{p}^{hydro}\tag{3}$$

$$\underline{p}^W \le p_t^W \le \overline{p}^W\tag{4}$$

$$0 \le p_t^{hydro} \le \overline{p}^{hydro}(1-x_t) \quad \forall x \in \{0,1\}\tag{5}$$

$$0 \le p_t^{pump} \le \overline{p}^{pump} x_t \quad \forall x \in \{0,1\}\tag{6}$$

$$p_t^{hydro} \le \min\left\{\left(\frac{E_t - \underline{E}}{\Delta t} + \eta_{pump}\, p_t^{pump}\right)\eta_{hydro}, \overline{p}^{hydro}\right\}\tag{7}$$

$$p_t^{pump} \le \min\left\{\left(\frac{\overline{E} - E_t}{\Delta t} - \frac{p_t^{hydro}}{\eta_{hydro}}\right)\frac{1}{\eta_{hydro}}, \overline{p}^{pump}\right\}\tag{8}$$

$$E_{t+1} = E_t + \Delta t\, \eta_{pump}\, p_t^{pump} - \frac{\Delta t}{\eta_{hydro}} p_t^{hydro}\tag{9}$$

$$E_1 = E^{initial}\tag{10}$$

$$E_{24} = E^{final}\tag{11}$$

$$\underline{E} \le E_t \le \overline{E}\tag{12}$$

$$\underline{E} \le E_t - \frac{\Delta t}{\eta_{hydro}} p_t^{hydro} \qquad (13)$$

$$E_t + \Delta t \, \eta_{pump} \, p_t^{pump} \le \overline{E} . \qquad (14)$$

In (2), the power output injected into the grid in each hour t is composed by three terms: (i) the amount of the available wind power producer, p_t^W; (ii) the amount of the hydro power generated that is delivered to the grid, p_t^{hydro}; and (iii) the amount of the power consumed by the pump operation of the plant. In (3), the power output injected into the grid is limited by the lower and upper capacities of hydro and wind power systems. In (4), the wind power of the wind farm must be within operating limits. In (5) and (6), the limits on the hydro and pump power are set, which include a 0/1 variable, x_t, to avoid enabling both hydro and pumping operation modes. In (7), the maximum value of p_t^{hydro} depends on two main physical limits: (i) the amount of available energy in the reservoir, E_t, due to wind energy stored by pumping; and (ii) the maximum generation capacity of the hydro turbines, $\overline{p}^{hydro}$. In (8), the pumping power is limited by the maximum power of pumping and the power associated with the energy available in the reservoir. In (7) and (8), η_{pump} and η_{hydro} are the efficiencies of pumping and hydro generation, respectively. In (9), the energy balance in the reservoir is computed. At the beginning of the $(t+1)$ hour, the reservoir energy level depends on the energy level in the previous hour as well as the pumped energy and the energy supplied to the grid by the hydro generation. The reservoir energy level is reduced by hydro generation and is reloaded by pumping operation. In (10) and (11), the initial, $E^{initial}$, and final, E^{final}, energy levels of the reservoir must be satisfied. These energy levels are assumed to be known, in order to obtain a consistent planning scheme for the reservoir. In (12), the limits of the reservoir energy levels are set. In (13), the energy level has to satisfy the constraint depending on two terms: (i) the reservoir energy level; and (ii) the energy generated by the hydro plant. In (14), the energy level upper limit has to satisfy the constraint depending on two terms: (i) the reservoir energy level; and (ii) the stored energy through by the pump operation of the plant.

5   Case Study

The proposed MILP approach has been tested on case studies based on a wind- hydro power system composed of eight wind turbines and two cascading river dams with pumping-storage capability. The time horizon considered is one day, divided into 24 hourly intervals. The input to the proposed approach includes the day-ahead hourly electricity price from the Iberian electricity market [19] and the forecasted available wind power. The electricity price and available wind power are shown in Fig. 2.

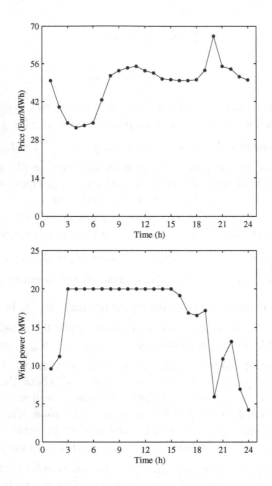

Fig. 2. Electricity prices (*left*) and available wind power (*right*)

The cost of the pumping operation assumes the same value as the electricity price at each hour. The computing time is less than 60 seconds on a 1.9-GHz-based processor with 2 GB of RAM using as a computing language the VBA for Microsoft Excel platform which is reasonable within a day-ahead decision making framework.

The operating data of the wind-hydro power system are shown in Table 1.

Table 1. Wind–hydro power system data

$\underline{p}^W$ (MW)	$\overline{p}^W$ (MW)	$\underline{p}^{hydro}$ (MW)	$\overline{p}^{hydro}$ (MW)	$\underline{p}^{pump}$ (MW)	$\overline{p}^{pump}$ (MW)	$\underline{E}$ (MWh)	$\overline{E}$ (MWh)	E^{final} (MWh)	η_{hydro}	η_{pump}
0	20	0	10	0	10	10	300	10	0.88	0.85

Two cases are presented in this section, considering different initial energy levels for the upper reservoir. The considered cases are as follows:

Case 1) The $E^{initial}$ is constrained to be equal to 20% of maximum energy level.

Case 2) The $E^{initial}$ is constrained to be equal to 75% of maximum energy level.

The two cases are analyzed with respect to the power producer's energy stored in the upper reservoir, optimal hourly power output injected into the grid and profit as follows.

The profiles of the energy stored in the upper reservoir over the 24 hour for both case studies are shown in Fig. 3.

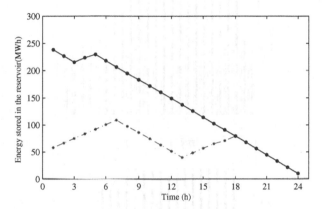

Fig. 3. Energy stored in the upper reservoir: Case 1 (*dashed-dot line*) and Case 2 (*solid line*)

In Fig. 3, the results for Case 1 show that the pumping operation occurs when the prices are low enough, i.e., for the hourly intervals between [1 , 7] h and [14 , 18] h. For the intervals [9 , 13] h and [19 , 24] h, the generating operation occurs in order to maximize the profit. In Case 2, the upper reservoir has an initial energy level greater than Case 1 then, the profit maximization is achieved with hydro production during 22 hours. Because the price is low in hours 4 and 5 the pumping operation occurs.

The hourly power output for both case studies is shown in Fig. 4.

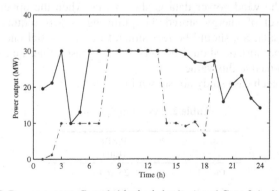

Fig. 4. Power output: Case 1 (*dashed-dot line*) and Case 2 (*solid line*)

In Fig. 4, a comparison between Case 1 and Case 2, as expected, shows that the wind-hydro power system tends to operate at a high production level when a high initial energy level in the reservoir is available. For a high initial energy level, the hydro system operates as a generator and sends the stored energy to the market until reaching the final value of energy required in the reservoir.

The power contribution of the wind farm and pumped-hydro system for both case studies in the day-ahead market is show in Fig. 5.

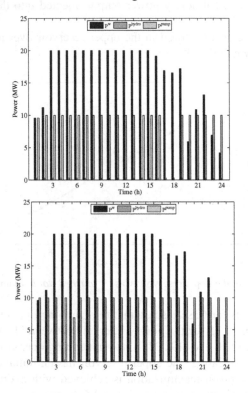

Fig. 5. Power contribution: Case 1 (*left*) and Case 2 (*right*)

In Fig. 5, for the wind power data is shown that when the upper reservoir starts with a higher level of energy stored the pumping is less required for the same electricity prices data. So, should be recommend that the initial energy stored in the starting of the time horizon should be kept at a highest value, i.e., the final energy stored should be aimed at this value.

The profits for each case study are shown in Table 2.

Table 2. Case study profits

	Profit (Eur)
Case 1	25,164
Case 2	29,666

In Table 2, a comparison of the profits shows, as expected by the last figures, that the profit is higher when the initial energy level is higher.

6   Conclusions

A MILP approach is proposed for the optimal operation planning of a wind-hydro power system, combining a wind farm with a pumped-hydro system to improve wind farm profits. The approach developed is illustrated considering two case studies. A comparison between these two case studies allows concluding that a coordination between the forecasted wind power and the level of the energy stored at the reservoir in the initial and final hours of the time horizon have to be decided by an upper level hierarchical management system. The approach is able to mitigate wind energy curtailments and avoids penalty risks related to energy deviations.

Acknowledgments. H.M.I. Pousinho thanks FCT for a post-doctoral grant (SFRH/BPD/ 52163/2013). This work was partially supported by FCT, through IDMEC under LAETA, Instituto Superior Técnico, Universidade de Lisboa, Portugal.

References

1. Al-Awami, A., El-Sharkawi, M.A.: Coordinated Trading of Wind and Thermal Energy. IEEE Trans. Power Syst. 2(3), 277–287 (2011)
2. Castronuovo, E.D., Peças Lopes, J.A.: On the Optimization of the Daily Operation of a Wind-Hydro Power Plant. IEEE Trans. Power Syst. 19(3), 1599–1606 (2004)
3. Angarita, J.L., Usaola, J., Martínez-Crespo, J.: Combined Hydro-Wind Generation Bids in a Pool-Based Electricity Market. Electr. Power Syst. Res. 79, 1038–1046 (2009)
4. Chen, C.L.: Optimal Wind-Thermal Generating Unit Commitment. IEEE Trans. Power Syst. 23(1), 273–279 (2008)
5. Maddaloni, J.D., Rowe, A.M., Van Kooten, G.C.: Network Constrained Wind Integration on Vancouver Island. Energy Policy 36(2), 591–602 (2008)
6. Collective Awareness Platforms for Sustainability and Social Innovation, http://ec.europa.eu/digital-agenda/collectiveawareness (accessed October 2013)
7. Lund, H.: Large Scale Integration of Wind Power into Different Energy Systems. Energy 30(13), 2402–2412 (2005)
8. Zalani, M.D., Mohamed, A., Hannan, M.A.: An Improved Control Method of Battery Energy Storage System for Hourly Dispatch of Photovoltaic Power Sources. Energy Convers. Manag. 73, 256–270 (2013)
9. Sebastián, R., Alzola, R.P.: Flywheel Energy Storage Systems: Review and Simulation for an Isolated Wind Power System. Renew. Sust. Energy Rev. 16(9), 6803–6813 (2012)
10. Karamanou, S.E., Papathanassiou, S., Papadopoulos, M.: Operating Policies for Wind-Pumped Storage Hybrid Power Stations in Island Grids. IET Renew. Power Gener. 3(3), 293–307 (2009)
11. Angarita, J.M., Usaola, J.G.: Combining Hydro-Generation and Wind Energy: Biddings and Operation on Electricity Spot Markets. Electr. Power Syst. Res. 77(5-6), 393–400 (2007)

12. Bakos, G.C.: Feasibility Study of a Hybrid Wind/Hydro Power-System for Low-Cost Electricity Production. Appl. Energy 72(3), 599–608 (2002)
13. Bueno, C., Carta, J.A.: Wind Powered Pumped Hydro Storage Systems, a means of Increasing the Penetration of Renewable Energy in the Canary Islands. Renew. Sustain. Energy Rev. 10(4), 312–340 (2006)
14. Anagnostopoulos, J., Papantoni, S.D.: Pumping Station Design for a Pumped-Storage Wind–Hydro Power Plant. Energy Convers. Manage. 48(11), 3009–3017 (2007)
15. Anagnostopoulos, J., Papantoni, S.D.: Simulation and Size Optimization of a Pumped-Storage Power Plant for the Recovery of Wind-Farms Rejected Energy. Renew. Energy 33(7), 1685–1694 (2008)
16. Korpaasa, M., Holen, A.T., Hildrumb, R.: Operation and Sizing of Energy Storage for Wind Power Plants. Elect. Power Energy Syst. 25, 599–606 (2003)
17. García-González, J., Ruiz de la Muela, R., Santos, L.M., González, A.: Stochastic Joint Optimization of Wind Generation and Pumped-Storage Units in an Electricity Market. IEEE Trans. Power Syst. 23(2), 460–468 (2008)
18. Bradwell, D.J., Kim, H., Sirk, A.H.C., Sadoway, D.R.: Magnesium-Antimony Liquid Metal Battery for Stationary Energy Storage. J. Am. Chem. Soc. 134, 1895–1897 (2012)
19. Red Eléctrica de España, S. A. Sistema de Información del Operador del Sistema, http://www.esios.ree.es (accessed October 2013)

Part XI
Monitoring and Control in Energy

Distributed Smart Metering
by Using Power Electronics Systems

Francisco M. Navas-Matos, Sara Polo-Gallego, Enrique Romero-Cadaval,
and Maria Isabel Milanés-Montero

Power Electrical and Electronic Systems (PE&ES)
School of Industrial Engineering (University of Extremadura), Spain
http://peandes.unex.es

Abstract. The main objective of this paper is to develop a monitor system implemented in existing power electronics equipment. The proposed algorithm is integrated in a previously developed Photovoltaic Array Emulator with the principal aim of monitoring important quality parameters that will be sent to a central repository (by using UDP protocol), where they will be available as Open data.

Keywords: Smart Metering, Power Quality, UDP Protocol and Open data.

1   Introduction

Historically, the electrical grid has been a "broadcast" grid, where generators cover the demand in a country or region. Whereas this model has served its purpose for the last century to some extent, the smartgrid [1] concept appears due to the growing need to reform the world's electrical grid and systems to address new societal challenges:

- The world is considering being more economical and environmentally friendly, and therefore the electrical grid is becoming more complex with the introduction of power electronics systems present in renewable energy generators.
- In a society where information and data are becoming increasingly important, information and communication technologies carry the conception of open data. Open data [2] refers to the notion that certain data should be unreservedly available to everyone for using and republishing as wished, free from restrictions imposed by copyright, patents or other mechanisms of control.

The European Platform Technology Smartgrids published 19 June 2013 The Summary of Priorities for Smartgrids Research Topics [3]. Specifically in D section with a priority 1 the sub-topic cluster Modelling Power System and ICT together is found.

In order to fulfill the research topic, we use the actual power electronic systems in order to be the host of the corresponding algorithms, able to get quality parameters on the base of the IEEE Standard 1459-2010 [4] and send it over the internet to an open grid of data, where everybody would be able to use the producer and consumer data

L.M. Camarinha-Matos et al. (Eds.): DoCEIS 2014, IFIP AICT 423, pp. 289–296, 2014.

for their work, as could be researching, madding statistical studies, matching generation and consumption as part of the grid control and so on.

The developed algorithm needs to be adaptable and embedded for every power electronics system, taking advantage of the electrical parameters already measured by these types of systems.

Therefore, our paper is built upon the development of the aforementioned algorithm and inclusion in a real system to prove its correct operation with the rapid prototyping tool based on xPC Target. We send over Internet to an open repository quality parameters calculated by an algorithm implemented in a photovoltaic array emulator connected to a commercial inverter.

2   Relationship to Collective Awareness

In order to contribute to collective awareness, the developed algorithm has the ability to send power quality data to a common repository, where they will be available as open data for other researchers that would be able to use them to develop advance strategies in order to operate the electric grid, for instance.

Besides in a vision of future, this kind of algorithms will avail of the increasing use of power electronics systems (for example, the converters used in distributed generators mainly based in renewable resources). These systems usually have electric sensors which the algorithm could use for monitoring the system, and enough resources for running the algorithm. Taking this into account, the power electronic systems will convert into sources of pre-processed data that can be used in high level intelligent systems that will optimize the operation of the grid.

3   System's Physical Description

The system is composed by a PV array emulator, an electronic converter working as a synchronous rectifier, which constitutes the interface circuit between the electrical grid by means of an autotransformer and the inverter. The real system is a SEMISTACK SKS 230F B8CI 190 V12 inverter by Semikron. We see the system's scheme in Fig. 1.

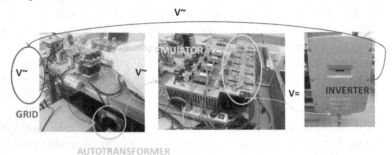

Fig. 1. Real Scheme of the System

The inverter, which returns the power, once reconverted to AC, to the grid, is a SMA sunny minicentral 6000A inverter.

Thus, we have a close-loop system consisting of a Photovoltaic array emulator controlled by Simulink (where we can control parameters, such as irradiance curve, DC voltage, number of panels in series and parallel, etc [5]), taking energy from the grid, and connect to a SMA, which returns it to the same point.

The system has a measurement conditioning board with a total of 10 Hall-effect sensors with the main characteristics shown in Table 1.

Table 1. Sensor characteristics

Magnitude	Sensor	Number of Sensors
Voltage	LV 25-P by LEM	5 (Vsa, Vsb, Vsc, Vdc ,Vga)
Current	LA 25-NP by LEM	5 (Isa, Isb, Isc, Idc, Iga)

4   Quality Parameters Defined on the Base IEEE Standard 1459-2010

Our system takes measurements in three different points:

1. Autotransformer to Emulator.
2. Emulator to Inverter.
3. Inverter to Grid.

In each one of these point, as if it is a single-phase system, only one voltage and current are measured thanks to a measurement board.

In single phase systems, the calculation of the root mean square (RMS) fundamental voltage U_1 and current I_1 is based on the instantaneous measured components voltage $u(k)$ and current $i(k)$. By getting the RMS fundamental magnitudes, the harmonic ones, U_H and I_H, can be obtained, since $U^2 = U_1^2 + U_H^2$ and $I^2 = I_1^2 + I_H^2$, where U and I are the RMS values of u and i, respectively.

The apparent power S is expressed as:

$$S^2 = (UI)^2 = S_1^2 + S_N^2 \tag{1}$$

The active power P can be calculated as the mean value of the power product of the instantaneous voltage and current. Afterwards, the non-active power is defined as:

$$N^2 = S^2 - P^2 \tag{2}$$

Other important quality rates measured are the total harmonic distortion of voltage THD_U and current THD_I:

$$THD_U = \frac{U_H}{U_1} \, , THD_I = \frac{I_H}{I_1} \tag{3}$$

Finally, the last parameters obtained are the power factor P/S at points 1 and 3, and the DC power at point 2.

Note that we get in our measurement block the three-phase values of power at the beginning and ending points thanks to three times the equivalent single-phase power.

Now, the measurement technique used in this work is based on the single phase Quality Meter proposed in [6]. Fig. 2 shows us the quality parameters algorithm.

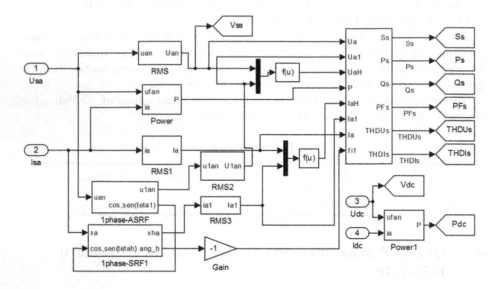

Fig. 2. Quality Parameters Algorithm (Points 1 and 2)

5   Creating a Real Time Quality Metering

5.1   Implementation Stages

The proposed algorithm will be implemented in two stages. First stage, that is the one described in following sections, is a Proof-of-the-concept. The second stage will be a real application implementation that will be done in future works. The resources needed for each step are:

- Proof-of-the-concept stage: A prototype of proof for the first implementation of the algorithm (mode used in the present paper). We establish a system structure based on three different PC's:
    - Host PC, used for the compilation of the algorithm to control the PV emulator and to calculate the required parameters.
    - Target PC, which executes in real time the code generated by the host PC to control the system.
    - Server PC, which receives the data from the target PC and sends them to an Open data.
- Real application stage: In this operation mode the Host PC is no longer needed, and only the control platform or PC, where the smart metering and power electronic controls are running, and the WebServer, where the Open data are available (that is a share PC for multiple services), are operating.

5.2   Quality Parameters Algorithm and Communication Implementation

The PV emulator was developed in the host PC with its corresponding control [5]; now to get the smart metering, the quality parameters algorithm is implemented inside of this model too, taking advantage of the inputs needed in the previous project.

Thus, once the host and target are connected via TCP/IP, inputs coming from the data acquisition target go through our measurement block, to get the desired quality parameters.

These parameters now have to end in the server computer for the data logging, being the chosen communication the UDP protocol [7].

The communication has been added to the host Simulink model, with the UDP send block and the previous pack block. The UDP send block needs the IP address to send to (server IP: 158.49.55.93), the IP port to send to (1992) and the sample time. We use an enabled subsystem, controlled by triggered function, which activates what is inside (UDP send Block, Pack and discrete function) every 5 seconds during the sample time. Before the pack block, data are treated by a discrete transfer function to calculate the average value of the five previous values and send it through the UDP send block. We can see this configuration in the Fig. 3.

The pack block is used to convert the Simulink signal of double type to a single vector of uint8 as required by the send block.

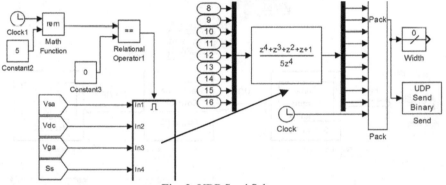

Fig. 3. UDP Send Scheme

Once the data has been sent, we need to receive it in the server PC. A simulink model with a UDP receive block and unpack has been created. The UDP receive block needs IP address to receive from (0.0.0.0 for accepting all) and the IP port to receive from (1992). Another important data is the output port width, known thanks to the width block. The Unpack Block changes again the uint8 data to the original state.

Therefore, we have to treat the data. We have configured an S- function to get the values of these data and write them in an array, which goes to a text format file. The S-function is inside of an enabled subsystem activated each 5 seconds (systems are synchronized). We can see the receive scheme in Fig. 4.

At the end, we obtain in our server PC a text format file with all the read quality parameters, ready to be uploaded to the Open data. The chosen format for this open data is an XML file [8], which is a widely used format for data exchange because it gives good opportunities to keep the structure of the data obtained from our original system (in our case text format file with the data array structure). In this way, new data could be added to the text file at the same time the XML format file is updated. In Fig. 5 we have a scheme of the communication system.

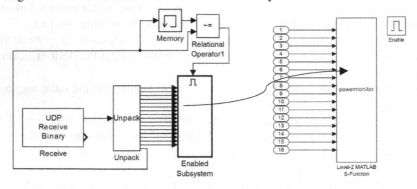

Fig. 4. Receive scheme

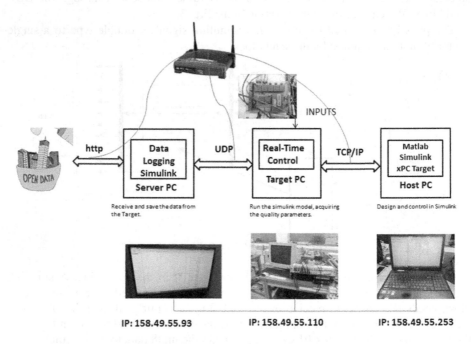

IP: 158.49.55.93   IP: 158.49.55.110   IP: 158.49.55.253

Fig. 5. Communication System Scheme

5.3   Real Time Guarantee

The real time behavior of the system is guaranteed because we only execute the S-function, which contains a clock function to write the current time and date to the corresponding data, comparing the mark of real time received from the target with the previous value, activating the function when they are different. This is because the clock block in charge of getting the real time is inside of an enabled subsystem activated each 5 seconds, so when a new data is received is because the time have changed from 5 to 10, for example.

6   Experimental Results

In table 2 the experimental test parameters are shown. The data obtained from an experimental test are shown in xPC target environment (Fig. 6a) and they can be accessed by conventional web navigator form in its corresponding repository (http://peandes.unex.es/powermonitor/) as it can be seen in Fig. 6b.

Table 2. Experimental Test Parameters

Parameter	Value
Kind of Solution	Euler
Simulation Step	1/3120
Basis Irradiance	1000 W/m2
Number of panels in series	8
Number of panels in parallel	1
V_{DC}	272V
V_{Supply}	75V

(a)   (b)

Fig. 6. (a) xPC Target Environment. (b) Data Repository.

With the experimental case we advise how the performance of the system has decreased. Note that after including the proposed algorithm in the initial power electronic system, the sample and switching frequency has been decreased from 8 to 3,120 kHz. Otherwise, the system will be stuck.

7   Conclusion and Future Works

A smart metering system able to generate Open data accessible in a central repository has been developed and tested using xPC Target. It has been used to get quality parameters from a system, which consists of a photovoltaic array emulator and a commercial inverter, saving them in an Open data in real time.

With this experiment we have proved how the metering algorithm could be implemented in an existing power electronics system. The inclusion of the proposed algorithm in the power electronics system causes the decrease of the sample frequency from 8 to 3,120 kHz. In the future, when we afford the real application stage, the metering algorithm will be included as a different execution tier to prevent affecting the original control algorithm. We consider that is possible to include these smart metering functions in the next generation of power electronics systems, which in some cases will required only the change of systems' firmware.

Besides other interesting work would be to develop a virtual lab [9] with the system, what would be meaningful and very popular in distance education. We will have with this proposal to implement the control in the opposite way, through the open data network to the real system. It will allow the students to see how the real system reacts when they change the control parameters.

To conclude, in the future we pretend to integrate the Open data generated by our algorithms in existing projects as Green Button Developer [10]. It will have to be done according to the ESPI scheme (Energy Services Provider Interface), which provides a way for Energy Usage Information (EUI) to be shared, in a controlled manner, between participants in the energy services markets.

References

1. European Commission, "European SmartGrids Technology Platform – Vision and Strategy for Europe's Electricity Networks of the Future", Directorate-General for Research – Sustainable Energy Systems (2006), http://www.smartgrids.eu
2. Auer, S., Bizer, C., Kobilarov, G., Lehmann, J., Cyganiak, R., Ives, Z.G.: DBpedia: A Nucleus for a Web of Open Data. In: Aberer, K., et al. (eds.) ISWC/ASWC2007. LNCS, vol. 4825, pp. 722–735. Springer, Heidelberg (2007)
3. Smartgrids SRA 2035. Summary of Priorities for Smartgrid research topics. V19 (June 2013)
4. IEEE Standard Definitions for the Measurement of Electric Power Quantities Under Sinusoidal, Nonsinusoidal, Balanced, or Unbalanced Conditions. IEEE Std 1459-2010 (Revision of IEEE Std 1459-2000), pp. 1–52 (March 2010)
5. Polo-Gallego, S., Roncero-Clemente, C., Romero-Cadaval, E., Miñambres-Marcos, V., Guerrero-Martínez, M.A.: Development of a Photovoltaic Array Emulator in a Real Time Control Environment using xPC Target. In: Camarinha-Matos, L.M., Tomic, S., Graça, P. (eds.) DoCEIS 2013. IFIP AICT, vol. 394, pp. 325–333. Springer, Heidelberg (2013)
6. Milanés, M.I., Miñambres, V., Romero, E., Barrero, F.: Quality Meter of Electric Power Systems based on IEEE Standard 1459-2000. CPE 2009 (2009)
7. MATLAB, xPC Target for Use with Real-time Workshop
8. http://opendatahandbook.org/
9. Masár, I., Bischoff, A., Gerke, M.: Remote Experimentation in Distance for Control Engineers. Control Systems Engineering Group, FernUniversität, Hagen
10. http://en.openei.org/wiki/Green_Button_Developer

An Innovator Nonintrusive Method for Disaggregating and Identifying Two Simultaneous Household Loads

Máximo Pérez-Romero[1,2], Enrique Romero-Cadaval[1], Adolfo Lozano-Tello[2],
João Martins[3], and Rui Lopes[3]

[1] Power Electrical & Electronics Systems R+D+i Group
University of Extremadura, Escuela de Ingenierías Industriales, Avda. de Elvas, s/n
06006 Badajoz, Spain
[2] Quercus Software Engineering Group
University of Extremadura, Escuela Politécnica, Avda. Universidad, s/n
10071 Cáceres, Spain
[3] Universidade Nova de Lisboa-FCT-DEE and UNINOVA-CTS
P-2829-516 Monte de Caparica, Portugal
mperez@peandes.net, {alozano,eromero}@unex.es,
jf.martins@fct.unl.pt, rm.lopes@campus.fct.unl.pt

Abstract. At the present time the monitoring systems are important in some areas, such as electric power supply industry and household environment because they provide useful information to energy storage and management tasks. A new nonintrusive monitoring method is proposed in this paper and it is able to disaggregate and identify two loads working simultaneously using a single measuring sensor and a least squares regression algorithm based on discrete form of the S-Transform.

Keywords: Monitoring, nonintrusive, disaggregation, identifying, simultaneously, S-Transform.

1 Introduction

The energy efficiency status report [1] showed by Joint Research Centre, or JRC, points out electricity consumption and efficiency trends in the EU-27 in 2012. According to the consumption statistics, final energy consumption in the EU-27 residential sector increased by 1.69%, furthermore the lowest consumption level of the last 20 years was reached in 2007. In terms of final residential energy consumption means 26,65% of total final energy consumption in the year 2010 showed in the breakdown of the Fig. 1. This sector was the second most consuming part of total consumption, so the residential field is important in energy efficiency programmes and policies.

Smart grid concept is a communication infrastructure to connect different distributed energy resources with three main targets, that is, by adapting the demand side to the distributed energy sources, transforming the distributed energy resources into systems providing active energy and supporting services, such as reactive power supply, to stabilize the distributed network and using storage devices in demand

L.M. Camarinha-Matos et al. (Eds.): DoCEIS 2014, IFIP AICT 423, pp. 297–304, 2014.

side [2]. Smart grids enable a two-way exchange of information and power between producers and consumers, and this leads to increase transparency and promote responsible energy saving measures on the consumer's side. It is important that consumers are involved to obtain effective smart electricity systems through trust, understanding and clear tangible benefits and consumers become proactive consumers or prosumers, being necessary to know individual loads on demand side.

Fig. 1. Final energy consumption breakdown into sectors in the EU-27, 2010 (Eurostat)

Load monitoring systems are important to manage smart grids because they provide accurate power and energy usage information to power usage prediction and management. There are two usual methods in literature to monitor appliances: intrusive and nonintrusive methods.

Intrusive methods use sensors in each appliance to measure electric signals, such as voltage and current waveform. This measuring manner is expensive because installation needs time and one sensor for each appliance. For overcoming this drawback researchers have studied other procedure to extract the individual energy information of each device. It is called Nonintrusive Load Monitoring (NILM) or Nonintrusive Appliance Load Monitoring (NIALM) and it consist of a single sensor placed in the main breaker level to measure current and voltage and an algorithm for disaggregate and identify each one of appliance. This question was firstly studied by George W. Hart [3] of Massachusetts Institute of Technology (MIT) and published in the 90´s.

The methods for disaggregating different appliances can be divided into two groups: the steady state analysis and the transient state analysis. Regarding first group, it is based on the change in steady state power and requires low sampling rates. Hart´s work was already mentioned at the end of the previous paragraph. Next, Cole y Albicki [4] have proposed an extension to the Hart method. It uses edges and slopes as appliance features together with steady-state power draw. Finally, Baranski and Voss have developed an approach [5], [6], [7] for their system consisting of a histogram made up by historical data. As far as transient state analysis is concerned, transient characteristics can be used to disaggregate different household devices because they are tied to type of a given appliance. One of the most important researchs belongs to Leeb [8], [9]. An extension of work mentioned above consists of investigating the line voltage transients appeared when appliances are switched on or off (Cox and Leeb, [10]). Finally Chang proposed a transient energy detector [11].

A new method for disaggregating and identifying different household loads based on the S-transform [12] is proposed in this paper (other S-transform method was used by [13]). This work is part of [14]. The novelty is the use of mathematical regression along with the S-transform and the development of algorithm to be implemented with the software *LabVIEW 2012*.

2   Relationship to Collective Awareness Systems

The Collective Awareness Platforms are ICT systems taking advantage of distributed knowledge and data from real environments, that is, Internet of things, to generate open databases. These types of platforms are scalable resulting in a big data source for different applications, such as real-time household power consumption data for managing energy demand of consumers, becoming collective tools.

In this way, this paper is based on such topic by disaggregating and identifying loads working individually or simultaneously and by sending such kind information in real-time to a decision central system.

3   S-Transform

The S-transform is a time-frequency analysis and representation technique proposed by Stockwell, Manshinha and Lowe in 1996. It is consists of a moving and scalable localizing Gaussian window leading to a time-frequency representation. Phase information is absolute because the origin of the time axis is taken as the fixed reference point.

The analysis begins with the target signal $h(t)$. $W(\tau,f)$ is the Continuous Wavelet transform (CWT) of the signal using a specific mother wavelet and its equation appears in (1).

$$W(\tau, f) = \int_{-\infty}^{\infty} h(t) w(t - \tau, f) dt.$$

(1)

where t is the time, f is the frequency and $w(t$-$\tau,f)$ is the specific mother wavelet. Then

$$S(\tau, f) = W(\tau, f) e^{-i2\pi f\tau} = \int_{-\infty}^{\infty} h(t) w(t - \tau, f) e^{-i2\pi f\tau} dt.$$

(2)

The mother wavelet is the first expression of (3).

$$w(\tau, f) = g(t) e^{-i2\pi ft}; g(t) = \frac{1}{\sigma\sqrt{2\pi}} e^{-t^2/2\sigma^2}; \sigma(f) = T = \frac{\lambda}{|f|} \qquad \lambda > 0.$$

(3)

The formula $g(t)$ is a Gaussian window and the parameter σ means the Gaussian window width ,where λ sets frequency resolution. In this paper the parameter λ has the value 1. As the window width is the inverse of frequency, S-Transform results in a multiresolution analysis of the signal. Therefore, a high value of frequency gives high resolution of time and a low value of frequency gives high resolution of frequency.

Finally, by combining the expressions (4) and (2) the complete definition of the continuous S-Transform is given by (5).

$$w(t - \tau, f) = \frac{|f|}{\sqrt{2\pi}} e^{-(t-\tau)^2 f^2/2} e^{-i2\pi f(t-\tau)}$$

(4)

$$S(\tau, f) = \int_{-\infty}^{\infty} h(t) \frac{|f|}{\sqrt{2\pi}} e^{-(t-\tau)^2 f^2/2} e^{-i2\pi f(t-\tau)} e^{-i2\pi f\tau} dt.$$

(5)

To efficiently implement on a computer the S-Transform it is necessary to develop the discrete form of the S-Transform. Stockwell *et al.* rewrote the continuous S-Transform by using Fourier transform $H(\alpha)$ of time series $h(t)$ as the equation (6) shows to get the discrete form. So, S-Transform can be run as computer algorithm by using the computational efficiency of Fast Fourier Transform and Inverse Fourier Transform in (6).

From the discrete time series $h[kT]$ corresponding to $h(t)$ with a sample time T the equation of the discrete Fourier transform is given by (7).

$$S(\tau, f) = \int_{-\infty}^{\infty} H(\alpha + f) e^{-\left(2\pi^2\lambda^2\alpha^2\right)\big/f^2} e^{i2\pi\alpha\tau} d\alpha \qquad f \neq 0. \tag{6}$$

$$H\left[\frac{n}{NT}\right] = \frac{1}{N} \sum_{k=0}^{N-1} h[kT] e^{-i2\pi nk\big/N} \qquad n = 0,1,...,N-1. \tag{7}$$

where k and $n = 0, 1, ..., N-1$.

Therefore, the discrete S-Transform is gotten by setting $f \rightarrow n/(NT)$ and $\tau \rightarrow jT$ resulting in (8) and (9), with λ equal to 1.

$$S\left[jT, \frac{n}{NT}\right] = \sum_{m=0}^{N-1} H\left[\frac{m+n}{NT}\right] e^{-2\pi^2 m^2\big/n^2} e^{i2\pi mj\big/N} \qquad n \neq 0. \tag{8}$$

$$S[jT, 0] = \frac{1}{N} \sum_{m=0}^{N-1} h\left(\frac{m}{NT}\right) \qquad n = 0. \tag{9}$$

where j, m and $n = 0, 1, ..., N-1$.

The S-Transform consists of a complex matrix (called S-matrix) with dimension N by M where rows correspond to frequency and columns correspond to time.

4   Case Study

The proposed system consists of parts showed in Fig. 2: a monitor, a light bulb, a single current sensor of the measuring board, a point of common coupling (PCC), a control platform *sbRIO 9631* of *National Instruments* consisting of a FPGA (hardware part) and a real-time processor (software part) and a notebook being only used to program the control platform. Overall system communicates with the central server called CEMIS.

The current sensor measures the current demanded by complete system and its information is sent to analog input of the control platform for disaggregating and identifying what loads are switched on or switched off. The two loads are connected in parallel to the PCC and one or both of them can be working simultaneously. The programming methodology of disaggregation and identification tasks, main topic of this paper, is based on codesign hardware-software and developed in *LabVIEW 2012*. The FPGA will get data from current sensor because the analog input is connected to it directly. It will also perform preprocessing tasks such as scale current data, convert from data type of fixed-point to unsigned word of 32 bits and write to a memory buffer. The sample frequency is 10 kHz and the number of points N is 201.

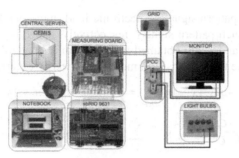

Fig. 2. System diagram

Fig. 3(a) shows the FPGA algorithm flowchart. The software part will read the data from buffer, convert from unsigned word of 32 bits to data type of double precision, extract one signal cycle, execute the S-transform and disaggregate and identify the loads by means of least squares regression as it is shown in the flowchart of Fig. 3(b).

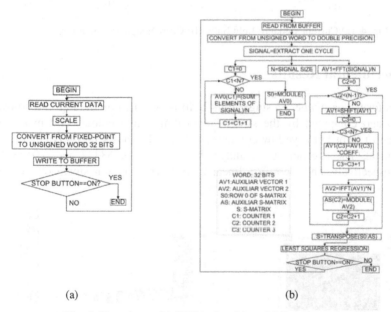

(a)   (b)

Fig. 3. Flowcharts: (a) FPGA algorithm, (b) Processor algorithm

The resultant matrix "S" is a transposed complex matrix with N rows and M = N/2+1 columns, that is, 201x102. The task of identification consists of the least squares regression (10).

$$\sum_{i=1}^{N}\left(z_i - \left(a_1 x_i + a_2 y_i\right)\right)^2 = e. \tag{10}$$

where z_i is the S-Transform of the overall measured current signal, x_i is the S-Transform of the pattern signal of linear load (light bulb), Fig. 4(a), y_i is the

S-Transform of the pattern signal of electronic load (monitor), Fig. 4(b), a_1 and a_2 are the amplitudes of such pattern signals, respectively, and e is the least squares error. The coefficients a_1 and a_2 are obtained in (12) from (11) and they can be used in the time domain because the S-Transform meets the linearity criteria.

$$\frac{\partial e}{\partial a_1} = \sum_{i=1}^{N} 2\left(z_i - a_1 x_i - a_2 y_i\right) x_i = 0; \quad \frac{\partial e}{\partial a_2} = \sum_{i=1}^{N} 2\left(z_i - a_1 x_i - a_2 y_i\right) y_i = 0. \tag{11}$$

$$a_1 = \frac{BD - CE}{AB - C^2}; a_2 = \frac{AE - CD}{AB - C^2}; A = \sum_{i=1}^{N} x_i^2; B = \sum_{i=1}^{N} y_i^2; C = \sum_{i=1}^{N} x_i y_i; D = \sum_{i=1}^{N} z_i x_i; E = \sum_{i=1}^{N} z_i y_i. \tag{12}$$

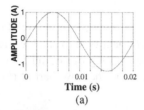 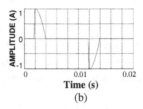

(a)   (b)

Fig. 4. (a) Current signal of light bulb. (b) Current signal of monitor.

5   Experimental Results

On one side, for the case where a single light bulb is switched on the Fig. 5(a) shows signal fitted and computed by the system algorithm (white colour) and measured current signal of light bulb (yellow colour) and Fig. 5(b) shows the S-Transform of measured signal. The coefficients a_1 and a_2 have values 0,368337 and 0,0163964 (Table 1), respectively, and identify a linear load as white colour signal of the Fig. 5(a) illustrates.

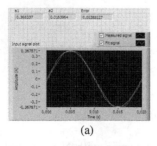 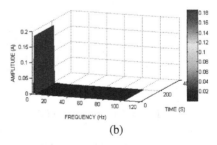

(a)   (b)

Fig. 5. (a) Measured and fitted current signal of light bulb (front panel of *LabVIEW 2012*). (b) S-Transform of measured current signal.

On the other hand, when a single electronic load or monitor is turned on the system discovers a nonlinear load currently working. The coefficients are shown in Table 1 (this table illustrates the coefficients and error for each appliance and the combination of both.) and its waveform appears in Fig. 6(a) whereas its S-Transform is shown by Fig. 6(b). Finally, the main case consists of one light bulb and monitor working simultaneously. The system calculates the values of coefficients a_1 and a_2 and, according to the gotten results, it can be said that two loads have been disaggregated and identified. Fig. 7(a) shows time series signals and Fig. 7(b) presents the S-Transform.

Table 1. Values of coefficients

LOAD	a_1	a_2	Error
Light bulb	0,368337	0,0163964	0,003888227
Monitor	0,0450099	0,309192	0,220845
Light bulb and monitor	0,409668	0,321245	0,181646

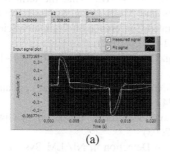 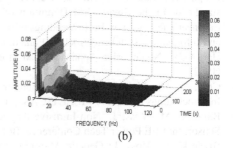

(a)   (b)

Fig. 6. (a) Measured and fitted current signal of monitor (front panel of *LabVIEW 2012*). (b) S-Transform of measured current signal.

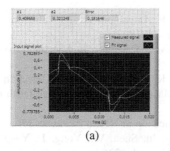 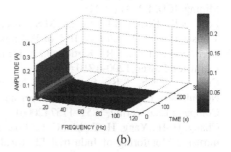

(a)   (b)

Fig. 7. (a) Measured and fitted current signal of light bulb and monitor (front panel of *LabVIEW 2012*). (b) S-Transform of measured current signals.

6   Conclusion and Further Work

The main feature of this new nonintrusive method is to disaggregate and identify two loads, a linear device and nonlinear device, by using S-Transform and least squares regression with one cycle of the time series signal. It is an advantage because the S-Transform uses computational efficiency of FFT and IFFT and other parameters are not necessary, such as harmonic analysis and reactive power, reducing computations. Furthermore, the system needs a single current sensor reducing cost. In the future, more loads will be connected to the system and the least squares regression will extend to detect such devices although the complexity will increase. There is also an open work to identify devices switched on by using neural networks.

Acknowledgements. This work has been developed under support of Telefónica Chair of University of Extremadura.

References

1. JRC. EESR (2012),
   http://iet.jrc.ec.europa.eu/energyefficiency/node/7250
2. Gregor, R., Simon, F., Ulf, J.J., Hahnel, P.B., Bernhard, W.-H.: What the term Agent stands for in the Smart Grid Definition of Agents and Multi-Agent Systems from an Engineer's Perspective. In: Proceedings of the Federated Conference on Computer Science and Information Systems, Wroclaw, Poland, pp. 1301–1305 (2012)
3. Hart, G.W.: Nonintrusive Appliance Load Monitoring. Proceedings of the IEEE, 1870–1891 (1992)
4. Cole, A., Albicki, A.: Algorithm for nonintrusive identification of residential appliances. In: Proceedings of the 1998 IEEE ISCAS, pp. 338–341 (1998)
5. Baranski, M., Voss, J.: Non-Intrusive Appliance Load Monitoring Based on an Optical Sensor. In: EEE Power Tech Conference, Bologna (2003)
6. Baranski, M., Voss, J.: Genetic Algorithm for Pattern Detection in NIALM Systems. In: IEEE International Conference on Systems, Man and Cybernetics, pp. 3462–3468 (2004)
7. Baranski, M., Voss, J.: Detecting Patterns of Appliances from Total Load Data Using a Dynamic Programming Approach. In: Fourth IEEE International Conference on Data Mining, ICDM 2004 (2004)
8. Leeb, S.B., Kirtley, J.L., Levan, M.S., Sweeney, J.P.: Development and Validation of a Transient Event Detector. AMP J. of Technology, 69–74 (1993)
9. Leeb, S.B., Shaw, S.R., Kirtley, J.L.: Transient Event Detection in Spectral Envelope Estimates. Power, 1200–1210 (1995)
10. Cox, R., Leeb, S.B., Shaw, S.R., Norford, L.K.: Transient Event Detection for Nonintrusive Load Monitoring and Demand Side Management Using Voltage Distortion. Computer Engineering, 1751–1757 (2006)
11. Chang, H.-H., Yang, H.-T., Lin, C.-L.: Load Identification in Neural Networks for a Nonintrusive Monitoring of Industrial Electrical Loads. In: Shen, W., Yong, J., Yang, Y., Barthès, J.-P.A., Luo, J. (eds.) CSCWD 2007. LNCS, vol. 5236, pp. 664–674. Springer, Heidelberg (2008)
12. Stockwell, R.G., Mansinha, L., Lowe, R.P.: Localization of the complex spectrum: the S transform. IEEE Trans. Signal Processing, 998–1001 (1996)
13. Martins, J.F., Lopes, R., Lima, C., Romero-Cadaval, E., Vinnikov, D.: A Novel Nonintrusive Load Monitoring System Based on the S-Transform. In: 13th International Conference on Optimization of Electrical and Electronic Equipment (OPTIM 2012), Brasov, Romania, May 24-26 (2012)
14. Pérez-Romero, M., Gallardo-Lozano, J., Romero-Cadaval, E., Lozano-Tello, A.: Local Energy Management Unit for Residential Applications. Electronics and Electrical Engineering 19(7), 61–64 (2013)

Distributed MPC for Thermal Comfort in Buildings with Dynamically Coupled Zones and Limited Energy Resources

Filipe A. Barata[1] and Rui Neves-Silva[2]

[1] Instituto Superior de Engenharia de Lisboa (ISEL), R. Conselheiro Emídio Navarro 1,
1959-007 Lisboa, Portugal
[2] Universidade Nova de Lisboa, Monte da Caparica,
2829-516 Caparica, Portugal
`fbarata@deea.isel.ipl.pt, rns@fct.unl.pt`

Abstract. This paper presents a distributed predictive control methodology for indoor thermal comfort that optimizes the consumption of a limited energy resource using a demand-side management approach. The building divisions are modeled using an electro-thermal modular scheme. For control purposes, this modular scheme allows an easy modeling of buildings with different plans where adjacent areas can thermally interact. The control objective of each subsystem is to minimize the energy cost while maintaining the indoor temperature in the selected comfort bounds. In a distributed coordinated environment, the control uses multiple dynamically coupled agents (one for each subsystem/zone) aiming to achieve satisfaction of available energy coupling constraints. The system is simulated with two zones in a distributed environment.

Keywords: Multi-zone thermal comfort, electro-thermal analogy, DMPC, limited energy resource.

1 Introduction

Buildings are responsible for a large share of our global energy use. Energy use is in fact the main determinant of a buildings global environmental footprint, considering its total life span.

Reducing energy consumption in the building stock is a trend in the world today, partly because of economic, partly because of environmental reasons. Energy dissipation depends on the construction of house, the materials from which the house is built, insulation during the year, outside temperature, and additional sources of energy. Consumption of energy is predominantly determined with a selection of materials and architectural solutions, and it can be further reduced with efficient management of heating or cooling. An effective heating/cooling management is provided in the framework of predictive control, particularly Model Predictive Control (MPC), has been granted to reduce and optimize the energy consumption in

L.M. Camarinha-Matos et al. (Eds.): DoCEIS 2014, IFIP AICT 423, pp. 305–312, 2014.

the residential sector namely to deal with temperature set points regulations [1, 2, 3] and load management [4, 5].

The distinct advantages of MPC based control solutions compared to classical controllers are: using also relevant future information in making control decisions using predicted profiles (e.g. ambient temperature and solar irradiation on outside walls), routine handling of multi-input-multi-output (MIMO) systems; routine respecting of system constraints (e.g. finite amount of heating/cooling power in room possible, desired room temperature spans); and explicit orientation of control actions towards the goal which can be using least energy possible, or spending least money possible, or causing least CO_2 emissions possible, or their combination.

Optimal control for indoor environment requires preservation of comfort conditions for buildings occupants and minimization of energy consumption and cost [6]. Basically, MPC makes a tradeoff between energy savings and thermal comfort. The MPC has also advantage when controlling distributed systems [7], [8]. Distributed Model Predictive Control (DMPC) algorithms are the state of the art in complex control problems with many interconnected subsystems. DMPC allows the distribution of decision-making while handling constraints in a systematic way. DMPC strategies can be characterized by the type of couplings or interactions assumed between constituent subsystems [9].

The method of subsystems sharing coupled constraints can be seen in [9], [10] being the strategy here presented a DMPC with coupled constraints (renewable energy must be shared by all divisions) and dynamically coupled zones. Thus, in a distributed coordinated environment, the control uses multiple dynamically coupled agents (one for each subsystem/division) aiming to achieve satisfaction of coupling constraints.

The desired approach here presented intends to take advantage from the innovative technology characteristics provided by future Smart Grids (SGs) [11]. In the smart world, simple household appliances, like dishwashers, clothes dryers, heaters, air conditioners will be fully controllable in order to achieve the network maximum efficiency.

Compared with the aforementioned literature, the novel contributions of this work are related with the existence of a system with coupled constraints and dynamically coupled zones in a cost function with distinct objectives allowing thermal comfort with a consumption to weatherize the divisions inside the available power constraints.

The remaining of this paper is organized as follows: Section 2 presents the technological contribution of this paper and Section 3 presents the system architecture. In Section 4 is introduced the house dynamic model and the model predictive controller is presented in Section 5. Some results and analysis are shown in Section 6 and conclusions are drawn in Section 7.

2   Relationship to Collective Awareness Systems

Collective awareness is an important issue to the use of groupware systems and virtual collaboration. From the Collective Awareness Systems (CAS) perspective, distributed networks with multi-agents may represent systems that are able to collectively contribute to a same objective. The analysis in this paper is on a line of technological contributions related with the CAS characteristics. Future appliances will be linked in the grid and will be fully controllable, monitored and regulated in real time in order to collaborate to achieve energy efficiency, balance between demand and supply, intelligent load control consumer comfort and CO_2 emissions reduction.

3   System Architecture

In this paper the large scenario considers a distributed network that involves a residential community, with electricity power source generated by their own renewable energy park. Hence, two kinds of energy are considered, the *green* that is from the renewable source and the *red* energy, from the grid. The energy from the grid is always available, although at a higher price, and it is only consumed when the *green* energy is not enough to satisfy the demand.

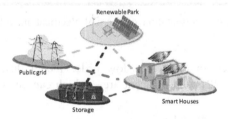

Fig. 1. Global scenario

The houses/divisions may have different plans, thermal loads, thermal characteristics, occupancy and comfort temperature bounds, and consequently with different energy needs for heating/cooling the spaces. The idea is to apply a predictive control law to maintain the temperature and power consumption inside their bounds. The *green* energy is limited and predictable and must be shared by all houses/divisions. The *red* resource consumed for comfort implies a penalty in the final cost function (6) due to the soft constraint violation. This penalization means that the maximum available *green* resource was exceeded. It is considered that the outside temperature, disturbances and daily comfort temperature bounds are known by each system inside the predictive horizon (*N*). The *green* resource that is not consumed at a certain instant is stored in batteries or delivered to the grid.

4   Thermal Model of the House

The idea here presented is to apply the principle of analogy between two different physical domains that can be described by the same mathematical equations. Thus, a linear electrical circuit represents the building and the state-space equations are obtained by solving that circuit. Here, the temperature is equivalent to voltage, the heat flux to current, the heat transmission resistance is represented by electrical resistance and the thermal capacity by electrical capacity. The equivalent circuit of the building is obtained by assembling models of the walls, windows, internal mass, etc. In the case of single-zone buildings, interior walls are being part of the internal thermal mass while exterior walls are forming the building envelope. Several approaches can be seen in [12, 5] where is shown that building models can be simpler or more complex depending on the objective. Fig. 2 shows the used thermal-electrical modular approach.

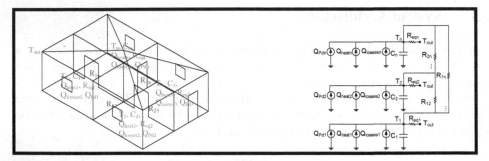

Fig. 2. Generic schematic representation of thermal-electrical modular analogy for several divisions

The model presented in (1-3) is a low order model describing the dominant dynamics for division i [2] with adjacent area, which can be considered suitable for control proposes.

$$\frac{dT_i}{dt} = \frac{1}{C_i}\left(Q_{heat_i} - Q_{losses_i} + Q_{Pd_i}\right), \tag{1}$$

$$Q_{losses_i} = \frac{T_{out} - T_i}{R_{eq_i}} + \sum_{j=1}^{n}\frac{T_j - T_i}{R_{ij}}, \tag{2}$$

$R_{eq_i} = R_{roof_i} \,//\, R_{window_i} \,//\, R_{wall_i} + R_{th_i}$, with $R_{window_i} = \sum R_{window_{materials}}$, $R_{roof_i} = \sum R_{roof_{materials}}$ and
$$R_{wall_i} = \sum R_{wall_{materials}} \tag{3}$$

For several adjacent zones (1) can be generically written as follows,

$$
\begin{bmatrix} \dot{T}_1 \\ \dot{T}_2 \\ \vdots \\ \dot{T}_n \end{bmatrix} =
\begin{bmatrix}
\frac{1}{R_{eq_1}C_1} - \frac{1}{R_{12}C_1} - \cdots \frac{1}{R_{1n}C_1} & \frac{1}{R_{12}C_1} & \cdots & \frac{1}{R_{1n}C_1} \\
\frac{1}{R_{21}C_2} & \frac{1}{R_{eq_2}C_2} \frac{1}{R_{21}C_2} - \cdots \frac{1}{R_{2n}C_2} & \ddots & \frac{1}{R_{2n}C_2} \\
\vdots & \vdots & \cdots & \vdots \\
\frac{1}{R_{n1}C_n} & \frac{1}{R_{n2}C_n} & \cdots & \frac{1}{R_{eqn}C_n} - \cdots \frac{1}{R_{n1}C_n} \frac{1}{R_{n2}C_n}
\end{bmatrix}
\begin{bmatrix} T_1 \\ T_2 \\ \vdots \\ T_n \end{bmatrix} +
$$

$$
\begin{bmatrix}
\frac{T_{out}}{R_{eq_1}C_1} + \frac{Pd_1}{C_1} \\
\frac{T_{out}}{R_{eq_2}C_2} + \frac{Pd_2}{C_2} \\
\vdots \\
\frac{T_{out}}{R_{eqn}C_n} + \frac{Pd_n}{C_n}
\end{bmatrix} +
\begin{bmatrix}
\frac{1}{C_1} & 0 & \cdots & 0 \\
0 & \frac{1}{C_2} & \cdots & \vdots \\
\vdots & \vdots & \ddots & 0 \\
0 & \cdots & 0 & \frac{1}{C_n}
\end{bmatrix}
\begin{bmatrix} u_1 \\ u_2 \\ \vdots \\ u_n \end{bmatrix} \tag{4}
$$

where in (1), Q_{losses_i} is heat and cooling losses (kW), T_i the inside temperature (°C), C_i the equivalent thermal capacitance (kJ/°C), and Q_{heat_i} the heat and cooling power (kW) and Q_{Pd_i} the external thermal disturbances (kW) (e.g. load generated by occupants, direct sunlight, electrical devices or doors and windows aperture to recycle the indoor air). In (2) T_{out} is the outdoor temperature (°C), R_{ij} the thermal resistance between divisions, R_{eq_i} the equivalent thermal resistance [3] and R_{th_i} the air thermal resistance to bulk of house.

The plant model representation (1) for one division with an adjacent zone can be approximated by a discrete model using Euler discretization [13] with a sampling time of Δt.

$$T_i(k+1) = A_i T_i(k) + B_i u_i(k) + D_i T_j(k) + d_i(k), \tag{5}$$

where $A_i = \left(1 - \dfrac{R_{eq_i} + R_{ij}}{R_{eq_i} C_i R_{ij}} \Delta t\right), B_i = \dfrac{\Delta t}{C_i}, D_i = \dfrac{T_j}{C_i R_{ij}} \Delta t, d_i = \dfrac{P_{d_i} \Delta t}{C_i} + \dfrac{T_{oa} \Delta t}{R_{eq_i} C_i}$, $u_l(k)$ is the

necessary heat/cooling power, $T_i(k)$ is the indoor temperature, $d_i(k)$ is a disturbance signal resulting from P_{di} the external disturbances (kW) (e.g. load generated by occupants, direct sunlight, electrical devices or doors and windows aperture to recycle the indoor air), and T_{oa}, the temperature of outside air (°C).

5   Model Predictive Control Cost Function

Fig. 3 shows the implemented MPC scheme. The controllers from the areas that are thermally coupled interchange information about their state prediction as can also be seen in (7).

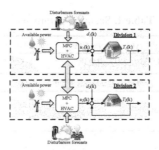

Fig. 3. Block diagram of the implemented MPC system in each division

At each time step, each one of the agents must solve his MPC problem. The cost function objectives are: minimize the energy consumption to heating and cooling; minimize the peak power consumption; maintain the zones within a desired temperature range and maintain the used power within the *green* available bounds. The generic problem to be solved by each agent, assumes the following form:

$$\min_{U,\bar{\varepsilon},\underline{\varepsilon},\bar{\gamma},\underline{\gamma}} \sum_{k=0}^{N-1} u_{t+k|t}^2 \Delta t + \phi \max\{u_{t|t}^2, ..., u_{t+N-1|t}^2\} + \rho \sum_{k=1}^{N} \left(\bar{\varepsilon}_{t+k|t}^2 + \underline{\varepsilon}_{t+k|t}^2\right) + \psi \sum_{k=1}^{N} \left(\bar{\gamma}_{t+k|t}^2 + \underline{\gamma}_{t+k|t}^2\right) \tag{6}$$

subject to the following constraints,

$$T_{t+k+1|t} = AT_{t+k|t} + Bu_{t+k|t} + DT_{j_{t+k|t}} + d_{t+k|t}, \tag{7}$$

$$\underline{T} - \underline{\varepsilon}_{t+k|t} \leq T \leq \bar{T} + \bar{\varepsilon}_{t+k|t}, \tag{8}$$

$$\underline{U}_{A_i} - \underline{\gamma}_{t+k|t} \leq U \leq \bar{U}_{A_i} + \bar{\gamma}_{t+k|t}, \tag{9}$$

$$\underline{\gamma}_{t+k|t}, \bar{\gamma}_{t+k|t}, \underline{\varepsilon}_{t+k|t}, \bar{\varepsilon}_{t+k|t} \geq 0. \tag{10}$$

In (6), u represents the power control inputs, ϕ is the penalty on peak power consumption, ρ is the penalty on the comfort constraint violation, ψ the penalty on the power constraint violation and N is the length of the prediction horizon. In (8), $\bar{\varepsilon}$ and $\underline{\varepsilon}$ are the vectors of temperature violations that are above and below the desired comfort zone defined by $\bar{T}$ and $\underline{T}$. In (9), $\bar{\gamma}$ and $\underline{\gamma}$ are the power violations that are above or lower the maximum, $\bar{U}_{A_i}$, and minimum, $\underline{U}_{A_i}$, available *green* power for heating/cooling the space, with $\underline{U}_{A_i} = -\bar{U}_{A_i}$.

6   Results

The presented results were obtained with an optimization *Matlab* routine. As a first approach towards developing a control structure it is considered an individual house with two divisions thermally coupled. The used parameters that characterize the house are showed in Table 1.

Table 1. Scenario parameters

C(kJ/ºC)	R_{roof}(ºC/W)	R_{walls}(ºC/W)	R_{d}(ºC/W)	$R_{windows}$(ºC/W)	R_{total}(ºC/W)	R_{r}(ºC/W)
9.2×10^3	0.192	0.031	0.038	0.023	0.050	0.030

Divisions	ρ	ψ	Φ	Δt	N	$T(0)$ (ºC)
D1						22
	100	500	2	1	24	
D2						23

It is considered that the divisions have the same thermal characteristics and penalties in the cost function. The planned thermal perturbations are, however, different for the two divisions (Table 1 and Fig. 4(b)). The division that uses the available *green* energy first is Division 1 (D1) and Division 2 (D2) uses only the remainder. By this reason, the maximum available energy to D1 is always the maximum *green* available stock, and for D2 is given by (11):

$$\bar{U}_{A2}(k:k+N) = \bar{U}_{A1}(k:k+N) - u_2(k:k+N). \tag{11}$$

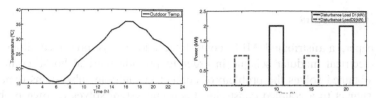

Fig. 4. (a) outdoor temperature forecasting (T_{oa}); (b) thermal disturbance profile (P_d)

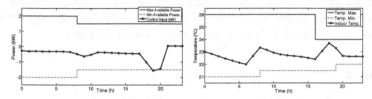

Fig. 5. Division 1 (a) power profile; (b) indoor temperature profile

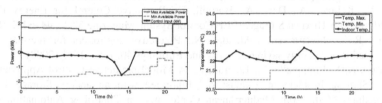

Fig. 6. Division 2 (a) power profile; (b) indoor temperature profile

The outdoor temperature forecasting (T_{oa}) has the profile present in [14]. The comfort limits and available renewable resource vary during the 24 period and it can be seen that both indoor temperature and consumed power are always maintained inside the constrained bounds (Fig. 5 and 6). Taking advantage of the predictive knowledge of the thermal disturbance and making use of the space thermal storage, it can also be seen that in both divisions the MPC changes the indoor temperature in anticipation to the thermal disturbance (Fig. 5(b) and 6(b)).

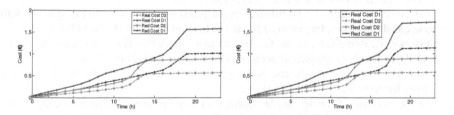

Fig. 7. (a) Cost profile with ψ=500 (b) Cost profile with ψ=1

For each one of the divisions it can be seen in Fig. 7 that the *"Real Cost"* is much lower than the cost of only consuming the *red* resource *"Red Cost"* at a higher fixed price. To illustrate the benefit of the power constraint penalization in (6), Fig. 7(b) shows that maintaining all the other features and changing only the penalty value D1 had a higher cost compared with Fig. 7(a).

7   Conclusions

In this paper, a distributed MPC control technique was presented along with a thermal-electrical modular scheme in order to provide thermal house comfort. The solution obtained solves the problem of control of multiple subsystems dynamically coupled subject to a coupled constraint. Each subsystem solves its own problem by involving its own and adjacent rooms state predictions and also the shared constraints. Changing the penalty values, the consumer can choose in each division between indoor comfort and lower costs. It could be observed through the simulations and results analysis that suitable dynamic performances were obtained.

References

1. Moroşan, P., Bourdais, R., Dumur, D., Buisson, J.: Building temperature regulation using a distributed model predictive control. Journal Energy and Buildings, 1445–1452 (2010)
2. Ma, Y., Kelman, A., Daly, A., Borrelli, F.: Predictive Control for Energy Efficient Buildings with Thermal Storage. IEEE Control System Magazine 32(1), 44–64 (2012)
3. Freire, R.Z., Oliveira, H.C., Mendes, N.: Predictive controllers for thermal comfort optimization and energy savings. Energy and Buildings 40, 1353–1365 (2008)
4. Giorgio, A.D., Pimpinella, L., Liberati, F.: A Model predictive Control Approach to the Load Shifting Problem in a household Equipped with an energy Storage Unit. In: Proceedings of the 20th Mediterranean Conference on Control & Automation (MED), Barcelona, Spain, pp. 1491–1498 (2012)
5. Zong, Y., Kullmann, D., Thavlov, A., Gehrke, O., Bindner, H.W.: Application of predictive control for active load management in a distributed power system with high wind penetration. IEEE Transactions on Smart Grid 3(2), 1055–1062 (2012)
6. Barata, F., Igreja, J., Neves-Silva, R.: Model Predictive Control for Thermal House Comfort with Limited Energy Resources. In: Proc. of the 10th Portuguese Conference on Automatic Control, pp. 146–151 (July 2012)
7. Negenborn, R.: Multi-Agent Model Predictive Control with Applications to Power Networks. PhD Thesis, Technische Universiteit Delft. Nederland (2007)
8. Scattolini, R.: Architectures for distributed and hierarchical Model Predictive Control – A review. Journal of Process Control 19, 723–731 (2009)
9. Trodden, P., Richards, A.: Distributed model predictive control of linear systems with persistent disturbances. International Journal of Control 83(8), 1653–1663 (2010)
10. Keviczky, T., Borrelli, F., Balas, G.: Decentralized Receding Horizon Control for Large Scale Dynamically Decoupled Systems. Automatica 42, 2105–2115 (2006)
11. Siano, P.: Demand response and smart grids—A survey. Renewable and Sustainable Energy Reviews 30, 461–478 (2014)
12. Hazyuk, I., Ghiaus, C., Penhouet, D.: Optimal temperature control of intermittently heated buildings using Model Predictive Control: Part I e Building modelling. Building and Environment 51, 379–387 (2012)
13. Luyben, W.: Process Modeling, Simulation and control for Chemicals Engineers, 2nd edn. McGrawHill
14. Barata, F., Neves-Silva, R.: Distributed Model Predictive Control for Thermal House Comfort with Auction of Available Energy. In: Proceedings of SG-TEP 2012: International Conference on Smart Grid Technology, Economics and Policies (2012)

Part XII

Modeling and Simulation in Energy

Amorphous Solar Modules Simulation and Experimental Results: Effect of Shading

Luis Fialho[1,3], Rui Melício[1,3], Victor M.F. Mendes[1,2], João Figueiredo[1,3],
and Manuel Collares-Pereira[1]

[1] Universidade de Évora, Évora, Portugal
ruimelicio@uevora.pt
[2] Instituto Superior of Engenharia de Lisboa, Lisbon, Portugal
[3] IDMEC/LAETA, Instituto Superior Técnico, Universidade de Lisboa, Lisbon, Portugal

Abstract. This paper focuses on the modeling of PV systems by the five parameters model, consisting on a current controlled generator, single-diode, a shunt and series resistances. An assessment for the identification of the parameters is used requiring data on open circuit, maximum power and short circuit tests. A simulation of a photovoltaic system on a parallel of two series connected amorphous solar modules under the effect of partial shading is presented. The estimated parameters are validated by a comparison with experimental measurements on photovoltaic modules.

Keywords: PV system, effect of shading, algorithm for parameter estimation, simulation, experimental results, shading on amorphous PV.

1 Introduction

The demand for energy, the foreseeable future scarcity and the price of fossil fuels coupled with the need for carbon footprint reduction turn out a political consciousness of the importance of energy savings and energy efficiency usage [1] and programs on the Demand-side Management have been developed in order to assist consumers on energy usage. Moreover, alternative sources of energy, for instance, wind and solar energy sources have turn out to be attractive for exploitation, not only for large scale systems, but also for micro and mini scale conversion systems [2], Disperse Generation owned by consumer. Disperse Generation significantly utilizes solar energy as a primary source of energy and have came into sight the utilization of Thermal and Photovoltaic (PV) systems. A PV system directly converts solar energy into electric energy. The main device of a PV system is a solar cell. Cells may be grouped to form arrays and panels. A PV array may be either a panel or a set of panels connected in series or parallel to form large PV systems without or with tracking systems in order to achieve higher values of energy conversion during sunny days due to the diverse perpendicular positions to collect the irradiation from the sun. The performance of a PV array depends on the operating conditions especially on solar irradiation, temperature, array configuration and shading. The shading on a PV array, for instances, due to a passing cloud or neighboring buildings causes not only power losses, but also further non-linear effects on the array V-I characteristics [3]. In order

L.M. Camarinha-Matos et al. (Eds.): DoCEIS 2014, IFIP AICT 423, pp. 315–323, 2014.

to protect the cells from destructive reverse voltages in case of shadowing or other abnormalities, a bypass diodes are utilized, for example, one bypass diode connected in parallel with each set of 18 cells [4] or with a panel is common practice as a compromise between protection and increase on the cost due to the extra pn-junction.

The impact of the shadow is important in case of large PV power system installations. Under partially shaded conditions, the PV characteristics of the system acquire complexity, appearing multiple peaks. Therefore, the importance to understand and to predict the shadow consequences in order not only to extract energy at the maximum possible power [3], but also to assess the protection on the systems.

2   Relationship to Collective Awareness Systems

Current trend in energy supply and usage are perceptibly economic, environmental and social unsustainable. There is a growing collective awareness of the urgent need to turn political statements and analytical work about the present unsustainable trend in energy supply and usage into concrete action in order to achieve what is anticipated as a sustainable development. The development of collective awareness platforms for a sustainable development allows the advancement not only of collective decision-making tools, but also of innovation mechanisms. These platforms allow individuals and groups to effectively react to environmental sustainability challenges, with an effective awareness of problems and possible solutions [5]. Solar energy is the most abundant form of energy on earth. PV systems are expanding due to supporting policies and cost reductions. Hence, PV systems are commercially available and a reliable technology with a potential for usage growth in almost everywhere. Although, can be argued that PV systems usage is not relevant in nowadays, PV systems are expected to be of relevant usage over the next decades [6].

3   Modeling

The equivalent circuit model used in this paper is the five parameters equivalent circuit, consisting in a current controlled generator, a single-diode, a shunt and series resistances as shown in Fig. 1.

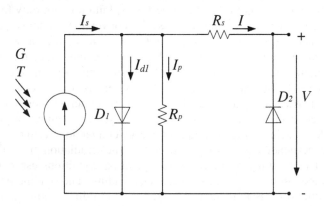

Fig. 1. Equivalent circuit of the solar module

In Fig.1, although not a part of the equivalent circuit, the anti-parallel bypass diode D_2 is shown for connection illustration purpose. G is the solar irradiance, T is the cell *p-n junction* temperature in [K], R_p is the equivalent shunt resistance, R_s is the equivalent series resistance, I is the output current, V is the output voltage, I_s is the photo generated electric current, I_{d1} is the current at diode D_1, I_p is the leakage current. If cells are associated in a module, an array or a panel and are equally subjected to the same irradiance, the junctions are at the same temperature T, then the circuit in Fig. 1 is also the equivalent circuit for the association, but with a suitable transformation for the parameters. The thermal voltage of a solar cell V_T is given by:

$$V_T = kT / q \tag{1}$$

where k is the Boltzman's constant, q is the electron charge.

The I-V characteristic is given by the implicit function associated with the model shown in Fig. 1 for the solar cell and is given by:

$$I = I_S - I_0 (e^{\frac{V+I R_S}{mV_T}} - 1) - \frac{V + I R_S}{R_p} \tag{2}$$

where I_0 is the diode reverse bias saturation current, and m is the diode ideality factor. (2). This equation has to be solved by an iterative method to determine for instance the output current in function of the output voltage. From (2), considering the short-circuit condition I_s is given by:

$$I_S = I_{SC} (1 + \frac{R_s}{R_p}) + I_0 (e^{\frac{I_{sc} R_s}{mV_T}} - 1) \tag{3}$$

The second term of (3) is usually small in comparison with the first term [7], [8]. So, the usual approximation considered for I_S is given by:

$$I_S \approx I_{SC} (1 + \frac{R_s}{R_p}) \tag{4}$$

Also from (2), considering the open circuit condition and due to the fact that the exponential at this condition is significant greater than one, I_S is given by:

$$I_S \approx I_0 \, e^{\frac{V_{oc}}{mV_T}} + \frac{V_{oc}}{R_p} \tag{5}$$

Equating (4) to (5), I_0 is approximated by the expression given by:

$$I_0 \approx [I_{SC} (1 + \frac{R_s}{R_p}) - \frac{V_{oc}}{R_p}] e^{- \frac{V_{oc}}{mV_T}} \tag{6}$$

Hence, I_0 can be computed by (6), if the parameters R_p, R_s, T, m are known.

Substituting (4) into (2) at Maximum Power Point (MPP) condition and due to the fact that the exponential at this condition is significant greater than one, I_0 is approximated by the expression given by:

$$I_0 \approx [(I_{SC} - I_{MP})(1 + \frac{R_s}{R_p}) - \frac{V_{MP}}{R_p}] e^{-\frac{V_{MP} + I_{MP} R_s}{mV_T}} \tag{7}$$

The electric power output equation and (2) allow for the determination of the derivative of the I in order to V at MPP, respectively given by:

$$\left(\frac{\partial I}{\partial V}\right)_{MP} = -\frac{I_{MP}}{V_{MP}}, \quad \left(\frac{\partial I}{\partial V}\right)_{MP} = -\frac{V_{MP} - I_{MP} R_s}{V_{MP}} \left(\frac{I_0}{mV_T} e^{\frac{V_{MP} + I_{MP} R_s}{mV_T}} + \frac{1}{R_p}\right) \tag{8}$$

Considering (7) and (8) I_{MP} is approximated by the implicit expression given by:

$$I_{MP} \approx \frac{(V_{MP} - I_{MP} R_s)}{mV_T} [(I_{SC} - I_{MP})(1 + \frac{R_s}{R_p}) - \frac{V_{MP} - mV_T}{R_p}] \tag{9}$$

Assuming (9) as having enough accurateness [8] for the parameter estimation R_p is approximated by the expression given by:

$$R_p \approx \frac{V_{MP} - I_{MP} R_s - mV_T}{(V_{MP} - I_{MP} R_s)(I_{sc}/I_{MP} - 1) - mV_T} \frac{V_{MP}}{I_{MP}} - R_s \tag{10}$$

and R_s from (6) of [8] is given by:

$$R_s \approx \frac{V_{oc} - V_{MP}}{I_{MP}} - \frac{mV_T}{I_{MP}} ln[\frac{V_{MP} - I_{MP} R_s + mV_T}{mV_T} \frac{I_{sc}(R_s + R_p) - V_{oc}}{I_{sc}(R_s + R_p) - 2V_{MP}}] \tag{11}$$

The ideality factor can be evaluated by (2) at maximum power point, replacing I_s given by (4), I_0 given by (6) and taking logarithms [8]. Hence m is given by:

$$m \approx \frac{1}{V_T} \frac{V_{oc} - V_{MP} - I_{MP} R_s}{ln[I_{sc}(R_s + R_p) - V_{oc}] - ln[(I_{sc} - I_{MP})(R_s + R_p) - V_{MP}]} \tag{12}$$

But $R_s > 0$, then (11) imposes the implicit relationship given by:

$$m < \frac{V_{oc} - V_{MP}}{V_T ln[\frac{V_{MP} - I_{MP} R_s + mV_T}{mV_T} \frac{I_{sc}(R_s + R_p) - V_{oc}}{I_{sc}(R_s + R_p) - 2V_{MP}}]} \tag{13}$$

Also to ensure that $R_p < \infty$, then the denominator in (10) imposes the relationship given by:

$$m < \frac{(V_{MP} - I_{MP} R_s)(I_{sc}/I_{MP} - 1)}{V_T} \tag{14}$$

The above expressions are capable of giving values for the five parameters equivalent circuit of a solar cell, using only data from open circuit, maximum power and short circuit tests. A Matlab/Simulink coding was carried out based on those expressions in order to simulate the behavior of PV systems.

4   Simulation

A simulation study concerning with the data measured from PV silicon amorphous solar modules Kaneka KA58 provided in [9] and placed in a photovoltaic facility at the Laboratório Nacional de Energia e Geologia (LNEG) in Lisbon, Portugal, is presented for illustration purpose. The coordinates for the PV modules site are: 38°46'18.50''N, 9°10'38.50''W.

The data for the silicon amorphous solar module Kaneka KA58 at STC [10] are shown in Table 1.

Table 1. Data for the Kaneka KA58 solar module at STC

Technology	V_m^*	I_m^*	V_{oc}^*	I_{sc}^*	β_{oc}	α_{sc}
Amorphous	63 V	0.92 A	85 V	1.12 A	-206 mV/ °C	1.3 mA/ °C

The characteristic curve for each tested PV module is measured outdoors quasi-simultaneously with the measurement of the reference unit I-V curve. The I-V curve is then converted to STC conditions by using the procedure described in IEC 60891 [11]. The I-V curve and the tracer used is shown in Fig. 2.

Fig. 2. Experimental I-V curves

The simulation results are compared with experimental observation carried out for an amorphous PV module technology with two modules connected in series.

The I-V curves simulated and experimental without shading are shown in Fig. 3.

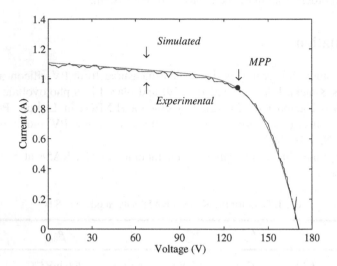

Fig. 3. I-V curves simulated and the experimental

The partial shading simulation was carried with twelve PV modules associating in parallel two series of six modules. The configuration in study with partial shading is shown in Fig. 4.

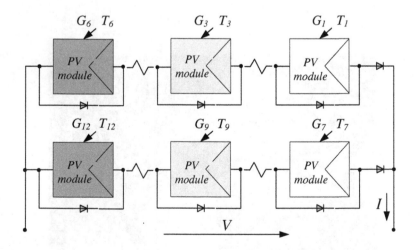

Fig. 4. Configuration for the simulation with partial shading

In Fig. 4, the partial shading is given by the following condition: $G = 1200W / m^2$ for the modules $\{1,2,4,5,7,8,10,11\}$; $G = 1000W / m^2$ for the modules $\{3,9\}$ and $G = 700W / m^2$ for the modules $\{6,12\}$. The I-V and P-V curves obtained by simulation of partial shading are respectively shown in Fig. 5 and Fig. 6.

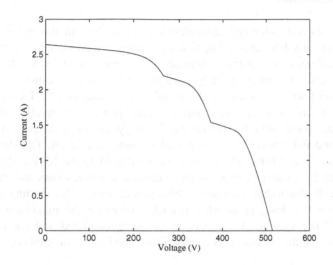

Fig. 5. I-V curve simulated for the PV modules with partial shading

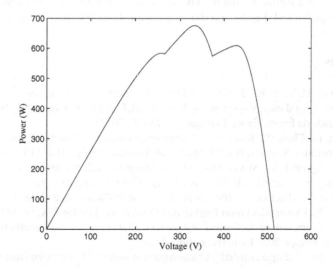

Fig. 6. P-V curve simulated for the PV modules with partial shading

Fig. 5 and Fig. 6 reveal the implication in the I-V and P-V curves due to the shading, i.e., due to the non-uniform illumination on the modules there are local MPPs, which can pose a difficult to the MPP tracking (MPPT) system and there is a loss on the expected energy conversion.

5   Conclusions

The five parameters solar cell equivalent circuit is used in this paper in order to acquire analytically I-V curves. The five parameters are estimated by a method using a set of equations and requiring information on open circuit, maximum power and short circuit tests. A simulation study using the five parameters estimation is reported for illustration purpose, presenting the simulate performance of a PV system before installation in sitú in what regard shading conditions on a parallel of two series connected amorphous solar PV modules technology, assessing the performance given by the I-V and P-V curves due to a partial shading simulation. The shading curves show multiple local MPPs, which can pose a difficult to the MPPT algorithm. The knowledge of these curves can be useful to minimize power losses and avoid damage to the solar cells due to hot spots during high power losses. This shading simulation is handy to help the designer in what regards extracting the maximum energy and protection of the system. A comparison between simulated and the experimental results shows that the estimation is a satisfactory approximation method.

Acknowledgments. This work was partially supported by Fundação para a Ciência e a Tecnologia, through IDMEC under LAETA, Instituto Superior Técnico, Universidade de Lisboa, Portugal. The authors gratefully acknowledge to Ms. L. Giacobbe for providind the measured data of amorphous PV modules at LNEG.

References

1. Seixas, M., Melicio, R., Mendes, V.M.F.: A simulation for acceptance of two-level converters in wind energy systems. In: Proc. of JIUE 2013 - 3ª Jornadas de Informática da Universidade de Évora, Évora, Portugal, pp. 75–79 (2013)
2. Blaabjerg, F., Chen, Z., Kjaer, S.B.: Power electronics as efficient interface in dispersed power generation Systems. IEEE Trans. Power Electron 19(5), 1184–1194 (2004)
3. Patel, H., Agarwal, V.: MATLAB-based modeling to study the effects of partial shading on PV array characteristics. IEEE Trans. Energy Conversion 23(1), 302–310 (2008)
4. Quaschning, V., Hanitsch, R.: Numerical Simulation of Photovoltaic Generators with Shaded Cells. In: 30th Universities Power Engineering Conference, London, UK, pp. 583–586 (1995)
5. European Commission: Collective Awareness Platforms for Sustainability and Social Innovation. Europe 2020 Initiative, retrieved (2013),
   `http://ec.europa.eu/digital-agenda/en/collective-awareness-`
   `platforms-sustainability-and-social-innovation`
6. International Energy Agency: Technology roadmap, solar photovoltaic energy (2010)
7. Carrero, C., Rodríguez, J., Ramírez, D., Platero, C.: Simple estimation of PV modules loss resistances for low error modelling. Renewable Energy 35, 1103–1108 (2010)

8. Carrero, C., Rodríguez, J., Ramírez, D., Platero, C.: Accurate and fast convergence method for parameter estimation of PV generators based on three main points of the I–V curve. Renewable Energy 36, 2972–2977 (2011)
9. Giacobbe, L.: Validação de modelos matemáticos de componentes de sistemas fotovoltaicos. Master Thesis, DEEC/IST (2005) (in Portuguese)
10. Kaneka Photovoltaic Products Information, http://www.pv.kaneka.co.jp
11. IEC 60891: Procedures for temperature and irradiance corrections to measured I–V characteristics of crystalline silicon photovoltaic devices (2009)

Simulation of Offshore Wind Turbine Link
to the Electric Grid through a Four-Level Converter

Mafalda Seixas[1,2,3], Rui Melício[1,3], and Victor M.F. Mendes[1,2]

[1] Universidade de Évora, Évora, Portugal
mafalda.seixas@gmail.com, ruimelicio@uevora.pt
[2] Instituto Superior de Engenharia de Lisboa, Lisbon, Portugal
vfmendes@deea.isel.pt
[3] IDMEC/LAETA, Instituto Superior Técnico, Universidade de Lisboa, Lisbon, Portugal

Abstract. This paper is on the modulation of offshore wind energy conversion systems with full-power converter and permanent magnet synchronous generator with an AC link. The drive train considered in this paper is a three-mass model which incorporates the resistant stiffness torque, structure and tower, in the deep water, due to the moving surface elevation. This moving surface influences the current from the converters. A four-level converter is considered with control strategies based on proportional integral controllers. Although more complex, this modulation is justified for more accurate results.

Keywords: Collective awareness, offshore wind turbine, multibody drive train, four-level converter, simulation.

1   Introduction

The demand for energy, the shortage of fossil fuels and the need for carbon footprint reduction have resulted in a global awareness of the importance of energy savings and energy efficiency [1]. The wind power industry and the construction of wind farms are undergoing rapid development [2], [3]. The installations and operation cost of OWT are still more expensive than that of onshore wind turbine, but an OWT situated sufficiently far away from the coast can capture more wind energy and will have a longer operation life [4], [5]. Offshore locations offer a number of advantages compared with onshore including a smoother, less turbulent wind and the ability to build larger turbines due to reduced concerns over visual impact [6]. Offshore floating structures are influenced by marine waves in coastal waters [7]. The coupling effect of the wind and wave dynamics are considered as an effect of the floating support structure motion on the blades and the shaft, as well the inertia induced by the motion of the blades [4]. The offshore wind energy conversion to mechanical energy over the rotor of an OWT is influenced by various forces acting on the blades and on the tower, i.e., centrifugal, gravity and varying aerodynamic forces acting on blades, gyroscopic forces acting on the tower and also by hydrodynamics depending on the foundation system and deep water. Consequently, the past experience gained regarding onshore wind turbines is not enough to be applied without further consideration to the development of OWT [8].

L.M. Camarinha-Matos et al. (Eds.): DoCEIS 2014, IFIP AICT 423, pp. 324–331, 2014.
© IFIP International Federation for Information Processing 2014

As wind energy is increasingly integrated into an electric grid, electric energy quality is becoming a concern of utmost importance, not only the stability of already existing power systems but also the total harmonic distortion (THD) must be assessed in order to ensure enough quality on the energy injected into the grid [9]. Some papers have been issued on OWT models, but mainly using simplified models to describe the drive train, the power converter or the control strategies. However, the increased wind power penetration, as nowadays occurs in Portugal, requires new models and system operation tools.

In this paper, an offshore variable-speed wind turbine in deep water is considered equipped with a permanent magnet synchronous generator (PMSG) using a full power converter, namely a back-to-back four-level converter topology, converting the energy of a variable frequency source in injected energy into the grid with constant frequency through an AC link. Although not imposed for OWT, the IEEE-519 imposes a THD not exceeding 5%. This THD limit is followed as a guideline for the filter design. The drive train considered is described by a three-mass model with an input stiffness torque due to the need to take in consideration the effects of the moving floating surface. The moving floating surface is modeled by one mass describing the tower and the floating structure. As a new contribution to earlier studies, a more realistic modeling of OWT with a complex multibody drive train is presented combined with a complex control strategy, and comprehensive simulation studies are carried out in order to adequately assert the system performance.

2   Relationship to Collective Awareness Systems

Awareness in a business reality implies a clear understanding of the enterprise, the operational context, the opportunities and risks: in essence, awareness in a business represents a solid base for sustainable enterprise competitiveness and acceptance of sustainability ideas. A collective awareness system (CAS) can thus be seen as a socio-technical artifact capable of openly linking objects, people, processes and knowledge for the benefit of the enterprise, and the society at large. Despite the marked social nature of this CAS is necessary public investment and advanced technological research [10]. Public investment and advanced technological research is critical to the still-emerging offshore business wind industry, the projects are expensive [11] and has a need to be supported by social acceptance. The awareness enabled by such a system can have very concrete impacts, for instance in empowering and motivating citizens to make informed decisions as consumers, or in fostering collective environmentally-savvy behavioral changes and a more direct democratic participation [12], regarding environmental and conversion of energy from alternative sources into electrical energy, ensuring sustainability. Only with a clear awareness of environmental problems will the people support their government's policies of investment in research in renewable energy, such as OWT. In this sense the creation of CAS supports the continuation of those policies which in turn supports the research in this area. Another important aspect in the creation of CAS is the dissemination of data on real physical systems that support the creation of mathematical models, as used in this work, closer to reality, thus allowing the study of concrete solutions.

3   Modeling

The configuration of the simulated WOT system with a four-level converter is shown in Fig. 1.

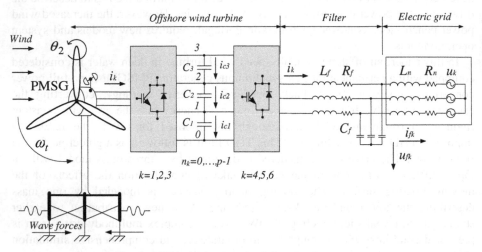

Fig. 1. OWT with back-to-back four-level converter

The configuration of the simulated drive train is shown in Fig. 2.

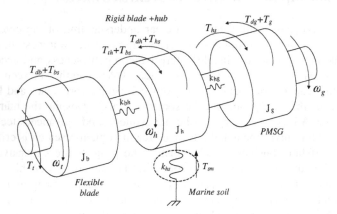

Fig. 2. OWT drive train

The modeling details regarding the wind speed, the wind turbine, the drive train can be seen, for instance, in [9]. The equations for modelling a PMSG are given by:

$$di_d / dt = 1 / L_d (v_d + p\omega_g L_q i_q - R_d i_d) \tag{1}$$

$$di_q / dt = 1 / L_q [v_q - p\omega_g (L_d i_d + M i_f) - R_q i_q] \tag{2}$$

where i_d, i_q are the stator dq currents, i_f is the equivalent rotor current, L_d, L_q are the stator inductances, M is the mutual inductance, R_d, R_q are the stator resistances, v_d, v_q are the stator voltages, p is the number of pair of poles, ω_g is the generator rotor speed. Due to the consideration of avowing demagnetization of the permanent magnet in the PMSG [9], a null stator current $i_d = 0$ has to be imposed.

The back-to-back four-level converter is an AC-DC-AC converter having eighteen unidirectional commanded IGBT's S_{ik} used as a rectifier, and with the same number used as an inverter. The rectifier is connected between the PMSG and a capacitor bank. The inverter is connected between this capacitor bank and a second order filter, which in turn is connected to an electric grid. The groups of six IGBT's linked to the same phase constitute the leg k of the converter. The converter has $p = 4$ levels. The voltage level is associated with the auxiliary variable n_k which range from 0 to $(p-1)$, used in the calculation of the u_{sk} generator output voltage. The voltage u_{sk} as a function of p and n_k is given by:

$$u_{sk} = \frac{1}{3}\frac{1}{(p-1)}(2n_k - \sum_{\substack{i=1 \\ i \neq k}}^{3} n_i)v_{dc} \qquad k \in \{1, 2, 3\}. \tag{3}$$

The current on each capacitor bank i_{cj} is given by:

$$i_{cj} = \sum_{k=1}^{3} \delta_{nk} i_k - \sum_{k=4}^{6} \delta_{nk} i_k \qquad j \in \{1, ..., p-1\} \tag{4}$$

where

$$\delta_{nk} = \begin{cases} 0 & j > n_k \\ 1 & j \leq n_k \end{cases}. \tag{5}$$

The voltage v_{dc} is modeled by the state equation given by:

$$\frac{dv_{dc}}{dt} = (\sum_{j=1}^{p-1}\frac{1}{C_j} i_{cj}). \tag{6}$$

A three-phase symmetrical circuit in series models the electric grid. The state equation for the injected current in the electrical grid is given by:

$$\frac{di_{fk}}{dt} = \frac{1}{L_n}(u_{fk} - R_n i_{fk} - u_k) \qquad k = \{4,5,6\} \tag{7}$$

where L_n and R_n are the electrical grid inductance and resistance, respectively, u_{fk} is the voltage at the filter, u_k is the voltage at the electric grid.

4   Control Method

The PI controllers are used for the control of the OWT. Pulse modulation by space vector modulation associated with sliding mode is used for controlling the converters. The sliding mode control is important for controlling the converters, by guaranteeing the choice of the most appropriate space vectors [9]. The power semiconductors present a finite switch frequency. Thus, for a value of the switching frequency, an error $e_{\alpha\beta}$ will exist between the reference value and the control value [9]. In order to guarantee that the system slides along the sliding surface $S(e_{\alpha\beta},t)$ is necessary that the state trajectory near the surfaces verifies in accordance with the stability condition [9] given by:

$$S(e_{\alpha\beta},t)\,\frac{dS(e_{\alpha\beta},t)}{dt} < 0. \tag{8}$$

A small error $\varepsilon > 0$ for $S(e_{\alpha\beta},t)$ exists in practice. Hence, the strategy is given by:

$$-\varepsilon < S(e_{\alpha\beta},t) < +\varepsilon. \tag{9}$$

Strategy (9) is taken by hysteresis comparators. The output voltage for each converter leg has four possible values $(0, \frac{1}{3}v_{dc}, \frac{2}{3}v_{dc}, v_{dc})$; therefore the three legs of the rectifier or of the inverter have sixty four possible combinations for the output voltages. The output voltage vectors in the $\alpha\beta$ coordinates for the four-level converter are shown in the Fig. 3.

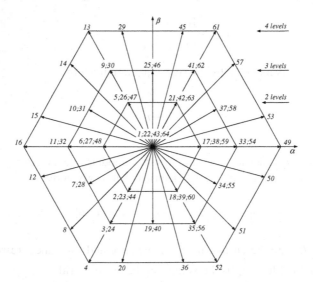

Fig. 3. Output voltage vectors for the four-level converter

The comparison of e_α and e_β errors with the values of the sliding surface $S(e_{\alpha\beta}, t)$ enables to find at each instant the variables σ_α and σ_β. The integer voltage variables σ_α and σ_β satisfy the following condition $\sigma_\alpha, \sigma_\beta \in \{-3, -2, -1, 0, 1, 2, 3\}$. These variables allow choosing the most appropriate vector.

5   Simulation

The mathematical model for the OWT with the four-level converter topology was implemented in Matlab/Simulink. The wind speed profile is shown in Fig. 4.

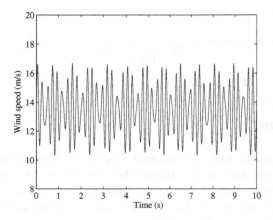

Fig. 4. Wind speed profile

The reference voltage and the voltage v_{dc} is shown in Fig. 5.

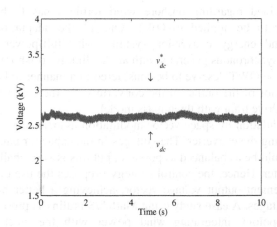

Fig. 5. The reference voltage v_{dc}

In Fig. 5, despite the wind speed profile the OWT system is capable of performing in a satisfactory manner, attenuating the influence of the wind speed on the system. The instantaneous current injected in the electric grid is shown in Fig. 6.

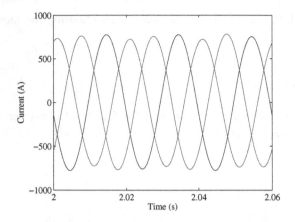

Fig. 6. Current injected in the electric grid

6   Conclusion

Europe has led offshore wind, for instance, due to the limited land available for onshore developments and so the offshore development is an actual option. Also offshore developments have a benefit in what regards the less variability and intermittence of the wind speed. The offshore wind energy conversion to mechanical energy over the rotor of an OWT is influenced by some actions different from the ones on onshore wind energy conversion systems, for instance, hydrodynamics depending on the foundation system and deep water, originating movements on the surface influencing the behavior of the output of the converters. Consequently, the past experience gained regarding onshore wind turbines has to be adapted and improved in order to be applied to OWT. One of the adaptations regards the modulation of wind energy conversion systems with full-power converter and permanent magnet synchronous generator with an AC link for offshore developments.

Hydrodynamics on OWT deserve to be considered in a manner to be more realistic to reveal the behavior of the output of the converters. So there is a justification for researching with a drive train with three-mass model.

Pulse width modulation by space vector modulation associated with sliding mode is used for controlling the converter. The voltages in the capacitor banks for the four-level converter should be as balanced as possible, but this is very challenging to do in a four-level converter. Hence, the control strategy proposes the use of four tables for selecting the convenient output voltage vector, achieving a better balanced for the capacitor banks voltages. A case study using Matlab/Simulink is presented for a four-level converter topology integrating wind power with the electrical grid. The simulation study revealed an overall good performance of the proposed OWT with the four-level converter. The wave form of the current injected in the electric grid seems purely sinusoidal, due to the strategy followed on the converters and the filtering used before the injection into the grid considered as a symmetric three phase sinusoidal voltage system.

Acknowledgments. This work was partially supported by Fundação para a Ciência e a Tecnologia, through IDMEC under LAETA, Instituto Superior Técnico, Universidade de Lisboa, Portugal.

References

1. Popovic-Gerber, J., Ferreira, J.A.: Power electronics for sustainable energy future–quantifying the value of power electronics. In: 3rd IEEE Energy Conversion Congress and Exposition, Atlanta, pp. 112–119 (2010)
2. Saheb-Koussa, D., Haddadi, M., Belhamel, M., Hadji, S., Nouredine, S.: Modeling and simulation of the fixed-speed WECS (wind energy conversion system): Application to the Algerian Sahara area. Energy 35, 4116–4125 (2010)
3. Fusco, F., Nolan, G., Ringwood, J.V.: Variability reduction through optimal combination of wind/wave resources – An Irish case study. Energy 35, 314–325 (2010)
4. Luo, N., Bottasso, C.L., Karimi, H.R., Zapateiro, M.: Semiactive control for floating offshore wind turbines subject to aero-hydro dynamic loads. In: International Conference on Renewable Energies and Power Quality – ICREPQ 2011, Las Palmas de Gran Canaria, pp. 1–6 (2011)
5. Musial, W., Butterfield, S., Boone, A.: Feasibility of floating platform systems for wind turbines. National Renewable Energy Laboratory – NREL/CP – 5oo-34874 (2003)
6. Wilkinson, M.R., Tavner, P.J.: Condition monitoring of wind turbine drive trains. In: 17th International Conference on Electrical Machines, Chania, pp. 1–5 (2006)
7. Holthuijsen, L.H.: Waves in Oceanic and Coastal Waters, pp. 145–196. Cambridge University Press, Cambridge (2007)
8. Bir, G., Jonkman, J.: Aeroelastic instabilities of large offshore and onshore wind turbines. Journal of Physics 75, 012069 (2007)
9. Melício, R.: Modelos dinâmicos de sistemas de conversão de energia eólica ligados à rede eléctrica: PhD Thesis: UBI/FE/DEE, Covilhã, Portugal (2010) (in Portuguese)
10. Future Internet Enterprise Systems, http://www.fines-cluster.eu/fines/jm/Documents/Download-document/410-FInES-Horizon-2020_Position-Paper-v2.00_Annex.html
11. Offshore Wind in Europe: Lessons for the US, http://theenergycollective.com/lewmilford/282836/offshore-wind-europe-lessons-us
12. Digital Agenda For Europe, http://ec.europa.eu/digital-agenda/en/collective-awareness-platforms-sustainability-and-social-innovation

Stochastic Modeling of Plug-In Electric Vehicles' Parking Lot in Smart Multi-Energy System

Maziar Yazdani Damavandi[1,2], M.P. Moghaddam[1], M.-R. Haghifam[1], Miadreza Shafie-khah[2], and João P.S. Catalão[2,3,4]

[1] Tarbiat Modares University (TMU), Tehran, Iran
[2] University of Beira Interior, Covilhã, Portugal
catalao@ubi.pt
[3] INESC-ID, Lisbon, Portugal
[4] IST, Univ. Lisbon, Portugal

Abstract. In this paper the role of Plug-In Electric Vehicles' (PIEVs) parking lot in operating Smart Multi-Energy System (SMES) has been investigated. SMES in this paper has been modeled as a multi-input multi-output model which consists of some storage and energy converters. In the proposed framework, the PIEV's parking lot behaves like an energy storage with selling energy price less than upstream network price and as manageable load when its purchase price is more than upstream network. On the other hand, traffic pattern of PIEVs in parking lot has an uncertain behavior and is modeled based on stochastic approach. In the stochastic model, two branches of scenarios for total state of charge and total capacity of parking lot in each hour are produced. The considered case studies show the effectiveness of the proposed model and the impact of PIEVs' parking lot in operation of SMES elements.

Keywords: Energy hub model, PIEVs' parking lot, stochastic modeling.

1 Introduction

Environmental aspects have been highlighted in development of societies by means of sustainable development. In this regard, sustainable energy development is the most important matter to take into account the applicable interaction of preserving the environment with providing the energy requirements [1]. Nowadays, integrated management of energy carriers and other energy related infrastructures (e.g. transportation system) is proposed as one of the approaches to achieve this goal [2].

Many researches have been oriented to model this new decision making environment and to propose management frameworks for Smart Multi-Energy Systems (SMES). Two pioneer models in this area are "energy hub system" and "matrix modeling". Both of the approaches consider the SMES as combination of operation centers (mostly co-generation or tri-generation units) and their interconnectors. Operation centers have been modeled by coupling matrix which converts input energy carriers to the output required energy services [3] and [4]; and Interconnectors transmit energy between operation centers based on energy carriers' physical constraints [5]. References [6] and [7] have modeled the operation

L.M. Camarinha-Matos et al. (Eds.): DoCEIS 2014, IFIP AICT 423, pp. 332–342, 2014.
© IFIP International Federation for Information Processing 2014

framework for operation centers and interconnectors in SMESs. On the other hand, mitigating environmental concerns by electrifying demand of carbon-based energy carriers are another important approach for sustainable development in energy sector.

Plug-In Electric Vehicles (PIEVs) are main tools for electrification in transportation system [8]. They reduce air pollution inside cities by consuming electricity which is supplied by renewable resources and the power plants located far from the cities. Moreover, PIEVs' batteries prepare bulk storage capacity for power system. PIEVs' parking lots are best opportunity for distribution system operators to utilize electric bulk storage facilities in demand side. It should be noted that, if these PIEVs' charge pattern as electric load are not controlled, it can worsen the network operation condition.

Hence, the main goal of this paper is to propose an operation framework for an energy hub which is equipped by PIEVs' parking lot to change the pattern of charging and discharging of PIEVs in parking lots to enhance the operation flexibility of system.

In [9] a single Plug-in Hybrid Electric Vehicle (PHEV) has been modeled as independent energy hub. Furthermore, [10] investigated the role of PHEVs as controllable loads in energy hub system operation and [11] utilized charging of PHEVs in SMES as flexible load for employing ancillary services (load frequency control) in energy market. However, PIEVs' parking lot has not been considered as the element of energy hub before the present paper.

In this paper, the PIEVs' parking lot in a smart energy hub is modeled as stochastic energy storage. The energy hub operator receives gas and electricity from upstream network and delivers energy services to the customers. Hence, energy hub consists of CHP unit, auxiliary boiler, heat storage and a PIEV's parking lot. In such a system, behavior of PIEV's parking lot follows uncertainties modeled by stochastic method. In the mentioned framework, the patterns of State of Charge (SOC) of connected PIEVs are included in the parking lot. The intention of the energy hub operator is to deliver energy to multi-energy demands in such a way that the benefit of operator being maximized. The numerical results show the role of PIEVs' parking lot for altering energy hub operation pattern and utilizing input energy carriers in better way.

2   Contribution to Collective Awareness Systems

Smartness in energy systems facilitates amendment of new resources in demand side. These new resources introduce interdependency in time and carrier domain which should be considered by integrated models. New highly dependent environment will increase the level of uncertainty in SMES which needs huge amount of information and non-deterministic models for appropriate decision making. Collective awareness system is a brilliant opportunity for SMES managers to openly link these uncertain smart energy centers and enhance their collaboration for the benefit of the enterprise. Although, participation of demand side players, enhance the system performance, imposes system with new human-centered layer. This human layer, authorize main portion of resources in the system e.g. PIEVs and small scale co- or tri-generation units. Collective awareness system will make new framework for implementing these resources and managing the behavior of human layer to act as the SMES managers' desire. In this paper the PIEVs traffic pattern in human layer of SMES system is modeled by stochastic models and their effectiveness on operation of other SMES elements and enhancement of SMES operation flexibility is discussed.

3   Stochastic Modeling of PIEVs' Parking Lot

A stochastic model is developed to quantify behavior patterns of PIEVs at a parking lot. The nominal capacity of parking and the sum of SOC of EVs plugged-in at the parking lot in each hour are the outputs of the model. Capacity of parking lot relies on the number and type of EVs parked at the parking lot. The hourly number of EVs connected to the grid at the parking lot is a probabilistic variable that is related to behavior of EV owners. In this paper, the pattern of available EVs at the parking lot is extracted from the real data that is obtained from number of vehicles parked at parking lots [12].

The energy storage capacity of each EV represents the total energy capacity and it is dependent to the EV class. For example, the energy storage capacity of plug-in hybrid electric vehicles (PHEVs) typically is between 6 kWh and 30 kWh; whereas, the capacity for BEVs varies from 30 to 50 kWh [13]. In [13], twenty four different classes have been considered for EV batteries. The probability distribution of the battery capacities in each EV class occurring in a market is illustrated in Fig. 1.

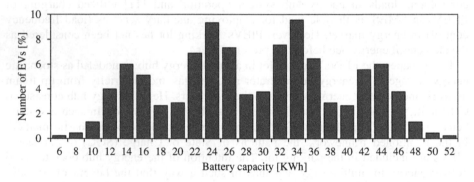

Fig. 1. Distribution of battery capacity

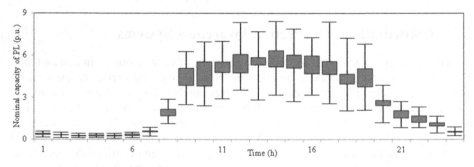

Fig. 2. The hourly nominal capacity of parking lot

In order to consider the market share of each EV class, the battery capacity of each class is considered. Taking into account the distribution of EV classes and probability number of EVs at parking lot, the hourly possibility of parking lot capacity is obtained as Fig. 2.

SOC of Parking lot is dependent to number of EVs parked at the parking lot, the type of each EV and the daily driven distance of each EV. The probabilistic traveled distance is applied as a parameter of calculating the SOC of parking lot. The lognormal distribution function is utilized to generate the probabilistic daily traveled distance [14]. The lognormal random variables are generated using standard normal random variable, N, and are computed using (1) [15].

$$M_d = \exp(\mu_{md} + \sigma_{md} N)$$ (1)

where M_d is the daily driven distance. μ_m and σ_m are the lognormal distribution parameters and are calculated from mean and standard variation of M_d based on the historical data, denoted as μ_{md} and σ_{md}, respectively [12]. μ_m and σ_m are calculated based on (2) and (3), respectively.

$$\mu_m = \ln\left(\frac{\mu_{md}^2}{\sqrt{\mu_{md}^2 + \sigma_{md}^2}}\right)$$ (2)

$$\sigma_m = \sqrt{\ln\left(1 + \frac{\sigma_{md}^2}{\mu_{md}^2}\right)}$$ (3)

Vehicles have been used in [12] made an average of 4.2 trips per day, yielding an average daily distance of 39.5 miles. On the other hand, an electric vehicle takes approximately 0.35 kWh to recharge for each mile traveling [12]. On this basis and according to the mentioned above discussion, the hourly SOC of Parking lot can be obtained as illustrated in Fig. 3.

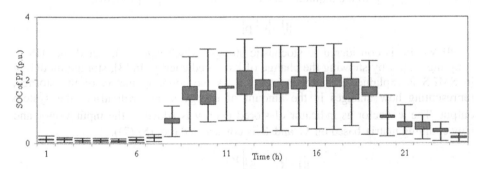

Fig. 3. The hourly SOC of parking lot

4   Operation Framework of SMES considering PIEVs' Parking Lot

SMES can consist of different elements and distributed energy resources. The focus of this paper is on the PIEVs' parking lot as flexible load in charging mode and large scale electric storage in discharge mode. Operator of SMES supply energy from

Electrical Distribution System (EDS) and Gas Distribution Network (GDN) and deliver required service in heat and electricity format to the Multi-Energy Demand (MED).

Fig. 4 shows a typical SMES which consists of Combined Heat and Power (CHP) unit, Auxiliary Boiler (AB), Heat Storage (HS), and PIEVs' Parking Lot (PL).

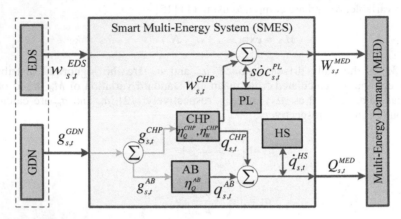

Fig. 4. A typical SMES schematic considering PIEVs' parking lot

4.1 Matrix Modeling of Smart Multi-Energy System

In depicted SMES schematic the coupling matrix $\mathbf{C}$, converts input energy vector, $\mathbf{p} = \begin{bmatrix} w_{s,t}^{EDS} & g_{s,t}^{GDN} \end{bmatrix}$, to the required services, $\mathbf{l} = \begin{bmatrix} W_t^{MED} & Q_t^{MED} \end{bmatrix}$ (eq. (4)).

$$[\mathbf{l}] = [\mathbf{C}][\mathbf{p}] \tag{4}$$

PIEVs' PL is considered as stochastic storage which can be modeled like HS in coupling matrix by showing the change of its stored energy. In [3], storage modeling in SMES is explained comprehensively. $\mathbf{S}$ is the coupling matrix of the storage, representing how changes in the amount of energy stored will affect the system output. The $\dot{\mathbf{e}}$ vector as indicator of stored energy is added to the input vector and new coupling matrix based on $\mathbf{C}$ and $\mathbf{S}$ is constructed (eq. (5)-(7)).

$$[\mathbf{l}] = [\mathbf{C} \quad -\mathbf{S}]\begin{bmatrix} \mathbf{p} \\ \dot{\mathbf{e}} \end{bmatrix} \tag{5}$$

$$\dot{E}_{\alpha,t} = E_{\alpha,t} - E_{\alpha,t-1} \approx e_\alpha U_\alpha \tag{6}$$

$$e_\alpha = \begin{cases} \eta_{\alpha,ch}^{Storage}, & \text{if } U_\alpha \geq 0 \quad (\text{Charge / Standby}) \\ 1/\eta_{\alpha,dis}^{Storage}, & \text{if } U_\alpha < 0 \quad (\text{Discharge}) \end{cases} \tag{7}$$

Moreover, the detailed matrix model of system is demonstrated in (8):

$$\begin{bmatrix} 1 & v_{G,s,t}^{CHP} \cdot \eta_W^{CHP} & 0 & 1/e_W^{PL} \\ 0 & v_{G,s,t}^{CHP} \cdot \eta_Q^{CHP} + v_{G,s,t}^{AB} \cdot \eta_Q^{AB} & 1/e_Q^{HS} & 0 \end{bmatrix} \cdot \begin{bmatrix} w_{s,t}^{EDS} & g_{s,t}^{GDN} & \dot{q}_{s,t}^{HS} & s\dot{o}c_{s,t}^{PL} \end{bmatrix}^T = \begin{bmatrix} W_t^{MED} \\ Q_t^{MED} \end{bmatrix} \tag{8}$$

4.2   Operational Optimization Problem

The objective of SMES operator is to maximize the benefit from energy trade between customers (MED and PIEVs' owners) and sellers (EDS, GDN and PIEVs' owners) considering the price of energy in peak hours.

Therefore, the objective function of optimization problem includes three terms; first and second terms are the benefit of operator from electricity and gas trade respectively, and the third one is the operator benefit from trading electricity to the PIEVs with contract price (which should be considered PIEV and operator cost, e.g. installation and battery degradation costs) and manipulation of PIEV's SOC in PL for demand management actions (eq. (9)).

$$
Maximizing \quad \sum_{s}\sum_{t}(w_{s,t}^{EDS}.\pi_{W,t}^{EDS} - W_t^{MED}.\pi_{W,t}^{MED}) + (g_{s,t}^{GDN}.\pi_{Q,t}^{GDN} - Q_t^{MED}.\pi_{Q,t}^{MED})
$$

$$
+ (w_{s,t}^{PL,in}.\pi_{W,t}^{PL,buy} - w_{s,t}^{PL,out}.\pi_{W,t}^{PL,sell})
\tag{9}
$$

Furthermore, due to physical characteristics of SMES's elements, the operation problem faces some constraints:

1) Input energy carriers limitation: EDS and GDN have some limitation for supplying required energy to the SMES. Moreover, the energy flow from EDS and GDN to the SMES is considered unidirectional (eq. (10)).

$$
0 \leq p \leq \overline{P}
\tag{10}
$$

2) CHP operational constraints: CHP unit operates in a predetermined operation zone which is based on its manufacturing characteristics.

$$
\underline{W}^{CHP} \leq w_{s,t}^{CHP} \leq \overline{W}^{CHP}
\tag{11}
$$

$$
\underline{Q}^{CHP} \leq q_{s,t}^{CHP} \leq \overline{Q}^{CHP}
\tag{12}
$$

$$
\lambda^{CHP} = q_{s,t}^{CHP} / w_{s,t}^{CHP}
\tag{13}
$$

3) AB operational constraints: AB heat output is constrained by upper and lower limits.

$$
\underline{Q}^{AB} \leq q_{s,t}^{AB} \leq \overline{Q}^{AB}
\tag{14}
$$

4) HS operational constraints: Interaction between HS and SMES is restricted and also the stored energy in the HS is limited by upper and lower limits.

$$
\left| \dot{q}_{s,t}^{HS} \right| \leq \gamma_Q^{HS}
\tag{15}
$$

$$
\underline{Q}^{HS} \leq q_{s,t}^{HS} \leq \overline{Q}^{HS}
\tag{16}
$$

5) Decision-making variables constraints: Utilizing same service from different energy vectors enhances the operator's degree of freedom. v is the decision making variable which determines this freedom in optimization problem.

$$
0 \leq v \leq 1
\tag{17}
$$

$$
v_{G,s,t}^{CHP} + v_{G,s,t}^{AB} = 1
\tag{18}
$$

4.3   PIEVs' Parking Lot Model in Operation Problem

PIEVs' parking lot behaves as an electrical load in charging mode and as a large scale storage when manage its discharge mode. In proposed model the SMES operator can manipulate the SOC of PIEVs to maximize its profit during operation period. Difference of PL with common storage in modeling is the variation of its capacity which is dependent to the arrival and departure time of PIEVs to the PL. Eq.s (19) and (20), determines the amount of changing in PL's SOC based on PL interaction with SMES and PIEVs' traffic in the parking.

$$\dot{soc}_{W,s,t}^{PL} = soc_{W,s,t}^{PL} - soc_{W,s,t-1}^{PL} + soc_{W,s,t}^{PL-ar} - soc_{W,s,t}^{PL-dep} \tag{19}$$

$$\dot{soc}_{W,s,t}^{PL} = l_{W,s,t}^{PL,in} - l_{W,s,t}^{PL,out} \tag{20}$$

In this model, it is assumed that, for arriving PIEVs to the PL in each hour the added SOC by variable $soc_{W,s,t}^{PL-ar}$ is based on the increase in SOC scenarios (eq. (21)) while for departed PIEVs from the PL, the loss in SOC by variable $soc_{W,s,t}^{PL-dep}$ is equal to portion of prior hour SOC considering the decrease in SOC scenarios (eq. (22)).

$$soc_{W,s,t}^{PL-ar} = \begin{cases} soc_{W,s,t}^{PL-Sc} - soc_{W,s,t-1}^{PL-Sc} & if \ soc_{W,s,t}^{PL-Sc} - soc_{W,s,t-1}^{PL-Sc} \geq 0 \\ 0 & if \ soc_{W,s,t}^{PL-Sc} - soc_{W,s,t-1}^{PL-Sc} < 0 \end{cases} \tag{21}$$

$$soc_{W,s,t}^{PL-dep} = \begin{cases} 0 & if \ soc_{W,s,t}^{PL-Sc} - soc_{W,s,t-1}^{PL-Sc} \geq 0 \\ ((soc_{W,s,t-1}^{PL-Sc} - soc_{W,s,t}^{PL-Sc}) / soc_{W,s,t-1}^{PL-Sc}).soc_{W,s,t-1}^{PL} & if \ soc_{W,s,t}^{PL-Sc} - soc_{W,s,t-1}^{PL-Sc} < 0 \end{cases} \tag{22}$$

Furthermore, the interaction amount of PL with SMES is restricted (eq. (23)) and the PL's SOC in each hour is limited by total capacity and minimum required SOC of PIEVs (eq. (24)).

$$\left| \dot{soc}_{W,s,t}^{PL} \right| \leq \gamma_W^{PL} \tag{23}$$

$$0 \leq \underline{SOC}_{W,s,t}^{PL} \leq soc_{W,s,t}^{PL} \leq \overline{SOC}_{W,s,t}^{PL} \leq Cap_{W,s,t}^{PL} \tag{24}$$

5   Numerical Results

SMES operator behaves like energy retailer to maximize its benefit by buying energy from wholesale market and selling it to customers through predetermined tariffs. But SMES operator has some physical asset to arbitrage between energy carriers for increasing its benefit. For considered SMES in this paper, the electricity prices in input and required service in output have been depicted in Fig. 5 and Fig. 6, respectively. Moreover, the GDN gas price is considered as 6 mu/p.u. and delivered heat price to the MED as 7 mu/p.u.

Two case studies have been produced to demonstrate the role of PIEVs' PL in the SMES operation. The first case is SMES normal operation without PL and the second one considers the stochastic model of PL. Furthermore, some results for showing the effectiveness of stochastic model and the comparison with deterministic one are reported. The case studies results depicted in Fig. 7.

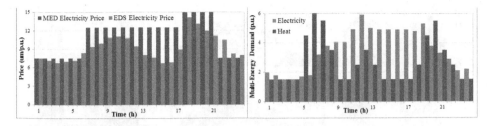

Fig. 5. MED and EDS price data   **Fig. 6.** Multi-energy demand data

Figure 7.(a) to 7.(c) show the results of first case study for input gas to SMES and multi energy electricity and heat supply mixture. As it can be seen, CHP unit produce between 7-12 and 17-22 hours while EDS electricity price is high and also MED has the simultaneous heat and electricity consumption. Furthermore, HS has stored exceeded energy of CHP in 9, 10, 18, 21, and 22 hours when the EDS' price is high and CHP generation is profitable; however, MED has low heat usage which can be solved by utilizing HS. These stored energy are given back to the SMES during 6, 13, 15, and 23 hours while the price is high and HS operation is more economical.

Figures 7.(d) to 7.(f) depict second case results which have considered the PL impact in operation of SMES elements. Figure 7.(d) shows less variation in output of CHP unit in hours 9-14. In these moments, PL as a stochastic storage enhances flexibility of the SMES and compensate the need for CHP variation by storing and injecting energy. Figures 7.(e) and 7.(f) demonstrate MED's electricity and heat usage; during low energy price (13-16), PL stores energy for injecting to SMES during peak period (18-21).

The SMES buys electricity from the PL in 12 mu/p.u. and sells it to PL in 10 mu/p.u. Therefore, it will be profitable for SMES operator to trade electricity with PL when the selling price is higher than EDS price and the buying price is lower than EDS price. As it can be seen in Fig. 7.(g), PL's SOC is less than scenarios amount between 9-13 and 18-21 and is more than scenarios amount between 14-17. In first discussed periods the electricity price is high and PL behave like storage and inject the energy to the SMES but in second period (14-27) the electricity price is low and PL behave as a manageable load which consume electricity to charge its PIEVs. Finally, in Fig. 7.(h) and 7.(i) the difference between stochastic and deterministic modeling for stored energy in HS and PL is depicted. The energy have been injected to the PL while the energy price is low for two reasons, first charging departed cars and secondly storing for consuming in peak hours. For the second reason, as a result of uncertain behavior of PIEVs, if operator charges the PIEVs' batteries and the PIEV departed from PL, operator miss part of its stored energy in PIEVs (PIEV's SOC) as loss. Therefore, as it can be seen in the stochastic case, the stored energy in PL is less compared to deterministic case because less stored energy means less loss and operation cost. However, less utilization of PL leads to more utilization of CHP which will result in more stored energy in HS (Fig. 7.(i)).

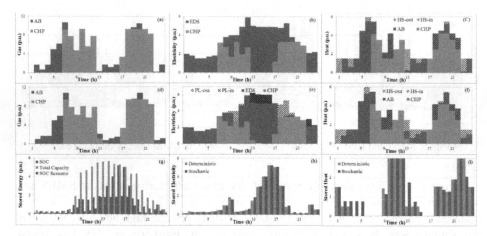

Fig. 7. Result of the proposed framework, (a) input gas without PL, (b) MED electricity consumption, (c) MED heat consumption mixture without PL, (d) input gas with PL, (e) MED electricity consumption mixture with PL, (f) MED heat consumption mixture with PL, (g) total capacity and SOC for PL during operation period, (h) stored electricity in PL for stochastic and deterministic case studies, (i) stored electricity in HS for stochastic and deterministic cases

6   Conclusion

This paper proposed a framework for modeling PIEVs' parking lot in SMES operation. The model considers the behavior of parking lots as storage and flexible loads dependent to the upstream network price. Stochastic approach is employed to consider uncertainties around traffic pattern in parking lot. The numerical results have showed parking lot change the SMES elements' operation condition and prepares more flexibility for SMES to deliver requiring services. Furthermore, comparing stochastic and deterministic results demonstrates more information for the operator to utilize SMES elements which resulted in less operation of parking lot and more operation of CHP and AB.

Acknowledgment. This work was supported by FEDER funds (European Union) through COMPETE and by Portuguese funds through FCT, under Projects FCOMP-01-0124-FEDER-020282   (Ref.   PTDC/EEA-EEL/118519/2010)   and   PEst-OE/EEI/LA0021/2013.   Also, the research leading to these results has received funding from the EU Seventh Framework Programme FP7/2007-2013 under grant agreement no. 309048.

References

1. Alanne, K., Saari, A.: Distributed energy generation and sustainable development. Renewable and Sustainable Energy Reviews 10, 539–558 (2006)
2. Galus, M.D., et al.: Integrating power systems, transport systems and vehicle technology for electric mobility impact assessment and efficient control. IEEE Trans. Smart Grid 3, 934–949 (2012)

3. Arnold, M., et al.: Distributed predictive control for energy hub coordination in coupled electricity and gas networks. In: Intelligent Infrastructures, pp. 235–273. Springer Netherlands (2010)
4. Chicco, G., Mancarella, P.: Matrix modelling of small-scale trigeneration systems and application to operational optimization. Energy 34, 261–273 (2009)
5. Hajimiragha, A., Cañizares, C., Fowler, M., Geidl, M., Andersson, G.: Optimal energy flow of integrated energy systems with hydrogen economy considerations. In: Proc. IREP Symposium Bulk Power System Dynamics and Control (2007), doi:10.1109/IREP.2007.4410517
6. Bozchalui, M.C., Hashmi, S.A., Hassen, H., Cañizares, C.A., Bhattacharya, K.: Optimal operation of residential energy hubs in smart grids. IEEE Trans. Smart Grid 3, 1755–1766 (2012)
7. Mancarella, P., Chicco, G.: Real-time demand response from energy shifting in distributed multi-generation. IEEE Trans. Smart Grid (2013) (in press)
8. Su, W., Chow, M.-Y.: Performance evaluation of an EDA-based large-scale plug-in hybrid electric vehicle charging algorithm. IEEE Trans. Smart Grid. Special Issues, Transportation Electrification and Vehicle-to-Grid Applications (2011)
9. Galus, M.D., Andersson, G.: Power system considerations of plug-in hybrid electric vehicles based on a multi energy carrier model. In: IEEE PES General Meeting, pp. 1–8 (2009)
10. Galus, M.D., Koch, S., Andersson, G.: Demand management of grid connected plug-in hybrid electric vehicles (PHEV). In: Energy 2030 Conference, Atalanta, pp. 1–8 (2008)
11. Galus, M.D., Koch, S., Andersson, G.: Provision of load frequency control by PHEVs, controllable loads, and a cogeneration unit. IEEE Trans. Industrial Electronics 58, 4568–4582 (2011)
12. van Haaren, R.: Assessment of electric cars' range requirements and usage patterns based on driving behavior recorded in the National Household Travel Survey of 2009. Columbia University, Fu Foundation School of Engineering and Applied Science, New York (2011)
13. Nemry, F., Leduc, G., Muñoz, A.: Plug-in hybrid and battery-electric vehicles: state of the research and development and comparative analysis of energy and cost efficiency. European Communities (2009)
14. Meliopoulos, S.: Power system level impacts of plug-in hybrid vehicles. Power Systems Engineering Research Center (PSERC) (2009)
15. Domínguez-García, A.D., Heydt, G.T., Suryanarayanan, S.: Implications of the smart grid initiative on distribution engineering. PSERC Document 11-05 (2011)

Appendix: SMES Elements' Characteristics and Nomenclature

Table 1. Data of SMES elements' characteristics

CHP Unit			Auxiliary Boiler			Heat Storage		
Output Energy (Min/Max)	η_e^{CHP}	η_h^{CHP}	Output Energy (Min/Max)	η_Q^{AB}	γ_Q^{HS}	Stored Energy (Min/Max)	$\eta_{Q,ch}^{HS}, \eta_{Q,dis}^{HS}$	
0/5 pu	0.35	0.45	0/10 pu	0.9	3 pu	0.5/3 pu	0.9	

Table 2. Definition of subscripts, parameters, and variables

Subscripts					
W	Electricity	G	Gas	Q	Heat
s	Scenario	t	Time	Sc	Scenario

Parameters and Variables					
E	Energy stored	η	Efficiency	G	Gas consumption
Q	Heat output	W	Electrical power	Cap	Parking lot total capacity
L	Energy demand	λ	Heat to power ratio	π	energy carrier price
γ	Maximum charge and discharge rate of heat storage	$\dot{e}$	(column vector) changes in stored energy	$\dot{h}$	Heat storage level difference in two consecutive time intervals

An underlined (overlined) variable is used to represent the minimum (maximum) value of that variable.
Capital letters denote parameters and small ones denote variables.

Part XIII

Optimization Issues in Energy - I

Decision Support in the Investment Analysis on Efficient and Sustainable Street Lighting

J.A. Lobão[1], T. Devezas[2], and João P.S. Catalão[2,3,4]

[1] Polytechnic Institute of Guarda, Portugal
[2] University of Beira Interior, Covilhã, Portugal
catalao@ubi.pt
[3] INESC-ID, Lisbon, Portugal
[4] IST, Univ. Lisbon, Portugal

Abstract. In recent years there has been a series of documents such as the European Strategy 20-20-20 to address the issue of energy efficiency in various sectors of activity. The objective is to reduce 20% of energy consumption, 20% of GHG emissions (Greenhouse Gases) and 20% of the energy consumed from renewable sources. Public lighting participates with 2.3% in global electricity consumption, so all contributions to the reduction in energy consumption will be relevant. Decision support in the investment analysis on efficient and sustainable street lighting allows a better use of the installed power. Hence, this paper deals with the reduction of losses in cables of a street lighting installation, depending on the luminaire used, presenting both simulation and experimental results. The economic choice of cables losses will allow improving the efficiency of the street lighting in general, providing also an optimal cost/benefit relationship. Moreover, real-time data acquisition systems of the equipment's consumption can be integrated into a collective awareness system.

Keywords: Street lighting, decision support, sustainable energy, efficiency lighting, losses.

1 Introduction

This study presents a new software application under development that compares and chooses the best investment and experimental validation in the solutions of installations of street lighting. The choice of efficient street lighting is related to the following factors: price, power consumption, reduction of losses in the conductors, useful life, and interest rate. The losses in the conductors will be analysed based on the current which passes throughout the electrical installation. The analysis allows various possibilities, allowing you to choose the analysis of a specific individual point of light, replace the existing technology on a street or in a selected group of streets or make replacement or control all luminaires installed simultaneously; investment analyses and advises more efficient.

Energy efficiency and consumption reduction in electrical installations and equipment have been the subject of investigation and research, from energy production

L.M. Camarinha-Matos et al. (Eds.): DoCEIS 2014, IFIP AICT 423, pp. 345–352, 2014.
© IFIP International Federation for Information Processing 2014

to final consumer, public (predominantly street) lighting participates with 2.3% in global electricity consumption [1], energy-efficient programs in this field are very welcome, since possibilities for energy savings in street lighting are numerous and since some of them enable reductions in electricity consumption of even more than 50%.

The cable losses analysed regularly throughout the transport and distribution of energy [2], [3], are often overlooked as a component of the cycle of lighting systems, and a means available to save energy and improve the overall performance of the installation in street lighting. Consumption reduction in electrical installations and lamps has been the subject of research, particularly in the aspects of economic choice of conductors section [4] and improving the efficiency lamps [5]. It is intended to connect these two aspects of the research, including on the economic analysis the influence of efficient lamps and losses caused by them in the installation of street lighting.

2   Relationship to Collective Awareness Systems

The electric grid is a massively interconnected network used to deliver electricity from suppliers to consumers. The electricity networks can intelligently integrate the behavior and actions of all users connected to it. In this sense, the parameterization of an installation of street lighting, associated with a data acquisition system in real-time consumption of the lighting, can be integrated in a collective awareness system. Associate software allows the decision maker to choose the purchase of efficient equipment to enable to improve energy efficiency, reduce carbon footprint, increase the security of energy supply and, last but not least, diminish the bill to be paid from production, transport, distribution and use of energy, thus linking objects, people and knowledge in order to foster new forms of social and business innovation.

The future development of the IoT will be especially relevant for energy efficient communication and energy-aware systems. In the service-based IoT applied to the production, transport, distribution and use of energy, all service providers are interconnected. A deregulated energy market, characterized by the growing use of decentralized energy systems and the increasing complexity of interactions between providers and consumers cannot be realized without an adequate IoT infrastructure,

Decision and policy makers will be able to base their actions on real-world, real-time data. Households and companies will be able to react to market fluctuations by increasing or decreasing consumption or production, thus directly contributing to increased energy efficiency, benefitting of the future collective awareness system in order to have better informed decision making and the effective involvement of the customer and sustainable energy systems.

3   Development

3.1   Identification of the Parameters

Physical parameters:
- Knot Connection (CK);
- Connections between knot Connection;

- Length of branch conductors in knot Connection;
- Section of branch conductors in knot Connection;

Load parameters:

- Power of the loads connected to the electrical installation;
- Efficiency of the loads;
- Power factor of the loads;
- Daily load diagram;
- Daily load diagram of the lamps and system control for economic analysis.

Operating parameters:

- Operating time of the street lighting installation;
- Monthly operating days (d);
- Months of annual operation (m);
- Cost of electricity (€);
- Interest rate.

3.2   Installation Characteristics

Fig. 1 shows a typical installation with the respective parameters.

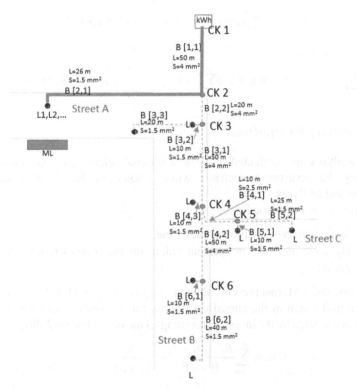

Fig. 1. Scheme of an installation

3.3   Calculations

After inputting the parameters and load diagrams, the following calculations are made:

- Determination of the load diagram associated to the branch knot Connection, adding the corresponding load diagrams.
- The currents in all conductors of the electrical installation due to:
    - Initial load diagram (I_1)
    - Load diagram of lamps efficient (I_2).
- Difference in cable losses (ΔP) in the conductors affected by the changed equipment (identified in bold in Fig. 1).

$$\Delta P[k,i] = \int_0^{24} R[k,i](I[k,i]_1)^2 d_t - \int_o^{24} R[k,i](I[k,i]_2)^2 d_t \qquad (1)$$

- Profits from the variation of cable losses (G1).

$$\Pr ofit1 = \sum_{j=1}^{n} (\Delta P[k,\, i]j) * d * m * \text{€} \qquad (2)$$

- Profits from the variation of power equipment (G2).

$$\Pr ofit2 = \sum_{j=1}^{n} [(P1[k,\, i]j - P2[k,\, i]j)] * d * m * \text{€} \qquad (3)$$

- Total profits.

$$R = \sum_{j=1}^{n} \Delta P[k,\, i]j * d * m * \text{€} + \sum_{j=1}^{n} [(P1[k,\, i]j - P2[k,\, i]j)] * d * m * \text{€} \qquad (4)$$

4   Economic Evaluation

Economic analyses are conducted to allow a rational selection of the solution to be taken during the investment decision, which should be based on a number of comparisons and analyses.

The methods can be grouped into:

- Static methods: simple payback time.
- Dynamic methods: net present value, internal rate of return and payback period.

In this work, the VAL (net present value) or the payback period (PP) is used, which is computed from the sum of the annual cash-flows for a given annual interest rate. The interest rate is indicated by the investor according to the desired profitability.

$$VAL = \sum_{k=0}^{n} \frac{R_k - D_k - I_k}{(1+a)^k} + \frac{V}{(1+a)^n} \qquad (5)$$

with:

R - Net profit;
D - Operation cost;
I - New investment;
n - Years of useful life;
V - Residual value for the old equipment;
a - Annual interest rate.

$$PP = ln\frac{100\,W_{el}C_e}{100\,W_{el}C_e - iC_{inv}} - ln\frac{100+i}{100} \tag{6}$$

with:

W_{el}- Electricity savings; Ce- Electricity cost; W_{el}Ce-Net profit
C_{inv} - New investment; i - Annual interest rate

5   Results

5.1   Software Developed

The developed software is intended to embrace a greater number of situations analyses. Thus, it is possible to analyze the efficiency simultaneously in the whole installation or in a particular street, as the example shown in Fig. 2. Here is analyzed the C Street, where it replaced luminaires 426.6W (lamp + ballast) by luminaires with using bi-power ballasts with an investment of € 51.5, decreasing to 310.6 W in half the time of use; this is due to flow reduction allowed by the use of the street in the late evening. It is also possible the analysis of specific individual lighting points as presented in the next section in terms of simulation and experimental verification.

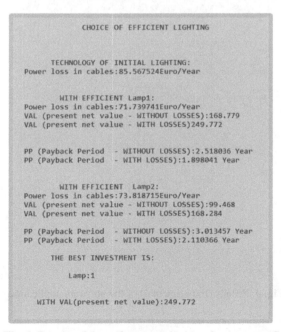

Fig. 2. Results of the software application for the street C

5.2   Simulation and Experimental Results

The load diagrams are shown in Fig. 3 with the power of each luminaire and lamps B[3,2], B[2,3], B[4,3], B[5,1], B[5,2], B[6,1], B[6,2], 100 W.

Fig. 4 presents the results of the new software application considering that the branch B[2,1] feeds a spotlight to illuminate and highlight the building's facade. The objective will be to replace this with a more efficient spotlight (LED) including in the analysis the losses in the conductor sections marked in Fig. 1 with the respective parameters, used as a likely example.

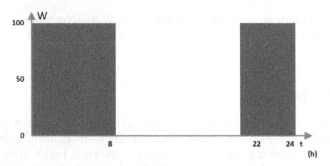

Fig. 3. Load diagram B[3,2], B[2,3], B[4,3], B[5,1], B[5,2], B[6,1], B[6,2]

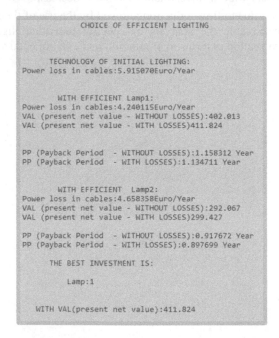

Fig. 4. Results (Experimental) of the software application

The results compare an initial situation, in the branch B[2,1] with one spotlight and halogen lamp of 240 W, with a spotlight LED of 30 W (90 €) and another spotlight and fluorescent compact lamp of 96 W (10€), equivalent in terms of lighting.

5.3   Experimental Validation

The experimental setup can be seen in Fig. 5.

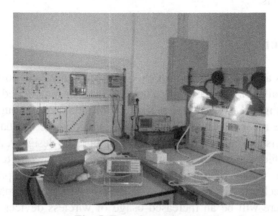

Fig. 5. Experimental setup

Laboratory measurements were performed at the beginning and end of the cables identified as bold in Fig. 1. With 240 W spotlight, 10 W losses were obtained. With 30 W LED spotlight (option 1), 5.83 W losses were obtained in B[1,1]. Figure 6 represents the measurements made at laboratory in cable B[1,1].

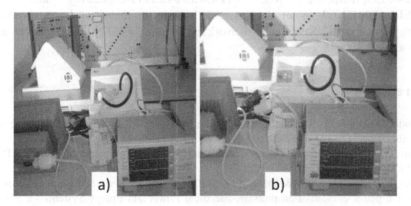

Fig. 6. Measure in the branch (cable) B[1,1], option 1 [a)begin; b) end]

6   Analysis of Results

The experimental results were analyzed based on the initial situation and the cost effective option indicated in the simulation results, during one year of operation and with a price of 0.10 €/kWh, considered normal on average in street lighting. From the data presented in the simulations using the new software application (Fig. 4), it can be seen that the total losses in branches affected by the substitution are equal to 1.6749 €. From the experimental results, the reduction of losses is equal to 4.13 W (41%). Considering that lighting works 11 hours during night period, operating under the same conditions of the simulations, the total losses are equal to 1.6742 €, validating the simulation results.

7   Conclusions

The work presents a software support in choosing luminaires and control systems for street lighting installations, in cases of projects of new or remodelled total, partial, or an individual spotlight. Losses in street lighting installations although small are not null and can make a considerable difference in the economic evaluation, supporting the investment decision. The use of software to support the designer, allows analysing and choosing effective solutions, thus avoiding the use of technologies and experiences from which there is no certainty of economic profitability. The incorporation of technologies in energy efficiency is a promising market segment in the future, and there will be an increased usage of wireless devices for remote data monitoring, where all devices will be interconnected and able to interact within collective awareness systems.

Acknowledgment. This work was supported by FEDER funds (European Union) through COMPETE and by Portuguese funds through FCT, under Projects FCOMP-01-0124-FEDER-020282 (Ref. PTDC/EEA-EEL/118519/2010) and PEst-OE/EEI/LA0021/2013. Also, the research leading to these results has received funding from the EU Seventh Framework Programme FP7/2007-2013 under grant agreement no. 309048.

References

1. Kostic, M., Djokic, L.: Recommendations for energy efficient and visually acceptable street lighting. Energy 34, 1565–1572 (2009)
2. Kaur, D.: Optimal conductor sizing in radial distribution systems planning. Electrical Power and Energy Systems 30, 261–271 (2008)
3. Pires, D.F., Antunes, C.H., Martins, A.G.: NSGA-II with local search for a multi-objective reactive power compensation problem. Electrical Power and Energy Systems 43, 313–324 (2012)
4. Vysotsky, V.S., Nosov, A.A., Fetisov, S.S., Shutov, K.A.: AC Loss and Other Researches with 5 m HTS Model Cables. IEEE Transactions on Applied Superconductivity 21, 1001–1004 (2011)
5. Orzáez, M.J.H., de Andrés Díaz, J.R.: Comparative study of energy-efficiency and conservation systems for ceramic metal-halide discharge lamps. Energy 52, 258–264 (2013)

Optimal Participation of DR Aggregators in Day-Ahead Energy and Demand Response Exchange Markets

Ehsan Heydarian-Forushani[1], Miadreza Shafie-khah[2],
Maziar Yazdani Damavandi[2], and João P.S. Catalão[2,3,4]

[1] Iran University of Science and Technology, Tehran, Iran
[2] University of Beira Interior, Covilhã, Portugal
catalao@ubi.pt
[3] INESC-ID, Lisbon, Portugal
[4] IST, Univ. Lisbon, Portugal

Abstract. Aggregating the Demand Response (DR) is approved as an effective solution to improve the participation of consumers to wholesale electricity markets. DR aggregator can negotiate the amount of collected DR of their customers with transmission system operator, distributors, and retailers in Demand Response eXchange (DRX) market, in addition to participate in the energy market. In this paper, a framework has been proposed to optimize the participation of a DR aggregator in day-ahead energy and intraday DRX markets. In this regard, the DR aggregator optimizes its participation schedule and offering/bidding strategy in the mentioned markets according to behavior of its customers. For this purpose, the customers' participation is modeled using a Supply Function Equilibrium (SFE) model. In addition, due to uncertainties of market prices and the behavior of consumers, an appropriate risk measurement, CVaR, is incorporated to the optimization problem. The numerical results show the effectiveness of the proposed framework.

Keywords: CVaR, day-ahead market, demand response exchange, DR aggregator, energy market, intraday market.

1   Introduction

In a competitive electricity market, demand response programs (DRPs) play an important role in improving market efficiency, reducing peak demand and price instability, and enhancing the reliability [1]. Most of independent market players (e.g. transmission system owner, distributors and retailers) can benefit from DR [2], [3], [4]. Moreover, by implementing the advanced smart grid infrastructure, DR share in system operation resources will be increased [5]. For this purpose, the regulatory bodies are changing market rules and regulations to support implementation of DR programs in electricity markets [6], [7]. From the market point of view, market players are divided into two sets: DR buyers and DR sellers.

DR buyers include retailers and distributors who need DR to improve their business and system reliability while DR sellers have the capacity to significantly modify electricity demand and sell DR to increase their profit [8].

L.M. Camarinha-Matos et al. (Eds.): DoCEIS 2014, IFIP AICT 423, pp. 353–360, 2014.
© IFIP International Federation for Information Processing 2014

DR sellers consist of large consumers which can meet the DR program requirements by themselves or a new market participant such as distribution system operators (DSOs), load service entities (LSEs), and DR aggregators which have the responsibility of managing customer responses. DR aggregators negotiate the amount of their consumers combined DR with TSO, distributors, and retailers.

As introduced in [8], DR can be treated as a tradable commodity in a market which is completely separated from energy market. In a DRX market, the DRX operator collects both the aggregated demand and individualized supply curves. Then, it balances the supply and demand at a common price to clear the market [8].

The literature contains some studies about demand-side players that bid to power markets [6], [7] and [9]. However, DR aggregation has not been discussed in the studies. It is obvious that simultaneous participation of DR aggregators in the energy and DRX markets has not been addressed thoroughly in the previous works.

The majority of the electrical energy is traded in day-ahead market. Hence, the market players have to submit their offers for entire hours of the day-ahead, several hours in advance. The offers have a degree of uncertainty due to the volatile nature of renewable based plants in the future smart grid. Therefore, participation in short timeframe markets for market players is crucial. In other words, from day-ahead market to the spot market, the market players can obtain some new data to update their preliminary offers in an intraday market [10], [11]. The intraday market is a corrector market that is closer to the operational hour; accordingly, that market allows market players to update their offers.

In this paper, the optimal participation of DR aggregator in intraday DRX market and day-ahead energy market is presented. For this purpose, the behavior of DR aggregator's customers has been modeled by Supply Function Equilibrium (SFE) method, unlike the previous studies that have considered constant DR demand curves [8]. On this basis, a new approach is developed for consumers' participating in DRX market. In addition, Conditional Value at Risk (CVaR) is incorporated to the model to tackle the uncertainties of market prices and the behavior of consumers.

The rest of the paper is organized as follows. In Section 2, the contribution to collective awareness systems is presented. Section 3 introduces the optimal participation of DR aggregator in DRX and energy markets. Section 4 is devoted to the case studies. Finally, Section 5 concludes the paper.

2   Contribution to Collective Awareness Systems

Recent advances in smart metering technology enable bi-directional communication between the utility operator and the consumers and facilitate the option of dynamic load adaptation. Toward this direction, DR provides incentives to major consumers, usually in the form of monetary rewards, to reduce their electricity consumption in peak periods. Since the importance of energy conservation and environmental protections are growing, DR can favorably affect the future smart grid [10], [11].

In this context, future collective awareness systems can positively affect future smart grid by obtaining precise information and the effective involvement of the consumers.

On this basis, improvements in collective awareness systems cause customers' behavior in demand side to play a crucial role in future smart grid. Although DR has been successfully applied in the industry sector [5], its application in the residential sector is a more challenging task. On this basis, some new market players (e.g. DR aggregators) should manage the customers' consumption.

Aggregators possess the technology to perform DR and are responsible for the installation of smart meters at end-user premises. Since each aggregator represents a significant amount of total demand in DR market, it can negotiate on behalf of home users with the operator more efficiently. Since these players are the link between customers and electricity markets, they have a critical role in moving towards future smart grid. In future smart grid, by supplementing the collective awareness systems and consequently increasing the participation of customers in DRPs, the players will have a more important role in the electricity markets.

3   Modeling the Optimal Behavior of DR Aggregator

DR aggregator aims to maximize its profit by participating in day-ahead energy and intraday DRX markets. On this basis, DR aggregator plays in DRX market as a DR seller and takes part in energy market as a *negative load* in demand side [12]. A schematic of DR aggregator presence in the mentioned markets is illustrated in Fig. 1. As it is shown, DR aggregator can participate both in intraday DRX market and day-ahead energy market. However, in the first case it has the role of a DR seller and in the second one it takes part on behalf of demand side.

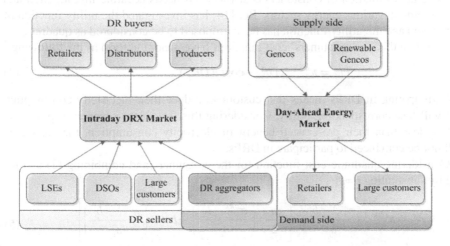

Fig. 1. Day-ahead energy and intraday DRX markets scheme

3.1   Modeling the Supply Function of DR Aggregator's Customers

In order to obtain the optimal behavior of the DR aggregator, the behavior of its customers is modeled first. For this purpose, a new approach based on SFE method is developed for customers' supply function in order to maximize their benefit.

Based on this, the clearing price of DRX market, π^{DRX}, can be formulated as follows:

$$\pi^{DRX} = a_i.DR_i + b_i.(1-\theta_i) \tag{1}$$

where DR_i is the amount of DR purchased from i-th customer. The coefficient θ is the *customer type* and represents a customer's willingness to participate in DRPs. It takes a value between 0 and 1. By increasing the amount of θ, the cost of DR decreases because the customer has more willingness to participate in DRPs. Additionally, a_i and b_i are SFE coefficients applied to all customers [13].

Since, a balance should exist between the amount of sold and purchased DR [8], the balance between electricity load and supply has been considered. On this basis and by introducing the Required DR, RDR, it will be as:

$$RDR = \sum_{i=1}^{l} DR_i = \sum_{i=1}^{l} \frac{\pi^{DRX} - b_i(1-\theta_i)}{a_i} \tag{2}$$

$$\pi^{DRX} = \left[RDR + \sum_{i=1}^{l} \frac{b_i(1-\theta_i)}{a_i} \right] \Big/ \sum_{i=1}^{l}(\frac{1}{a_i}) \tag{3}$$

In this approach, each aggregator should maximize its benefit in the worst case. The worst condition for aggregators occurs when θ tends to 0. In this condition, aggregators have the least capacity to participate in DR, and their profit will be low.

Accurate estimation of consumers cost functions needs accurate investigation and data mining in various energy sectors. Ref [14] has investigated the utility function of consumers and the utility function has been proposed to be considered as quadratic.

Based on this, the consumers' cost functions have quadratic form as the following:

$$Pf_i = \pi^{DRX} \times DR_i - \text{cost}_i(DR_i) \tag{4}$$

Participating in DRPs means that customers reduce their electricity consumption and will lose corresponding utility. Considering this fact, if the revenue of providing DR be less than their pre-existed benefit of electricity consumption, the customers will not be convinced to participate in DRPs.

Considering quadratic cost function for the consumers and combining (2) and (4) and by substituting $\theta=0$:

$$pf_i = \pi^{DRX} \times (\frac{\pi^{DRX} - b_i}{a_i}) - \left[\frac{am_i}{2} \times \left(\frac{\pi^{DRX} - b_i}{a_i} \right)^2 + bm_i \times \left(\frac{\pi^{DRX} - b_i}{a_i} \right) \right] \tag{5}$$

where am_i and bm_i are the customers marginal cost function coefficients. As can be observed from (5), SFE model is utilized for offering strategy of consumers [15].

On this basis, each seller can offer its a_i and b_i to increase its profit. Based on this, the expected clearing price of demand response can be formulated as follows:

$$\pi^{DRX} = \left[RD + \sum_{i=1}^{N^{DRS}} \frac{b_i}{a_i} \right] \bigg/ \sum_{i=1}^{N^{DRS}} \left(\frac{1}{a_i} \right) = \left[RD + \frac{b_i}{a_i} + \sum_{i \neq j}^{N^{DRS}} \frac{b_j}{a_j} \right] \bigg/ \sum_{i=1}^{N^{DRS}} \left(\frac{1}{a_i} \right) \qquad (6)$$

3.2   Modeling the Uncertainty of Market Prices

In order to successfully participate in electricity market, DR aggregators have to forecast market prices. In this paper, two uncertain market prices are considered: day-ahead and intraday. For this purpose, Roulette Wheel Mechanism (RWM) technique is applied for scenario generation in each hour. In order to develop an accurate and appropriate model, market prices have been characterized by log-normal distribution in each hour [16]. Thus, considering μ and σ represent mean value and standard-deviation, respectively, the PDF of market prices is represented by (7):

$$f_{Pr}(Pr, \mu, \sigma) = \frac{1}{Pr \sigma \sqrt{2\pi}} \exp\left[-\frac{(\ln Pr - \mu)^2}{2\sigma^2} \right] \qquad (7)$$

3.3   Incorporating Risk Management

Since the profit of DR aggregator is related to the uncertain behavior of its customers, it should manage its related risk. Conditional value-at-risk (CVaR) can be an appropriate technique to incorporate risk management into the problem. The formulation of CVaR is indicated in (8)-(9):

$$Max: \quad \xi - \frac{1}{1-\alpha} \sum_{s=1}^{S_N} \rho_s . \eta_s \quad , \quad \eta_s \geq 0 \qquad (8)$$

$$-B_s + \xi - \eta_s \leq 0 \qquad (9)$$

The parameter α is usually assigned within the interval of 0.90 to 0.99, which in this work is set to 0.95. If the profit of scenario s is higher than ξ, the value of η_s is set to 0. Otherwise, η_s is assigned to the difference between ξ and the related profit. The above formulated constraint is applied to unify the risk-metrics CVaR.

3.4   Objective Function of the DR Aggregator

In the pool based model, each DR aggregator manages its customers' responses and offers its price-quantity. The uncertain characteristic of the day-ahead energy and the intraday DRX market prices is fully considered. The aggregator offers a specified quantity in day-ahead market and gets accepted level of energy from day-ahead market for each hour. Then, it can update the offers in intraday market. The price scenarios utilized at the DRX market become more accurate. Moreover, in the proposed stochastic framework, risk aversion is implemented by restricting deviations of expected profit using CVaR technique. According to the above mentioned description, the objective function can be expressed as (10):

$$\text{Max } EP = \beta\left(\xi - \frac{1}{1-\alpha}\sum_{s=1}^{S_N}\rho_s.\eta_s\right) + \sum_{s=1}^{S_N}\rho_s\sum_{t=1}^{T}\left[\pi_{ts}^{D}.P_t^{D} + \pi_{ts}^{I}.P_{ts}^{I,sell} - \sum_{d=1}^{ND}\pi_{td}^{DRX}.DR_{td}\right] \quad (10)$$

where, β is the weighting factor to achieve a tradeoff between profit and CVaR.

The first term indicates the CVaR multiplied by β. The next two terms represent the incomes achieved from selling energy in the day-ahead and intraday markets, respectively. Finally, the cost of buying energy from DRX market is represented in the last term. Eq. (10) is maximized considering the constraints described below:

$$P_{ts}^{Sch} = P_t^{D} + P_{ts}^{I,sell} - \sum_{d=1}^{ND}DR_{dts} \quad (11)$$

$$-\sum_{t=1}^{T}\left[\pi_{ts}^{D}.P_t^{D} + \pi_{ts}^{I}.P_{ts}^{I,sell} - \sum_{d=1}^{ND}\pi_{td}^{DRX}.DR_{td}\right] + \xi - \eta_s \le 0 \quad (12)$$

The total scheduled energy of the aggregator in both day-ahead and intraday markets is given in (11). Eq. (12) is related to incorporating the risk into the problem.

4   Numerical Study

In order to illustrate effectiveness of the model, some numerical studies are accomplished. It is assumed that DR aggregator's customers are clustered into four DR sellers which offer to the aggregator. The values of θ are considered as shown in Table 1 for each DR seller. A common load curve of a real-world system is considered [17]. The peak of typical load curve is considered to be 100MW.

Table 1. θ coefficient for DR sellers

DR sellers	Seller 1	Seller 2	Seller 3	Seller 4
θ	0.9	0.7	0.5	0.3

When the hourly required DR is less than the sellers' capacity, a competition has been raised between sellers to sell DR. This competition occurs in a pool-based market. SFE coefficients (a_i, b_i) for each seller are obtained using the approach expressed in section 3.1. The amount of traded DR by each customer is indicated in Fig. 2. As it can be seen, each seller wins an amount of DR that is related to its willingness coefficient. Sellers 1, 2, and 3 can participate in DRX market in most hours, while seller 4 can participate only in peak times. Furthermore, in off-peak intervals, the traded DR between players is decreased.

Table 2 presents the effect of participation in the intraday DRX market on the DR aggregator's expected profit. The capacity of DR for offering to DRX market is supposed to limit up to 20%. As it can be seen in Table 2, an increase in DR aggregator's risk causes a reduction in its expected profits. Moreover, the participation in intraday DRX market can increase its profit.

The effect of DR participation level on the cost and income of DR aggregator is presented in Table 3. The maximum capacity of DR for participating in the DRX market is supposed to be 10%. As it can be seen in Table 3, the intraday DRX market encourages the aggregator to bid more quantities to the day-ahead market, because it expects to make the uncertainties up. On other hand, the prices of DRX market are more stable than those of the intraday market. As can be observed, the increases in DR participation level cause linear increases in the expected profits of the aggregator up to about 20% of DRP participation level. After this point, the impact of the intraday DRX market capacity on the aggregator's profit is decreased, thus its tendency for participating in the DRX market is saturated.

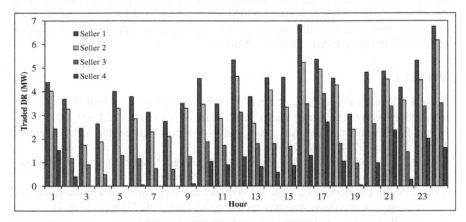

Fig. 2. Traded DR in DRX market

Table 2. Effect of intraday DRX market on DR aggregator's profit and CVaR

Case	Risk level (β)	CVaR ($)	Expected profit ($)
Without participation in intraday DRX market	0	8861.7	9421.5
	1	8989.8	9159.4
With participation in intraday DRX market	0	10059.6	10565.4
	1	10115.0	10126.1

Table 3. Effect of DR participation level on DR aggregator's costs and incomes

DR participation level (%)	10	20	30	40	50
Income from day-ahead market ($)	6359.3	6932.7	7249.4	7278.2	7287.4
Income from intraday market ($)	4677.3	5775.9	5974.5	6043.1	6171.7
Cost from DR sellers ($)	2030.7	2582.5	2689.9	2711.6	2778.0
Expected profit ($)	9005.9	10126.1	10534	10609.7	10681.1

5   Conclusion

This paper investigated the impacts of intraday DRX market on optimal trading of a DR aggregator in a market environment. In this regards, the aggregator can participate in intraday market and use demand response resources a pool based DRX market in order to reduce its risk and maximize its profit. Behavior of consumers in selling DR was modeled using a new approach based on SFE method to maximize their benefit considering the amount of hourly required DR. In addition, the uncertain natures of day-ahead and intraday market price were modeled using RWM method. Furthermore, CVaR was applied as a risk measure that DR aggregator can specify its desirable weighting between the expected profit and risk. The results showed that establish of an intraday DRX market as an adjustment market may provide more opportunities for DR aggregators. The more application of DR could significantly increase the expected profit of DR aggregators and reduce their risks.

Acknowledgment. This work was supported by FEDER funds (European Union) through COMPETE and by Portuguese funds through FCT, under Projects FCOMP-01-0124-FEDER-020282 (Ref. PTDC/EEA-EEL/118519/2010) and PEst-OE/EEI/LA0021/2013. The research also received funding from the EU Seventh Framework Programme FP7/2007-2013 under grant agreement no. 309048.

References

1. Shao, S., Pipattanasomporn, M., Rahman, S.: Demand response as a load shaping tool in an intelligent grid with electric vehicles. IEEE Trans. Smart Grid. 2, 624–631 (2011)
2. Siano, P.: Demand response and smart grids—A survey, Renewable and Sustainable. Energy Reviews 30, 461–478 (2014)
3. Nguyen, D.T., Negnevitsky, M., Groot, D.: Walrasian Market Clearing for Demand Response Exchange. IEEE Trans. Power Sys. 27, 535–544 (2012)
4. Yousefi, S., Moghaddam, M.P., Majd, V.J.: Optimal real time pricing in an agent-based retail market using a comprehensive demand response model. Energy 36, 5716–5727 (2011)
5. Palensky, P., Dietrich, D.: Demand side management: demand response, intelligent energy systems, and smart loads. IEEE Trans. Ind. Informat. 7, 381–388 (2011)
6. Goel, L., Aparna, V.P., Wang, P.: A framework to implement supply and demand side contingency management in reliability assessment of restructured power systems. IEEE Trans. Power Sys. 22, 205–212 (2007)
7. Aalami, H.A., Moghaddam, M.P., Yousefi, G.R.: Demand response modeling considering interruptible/curtailable loads and capacity market programs. Applied Energy 87, 243–250 (2010)
8. Nguyen, D.T., Negnevitsky, M., de Groot, M.: Pool-based demand response exchange—concept and modeling. IEEE Trans. Power Sys., Power Systems 26, 1677–1685 (2011)
9. Moghaddam, M.P., Abdollahi, A., Rashidinejad, M.: Flexible demand response programs modeling in competitive electricity markets. Applied Energy 88, 3257–3269 (2011)
10. Hajati, M., Seifi, H., Sheikh-El-Eslami, M.K.: Optimal retailer bidding in a DA market e a new method considering risk and demand elasticity. Energy 36, 1332–1339 (2011)
11. Aazami, R., Aflaki, K., Haghifam, M.R.: A demand response based solution for LMP management in power markets. Elec. Power & Energy Sys. 33, 1125–1132 (2011)
12. Megawatts vs. negawatts: how a little can do a lot. Electricity Journal 21, 5–5, (2008)
13. Kirschen, D.S.: Demand-side view of electricity markets. IEEE Trans. Power Sys. 18, 520–527 (2003)
14. Yusta, J.M., Khodr, H.M., Urdaneta, A.J.: Optimal pricing of default customers in electrical distribution systems: Effect behavior performance of demand response models. Electr. Power Sys. Res. 548, 58–77 (2007)
15. Li, T., Shahidehpour, M.: Strategic bidding of transmission-constrained GENCOs with incomplete information. IEEE Trans. Power Sys. 20, 437–447 (2005)
16. Conejo, A.J., Nogales, F.J., Arroyo, J.M.: Price-takerbiddingstrategyunderprice uncertainty. IEEE Trans. Power Sys. 17, 1081–1088 (2002)
17. Shayesteh, E., Yousefi, A., Moghaddam, M.P.: A probabilistic risk-based approach for spinning reserve provision using day-ahead demand response program. Energy 35, 1908–1915 (2010)

Renewable Power Forecast to Scheduling
of Thermal Units

Pedro M. Fonte[1,3], Bruno Santos[2], Cláudio Monteiro[2,3], João P.S. Catalão[4],
and Fernando Maciel Barbosa[3,5]

[1] ISEL – Lisbon Superior Engineering Institute, Lisbon, Portugal
[2] Smartwatt,S.A, [3]University of Porto, [5]INESC TEC PORTO, Porto, Portugal
[4] University of Beira Interior, Covilhã, Portugal
pfonte@deea.isel.pt, brunosantos@smartwatt.com, catalao@ubi.pt
{cdm,fmb}@fe.up.pt

Abstract. In this work is discussed the importance of the renewable production forecast in an island environment. A probabilistic forecast based on kernel density estimators is proposed. The aggregation of these forecasts, allows the determination of thermal generation amount needed to schedule and operating a power grid of an island with high penetration of renewable generation. A case study based on electric system of S. Miguel Island is presented. The results show that the forecast techniques are an imperative tool help the grid management.

Keywords: Probabilistic renewable power forecast, kernel density estimator, Power scheduling.

1 Introduction

The increasing introduction of electric energy production with renewable sources, mainly those with high variability, has created several challenges to the energy networks operators, predominately in the scheduling chapter. This problem is potentiated in the low power networks, especially in island without any connection to continental networks. Due to its large implementation, generally, wind generation is considered the main source of variability on renewable generation but there are other sources that can introduce much faster variations, as solar generation [1]. On the other hand, hydro generation, although being easier to control its operation can depend on economic strategies, disconnected from available resource. If there is a small storage capacity, the production can be temporally disconnected form the rainfall. In an island context, a large variation on renewable production can introduce stability problems in the network, which can originate generation and/or load shed and, at limit, black-outs. By this sense and for security, the scheduling is generally done by a conservative way, with low risk but far away from an optimal operation.

In this work is proposed an approach to a renewable power forecast method based on kernel density estimator (KDE).The main objective is to develop a short-term renewable power prediction to allow in the future the creation of a scheduling

L.M. Camarinha-Matos et al. (Eds.): DoCEIS 2014, IFIP AICT 423, pp. 361–368, 2014.

approach for insular electricity networks. The developed methodology was tested in the S. Miguel Island in at Azores, Portugal.

2   Relationship to Collective Awareness Systems

Following the digital agenda for Europe "A Europe 2020" Initiative, the Collective Awareness Platforms (CAPS) are expected, among others, to support environmentally aware. By this sense, this work is based on the necessity of defining tools to allow a growth implementation of "clean" energy sources to produce electric energy. In spite of being renewable and "clean" techniques, due to their uncertainty creates great technical challenges to the power systems. In this work we present a set of tools in order to minimize the power production with pollutants fossil fuels and on other hand maximize the production with renewable and non-pollutants sources maintaining the technical requirements.

3   Problem Discussion

To decrease the fuel consumption by thermal generation, and consequently to decrease the costs of power production in S. Miguel Island, a large amount of renewable energy sources (RES), mainly wind generation, have been integrated. This is a reality in peak load periods but during off-peak periods the system is already saturated with RES. So, it's necessary, most of the times, to limit the wind power output in order to maintain the thermal generators operating above their technical minimum. Therefore, an efficient use of accurate short-term probabilistic forecast of RES can allow optimal committed/dispatched thermal generators, mainly during off-peak to minimize the necessity of renewable power derate. The production mix is composed by 1 thermal power plant with 8 units (divided in 2 groups with different rated power) 2 geothermal power plants with 5 units, 7 small hydro and 1 wind power plant. In table 1 is shown the rated power to each power source.

Table 1. Rated power of each power source

Source	(# units) total power
Fuel	(4) 28 MW
	(4) 64 MW
Wind	(9) 9 MW
Small hydro	(7) 5 MW
Geothermal	(5) 29,6 MW
other	*(2) < 1 MW*

During 2012 several measurements were done and it was concluded that load varied between a minimum of 17,3 MW and a maximum of 70,2 MW, while the maximum renewable production reached 62,2 MW. By operational security reasons it is mandatory that, at least, two thermal units must be on-line to avoid the loss of thermal production. It means that several times, especially during off-peak periods, it was necessary to derate renewable production and, in some cases, work with thermal units below theirs minimum technical limits (12,26 MW). In these cases the thermal units are forced to work with poor efficiency and high fuel consumption whereas renewable sources are wasted. In 2012 these episodes occurred during 745 hours. In figure 1 is shown the thermal production as well as the sum of minimum technical limits of 2 units for 30 min intervals since 0h00, 1 Dec. up to 24h00, 31 Dec. During all over off-peak periods the thermal units had to work below their minimum limits.

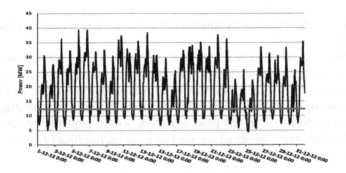

Fig. 1. Thermal production and technical limits

With an efficient forecast tool is possible to have an idea of expected renewable production in a forecast horizon and manage the mix of production, reducing with a certain degree of confidence the thermal production.

4   Renewable Power Probabilistic Forecasts

To ensure an optimized and secured scheduling operation of power grids, is mandatory to have an idea of load forecast, as well as the total renewable power forecast. With this information, is possible to know, under a certain degree of confidence, the thermal capacity needed to each time interval. Therefore, the wind and hydro and geothermal power forecast must be known as well as the load. Traditional forecasts techniques only forecast one value to each variable (*point forecast*) which doesn't afford any probability or confidence interval. Recent techniques present forecasts with uncertainties as densities estimations [2], probabilistic forecasts [3-7], quantiles regressions [8], among others. In this work it's proposed the use of KDE for the computing of predictive probability density function for each time-step of the prediction horizon.

4.1   Forecasting Formulation Modeling

In forecasting problems, the estimative of the future conditional density function has a very important role, since it describes the relation between explanatory and target variables. The conditional density estimation can be seen as a generalization of regression, since conditional density estimation aims at obtaining the full probability density function $f_{y/x}(y/x)$, while the regression aims at estimating the conditional mean $E(y/x)$ [9]. In probabilistic forecasting problems the obtained *pdf* can be used to represent the uncertainty. In this work, we used a Nadaraya-Watson estimator represented by (1) which allow to estimating a random variable Y, when the explanatory random variable X is equal to x [2][10],

$$f_{Y|X}\left(y|x\right)=\frac{f_{Y,X}\left(y,x\right)}{f_X\left(x\right)}.$$

(1)

Where $f_{Y,X}(y,x)$ is the joint density function of (Y,X) and $f_X(x)$ is the marginal density function of X.

In the case of power and load forecast it consists in the estimation the future conditional *pdf* of a random variable for each look-ahead time step $t+k$, given a set with N pairs of samples (p_n, x_n) summarizing all information available up to instant t. Each pairs consists on a set of explanatory variables X_n and corresponding value of variable to be predicted p_n. In this process is assumed the explanatory variables $x_{t+k/t}$ are known for each time-step ahead we want to forecast, resulting (2) where p_{t+k} is the power forecasted for look ahead time $t+k$,

$$\hat{f}_P\left(p_{t+k}\left|x_{t+k|t}\right.\right)=\frac{\hat{f}_{P,X}\left(p_{t+k},x_{t+k|t}\right)}{\hat{f}_X\left(x_{t+k|t}\right)}.$$

(2)

Since the joint and marginal densities aren't known, a nonparametric kernel estimation of the regression function can be used [10].

As renewable power production can depend on several variables (for instance wind speed and direction for wind power forecast or solar radiation and temperature for solar power forecast) a multivariate KDE was used [2][10]. For a given independent and identical distributed multivariate data $(X_{1d}, \ldots, X_{nd})$ from d different variables from an unknown multivariate density function f, the multivariate KDE is given by (3)

$$\hat{f}\left(x_1, \ldots, x_d\right)=\frac{1}{n}\sum_{i=1}^{n}\prod_{j=1}^{d}\frac{1}{h_j}K_j\left(\frac{x_j-X_{ij}}{h_j}\right),$$

(3)

where n is the number of samples, d the number of variables and K_j is the kernel function to each variable j. In (3) h_j is the bandwidth (smoothing parameter) of each kernel around each sample X_{ij}.

Using Nadaraya-Watson estimator, the conditional density is given by (4)

$$\hat{f}_{P_{t+k}|X_{t+k|t}}\left(p,x\right)=\sum_{i=1}^{n}K\left(\frac{p-P_i}{h_p}\right)\cdot\frac{\prod_{j=1}^{d}K\left(\frac{x_j-X_{ij}}{h_j}\right)}{\sum_{i=1}^{n}\left[\prod_{j=1}^{d}K\left(\frac{x_j-X_{ij}}{h_j}\right)\right]}.$$

(4)

In the case of deterministic forecast, to estimate the conditional mean $P_{t+k}=E(p_{t+k}/x_{t+k/t})$, Nadaraya-Watson is also used to estimate P_{t+k} as a locally weighted average (5)

$$\hat{P}_{t+k} = E\left(p_{t+k}\left|x_{t+k|t}\right.\right) = \sum_{i=1}^{n} \frac{\prod_{j=1}^{d} K\left(\frac{x_j - X_{ij}}{h_j}\right)}{\sum_{i=1}^{n}\left[\prod_{j=1}^{d} K\left(\frac{x_j - X_{ij}}{h_j}\right)\right]} . P_i \tag{5}$$

The first step is to choice the kernel, and in the literature there are several possible kernels, namely, Normal, Biweight, Epanechnikov, Logistic, among others, but the most important step is the chosen of bandwidth h. Small bandwidth values lead to an over fitted prediction function, while high values generalize too much [10]. In this work we used Normal Kernel and the bandwidths were calculated by the leave-one-out cross validation (LOO-CV) [10] with the hill climbing optimization algorithm.
After the definition of the expected value by (5) and standard deviation it is necessary to define the probability distribution. Following [2] *Beta pdf* (6) was found to be a good approximation for modelling variables with minimum and maximum limits, beyond to be very flexible.

$$f(x,\alpha,\beta) = \frac{1}{B[\alpha,\beta]}\left(x^{\alpha-1}(1-x)^{\beta-1}\right) \tag{6}$$

Where $B[\alpha,\beta]$ is a normalization constant to ensure that the probability integrates to one, x is defined between $[0,1]$ and shape parameters α and β are higher than 0. Allowing the evaluation of the distribution in a continuous way, this distribution fitting also makes possible to approximately simulate the obtained distribution using only two parameters. To define the distribution parameters α and β, calculated for every time-step of the forecast the moment's method was used.

4.2   Wind Power Forecast

Although depending on several variables, typically wind speed and direction are used as explanatory variables to wind power forecast. In this work the explanatory variables, wind speed and direction, are provided as spot forecast from numerical weather predictions (NWP) tools. As the NWP are one of the largest sources of uncertainty in wind power forecast, each kernel (Normal) is centered in spot forecast with a bandwidth h_x obtaining the prediction from (5).

4.3   Hydro and Geothermal Power Forecast

The geothermal power plant although being renewable, isn't characterized by its uncertainty, once the resource is relatively easy to control. Due to that, the geothermal power production remains relatively constant all over the year. It is assumed that the uncertainties are small oscillations around a specific value. Thus, the expected value is modeled from a 24 hours moving average of past productions. To model the uncertainty, centered in the set point it's applied a kernel to obtain the uncertainty.

In the case of hydro power, the point forecast of the power is obtained by a tool named H4C proposed by [11]. The inputs of H4C are, data of hydro power production and precipitation forecast provided by NWP. Finally, to model the uncertainty the output variable of H4C will be used as input of KDE method.

4.4   Load Forecast

As in the generality of power grids, the power consumption in S. Miguel is affected by environmental factors, as temperature and social factors, as hour of the day, weekends, and holidays, among others. As previous forecasts, also in load forecast multivariable KDE was used. In this case, the explanatory variables are ambient temperature, hour of the day, day of the week and week of the year.

4.5   Aggregation of Forecasts

After the definition of all forecasts and distributions, is necessary to define the equivalent probabilistic model of the aggregations. Assuming that all renewable sources are statistically independent, their sum can be obtained by the convolution of all *pdf* [12] which resulted from RES forecasts. Finally, the convolution (subtraction) between aggregated RES forecast and load forecast (L-RES), results on the power which should be produced by thermal units. These values represent the remaining load to be feed by thermal units under a level of uncertainty defined by a *Beta* distribution.

5   Case Study

To perform the case study there is an off-line hourly data set of all variables necessary to the forecasts as well as the on-line thermal power units and total demand during 2012. In figure 2 is shown the forecast (24 hours ahead, refreshed at 0h00 of each day) of thermal power production resulting from (L-RES) and its confidence (uncertainty) interval from 5% up to 95%, since 0h00 November 15th up to 23h00 November 21st of 2012. In the same figure is shown the maximum and minimum limits of on-line thermal units committed by transmission system operator (TSO).

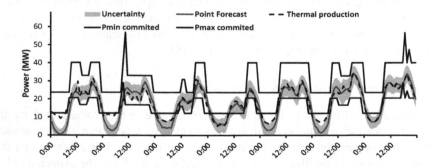

Fig. 2. Measured and forecasted thermal power production

It is shown that there is a good fitting between measured and forecasted thermal power. The large differences happen when the forecasts are below the lower technical limits of on-line thermal units. As a result of a rule of thumb, there are always two generators on-line whose sum of technical minima are 12,26 MW. So it means there was the necessity of wind power derating. In some of that cases, even with the derating there were the necessity of thermal units working below the minimum. In these cases the thermal units are working with low efficiency and high consumption of fuel and wasting "clear" energy.

In figure 3 is shown the result of a possible unit commitment based on thermal power forecast under uncertainty. Comparing with figure 2 is clear the decrease of thermal production mainly in off-peak periods. In the future works is intended to add more restrictions to the problem as contingencies and start/stop times.

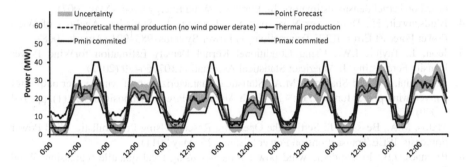

Fig. 3. Unit commitment and thermal forecast with uncertainty

Due to the real necessity of wind power derrating another source of uncertainty was introduced in the available off-line data set. There are pairs of explanatory variables which do not correspond to the real wind power available due to the derrating. To overcome this problem a real power curve of wind farm was modeled and during the periods where the power is limited or at least suspicious of limitation the obtained power curve was used as an estimation of wind power production. By this way another characteristic is shown in figure 3, a theoretical thermal production (without derrating).

6   Conclusions

There are vast consensuses on the need of forecasting techniques when there is a large penetration of renewable power plants in the electric grid. Nevertheless this issue becomes more important in the context of an island manly when there is low rated capacity. In this work was shown that accurate forecast techniques have an important role given important information to the scheduling and operation of power grid. Based on this work a short-term scheduling methodology based on risk analysis can be developed in order to create a tool which can optimize de unit commitment and due to that minimize the fuel consumption and maximize the weight of renewable production.

Acknowledgments. This work and methodology was developed in Smartwatt S.A. under SINGULAR Project – *Smart and Sustainable Insular Electricity Grids Under Large-Scale Renewable Integration* financed by European Commission.

References

1. Sengupta, M., Keller, J.: PV ramping in a distributed generation environment: A study using solar measurements. In: 2012 38th IEEE Photovoltaic Specialists Conference (PVSC), June 3-8, pp. 586–589 (2012)
2. Bessa, R.J., Miranda, V., Botterud, A., Wang, J.: Time Adaptive Conditional Kernel Density Estimation for Wind Power Forecasting. IEEE Transactions on Sustainable Energy (October 2012)
3. Juban, J., Fugon, L., Kariniotakis, G.: Probabilistic short-term wind power forecasting based on kernel density estimators. In: European Wind Energy Conf. (May 2007)
4. Bludszuweit, H., Dominguez-Navarro, J.A., Llombart, A.: Statistical Analysis of Wind Power Forecast Error. IEEE Transactions on Power Systems, 983–991 (August 2008)
5. Jeon, J., Taylor, J.W.: Using Conditional Kernel Density Estimation forWind Power Density Forecasting. J. American Statistical Association 107, 66–79 (2012)
6. Al-Awami, A.T., El-Sharkawi, M.A.: Statistical characterization of wind power output for a given wind power forecast. In: 41st North American Power Symposium, vol. (1), pp. 1–4 (2009)
7. Matos, M.A., Bessa, R.J.: Setting the Operating Reserve Using Probabilistic Wind Power Forecasts. IEEE Transactions on Power Systems 26 (May 2011)
8. Bremnes, J.B.: Probabilistic wind power forecasts using local quantile regression. Wind Energy 1(1), 47–54 (2004)
9. Fu, G., Shih, F.Y., Wang, H.: A kernel-based parametric method for conditional density estimation. Pattern Recognition 44(2), 284–294 (2011)
10. Demir, S.: On the Adaptive Nadaraya-Watson. Journal of Mathematic and Statistics 39(3), 429–437 (2010)
11. Monteiro, C., Ramirez-Rosado, I.J., Fernandez-Jimenez, L.A.: Short-term forecasting model for electric power production of small-hydro power plants. Renewable Energy (6), 387–394 (2013)
12. Williamson, R.C.: Probabilistic Arithmetic, University of Queensland, Ph.D. Thesis (1989)

Part XIV

Optimization Issues in Energy - II

Optimum Generation Scheduling Based Dynamic Price Making for Demand Response in a Smart Power Grid

Nikolaos G. Paterakis[1], Ozan Erdinc[1], João P.S. Catalão[1,2,3],
and Anastasios G. Bakirtzis[4]

[1] University of Beira Interior, Covilhã, Portugal
catalao@ubi.pt
[2] INESC-ID, Lisbon, Portugal
[3] IST, Univ. Lisbon, Portugal
[4] Aristotle University of Thessaloniki, Greece

Abstract. Smart grid is a recently growing area of research including optimum and reliable operation of bulk power grid from production to end-user premises. Demand side activities like demand response (DR) for enabling consumer participation are also vital points for a smarter operation of the electric power grid. For DR activities in end-user level regulated by energy management systems, a dynamic price variation determined by optimum operating strategies should be provided aiming to shift peak demand periods to off-peak periods of energy usage. In this regard, an optimum generation scheduling based price making strategy is evaluated in this paper together with the analysis of the impacts of dynamic pricing on demand patterns with case studies. Thus, the importance of considering DR based demand pattern changes on price making strategy is presented for day-ahead energy market structure.

Keywords: Demand response, Home energy management, Optimal scheduling, Real-time pricing.

1   Introduction

Smart grid concept refers to operating bulk power system in a more efficient, reliable, secure, environment friendly and economic way together with the utilization of advanced monitoring, protection and control systems by a two-way information and energy flow in all nodes of the power grid. Smartly operated grid structure is envisioned to support high levels of renewable energy penetration, new loads like electric vehicles (EVs), consumer participation with increased level of awareness and to optimize the operation of production, transmission and distribution systems. Especially, the issue of enabling consumer participation in demand side is a pivotal advantage of the smart grid idea providing a smoother power profile to be faced by utilities and all parts of the power grid. In this regard, smart households that can monitor their use of electricity in real-time and take actions to lower their electricity bills have also been given specific importance for the research of possible demand side actions [1]. Demand side actions in a smart grid generally focus on demand response (DR) strategies creating a two-side game between utility and consumers [2].

As one of the most important contribution and interest area for smart grid idea, DR activities for controlling the demand side of the power production/demand balance

L.M. Camarinha-Matos et al. (Eds.): DoCEIS 2014, IFIP AICT 423, pp. 371–379, 2014.

instead of mature grid structure just dealing with the production side of the equation include a major portion in the latest literature dealing with smart grid applications.

Within the given demand side actions topic, smart home structures together with smart home energy management (HEM) systems capable of controlling home size distributed energy production facilities, electric vehicle (EV) based storage production options, controllable new generation smart appliances have also been the specific topic of some research activities in this area. The dynamic retail price variation for DR activities in end-user premises follows the wholesale price dynamics resulting from the short-term scheduling of the generation side facilities. In this study, a production side scheduling including dispatchable sources of energy will be provided in order to obtain a dynamic price variation to be presented to end-users in a smart grid environment. Then, these price variations will be considered as an input to the HEM of a sample residential end-user in order to provide the relevant demand response activities considering controllable/non-controllable load facilities. Thus, the impact of price variations on end-user load shapes can be easily examined with the provided structure.

The paper is organized as follows: Section 2 gives contribution to Collective Awareness Systems. Section 3 presents the methodology and the obtained results are discussed in Section 4. Finally, concluding remarks and future works are summarized in Section 5.

2   Contribution to Collective Awareness Systems

In the literature, there are several papers dealing with optimum scheduling of production facilities for power systems. Simoglou et al. [3] proposed a detailed model for solving the hydrothermal scheduling problem for a day ahead energy and reserve market. Morales et al. [4] provided a work on evaluating reserves in a power system under high penetration of wind power under the scope of two-stage stochastic programming. They also consider demand side as a bounded or involuntarily shed resource, providing some elasticity but not explicitly referring to DR.

There are also many recent studies dealing with DR strategies for smart households. Chen et al. [5] and Tsui and Chan [6] developed an optimization strategy for the effective operation of a household with a price signal based DR. Pipattanasomporn et al. [7] and Kuzlu et al. [8] presented a HEM considering peak power limiting DR strategy for a smart household, including both smart appliances and EV charging. Shao et al. [9] also investigated EV for DR based load shaping of a distribution transformer serving a neighborhood. Angelis et al. [10] performed the evaluation of a HEM strategy considering the electrical and thermal constraints imposed by the overall power balance and consumer preferences. Chen et al. [11] provided an appliance scheduling in a smart home considering dynamic prices and appliance usage patterns of consumer.

The studies referred above together with many other studies not referred here have provided valuable contributions to the application of smart grid concepts in household areas. However, the papers dealing with scheduling issue generally neglect demand side uncertainties caused by possible DR activities in a smart power grid environment. Besides, from the DR points of view, many of the mentioned papers referred above failed to address either production side of the game for providing the necessary price variations or the demand side in terms of vehicle-to-grid (V2G) option of EVs for lowering the demand peak periods together with different DR strategies.

3   Methodology

In this section, the methodology for the optimum operation strategy combined with DR activities is presented. First, a short-term generation scheduling algorithm that covers next day's load profile with the least cost is presented. Then, utilizing the mentioned scheduling based price variation, a DR strategy with considering different case studies of utilizing the EV in different modes of operation is performed. The relevant details of the employed methodology are as follows:

3.1   Scheduling Model for Obtaining Price Variation

For the sake of simplicity, in the presented scheduling model we neglect the need of considering reserves. The overall objective of the day-ahead energy market clearing procedure is to minimize the total costs associated with electricity production. The objective function given in Eq. (1) could be easily extended to comprise other costs such as generator's no load cost or the cost of energy not served.

$$Minimize \sum_{t \in T} \sum_{i \in I} (SUC_i \cdot y_{i,t} + SDC_i \cdot z_{i,t}) + \sum_{t \in T} \sum_{i \in I} \sum_{f \in F} c_{i,f} \cdot b_{i,f,t} \tag{1}$$

The objective of the system should be achieved subject to several constraints presented in (2)-(12):

$$\sum_{i \in I} p_{i,t} = D_t, \forall t \in T \tag{2}$$

$$p_{i,t} \geq P_i^{min} \cdot u_{i,t}, \forall t \in T, \forall i \in I \tag{3}$$

$$p_{i,t} \leq P_i^{max} \cdot u_{i,t}, \forall t \in T, \forall i \in I \tag{4}$$

$$\sum_{f \in F} b_{i,f,t} = p_{i,t}, \forall i \in I, \forall t \in T \tag{5}$$

$$0 \leq b_{i,f,t} \leq B_{i,f,t}, \forall i \in I, \forall f \in F, t \in T \tag{6}$$

$$p_{i,t} + p_{i,(t-1)} \leq RU_i \cdot 60, \forall i \in I, \forall t \in T \tag{7}$$

$$p_{i,(t-1)} - p_{i,t} \leq RD_i \cdot 60, \forall i \in I, \forall t \in T \tag{8}$$

$$\sum_{\tau=t-UT_i+1}^{t} y_{i,t} \leq u_{i,t}, \forall i \in I, \forall t \in T \tag{9}$$

$$\sum_{\tau=t-DT_i+1}^{t} z_{i,t} \leq 1 - u_{i,t}, \forall i \in I, \forall t \in T \tag{10}$$

$$y_{i,t} - z_{i,t} = u_{i,t} - u_{i,(t-1)}, \forall i \in I, \forall t \in T \tag{11}$$

$$y_{i,t} + z_{i,t} \leq 1, \forall i \in I, \forall t \in T \tag{12}$$

Constraints (2)-(4) enforce system power balance and generating unit technical limits. The cost for generating power for each unit is described with a step-wise non-decreasing marginal cost function like the one presented in Ref. [3] and is expressed by the second term of (1) and the constraints (5) and (6). Constraints (7) and (8) enforce the unit ramp rate limits. The unit minimum up and down time constraints are considered by (9) and (10), respectively. Constraint (11) considers the start-up and shut-down status change logic while (12) states that a unit cannot be simultaneously started-up and shut-down.

3.2   Demand Response Activities

A HEM system for DR strategies regulates the operation of such a smart household considering price based and other signals from the utility, production of small scale own facilities, load consumption of smart appliances, etc. together with different consumer preferences. In this paper, the smart EV operation considering smart charging and possible V2G mode is evaluated with an optimization based HEM strategy. A bi-directional EV grid connection is considered for the analyzed household structure. The specifications of a Chevy Volt with a battery rating of 16 kWh is taken into account [12]. The Chevy Volt is employed with a charging station limited to a charging power of 3.3 kW. The same power limit is also assumed to be valid for discharging operation in V2G mode. The charging and discharging efficiencies are considered as 0.95. The EV is modeled with a state-of-energy equation, as follows:

$$State - of - Energy = E_{bat,in} + \frac{\int_0^t P_{bat} dt}{E_{bat,cap}} \tag{13}$$

There are many DR strategies for load demand management for households. In this study, a dynamic price based DR strategy is considered for the interaction with smart grid operator utility. The time-varying price signal available for the consumer via smart meter is obtained by the scheduling mentioned above in Section 3.1, adding a flat rate that reflects the fact that the retail prices are the wholesale market prices plus surcharges for transmission and distribution networks usage plus taxes. The DR action is provided as leading to the optimized operation of household appliances with lowest daily price under limitations of power supply guarantee.

4   Test and Results

The load demand considered in this study is created using real demographic data from Crete Island, Greece and is separated in two components [13]. The first component, comprising the industrial and commercial loads is shown in Fig. 1 and reflects the fact that Crete´s economy is primarily based on agricultural and light commercial sector, rather than heavy industry.

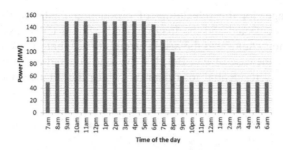

Fig. 1. Industrial load demand

The second component of the load demand, residential load profiles are created by considering general daily habits of inhabitants and possibility of owning EV and smart DR opportunities. There are approximately 200,000 households in Crete. The general load profile of a household is obtained using the home appliance data in Table 1. In this table, the cycle of utilization presents how many cycles the appliance face in separate hours.

Table 1. Considered household appliances

Appliance	Power [kW]	Cycle of utilization	Appliance	Power [kW]	Cycle of utilization
Refrigerator	0.3	24	Toaster	0.85	3
TV	0.125	17	Oven	2	3
Coffee Maker	0.85	4	Washing-machine	2.3	1
Computer	0.15	15	Dishwasher	1.3	3
Water heater	2	1	Lighting	0.15	7

The load demand of a sample household without and with EV using the data given in Table 1 is shown in Fig. 2. It should be noted that the household owners arrive home at 5 pm and directly plug their EVs. Besides, it should be stated that the percentage of households for cases (1) without EV, (2) with EV, (3) with DR, EV and without V2G, and (4) with EV, DR and V2G are assumed as 70%, 10%, 10% and 10%, respectively.

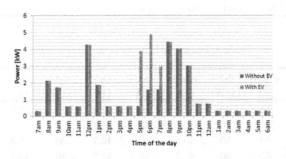

Fig. 2. Household load demand with and without EV

In the first stage, for the scheduling of production facilities, the DR and V2G opportunities are neglected and the load demand without EV is assumed to contain 70% of the total households and the rest is taken into account to cover the load profile given in Fig. 2 for the case of EV. Besides, the load demand given in Fig. 1 is added to the obtained total residential load demand. Thus, the total load for scheduling is given in Fig. 3.

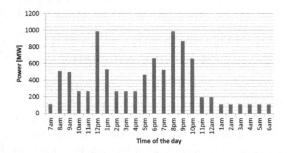

Fig. 3. Considered total load demand for scheduling

The test power system consists of five units of different technologies, a fact that explicitly affects the associated costs and technical features. Detailed technical and economic data are given in Table 2 and Table 3. After the scheduling procedure, the obtained price variation for analyzing the DR activities is shown in Fig. 4, corresponding to the wholesale market prices plus the extra charges described above, considered a flat 5cents/kWh surcharge. As seen, for the peak power periods of total demand given in Fig. 3, the scheduling algorithm provides higher prices compared to off/peak periods as expected.

Table 2. Technical Data of Generating Units

	Pmax [MW]	Pmin [MW]	SUC [€]	SDC [€]	RU [MW/min]	RD [MW/min]	UT [h]	DT [h]	Tech.
Unit1	180	40	45500	10000	1.8	1.8	8	4	Lignite-fired
Unit2	200	50	60000	10000	3	3	8	4	Lignite-fired
Unit3	250	100	16000	5000	20	20	4	3	CCGT
Unit4	300	90	19800	5000	24	24	4	3	CCGT
Unit5	120	30	2600	500	8	8	1	1	OCGT

Table 3. Economic Data of Generating Units

	b1 [MW]	c(b1) [€]	b2 [MW]	c(b2) [€]	b3 [MW]	c(b3) [€]	b4 [MW]	c(b4) [€]	b5 [MW]	c(b5) [€]
Unit1	80	32	40	32.2	30	32.4	20	32.6	10	32.8
Unit2	100	32.7	60	32.9	40	33.1	-	-	-	-
Unit3	120	40	50	40.5	40	40.7	30	40.8	10	41.3
Unit4	130	42	60	42.5	50	42.9	40	43.2	20	43.5
Unit5	70	60	35	61	15	62	-	-	-	-

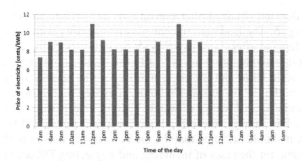

Fig. 4. Obtained price variation

The HEM system in the sample household considers the daily electricity prices in Fig. 4 together with regular load demand patterns of the household to decide for the best operating strategy for EVs. The smart EV charging by optimization based HEM strategy with and without V2G option is evaluated and the results are given in Fig. 5. The HEM strategy shifts the EV charging to off-peak hours with lower prices. This type of operation leads to a lower total daily electricity cost than the case where consumers manually decide the charging time of their EVs without V2G option.

It can also be seen from Fig. 5 that V2G option decreases the energy procurement from the utility during peak power and price periods. It is considered in this case that the EV is plugged-in and the household load demand is supplied by the EV until the battery energy reaches to the restricted lower battery energy limit for the periods determined by the optimum DR strategy. After reaching this limit, procurement of energy from utility starts again to fully supply the load demand in all conditions.

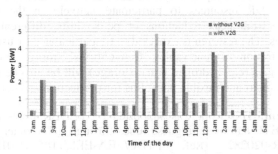

Fig. 5. Household load demand with DR while neglecting and considering V2G

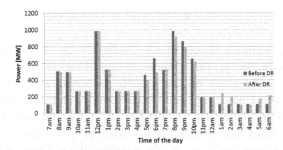

Fig. 6. Compared total load demand profiles with and without DR activities

Total load demand obtained after considering also DR strategies is comparatively presented in Fig. 6. The load pattern changes considerably with shifting some portion of peak load demand after 5 pm to off-peak periods especially after midnight hours. Also, the load demand before 5 pm is totally the same as only EVs are considered as controllable load in HEM and EV charging/discharging activities are assumed to start after 5 pm when the EV owners arrive home. It is sure that if more high power loads (washing machines, dishwashers, water heaters, etc.) were also considered as controllable in DR strategy, the other hours before 5 pm would be affected by this issue. Besides, it should be stated that the total consumed energy in the evaluated 24-h period is the same for the cases of including and neglecting DR activities as expected.

5   Conclusion

In this paper, a sample generation scheduling with neglecting reserve requirements for the day-ahead market operation was provided. The obtained prices with the scheduling strategy were employed for a smart household structure to better observe the impacts of dynamic pricing on DR activities, according to the total load profile. It is clear that the change in load demand with DR activities would have a significant impact on the price making structure, not only for the reason of load pattern change but also for the reason of reserve requirements and other market based actions in the real time market. This study is an innovative first step towards a thorough market based evaluation of DR as a system resource. Future studies will examine the interaction between DR activities and generating side in several market structures such as day-ahead, intra-day and real-time market. The basic outlook will be the investigation of DR resources to participate actively in the wholesale market providing, apart from peak reductions or peak shifts, different types of critical ancillary services. The research is expected to be particularly focused on insular grids with high penetration of renewables. Conventional storage units and upcoming storage technologies (e.g. hydrogen based storage) will be also be taken into account in order to form smart and flexible portfolios for production and demand facilities.

Acknowledgment. This work was supported by FEDER funds (European Union) through COMPETE and by Portuguese funds through FCT, under Projects FCOMP-01-0124-FEDER-020282   (Ref.   PTDC/EEA-EEL/118519/2010)   and   PEst-OE/EEI/LA0021/2013.   The research also received funding from the EU Seventh Framework Programme FP7/2007-2013 under grant agreement no. 309048.

Nomenclature. $B_{i,f}$ is size of step f of unit i marginal cost function; D_t is demand in hour t; DT_i is minimum down time of unit i; $E_{bat,cap}$ is battery energy storage capacity; $E_{bat,in}$ is initial battery energy while EV arrives home; P_{bat} is battery charging/discharging power; RU_i is ramp- up rate of unit i; RD_i is ramp-down rate of unit i; SUC_i is start-up cost of unit i; SDC_i is shut-down cost of unit i; UT_i is minimum up time of unit i; $c_{i,f}$ is marginal cost of step f of unit I marginal cost function; $b_{i,f,t}$ is portion of step f of the i-th unit's marginal cost function loaded in hour t; $f(F)$ is index (set) of steps of the marginal cost function of unit i, $i(I)$ is index (set) of generating units; $p_{i,t}$ is power output of unit i in hour t limited between

P_i^{min} and P_i^{max}; $t(T)$ is index (set) of hours of the planning period; $u_{i,t}$ is binary variable which is 1 if unit i is committed during hour t; $y_{i,t}$ is binary variable which is 1 if unit i is started-up during hour t; $z_{i,t}$ is binary variable which is 1 if unit i is shut-down during hour t.

References

1. Gellings, C.W.: The Smart Grid: Enabling Energy Efficiency and Demand Response. CRC Press (2009)
2. Borlease, S.: Smart Grids: Infrastructure, Technology and Solutions. CRC Press (2013)
3. Simoglou, C.K., Biskas, P.N., Bakirtzis, A.G.: A MILP approach to the short term hydrothermal self-scheduling problem. In: IEEE Bucharest Power Tech Conference (2009)
4. Morales, J.M., Conejo, A.J., Ruiz, J.P.: Economic valuation of reserves in power systems with high penetration of wind power. IEEE Trans. Power Systems 24, 900–910 (2009)
5. Chen, Z., Wu, L., Fu, Y.: Real-time price-based demand response management for residential appliances via stochastic optimization and robust optimization. IEEE Trans. Smart Grid 3, 1822–1831 (2012)
6. Tsui, K.M., Chan, S.C.: Demand response optimization for smart home scheduling under real-time pricing. IEEE Trans. Smart Grid 3, 1812–1821 (2012)
7. Pipattanasomporn, M., Kuzlu, M., Rahman, S.: An algorithm for intelligent home energy management and demand response analysis. IEEE Trans. Smart Grid 3, 2166–2173 (2012)
8. Kuzlu, M., Pipattanasomporn, M., Rahman, S.: Hardware demonstration of a home energy management system for demand response applications. IEEE Trans. SmartGrid 3, 1704–1711 (2012)
9. Shao, S., Pipattanasomporn, M., Rahman, S.: Demand response as a load shaping tool in an intelligent grid with electric vehicles. IEEE Trans. Smart Grid 2, 624–631 (2011)
10. De Angelis, F., Boaro, M., Squartini, S., Piazza, F., Wei, Q.: Optimal home energy management under dynamic electrical and thermal constraints. IEEE Trans. Industrial Informatics 9, 1518–1527 (2013)
11. Chen, X., Wei, T., Hu, S.: Uncertainty-aware household appliance scheduling considering dynamic electricity pricing in smart home. IEEE Trans. Smart Grid 4, 932–941 (2013)
12. GM Chevy Volt specifications, http://gm-volt.com/full-specifications/
13. Hellenic Statistical Authority, http://www.statistics.gr

Application of NSGA-II Algorithm to Multiobjective Optimization of Switching Devices Placement in Electric Power Distribution Systems

António Vieira Pombo[1,2], Vitor Fernão Pires[1,2], and João Murta Pina[1]

[1] Faculdade de Ciências e Tecnologia
Universidade Nova de Lisboa
Monte da Caparica, Portugal
jmmp@fct.unl.pt
[2] Escola Superior de Tecnologia
Instituto Politécnico de Setúbal
Setúbal, Portugal
{antonio.pombo,victor.pires}@estsetubal.ips.pt

Abstract. The Electric Utility Industry all around the world is facing numerous challenges, which include amongst others, the optimal use of expensive assets and resources and the maintenance of electric grid and customer quality service levels. The optimal placement of switches in electrical distribution networks will allow the control over service quality levels and the maximization of investments in equipments. This work proposes a genetic evolutionary algorithm NSGA-II for the optimization between the maximal return of investments on existing assets, while maintaining the quality of service provided. The trade off between total cost of investments and service quality levels SAIDI (System Average Interruption Duration Index) and SAIFI (System Average Interruption Frequency Index) is analyzed, to choose the optimal placement of switches in the distribution electrical networks. The proposed method was tested with a Portuguese real distribution network. The obtained results allowed to verify the performance of the adopted approach.

Keywords: Switch Placement, Reliability, Genetic Algorithm, Electrical Distribution Networks.

1 Introduction

Planning and operation in Distribution system consists on satisfying the system load and energy demand, as economically as possible, while assuring supply continuity and contracted quality levels. Normally, the approaches that public owned utilities tend to implement consists on designing and operating electrical distribution infrastructures to minimize the total cost to the society. Generally this approach is referred to as value based planning.

On the other hand, private owned utilities tend to defer in time network investments and maximize their return. Their goal consists on supplying safe and reliable electric

L.M. Camarinha-Matos et al. (Eds.): DoCEIS 2014, IFIP AICT 423, pp. 380–387, 2014.

power at a reasonable cost. Recent trends however show that the future focus will not be only on the companies' economic performance, but rather on successfully combining economic performance with business impacts on social welfare and environmental factors (Fig. 1). Aging infrastructures and reliability are the most critical issues that preoccupy utility CEOs. And these two are closely related. A decrease in infrastructure investment decreases in time the reliability of the network. On the other hand an increase in network investment increases system reliability. But why focus on reliability? The effects of power outages go beyond the inconvenience experienced from the outage itself. Power outages cost local businesses thousands of euros in lost sales, interrupted manufacturing and lost data. The economic cost of outages reaches beyond lost productivity. A multi-day outage can cost residents hundreds of dollars in lost food and can damage other personal items when sump pumps stop working. Under this context, in electric power distribution system planning, the optimal placement of switches is one of the areas that contribute to the utility goal achievement of operating in the social welfare area. By choosing the optimal place and number of switches in the distribution electrical network, it is possible to maximize the quotient between the benefits to the community and the investments made by the utility on the network. By changing the state of the switches (open/close) the distribution network can be reconfigured, after a fault, thus minimizing the number of customers affected, fault duration and frequency. Although with different objectives, all reconfiguration problems are a combinatorial constrained problem, described by a non-linear and non-differential objective function.

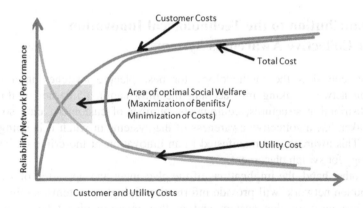

Fig. 1. Value Based Planning and Social Welfare operation

To address this optimization problem, where the objectives are generally conflicting, preventing simultaneous optimization of each objective, several algorithms have been developed. Due to the importance of the problem, several research articles related to optimal placement of switches in distribution networks have been written. In these works, different mathematical models and optimization methodologies have been used. Some of them present this issue as a single objective problem. Others use several objectives functions, but they are agglomerated into a new single objective function. Finally, some authors deal with this problem with several objective functions that are not agglomerated, allowing in this way to obtain a Pareto Optimality. Several optimization methods have been used to address the optimal switch placement problem such as Genetic Algorithms (GA) [1], [2], Particle Swarm Optimization (PSO) [3], [4], Ant Colony Systems (ACS) [5], [6] and Tabu Search Algorithm (TSA) [7]. In [8] is used a modified shuffled frog leaping algorithm and in [9] a hybrid algorithm, both using a multi-objective fuzzy logic (FL) approach.

In this work, a new algorithm applied to this problem is proposed. Due to the characteristics of this kind of problem, two important reliability measures generally used by electric utilities were selected for the objective functions [10]. These are SAIFI, which is an indicator of utility network performance and SAIDI, which measures the operating performance of the utility in restoring customer interruptions. The evolutive genetic algorithm NSGA-II was adopted for the optimization of this problem. This kind of genetic Algorithm is a popular meta-heuristic algorithm that have been applied to multiobjective optimization in power systems such as environmental/economic power dispatch [11] and capacitor placement [12], and as can be seen by this work is well suited for this type of problem. In this multi-objective optimization the goal is not to find a single optimal solution, but rather a trade off between different objective functions, called the "Pareto Set", which is the set of all feasible solutions whose vector of the various objectives is not dominated by any other solution. Results of a real distribution feeder are presented in order to verify the effectiveness of the proposed approach.

2   Contribution to the Technological Innovation for Collective Awareness Systems

This paper considers the methodology for best placing switches in a electrical distribution network, taking into account the information ("internet of things") of existing distribution structures, equipments, number of customers, etc., so that the decision taken has a collective awareness of the systems in witch it is going to have an effect. This awareness is manifested in an innovation of the criteria's behind the methodology for switch placement.

On the other hand, the implications of the placement and operation of switches in the distribution network, will provide information to higher systems, so that they too have an awareness of this system and in that measure also take an innovative perspective on other solutions affecting the planning and operation of electrical distribution networks.

3   Problem Formulation

As said before, in multi-objective optimization we are not interested in finding a single optimal solution, but rather a trade off between the two different objective functions, called the "Pareto Set", which is the set of all feasible solutions whose vector of the various objectives is not dominated by any other solution. Considering that which preoccupies most the CEOs of electric utilities is aging infrastructures and reliability, we chose system reliability (SAIDI and SAIFI) [10] and investment in equipment and installation costs for the objective functions to be minimized.

SAIDI (System Average Interruption Duration Index) is considered one the most widely spread indices used by electrical utilities to measure reliability. The mathematical representation of this index is presented below:

$$SAIDI = \frac{\sum r_i N_i}{N_t} \tag{1}$$

where:

r_i is the average outage time per interruption of load point i due to outages in section s;

N_i is the number of clients in load point i;

N_t is the total number of clients in the network;

Another reliability system index that is used is the SAIFI (System Average Interruption Frequency Index). The difference between this index and the previous one (SAIDI), is that instead of considering the average interruption duration, in this index is considered the average interruption frequency. Equation (2) presents the mathematical representation of the SAIFI where λ_i is the average outage time per interruption of load point i due to outages in section s. Comparing equation (1) and (2) it is possible to verify that the difference variables r_i and λ_i. However r_i represents interruption time and λ_i represents interruption frequency, which gives a completely different concept.

$$SAIFI = \frac{\sum \lambda_i N_i}{N_t} \tag{2}$$

Beside the minimization of the reliability indexes SAIDI and SAIFI, a third minimization function is considered. This last function is the Total Costs (TC) in which is considered the investment in buying and installation of equipment cost associated with each solution for a given reliability (3).

$$TC = (Equipment\ Cost) + (Installation\ Cost) \tag{3}$$

where:

Equipment Cost is the cost of switches;

Installation Cost is the cost of the installations of switches in the network;

A radial electrical distribution networks characterized by several feeders coming out of a substation and its branches was considered in this work.

The GA used to generate the "Pareto Set" of feasible solutions uses a binary representation representing the place of the switches in each branch.

4   The NSGA-II Genetic Algorithm

The chosen Genetic Algorithm NSGA-II (Fast Non-dominated Sorting Genetic Algorithm) is an elitist multiobjective evolutionary algorithm (MOEA). This algorithm it is characterized by a fast nondominated sorting and by an efficient crowding-distance assignment approach. Due to its characteristics, it has been used in many applications and its performance tested in many comparative studies.

It consists on finding a vector "x" that minimizes a given set of "n" objective functions. The solutions are normally restricted by a series of constrains and consists on a set of results that satisfies the objective at an acceptable level without being dominated by any other solution. In this case, a solution is said to be Pareto Optimal solution and each solution cannot be improved with respect to any other objective without worsening at least one other objective.

The algorithm NSGA-II achieves the three conflicting goals expected for a GA, that is, the best known Pareto Front: 1) should be as close as possible to the true Pareto Front; 2) should be uniformly distributed and diverse; 3) should capture the all spectrum of the Pareto Front.

The general solving procedure of a GA is as following:

1. Generation of the initial N population, each of which is randomly initialized. Each solution of the population is called a Chromosome;
2. Generation of new Q solutions from existing ones, using two operators: Crossover and Mutation. In Crossover two chromosomes, called Parents are combined together to form a new chromosome, called offspring. The selection of the Parents considers the preference towards fitness values. Mutation introduces random changes and is applied at gene level. Typically the probability of this change is very small, less then 1%. Crossover leads to the convergence of the population and mutation reintroduces genetic diversity to the population.
3. Evaluation and assignment of a fitness value to each solution;
4. Selection of N solutions from the Q set of solutions based on their fitness;
5. Continuation of the anterior four steps until the stopping criteria is met.

The fitness assignment function of the algorithm uses the Crowding Distance approach. This concept aims to obtain a uniform spread of solutions along the best-known Pareto Front without using a fitness sharing parameter. This is done by: 1) ranking the population and identify non-dominated fronts; 2) for each objective function sort the solutions in the ascending order; 3) finding the total crowding distance of a solution. The main advantage of this approach is that a measure of population density around a solution is compounded without requiring a user-defined parameter. Considering two random solutions, if they are in the same non-dominated front, the solutions with a higher crowding distance wins, otherwise, the solution with the lowest rank is selected.

The problem of Elitism, (the best solution found so far during the search has immunity against selection and always survives in the next generation), is obtained by the NSGA-II algorithm, using a fixed size N population. In step 2 of the general solving procedure, the generation of the Q solutions, considers the offspring solutions and the parent solutions. The fitness assignment considers the two populations and this makes sure that all the non-dominated solutions are included in the next populations N and the selection based on the crowding distance will promote diversity.

Regarding the genetic encoding, a matrix with N lines and M columns is used. The lines of the matrix represent the population. The branches are numerated in a sequential form. Thus, the index of each column is directly related with the beginning of a branch. Inside on each element of the matrix it is used a binary codification, where 1 represents the allocation of a switch and 0 a branch without a switch in the beginning.

5   Case Study

In order to verify the effectiveness of the adopted algorithm, tests over a real case were implemented. Fig. 3 shows the 94-Bus Portuguese distribution feeder that was used in this work. This is a 15 kV radial network with 94-bus feeder.

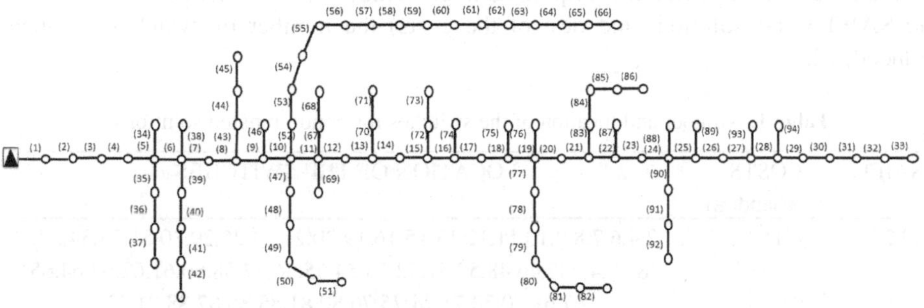

Fig. 2. Distribution network example for network representation

In order to evaluate the effectiveness of the proposed approach, two different tests have been made. SAIFI versus Cost was used as objective function in the first one. For the switch cost it was considered 2500 € and for the installation 1000 €.

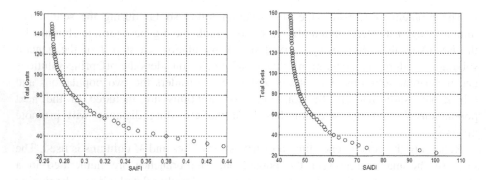

Fig. 3. Nondominated solutions obtained from de NSGA II algorithm considering as objective functions – Total Costs vs. SAIDI and Total Costs vs. SAIFI

The output of the NSGA II algorithm is presented in Fig. 4. In the figure the total costs is given in thousands of euros. As can be seen by the output of the algorithm, the final set of solutions is represented by the Pareto front. Each solution of the Pareto front represents a specific solution that is optimal in its own way. In fact, from these results it is possible to verify that when the costs improves the SAIFI decreases and vice-versa. The decision maker can select a particular solution from these multiple solutions. Table 1 shows the number and location of the switches for the best solution considering only SAIFI and the best solution considering only costs. As expected, for the SAIFI best solution it is required a huge number of switches, 60. However, for the SAIFI worst solution (and best of the costs) the number of switches is much reduced, 12.

Table 1. Number and location of the switches for nondominated solutions

SAIFI	COSTS (thousand(€))	LOCATION OF THE SWITCHES
0.27	150	2,4,6,7,8,9,10,11,12,13,15,16,19,20,21,23,25,29,30,31,33,34,37, 38,40,42,45,46,48,50,51,52,53,54,55,56,57,58,60,61,62,63,64,65 67,68,70,72,73,74,75,76,80,81,85,86,87,88,91,92
0.44	30	6,11,20,34,35,39,43,46,47,55,77,89

For the second test SAIDI versos cost was used as objective function. Fig. 4 shows the output of the algorithm using this new objective function. As expected when the cost improves the SAIDI decreases and vice-versa.

6   Conclusions

The allocation of switches in an electric distribution network represents an important measure to improve the power quality. Under this context, this paper presents a multi objective optimization technique to determine the optimal number and place of switches in an electrical distribution feeder. Reliability index SAIDI and SAIFI and cost of equipment's are the objective functions that are minimized to obtain the number and location of switches in an electrical distribution feeder. The algorithm used to find the optimal placement of switches is a Genetic Algorithm NSGA-II. The multi objective problem uses the Pareto optimal front to obtain the set of non dominated solutions. The adopted algorithm was applied to a 94-Bus Portuguese network to illustrate its performance. The number and location of switches in the distribution feeder have been determined in order to minimize the equipments cost and the reliability indexes SAIDI and SAIFI. This algorithm also shows the advantage of multiobjective optimization over single objective ones, since provides a superior balance between cost and reliability indexes. The proposed planning approach can be useful tool for a practical distribution system planning.

References

1. Logrono, D.O., Wu, W.-F., Lu, Y.-A.: Multi-Objective Optimal Placement of Automatic Line Switches. In: Proceedings of the Institute of Industrial Engineers Asian Conference, pp. 471–478 (2013)
2. Ma, L., Lv, X., Wang, S., Miyajima, H.: Optimal switch placement in distribution networks under different conditions using improved GA. In: Second International Conference on Computational Intelligence and Natural Computing Proceedings (CINC), pp. 236–239 (2010)
3. Barroso, G.C., Leao, R.P.S.: Switch placement algorithm for reducing customers outage impacts on radial distribution networks. In: TENCON - IEEE Region 10 Conference (2012)
4. Moradi, A., Fotuhi-Firuzabad, M.: Optimal Switch Placement in Distribution Systems Using Trinary Particle Swarm Optimization Algorithm. IEEE Transactions on Power Delivery 23, 271–279 (2008)
5. Teng, J., Liu, Y.: A novel ACS-Based Optimum Switch Relocation Method. IEEE Transactions on Power Systems 18, 113–120 (2003)
6. Tippachon, W., Rerkpreedapong, D.: Multiobjective Optimal Placement of Switches and Protective Devices in Electric Distribution Systems using Ant Colony Optimation. Electric Power System Research 79, 1171–1178 (2009)
7. Silva, L.G.W., Pereira, R.A.F., Mantovani, J.R.S.: Opimized Placement of Control and Protective Devices in Electric Distribution Systems Through Reactive Tabu Search Algorithm. Electric Power System Research 78, 372–381 (2008)
8. Goroohi Sardou, I., Banejad, M., Hooshmand, R., Dastfan, A.: Modified shuffled frog leaping algorithm for optimal switch placement in distribution automation system using a multi-objective fuzzy approach. IET Generation, Transmission & Distribution 6, 493–502 (2012)
9. Nascimento Alves, H.: A hybrid algorithm for optimal placement of switches devices in electric distribution systems. IEEE Latin America Transactions 10, 2118–2223 (2012)
10. Third Benchmarking Report on Quality of Electricity Supply – CEER - Council of European Energy Regulators. Technical report (2005)
11. Abido, M.A.: Environmental/economic power dispatch using multiobjective evolutionary algorithms. IEEE Transactions on Power Systems 18, 1529–1537 (2003)
12. Pires, D.F., Antunes, C.H., Martins, A.G.: NSGA-II with local search for a multi-objective reactive power compensation problem. International Journal of Electrical Power & Energy Systems 43, 313–324 (2012)

Stochastic Unit Commitment Problem with Security and Emissions Constraints

Rui Laia[1,2], Hugo M.I. Pousinho[1,2], Rui Melício[1,2], Victor M.F. Mendes[1,3], and Manuel Collares-Pereira[1]

[1] University of Évora, Évora, Portugal
{rui.j.laia,hpousinho}@gmail.com,
{ruimelicio,collarespereira}@uevora.pt
[2] IDMEC/LAETA, Instituto Superior Técnico, Universidade de Lisboa, Lisbon, Portugal
[3] Instituto Superior of Engenharia de Lisboa, Lisbon, Portugal
vfmendes@deea.isel.pt

Abstract. This paper presents a stochastic optimization-based approach for the unit commitment (UC) problem under uncertainty on a deregulated electricity market that includes day-ahead bidding and bilateral contracts. The market uncertainty is modeled via price scenarios so as to find the optimal schedule. An efficient mixed-integer linear program is proposed for the UC problem, considering not only operational constraints including security ones on units, but also emission allowance constraints. Emission allowances are used to mitigate carbon footprint during the operation of units. While security constraints settle on spinning reserve are used to provide reliable bidding strategies. Numerical results from a case study are presented to show the effectiveness of the approach.

Keywords: Emission allowances, stochastic optimization, security constraints, unit commitment.

1 Introduction

The UC is one of the most challenging optimization problems in power system operation, which has deserved an increasing interest due to today's energy shortage [1]. In a deregulated electricity market, the generation companies (GENCOs) operate under a high competition degree due to the nodal variations of electricity prices [2] in order to obtain the best profit bidding in the day-ahead market and bilateral contracting. So, the optimal schedule of the thermal units must consider the electricity prices uncertainty and other requirements, such as: technical operating constraints including security ones on the units and environmental constraints required to ensure admissible emission allowance levels. Multiple deterministic approaches have been proposed to solve the UC problem, which can be categorized into: priorities list, classical mathematical programming methods and intelligent methods [3]. Within the classical mathematical programming methods, mixed-integer linear programming (MILP) has been broadly applied for solving the UC problem [4], due to the suitable

L.M. Camarinha-Matos et al. (Eds.): DoCEIS 2014, IFIP AICT 423, pp. 388–397, 2014.

proprieties to guarantee global optimality and support the management decision. However, the literature about UC problem tends to address the problem without security and emission constraints uncertainty [5]. Security constraints provide reliable bidding strategies, for instance, spinning reserve levels recommended by the Union for the Coordination of the Transmission of Electricity [6]. Hence, this paper proposes a stochastic MILP approach to handle the electricity price uncertainty and solve the UC problem considering appropriate constraints to address a more realistic and feasible results for the management of thermal units in a competitive electricity market environment. A case study is presented for a schedule over a time horizon of 24 hours with hourly periods.

2   Relationship to Collective Awareness Systems

The technological evolution on electric power system encouraged by the expansion of distributed generation has been crucial to create collective awareness systems useful to define new energy consumption and production patterns. A collective awareness system can result from the development of powerful optimization approaches for the management of energy systems, helping to make decisions. The collective awareness system not only promotes the sustainable use of energy resources in favor of an effective low-carbon economy [7], but also processes the optimal decision. In order to achieve this optimal decision, collective tools are essential to provide real-time information on market data such as, the electricity prices, bilateral contracts and admissible emission allowance levels, allowing a GENCO to maximize the expected profit. Hence, research on technological innovation for collective tools based on approaches for solving the UC problem of a GENCO is crucial to achieve guidelines for the best bidding in an electricity market.

3   State of the Art

A review of literature describing the UC problem of a GENCO reveals that this problem has been treated in some way by avowing stochastic modeling, i.e., ignoring the random event on the electricity market [5]. Such treatment cannot provide a convenient level of precautions on the decision. This treatment is not appropriated, because in nowadays the most certainty thing for a GENCO is the uncertainty. Optimization methods for solving the UC problem have been addressed since the old priorities list method [8] to the classical mathematical programming methods until the more recently reported artificial intelligence methods [9]. Although, easy to implement and requiring a small computation time, the priority list method does not ensures an economic convenient solution near a global optimal one, implying a higher operation cost [8]. Within the classical methods are included dynamic programming (DP), linear programming, nonlinear programming and Lagrangian relaxation-based techniques [10]. DP methods are flexible but suffer from the "curse of dimensionality", due to the increase in the problem size related with the number of thermal units to be committed and the number of states considered for modeling the thermal behavior of each unit, implying an eventually huge use of computation memory and processing time. Although the Lagrangian relaxation [11] can overcome the previous limitation, does not always lead to a conveniently feasible solution, requiring in order to set a feasible solution the satisfaction of some violated constraints using heuristics, undermining the optimality. Artificial intelligence (AI) methods based on artificial neural networks [12], genetic algorithms [13], and evolutionary algorithms [9] have also been applied. However, the major limitation of the AI methods is the likelihood to obtain a convenient solution near global optimum, especially with a few thermal units. MILP has been applied with success for solving the UC problem [14]. Although, nonlinear constraints have to be converted into linear ones by piecewise linear approximation, MILP allows an easily inclusion of new constraints that makes the formulation of the problem more appropriated in order to conveniently support the management decision.

The literature about the UC problem of a GENCO tends to address the problem without security and emission constraints uncertainty. The emission constraints that can significantly affect the solutions of UC problem cannot be disregarded in the present context of the regulation. For instance, the authors in [5] address the problem of modeling the emission constraints into the UC, but with no uncertainty modeled. Hence, this paper as a contribution proposes a stochastic MILP approach to handle the electricity price uncertainty and solve the UC problem considering appropriate constraints, addressing a more realistic and feasible results for the management of thermal units in a competitive electricity market environment.

4   Problem Formulation

The UC problem can be stated as to find the schedule on status and the power generated for each thermal unit i at each time period t that optimizes performance criterion, involving market trading revenue and costs subject to a set of constraints on security, emissions and operation of the units.

4.1   Objective Function

The UC problem of a price-taker GENCO on price uncertainty has an objective function given by a measure of the expected profit attained by the sales of energy in a day-ahead market with bilateral contracts. The objective function to be maximized can be stated as:

$$\sum_{t=1}^{T}\sum_{m=1}^{M}\lambda_{mt}^{bc}\,p_{mt}^{bc} + \sum_{\omega=1}^{\Omega}\rho_{\omega}\left\{\sum_{t=1}^{T}\lambda_{\omega t}^{b}\,p_{\omega t}^{b} - \sum_{i=1}^{I}F_{\omega it}\right\} \tag{1}$$

The objective function in (1) is composed of two terms, namely: the revenue from selling through bilateral contracts between the GENCO and other market entities; the expected profit obtained by GENCO from selling its production in the day-ahead market minus the incurred operating costs. In (1), λ_{mt}^{bc} is the electricity price at period t for the bilateral contract m; p_{mt}^{bc} is the power at period t for the bilateral contract m; ρ_ω is the probability of occurrence for the scenario ω; $\lambda_{\omega t}^b$ is the electricity price at period t for the scenario ω; and $p_{\omega t}^b$ is the power to bid in the day-ahead market at period t for the scenario ω. At each period, the operating costs, $F_{\omega it}$ can be stated as:

$$F_{\omega it} = A_i u_{\omega it} + d_{\omega it} + b_{\omega it} + C_i z_{\omega it} \quad \forall \omega, \quad \forall i, \quad \forall t \tag{2}$$

The operating costs in (2) are composed of four terms, namely: the fixed cost, A_i, variable cost, $d_{\omega it}$, start-up cost, $b_{\omega it}$, and shut-down cost, C_i, of the units.

4.2   Constraints

The optimization problem is subject to a set of constraints due to the modeling. The modeling for the variable cost function by piecewise linear approximation introduces the constraints stated as:

$$d_{\omega it} = \sum_{l=1}^{L} F_i^l \delta_{\omega it}^l \quad \forall \omega, \quad \forall i, \quad \forall t \tag{3}$$

$$p_{\omega it} = p_i^{min} u_{\omega it} + \sum_{l=1}^{L} \delta_{\omega it}^l \quad \forall \omega, \quad \forall i, \quad \forall t \tag{4}$$

$$(T_i^1 - p_i^{min}) t_{\omega it}^1 \leq \delta_{\omega it}^1 \quad \forall \omega, \quad \forall i, \quad \forall t \tag{5}$$

$$\delta_{\omega it}^1 \leq (T_i^1 - p_i^{min}) u_{\omega it} \quad \forall \omega, \quad \forall i, \quad \forall t \tag{6}$$

$$(T_i^l - T_i^{l-1}) t_{\omega it}^l \leq \delta_{\omega it}^l \quad \forall \omega, \quad \forall i, \quad \forall t, \quad \forall l = 2,...,L-1 \tag{7}$$

$$\delta_{\omega it}^l \leq (T_i^l - T_i^{l-1}) t_{\omega it}^{l-1} \quad \forall \omega, \quad \forall i, \quad \forall t, \quad \forall l = 2,...,L-1 \tag{8}$$

$$0 \leq \delta_{\omega it}^L \leq (p_i^{max} - T_{\omega it}^{L-1}) t_{\omega it}^{L-1} \quad \forall \omega, \quad \forall i, \quad \forall t \tag{9}$$

In (4), the power generation of the unit i is given by the minimum power generation plus the sum of the power $\delta_{\omega it}^l$ associated with each segment l. The binary variable $u_{\omega it}$ ensures that the power generation is equal to 0 if the unit i is offline. In (5)–(9), the limits of the power generated in each segment are set. This power must be between zero and the maximum size of each segment. This is assured with a binary variable, $t_{\omega it}^l$, which is equal to 1 if the power generation of the unit at period t has exceeded segment l.

The modeling for the start-up cost is given by a stairwise linear approximation [15]. This linear approximation introduces the constraints stated as:

$$b_{\omega it} \geq K_i^{\beta}\left(u_{\omega it} - \sum_{r=1}^{\beta} u_{\omega it-r}\right) \quad \forall \omega, \quad \forall i, \quad \forall t \tag{10}$$

$$b_{\omega it} \geq 0 \quad \forall \omega, \quad \forall i, \quad \forall t \tag{11}$$

In (10), the expression in parentheses is equal to 1 if the unit i is online at period t and has been offline β preceding hours.

The modeling to limit the power of unit i introduces the constraints stated as:

$$p_i^{\min} u_{\omega it} \leq p_{\omega it} \leq p_{\omega it}^{\max} \quad \forall \omega, \quad \forall i, \quad \forall t \tag{12}$$

$$p_{\omega it}^{\max} \leq p_i^{\max}(u_{\omega it} - z_{\omega it+1}) + SD_i\, z_{\omega it+1} \quad \forall \omega, \quad \forall i, \quad \forall t \tag{13}$$

$$p_{\omega it}^{\max} \leq p_{\omega it-1}^{\max} + RU_i\, u_{\omega it-1} + SU_i\, y_{\omega it} \quad \forall \omega, \quad \forall i, \quad \forall t \tag{14}$$

$$p_{\omega it-1} - p_{\omega it} \leq RD_i\, u_{\omega it} + SD_i\, z_{\omega it} \quad \forall \omega, \quad \forall i, \quad \forall t \tag{15}$$

In (12), the generating limits of the units are set. In (13)–(15), the relation between the start-up and shut-down variables of the unit are given, using binary variables and their weights. In (13) and (14), the upper bound of $p_{\omega it}^{\max}$ is set, which is the maximum available power of the unit. These constraints involve: unit's actual capacity, start-up, SU_i, and shut-down, SD_i, ramp rate limits, and ramp-up, RU_i, limit. In (15), the ramp-down, RD_i, and shut-down ramp rate limits are considered.

The modeling for the minimum down time in a linear formulation introduces the constraints stated as:

$$\sum_{t=1}^{J_i} u_{\omega it} = 0 \quad \forall \omega, \quad \forall i \tag{16}$$

$$\sum_{t=k}^{k+DT_i-1}(1 - u_{\omega it}) \geq DT_i z_{\omega it} \quad \forall \omega, \quad \forall i, \quad \forall k = J_i + 1 \dots T - DT_i + 1 \tag{17}$$

$$\sum_{t=k}^{T}(1 - u_{\omega it} - z_{\omega it}) \geq 0 \quad \forall \omega, \quad \forall i, \quad \forall k = T - DT_i + 2 \dots T \tag{18}$$

$$J_i = \min\{T, (DT_i - s_{\omega i0})(1 - u_{\omega i0})\}$$

In (16)–(18), the minimum down time DT_i imposes that unit i have to be down by at least the minimum down time before startup.

The modeling for the minimum up time is introduced by linear constraints stated as:

$$\sum_{t=1}^{N_i}(1 - u_{\omega it}) = 0 \quad \forall \omega, \quad \forall i \tag{19}$$

$$\sum_{t=k}^{k+UT_i-1} u_{\omega it} \geq UT_i y_{\omega it} \quad \forall \omega, \quad \forall i, \quad \forall k = N_i+1\dots T - UT_i+1 \tag{20}$$

$$\sum_{t=k}^{T} (u_{\omega it} - z_{\omega it}) \geq 0 \quad \forall \omega, \quad \forall i, \quad \forall k = T - UT_i+2\dots T \tag{21}$$

$$N_i = \min\{T, (UT_i - U_{\omega i0}) u_{\omega i0}\}$$

In (19)–(21), the minimum up time UT_i imposes that unit i have to be on by at least the minimum up time before shutdown.

The relation between the binary variables to identify start-up, shutdown and prohibited operating zones is stated as:

$$y_{\omega it} - z_{\omega it} = u_{\omega it} - u_{\omega it-1} \quad \forall \omega, \quad \forall i, \quad \forall t \tag{22}$$

$$y_{\omega it} + z_{\omega it} \leq 1 \quad \forall \omega, \quad \forall i, \quad \forall t \tag{23}$$

The modeling for the emission function, $E_{\omega it}$, used to quantify the emission of a thermal unit, introduces the constraint stated as:

$$E_{\omega it} = Ae_i u_{\omega it} + \sum_{r=1}^{R} Fe_i^r \delta e_{\omega it}^r \quad \forall \omega, \quad \forall i, \quad \forall t \tag{24}$$

The modeling for the emission allowance over the time horizon introduces the constraint stated as:

$$\sum_{i=1}^{I}\sum_{t=1}^{T} E_{\omega it} \leq EMS \quad \forall \omega \tag{25}$$

In (25), the sum of the emissions associated with the committed units during the time horizon is set to a value not greater than the total emission allowance, EMS.

In order to ensure that GENCO is capable to accomplish their agreements, the balance of trade constraint can be stated as:

$$\sum_{i=1}^{I} P_{\omega it} = p_{\omega t}^b + \sum_{m=1}^{M} p_{mt}^{bc} \quad \forall \omega, \quad \forall t \tag{26}$$

In (26), the total power generated by the units is given by the total contracted power plus the day-ahead market power in each period.

The system security is included considering the European network codes [6], which recommend a spinning reserve level available at all times to cover eventual contingencies.

Hence, the spinning reserve, $SR_{\omega t}$, necessary to ensure reliability, is stated as:

$$SR_{\omega t} + \sum_{i=1}^{I} P_{\omega it} \leq \sum_{i=1}^{I} p_{\omega it}^{\max} \quad \forall \omega, \quad \forall t \tag{27}$$

5   Case Study

The proposed stochastic MILP approach has been tested on a representative case study based on a GENCO with ten thermal units. The time horizon considered is one day, divided into 24 hourly intervals. The approach Wavelet-Neuro-Fuzzy hybrid [16] is elected to generate 30 price scenarios, using historical data from the Iberian electricity market. The GENCO has to satisfy a bilateral contract with power and prices at each period. The price scenarios and the bilateral contract are shown in Fig. 1.

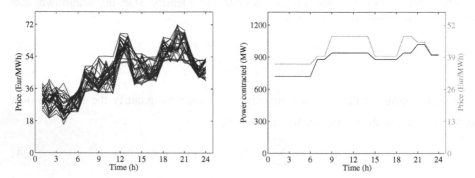

Fig. 1. Price scenarios (*left*) and power contracted and prices for bilateral contract (*right*)

The ten thermal units data, the minimum and maximum power, ramp up/down values, start-up and shut-down ramp rate values, minimum up/down time, fixed and shut-down costs are shown in Table 1.

Table 1. Thermal units' data

Unit	p_i^{min} (MW)	p_i^{max} (MW)	RU_i (MW)	RD_i (MW)	SU_i (MW)	SD_i (MW)	UT_i (h)	DT_i (h)	A_i (Eur/h)	C_i (Eur/h)
U1	45	85	35	35	60	70	8	3	2450	100
U2	70	125	45	40	100	95	5	4	2900	170
U3	110	160	60	50	125	140	8	4	3150	215
U4	60	125	55	55	90	80	5	3	3060	120
U5	90	170	40	60	100	100	6	3	2995	155
U6	90	170	40	60	100	100	6	3	2995	155
U7	80	145	35	40	90	105	9	6	3225	120
U8	145	215	45	70	160	170	6	4	3810	110
U9	200	380	60	50	230	250	10	6	4235	160
U10	220	330	70	60	230	245	10	6	4490	135

The variable costs of the thermal units have been modeled through piecewise linear approximations with three segments. The start-up costs are modeled through stairwise approximations with ten intervals. The thermal unit data associated with both models are available at [17].

The optimal hourly power generation for EMS = 200 tons and EMS = 300 tons for scenario 23 (#23) is shown in Fig. 2.

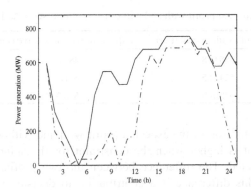

Fig. 2. Bidding power #23: EMS = 200 tons (*dashed–dotted line*), EMS = 300 tons (*solid line*)

In Fig. 2, a comparison between EMS = 200 tons and EMS = 300 tons, as expected, shows that the thermal system tends to operate at a high production level when a high emission allowance level is available. This operation is a compromise between the economic favoring and the level of the emission allowance in order to optimize the decision: when the emission allowance level is at a high value the production tends to follow the hourly price; otherwise, the production tends to be allocated in the hour with higher prices.

The hourly UC for the units at profiles EMS = 200 tons and EMS = 300 tons for #23 are shown in Fig. 3.

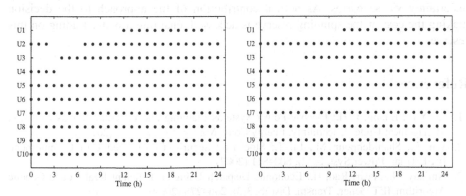

Fig. 3. Schedules #23: EMS = 200 tons (*left*), EMS = 300 tons (*right*)

In Fig. 3 is shown how the UC problem based on maximization of the expected profit is affected by the emission constraints. The UC of units U1, U2, U3 and U4 are affected by the different emission allowance levels. The number of units committed at each hour increases as the emission allowance level increases, as was expected.

The expected profit with and without spinning reserve for different emission allowance levels are shown in Table 2.

Table 2. Expected profit with and without spinning reserve and CPU time

	Exp. profit with reserve (Eur)	Exp. profit without reserve (Eur)	CPU time (min)
EMS = 200	485,988	487,086	5.53
EMS = 250	524,355	531,325	6.22
EMS = 300	528,168	536,310	3.32

In Table 2, a comparison of the expected profit with and without spinning reserve shows that the profit is higher when the constraint for the spinning reserve is not considered, as expected. The difference between the two profits is the cost of the spinning reserve. This difference is a contribution to decision-making in order to value this reserve.

6   Conclusions

A stochastic MILP approach for solving the UC problem of a price-taker thermal and emission constrained GENCO is presented, giving as main results: the short-term bidding strategies and the optimal schedule of the thermal units for different emission allowance levels. The proposed stochastic MILP approach proved both to be accurate and computationally acceptable, since the computation time scales up linearly with number of price scenarios, units and hours on the time horizon. This computational acceptance is due to the stochastic MILP being suitable to address parameter uncertainty via scenarios. As a final contribution of the approach to the decision making the cost of the spinning reserve is accessed, enabling a better trading on this reserve.

References

1. Li, Y.P., Huang, G.H.: Electric-Power Systems Planning and Greenhouse-Gas Emission Management Under Uncertainty. Energy Conv. Manag. 57, 173–182 (2012)
2. Wu, L., Shahidehpour, M., Li, T.: Stochastic Security-Constrained Unit Commitment. IEEE Trans. Power Syst. 22(2), 800–811 (2007)
3. Amjady, N., Nasiri-Rad, H.: Economic Dispatch Using an Efficient Real Coded Genetic Algorithm. IET Gener. Transm. Distrib. 3(3), 266–278 (2009)
4. Morales-España, G., Latorre, J.M., Ramos, A.: Tight and Compact MILP Formulation of Start-Up and Shut-Down Ramping in Unit Commitment. IEEE Trans. Power Syst. 28(2), 1288–1296 (2013)

5. Yamin, H., Shahidehpour, M.: Self-Scheduling and Energy Competitive Electricity Markets. Electr. Power Syst. Res. 71(3), 203–209 (2004)
6. UCTE. Operation handbook: G3 Recommended Secondary Control Reserve (2004)
7. Collective Awareness Platforms for Sustainability and Social Innovation, http://ec.europa.eu/digital-agenda/collectiveawareness (accessed October 2013)
8. Senjyu, T., Shimabukuro, K., Uezato, K., Funabashi, T.: A Fast Technique for Unit Commitment Problem by Extended Priority List. IEEE Trans. Power Syst. 18, 882–888 (2003)
9. Dhillon, J.S., Kothari, D.P.: Economic-Emission Load Dispatch Using Binary Successive Approximation-Based Evolutionary Search. IET Gener. Trans. Distrib. 3(1), 1–16 (2009)
10. Chandrasekaran, K., Hemamalini, S., Simon, S.P., Padhy, N.: Thermal UC Using Binary/Real Coded Artificial Bee Colony Algorithm. Electr. Power Syst. Res. 84(1), 109–119 (2012)
11. Dieua, V.N., Ongsakul, W.: Augmented Lagrange Hopfield Network Based Lagrangian Relaxation for Unit Commitment. Int. J. Electr. Power Energy Syst. 33(3), 522–530 (2011)
12. Ouyang, Z., Shahidehpour, M.: A Hybrid Artificial Neural Network-Dynamic Programming Approach to Unit Commitment. IEEE Trans. Power Syst. 7(1), 236–242 (1992)
13. Kumar, V.S., Mohan, M.R.: Solution to Security Constrained Unit Commitment Problem Using Genetic Algorithm. Electr. Power Energy Syst. 32(2), 117–125 (2010)
14. Carrion, M., Arroyo, J.M.: A Computationally Efficient Mixed-Integer Linear Formulation for the Thermal Unit Commitment Problem. IEEE Trans. Power Syst. 21, 1371–1378 (2006)
15. Ostrowski, J., Anjos, M.F., Vannelli, A.: Tight Mixed-Integer Linear Programming Formulations for the Unit Commitment Problem. IEEE Trans. Power Syst. 27, 39–46 (2012)
16. Catalão, J.P.S., Pousinho, H.M.I., Mendes, V.M.F.: Short-Term Electricity Prices Forecasting in a Competitive Market by a Hybrid Intelligent Approach. Energy Conv. Manag. 52(2), 1061–1065 (2011)
17. Thermal Units Data, http://hmi-21.wix.com/thermalunits (accessed October 2013)

Part XV

Operation Issues in Energy - I

Operation Modes of Battery Chargers
for Electric Vehicles in the Future Smart Grids

Vítor Monteiro, João C. Ferreira, and João L. Afonso

Centro Algoritmi – University of Minho – Guimarães, Portugal
{vitor.monteiro,joao.ferreira,
joao.l.afonso}@algoritmi.uminho.pt

Abstract. This paper presents an on-board bidirectional battery charger for Electric Vehicles (EVs), which operates in three different modes: Grid-to-Vehicle (G2V), Vehicle-to-Grid (V2G), and Vehicle-to-Home (V2H). Through these three operation modes, using bidirectional communications based on Information and Communication Technologies (ICT), it will be possible to exchange data between the EV driver and the future smart grids. This collaboration with the smart grids will strengthen the collective awareness systems, contributing to solve and organize issues related with energy resources and power grids. This paper presents the preliminary studies that results from a PhD work related with bidirectional battery chargers for EVs. Thus, in this paper is described the topology of the on-board bidirectional battery charger and the control algorithms for the three operation modes. To validate the topology it was developed a laboratory prototype, and were obtained experimental results for the three operation modes.

Keywords: Battery Charger, Grid to Vehicle (G2V), Vehicle to Grid (V2G), Vehicle to Home (V2H), Electric Vehicles, Smart Grids.

1 Introduction

Nowadays, the electric mobility is the main alternative to the traditional transportation system in order to reduce the greenhouse gases emission and to help to address environmental and energy issues [1]. It is also important to reduce the oil consumption in this sector [2], where it is predictable that 55% of the total oil consumption in the world will be allotted by this sector in 2030 [3]. This new paradigm has been supported by the several alternatives that are already available, mainly Electric Vehicles (EVs), as the Nissan Leaf. Nevertheless, the impact of the electric mobility cannot be neglected [4], [5], [6]. The uncontrolled EVs proliferation, which represents extra loads to the power grids, can worsen some power quality problems, as power losses and high values of Total Harmonic Distortion (THD) in the currents and voltages [7]. These problems are caused by the EV battery chargers that are implemented with static power converters. Considering this scenario, it is extremely necessary equip the EVs with battery chargers with sinusoidal current consumption and controlled power factor, aiming to preserve the power quality [8], [9].

L.M. Camarinha-Matos et al. (Eds.): DoCEIS 2014, IFIP AICT 423, pp. 401–408, 2014.

Currently, the energy required to the battery charging process follows from the power grid to the EVs. This operation mode is identified in the literature as Grid-to-Vehicle (G2V) [10]. Nevertheless, considering that the energy can follow in opposite sense, arises the operation mode Vehicle-to-Grid (V2G) [11]. Taking into account the future smart grids, where the EVs will be connected to a collaborative broker [12], the aforementioned operation modes can be used to help the power grid, especially for load-shedding and compensation of renewable production intermittency (providing both backup and storage). Besides the G2V and V2G operation modes, the energy stored in the batteries can also be used to feed other loads, typical at home. This operation mode is denominated as Vehicle-to-Home (V2H) [13]. The selection of each one of the G2V, V2G and V2H operation modes, will be optimized taking into account the EV driver profile and benefits, and the power grid capabilities. These operation modes are further described in detail. Fig. 1 illustrates the operation modes of the EVs in the future smart grids. As shown in this figure, the G2V and V2G operation modes can be performed at private (homes) or public places, and the V2H is performed at private (homes) places.

This paper results from a preliminary studies conducted in the first year of a PhD work, and intents contribute to the technological innovation of the electric mobility in smart grids. It is presented an on-board bidirectional battery charger with respective control algorithms, which enables the G2V, V2G and V2H operation modes. In order to evaluate these operation modes are presented some experimental results.

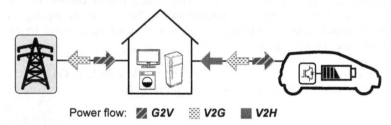

Power flow: ▨ *G2V*   ▧ *V2G*   ■ *V2H*

Fig. 1. Representation of Electric Vehicle with G2V, V2G and V2H operation modes

2   Relationship to Collective Awareness Systems

The future smart grids will bring a set of advantages to the end-user. This is more relevant taking into account that it involves the introduction of several technologies aiming to establish a bidirectional communication between the power grid collaborative broker and the users. Knowing that the electric mobility is a topic of highest importance in smart grids, the end-user will have the opportunity to participate actively in the energy market. Thereby, to the smart grids will be possible predict the energy demand and control the energy production [14], [15]. This interactivity, including supervision, control and communication applications, will be the final topic that will be addressed in the PhD work. For such purpose it will be developed an Information and Communication Technologies (ICT) application.

The main goal of the G2V, V2G and V2H operation modes that are described in this paper is establish a bidirectional flux of energy and information between the EVs and the power grids. This collaboration will strengthen the collective awareness systems, contributing to solve and organize issues related with the energy resources and power grids.

3   System Architecture and Operation Modes

The on-board bidirectional battery charger that is presented in this paper is composed by two power converters, one ac-dc and other dc-dc. The ac-dc converter is a full-bridge bidirectional converter that works in three distinct ways according to the operation mode. The dc-dc converter is a buck-boost bidirectional converter. Fig. 2 shows the electric diagram of the on-board bidirectional battery charger.

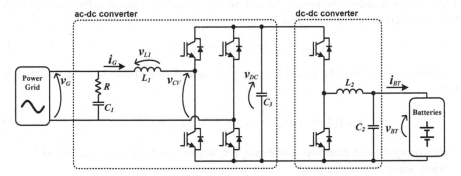

Fig. 2. On-board bidirectional battery charger composed by two power converters: ac-dc full-bridge bidirectional converter and dc-dc buck-boost bidirectional converter

3.1   Grid-to-Vehicle (G2V) Operation Mode

During the G2V operation mode the energy flows from the power grid to the batteries. In this operation mode the ac-dc full-bridge bidirectional converter works as active rectifier with sinusoidal current consumption and unitary power factor. The dc-dc buck-boost bidirectional converter works as buck converter aiming to charge the batteries with different stages of current and voltages. In this operation mode, the current reference (i_G^*) is obtained through:

$$i_G^* = \frac{P_{DC}^* + i_{TB}v_{TB}}{V_G^2} \, v_G \, ,\tag{1}$$

where, V_G is the RMS value of the power grid voltage, v_G the instantaneous value of the power grid voltage, i_{TB} and v_{TB} the current and voltage in the batteries, and P_{DC}^* is the power reference obtained through the dc-link voltage. The dc-link voltage is controlled by a Proportional-Integral (PI) control:

$$P_{DC}^* = k_p(v_{DC}^* - \overline{v_{DC}}) + k_i \int (v_{DC}^* - \overline{v_{DC}})dt \, ,\tag{2}$$

where, $v_{DC}{}^*$ is the dc-link voltage reference and $\overline{v_{DC}}$ the average value of the dc-link voltage during one cycle. In this controller is used the $\overline{v_{DC}}$ instead of v_{DC} because it avoid introduce the dc-link voltage oscillation into the control. For such purpose, $\overline{v_{DC}}$ in discrete samples is obtained by:

$$\overline{v_{DC}}[n] = \frac{1}{T} \sum_{n=1}^{T} v_{DC}[n] , \tag{3}$$

where, T corresponds to one cycle of the power grid voltage (50 Hz), and taking into account that the sampling frequency is 40 kHz in digital control corresponds to 800 samples. Analyzing the circuit presented in Fig. 2 it can be established:

$$v_G(t) = v_{L_1}(t) + v_{CV}(t) , \tag{4}$$

where, $v_G(t)$, $v_{L_1}(t)$ and $v_{CV}(t)$ are, respectively, the instantaneous values of the power grid voltage, inductance voltage, and voltage produce by the ac-dc converter. With the instantaneous values of the current $i_G(t)$, and the current reference ($i_G{}^*$) can be determined the current error by:

$$i_{Gerror}(t) = i_G{}^*(t) - i_G(t) . \tag{5}$$

Combining (4) and (5), and substituting the inductance voltage can be established that:

$$v_{CV}(t) = v_G(t) - L_1 \frac{i_G{}^*(t)}{dt} + L_1 \frac{i_{Gerror}(t)}{dt} . \tag{6}$$

Rewriting (6) in terms of discrete samples, where k is the actual sample and k-1 the previous sample, it is obtained:

$$v_{CV}[k] = v_G[k] - \frac{L_1}{T} (2i_G{}^*[k] - i_G{}^*[k-1] - i_G[k] - i_{Gerror}[k-1]) . \tag{7}$$

With this voltage (v_{CV}) is implemented a predictive current control. With the help of a unipolar sinusoidal Pulse Width Modulation (PWM) strategy with a 20 kHz are calculated the gate pulse patterns. In order to avoid the effects of the deadtime in the produced current, the voltage reference ($v_{pwm}{}^*[k]$) that is compared with the triangular carrier is given, at each k sample, by:

$$v_{pwm}{}^*[k] = v_{CV}[k] + u\Delta v , \tag{8}$$

where, Δv is the voltage that is added to the voltage $v_{CV}[k]$, and u assumes the value 1 when the reference current ($i_G{}^*$) is greater than zero, and -1 when is lesser than zero. To the dc-dc bidirectional converter, which works as buck converter during this operation mode, the voltage reference that the converter must produce during the constant current ($v_{pwm\_CI}{}^*$) and constant voltage ($v_{pwm\_CV}{}^*$) stages is given by PI controllers and are, respectively:

$$v_{pwm\_CI}{}^* = k_p(i_{BT}{}^* - i_{BT}) + k_i \int (i_{BT}{}^* - i_{BT})dt \tag{9}$$

$$v_{pwm\_CV}{}^* = k_p(v_{BT}{}^* - v_{BT}) + k_i \int (v_{BT}{}^* - v_{BT})dt \tag{10}$$

3.2   Vehicle-to-Grid (V2G) Operation Mode

During the V2G operation mode the energy flows from the batteries to the power grid. In this operation mode the ac-dc full-bridge bidirectional converter works also as an inverter, however, with sinusoidal current injection. The dc-dc buck-boost bidirectional converter works as boost converter aiming to discharge the batteries with constant power (as presented in this paper, however, can also be constant current). In this operation mode, the ac-dc converter control is similar to the one used in the G2V operation mode. To synthesize the reference current it was also used the predictive current control. The main difference is the dc-dc operation, which, in this case, operates as boost converter. The power to be delivered to the power grid is established as external input parameter. In the context of this paper, the value of the power is received through a serial port, however, it is predictable that it will be received through a wireless communication aiming to enable the collaborative integration of the EV in the future smart grids. With the reference power ($P_{BT}{}^*$) and the batteries voltage (v_{BT}) is determined the current reference ($i_{BT}{}^*$):

$$i_{BT}{}^* = \frac{P_{BT}{}^*}{v_{BT}}, \tag{11}$$

and the voltage reference ($v_{pwm\_CP}{}^*$) that the converter must produce is obtained by:

$$v_{pwm\_CP}{}^* = k_p(i_{BT}{}^* - i_{BT}) + k_i \int (i_{BT}{}^* - i_{BT})dt . \tag{12}$$

3.3   Vehicle-to-Home (V2H) Operation Mode

During the V2H operation mode the energy flows from the batteries to feed home loads during power outages. It also can be used to feed loads in places without connection to the power grid. In this operation mode the ac-dc full-bridge bidirectional converter works as inverter aiming to produce a true sine wave voltage output to feed the home loads. The dc-dc buck-boost bidirectional converter also works as boost converter aiming to increase the batteries voltage to an appropriated value to the operation of the ac-dc converter. In this operation mode the ac-dc converter works as voltage source inverter. Thus, analyzing the circuit presented in Fig. 2 it can be established:

$$v_{CV}(t) = v_G{}^*(t) - v_{L_1}(t), \tag{13}$$

where, $v_{CV}(t)$, $v_G{}^*(t)$, and $v_{L_1}(t)$ are, respectively, the instantaneous values of the voltage produce by the ac-dc converter, the reference voltage, and the inductance voltage. Substituting $v_{L_1}(t)$ it can established:

$$v_{CV}(t) = v_G{}^*(t) - L_1 \frac{di_{L_1}(t)}{dt}, \tag{14}$$

Also in this control was used the strategy presented to the G2V operation mode in order to avoid the effects of the deadtime.

In this operation mode the dc-dc converter operates as boost converter in order to maintain the dc-link voltage regulated to the proper functioning of the ac-dc converter. For such purpose, the dc-link voltage is controlled through a PI and the voltage reference ($v_{pwm\_DC}{}^*$) that the converter must produce is given by:

$$v_{pwm\_DC}{}^* = k_p(v_{DC}{}^* - v_{DC}) + k_i \int (v_{DC}{}^* - v_{DC})dt \tag{15}$$

4   Experimental Results

In order to evaluate the on-board bidirectional battery charger working in the G2V, V2G and V2H operation modes it was developed a laboratory prototype. Fig. 3 shows part of the developed prototype.

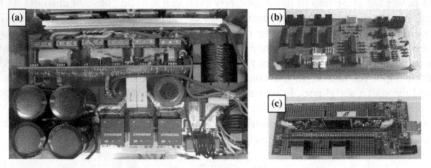

Fig. 3. Parts of the developed prototype of the on-board bidirectional battery charger: (a) Power converters; (b) Signal conditioning circuit; (c) Digital control platform

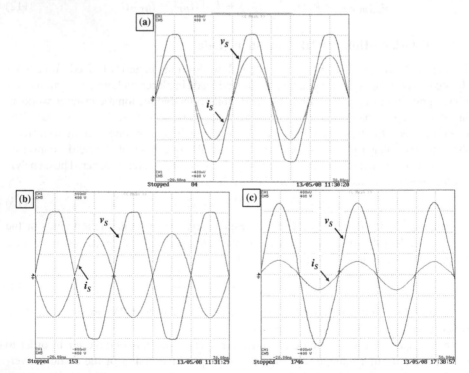

Fig. 4. Experimental results of the power grid voltage (v_G – 100 V/div) and current (i_G – 10 A/div): (a) G2V operation mode; (b) V2G operation mode; (c) V2H operation mode

The experimental results were obtained with the on-board bidirectional battery charger and with a set of 24 sealed 12 V 33 Ah Absorbed Glass Mat (AGM) batteries. Fig. 4 shows the obtained results. More specifically: Fig. 4 (a) shows the power grid voltage (v_G) and the current (i_G) during the G2V operation mode; Fig. 4 (b) presents the power grid voltage (v_G) and the current (i_G) during the V2G operation mode; and Fig. 4 (c) shows the output voltage (v_G) and the load current (i_G), during the V2H operation mode.

5   Conclusions and Further Work

This paper presents the development of an on-board bidirectional battery charger for Electric Vehicles (EVs), which can work in Grid-to-Vehicle (G2V), Vehicle-to-Grid (V2G), and Vehicle-to-Home (V2H) operation modes, targeting the future smart grids scenario. Taking into account that the introduction of EVs in smart grids will be an important subject, the bidirectional communication between EV drivers and the collaborative brokers of smart grids will strengthen the collective awareness systems. Therefore, it will contribute to solve and organize issues related with energy resources and power grids. The developed laboratory prototype, functioning in the three operation modes, was validated through experimental results.

Currently, the selection of the operation modes (G2V, V2G and V2H) is performed manually by the driver, through the user interface of the EV. Nevertheless, until the final of the PhD work where this paper is encompassed, this selection will be wireless controlled. Besides, it will be developed an Information and Communication Technologies (ICT) application aiming to allow the communication with a collaborative broker of the smart grids.

Acknowledgment. This work is financed by FEDER Funds, through the Operational Programme for Competitiveness Factors – COMPETE, and by National Funds through FCT – Foundation for Science and Technology of Portugal, under the project FCOMP-01-0124-FEDER-022674, and QREN project AAC n.º36/SI/2009 – 13844. Mr. Vítor Monteiro was supported by the doctoral scholarship SFRH/BD/80155/2011 granted by the FCT agency.

References

1. Boulanger, A.G., Chu, A.C., Maxx, S., Waltz, D.L.: Vehicle Electrification: Status and Issues. Proceedings of the IEEE 99(6), 1116–1138 (2011)
2. Camus, C., Farias, T.: Electric vehicles as a mean to reduce, energy, emissions and electricity costs. In: IEEE International Conference on the European Energy Market, pp. 1–8 (May 2012)
3. International Energy Outlook 2009. U.S. Department of Energy Washington DC (May 2009)
4. Dyke, K., Schofield, N., Barnes, M.: The Impact of Transport Electrification on Electrical Networks. IEEE Trans. Ind. Electron. 57, 3917–3926 (2010)

5. Jian, L., Xue, H., Xu, G., Zhu, X., Zhao, D., Shao, Z.Y.: Regulated Charging of Plug-in Hybrid Electric Vehicles for Minimizing Load Variance in Household Smart Micro-Grid. IEEE Trans. Ind. Electron. 60, 3218–3226 (2013)
6. Clement-Nyns, K., Haesen, E., Driesen, J.: The impact of charging Plug-In hybrid electric vehicles on a residential distribution grid. IEEE Transactions on Power Systems 25, 371–380 (2010)
7. Basu, M., Gaughan, K., Coyle, E.: Harmonic distortion caused by EV battery chargers in the distribution systems network and its remedy. UPEC International Universities Power Engineering Conference, pp. 869–873 (2004)
8. Monteiro, V., Ferreira, J.C., Meléndez, A.A.N., Afonso, J.L.: Electric Vehicles On-Board Battery Charger for the Future Smart Grids. In: Camarinha-Matos, L.M., Tomic, S., Graca, P. (eds.) Technological Innovation for the Internet of Things, ch. 38, 1st edn., pp. 351–358. Springer (2013)
9. Vítor Monteiro, H., Goncalves, J.L.: Afonso, "Impact of Electric Vehicles on power quality in a Smart Grid context. In: IEEE EPQU 11th International Conference on Electrical Power Quality and Utilisation, pp. 1–6 (2011)
10. Haghbin, S., Lundmark, S., Alaküla, M., Carlson, O.: Grid-Connected Integrated Battery Chargers in Vehicle Applications: Review and New Solution. IEEE Transactions on Industrial Electronics 60(2), 459–473 (2013)
11. Kramer, B., Chakraborty, S., Kroposki, B.: A review of plug-in vehicles and vehicle-to-grid capability. In: IECON 2008 - 34th Annual Conference of IEEE Industrial Electronics, pp. 2278–2283 (2008)
12. Ferreira, J.C., Santos, R., Monteiro, V., Afonso, J.L.: Cloud Collaborative Broker for Distributed Energy Resources. In: IEEE Iberian Conference on Information Systems and Technologies, Lisbon Portugal, June 19-22, vol. 1, pp. 33–40 (2013)
13. Green Car Congress, Nissan to launch the "LEAF to Home. V2H power supply system with Nichicon" EV Power Station (June), http://www.greencarcongress.com/2012/05/leafvsh-20120530.html
14. Meliopoulos, A., Cokkinides, G., Huang, R., Farantatos, E., Choi, S., Lee, Y., Yu, X.: Smart Grid Technologies for Autonomous Operation and Control. IEEE Transactions on Smart Grid 2(1) (March 2011)
15. Sahin, D., Kocak, T., Ergut, S., Buccella, C., Cecati, C., Hancke, G.P.: Smart Grid and Smart Homes: Key Players and Pilot Projects. IEEE Industrial Electronics Magazine 6, 18–34 (2012)

Power Outage Detection Methods for the Operation of a Shunt Active Power Filter as Energy Backup System

Bruno Exposto, J.G. Pinto, and João L. Afonso

Centro Algoritmi, University of Minho, Guimarães, Portugal
{bruno.exposto,gabriel.pinto,
joao.l.afonso}@algoritmi.uminho.pt

Abstract. This paper presents the study of power outage detection methods that can be applied to a Shunt Active Power Filter (SAPF) with energy backup capability. SAPFs can successfully compensate Power Quality problems related with distorted or unbalanced currents and low power factor. Future Smart Grids will combine devices, control strategies and functionalities to increase the grid reliability and the power management capability. One of the main tools necessary to enable these features is the information of what is occurring in all the smart grid parts. In this context the fast detection of power outages is critical, so this paper also contributes for the discussion of the best ways to extract information in the context of future smart grids. The combination of information and flexible devices in a smart grid will enable the implementation of collective awareness systems, which can deal with different electrical grid problems and situations in an organic manner.

Keywords: Shunt Active Power Filter, Energy Backup, Power Quality, Power Outage Detection, Kalman Filter.

1   Introduction

Smart Grids will become a reality in the future. Efforts are being made to pursue this objective in several countries of the world [1], [2]. The implementation of this new technology implies massive resources deployment, i.e., it must be implemented, study and regulated different parts that will constitute the smart grid. In the technical point of view, Smart Grids generically will integrate three major systems: infrastructure system, smart management system and smart protection system [2]. This technical global vision, although that can change according with the definition of each one point of view and work scope, provides a framework for what can be studied in the Smart Grid context. In terms of infrastructure it must be studied new ways of delivering energy, with the maximum quality, efficiency, reliability and controllability. To achieve these goals, it must be studied new types of power electronics devices, extend the existing power electronics devices capabilities and inherently study new types of control systems.

To achieve good Power Quality indexes, future Smart Grid probably will have to integrate different Power Quality Conditioners [3], [4]. Today are available several

L.M. Camarinha-Matos et al. (Eds.): DoCEIS 2014, IFIP AICT 423, pp. 409–416, 2014.

types of Power Quality Conditioners, namely shunt conditioners [5], series condition-ers [4], and the integrated approach known as Unified Power Quality Conditioners (UPQC) [4], [6]. Besides these types of Power Quality Conditioners, exist also in the literature, several references to other devices that have Power Quality compensation capabilities [7].

Shunt Active Power Filters (SAPFs) are shunt connected with the electrical power grid to compensate current problems. This type of Power Quality Conditioners usual-ly is designed to mitigate the current harmonics, the current unbalances, low power factor and the neutral currents [8].

SAPFs are generally constituted by an inverter that can be single-phase or three-phase, voltage-source or current source. There are more defining characteristics that can be applied to the inverters that are used in SAPF, but the focus in this work is three-phase voltage-source inverters [5].

In normal grid operation, the SAPF is compensating the load currents, but when a power outage occurs, the load and the SAPF will be disconnected due to lack of ener-gy. If the process that is being executed by the load is critical, the placement of an SAPF upstream of the load will not useful to deal with the situation of a power out-age. In this situation will be necessary to deploy backup equipment like Uninterrupti-ble Power Supplies (UPS). If the power line problems are outages, blackouts or losses of the utility line with time bigger than 10 ms, sags or dips, or even short un-der-voltages with lasting time less than 16 ms or even overvoltages with lasting time inferior to 16 ms, the Off-line UPS can be a good solution [9]. Considering that the power converters of a three phase offline UPS is very similar to the SAPF hardware, it would be advantageous combine the two functionalities in the same equipment.

Given this, in this paper is proposed a topology of SAPF that combines the func-tionalities of current compensation, when connected to the power grid, with the UPS functionalities, to feed the loads that are downstream in case of an power outage.

The combination of the two functions can result in a shorter payback time, so this solution has several advantages.

This paper presents some preliminary results from a PhD work related with the de-velopment of devices to the compensate Power Quality problems. The paper describes the hardware topology of the Shunt Active Power Filter, the control system as well as the energy backup control scheme. Also it describes power outage detection schemes. The behavior of the SAPF with energy backup capability is analyzed, and simulation results are presented.

2   Relationship to Collective Awareness Systems

Future Smart Grids will have several features: digitalization, flexibility, intelligence, resilience, sustainability, and customization [10]. Those combined features will allow extending the capabilities of the conventional electrical grids, to levels never thought before. The intelligence of Smart Grids will rely always on data. In fact, to reach col-lective awareness in a complex system is necessary have large data provided by a network of sensors and devices [11]. The Smart Grid will have these two providers of data associated with the capability of change in a real time the behavior of the devices. Given this, the study of methods to obtain information, at the same time that are used devices that have the capability to interact with the grid present several advantages in the future Smart Grids. If we consider that the topology of an off-line UPS has some touching points with the topology of the SAPF, combining the two features will result in a device that is capable of increasing the overall Power Quality at the same time that protects the load of power outages and other aforementioned events.

The collection of data, the increase of Power Quality in the grid and the change of behavior of the devices connected to the grid will be addressed in the PhD work so this is a good contribution to the Smart Grid issue and to the collective awareness systems.

3   Shunt Active Power Filter with Energy Backup

The topology of the SAPF is composed by a three-phase four wire shunt converter, with three IGBT legs and a split capacitor in the DC-link (Fig. 1). This topology was chosen because enables the compensation of the three-phases plus the neutral, using only six IGBTs. The main disadvantage is the difficulty in the correct regulation of the DC-link voltages i.e. the difficulty in maintain the voltage of the two capacitors (C_1 and C_2) equal. The power inverter is connected to the power grid trough three LCR passive output filters to eliminate the switching noise. In the DC-link of the SAPF is connected a reversible DC-DC power converter. This power converter provides an interface with the backup battery pack. When the SAPF is operating in a normal mode, the DC-DC converter charges the battery pack. If a power outage occurs and the energy backup mode is triggered, the DC-DC converter transfers energy from the battery pack to the DC-Link and insures that the voltage in this point is always equal to the predefined reference value.

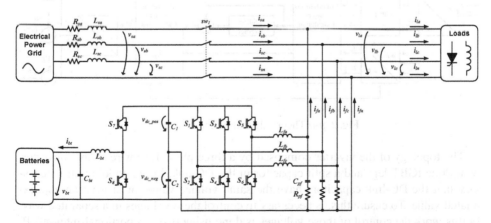

Fig. 1. Topology of the proposed Shunt Active Power Filter with energy backup

The control of the SAPF is divided in several independent algorithms interacting in an integrated way. To the operation as SAPF, is implemented a control algorithm based in the *p-q* Theory. Also it was implemented a digital phase locked loop (PLL), that is used in conjunction with the *p-q* Theory. The detection of the power outage sag or dips is done using a Kalman filter based RMS calculator. The control system is also in charge of the voltage and current control, in both of the operating modes of the SAPF, and the resynchronization after the normal power restoration.

3.1 Control System

In the SAPF mode the determination of the compensation currents is done using the *p-q* theory. This time domain approach has been largely used in SAPFs and Unified Power Quality Conditioners and has the advantages of requiring less processing capability [12], [13].

If the grid voltages are distorted, it is not possible generate compensating reference currents that aim to have sinusoidal source currents, because the *p-q* Theory will generate reference currents that when injected in the grid and in conjunction with the load currents, will ensure constant power at the source. In this case exists a modification of the *p-q* Theory that instead of constant instantaneous power at source, results in sinusoidal currents at the source, even with distorted power grid voltages. This modification consists in using the positive sequence of the fundamental voltages instead of the real voltages [13], [14]. To determine the positive sequence of the voltages, in this work was implemented a digital PLL in the *α-β* reference frame. In sum, the use of the digital PLL in conjunction with the control algorithm based in the *p-q* Theory results in sinusoidal source currents. In Fig. 2 is showed the implementation of the *p-q* Theory algorithm.

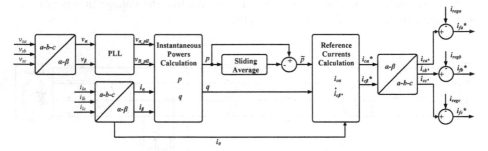

Fig. 2. *p-q* Theory algorithm implementation

The topology of the inverter composed by a three-phase four wire shunt converter, with three IGBT legs and a split capacitor in the DC-link. In this topology it's necessary that the DC-link capacitors have the same voltage value, or at least an approximated value. To ensure this, is necessary to control these voltages in a separated way. In this work the control of those voltages is done using two (proportional-integral) PI controllers (1). If, for example, $v_{a\_pll} \geq 0$ (calculated through the inverse Clark transform of v_{α_pll} and v_{β_pll}) the top capacitor voltage (V_{DC_pos}) is regulated using i_{rega}, otherwise is the regulated the bottom capacitor voltage (V_{DC_neg}). This is done for all the three-phases of the SAPF. In (2) and (3) is possible to see how is done this regulation. Finally, the reference compensation current generated by the p-q Theory algorithm (i_{cx}^*) and the regulation current (i_{rx}) are used to calculate the inverter reference current (3).

$$p_{reg\_x} = k_p (V_{DC\_x}^{*} - V_{DC\_x}) + \frac{1}{k_i} \int (V_{DC\_x}^{*} - V_{DC\_x})$$
(1)

$$i_{rx} = \begin{cases} p_{reg\_pos} \dfrac{v_{x\_pll}}{v_{a\_pll}^2 + v_{b\_pll}^2 + v_{c\_pll}^2} & , \quad v_{x\_pll} \geq 0 \\[4mm] p_{reg\_neg} \dfrac{v_{x\_pll}}{v_{a\_pll}^2 + v_{b\_pll}^2 + v_{c\_pll}^2} & , \quad v_{x\_pll} < 0 \end{cases}$$
(2)

$$i_{ref\_x} = i_{cx}^* - i_{rx}$$
(3)

3.2 Outage Detection Methods

In this work, the power outage detectors are responsible for evaluate the RMS values of the grid voltages and determine SAPF operation mode. If the system doesn't detect sag, swell or a power outage, the SAPF will operate in the normal mode compensating the load currents. Otherwise, the SAPF will stop the compensation of the loads and operate n backup mode.

The event detection algorithm must be as fast as possible so that the transfer time between normal mode and backup mode be very short. In *Moschitta, et al.* [15] the authors make a comparison of several sag and power outage detection methods. Their comparison showed that a Kalman filter can be used with success in the detection of sags and power outages decreasing the detection time. The authors propose and discuss other methods but the Kalman filter method has a good performance, is relatively simple to implement and is fast. In this work were implemented three Kalman filters, one per phase. The estimation of the RMS values of the voltages is done using the values obtained in the Kalman filters outputs through the following equation:

$$V_{RMS} = \sqrt{\frac{\hat{s}[n]^2 + \hat{q}[n]^2}{2}}.$$
(4)

Where $\hat{s}[n]$ and $\hat{q}[n]$ are the estimated values, of voltage and its quadrature respectively. Using the RMS values calculated using the Kalman filters the control system monitors the grid voltages continuously, to determine the operating mode of the SAPF. The control system of the SAPF changes its behavior in a situation where exists a sag higher than 20% of the nominal voltage, or a power outage. In this situation is activated the energy backup mode of the SAPF and the current control of the inverter is changed to voltage control (the reference voltages are generated using the PLL). Also the loads and the SAPF are isolated from the power grid.

After the restoration of the normal state of the grid, the control system resynchro-nizes with the grid and after this, reconnects the load to the grid and returns to the nor-mal mode. In Fig. 3 is possible to see the full operating diagram of the SAPF.

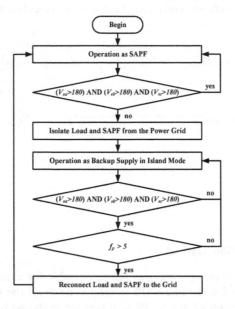

Fig. 3. Shunt Active Power Filter operation diagram

4   Simulation Models and Simulation Results

The simulation model is composed by the topology showed in Fig. 1. The simulations were performed using *PowerSim* PSIM software. In Table 1 is possible to see the simulation parameters values of the model.

Table 1. Shunt Active Power Filter simulation model parameters

Parameters	Values	Units
R_{sx}	10	mΩ
L_{sx}	100	μH
C_{fx}	25	μF
R_{fx}	4	Ω
C_x	10	mF
L_{bt}	2	mH
C_{bt}	1	mF

In the simulations, the SAPF starts to compensate the loads (Fig. 4 a) and is possi-ble to see that the source currents (i_{sx}) are sinusoidal and in phase with the voltages (v_{sx}). At t=0.5 s the source voltages are removed and then the SAPF acts as energy backup (Fig. 4 b) maintaining the load voltages (v_{lx}) with the nominal RMS value. The load stays some time being fed by the SAPF for a certain period of time. Then at t=0.8 s (Fig. 4 c) the source voltages are restored and the SAPF starts the resynchronization of the load voltages (v_{lx}) and the source voltages (v_{sx}). When the resynchronization is complete (t=0.815 s), the load and the SAPF is reconnected to the grid and the SAPF returns to normal mode.

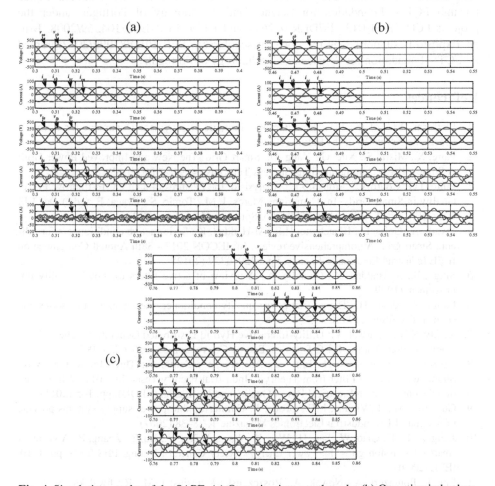

Fig. 4. Simulation results of the SAPF: (a) Operating in normal mode; (b) Operating in backup mode; (c) During transference to backup mode; (d) During transference to normal SAPF mode

5   Conclusions and Future Work

This paper presents a Shunt Active Power Filter (SAPF) that is able to compensate Power Quality problems and that also serves as energy backup system during power outages. The papers also discuss some methods of power outage detection. The simulation results show that the SAPF can successfully compensate the current harmonics, the current unbalances and the power factor. The simulation results also show that the SAPF can be used as energy backup system in case of a power outage. Also the power presented outage detection scheme performs correctly. Nevertheless further tests are needed, namely with different loads and power outage conditions.

Acknowledgment. This work is financed by FEDER Funds, through the Operational Programme for Competitiveness Factors – COMPETE, and by National Funds through FCT – Foundation for Science and Technology of Portugal, under the projects FCOMP-01-0124-FEDER-022674 and PTDC/EEA-EEL/104569/2008. Bruno Exposto is supported by the doctoral scholarship SFRH/BD/87999/2012 granted by FCT – Foundation for Science and Technology of Portugal.

References

1. Platform, A.C. of the E.T.: SmartGrids - Strategic Deployment Document for Europe's Electricity Networks of the Future (2010)
2. Fang, X., Misra, S., Xue, G., Yang, D.: Smart Grid, The New and Improved Power Grid: A Survey. IEEE Communications Surveys Tutorials 14, 944–980 (2012)
3. Pinto, J.G., Couto, C., Afonso, J.L.: Analysis of the Features of a UPQC to Improve Power Quality in Smart Grids. In: Camarinha-Matos, L.M., Tomic, S., Graça, P. (eds.) DoCEIS 2013. IFIP AICT, vol. 394, pp. 299–306. Springer, Heidelberg (2013)
4. Javadi, A., Geiss, N., Blanchette, H.F., Al-Haddad, K.: Series active conditionners for reliable Smart grid: A comprehensive review. In: IECON 2012 - 38th Annual Conference on IEEE Industrial Electronics Society, pp. 6320–6327 (2012)
5. Singh, B., Al-Haddad, K., Chandra, A.: A review of active filters for power quality improvement (1999)
6. Fujita, H., Akagi, H.: The unified power quality conditioner: the integration of series and shunt-active filters (1998)
7. Reddy, K.N., Agarwal, V.: Utility-Interactive Hybrid Distributed Generation Scheme With Compensation Feature. IEEE Trans. on Energy Conversion 22, 666–673 (2007)
8. Exposto, B., Gonçalves, H., Pinto, J., Afonso, J.L., Couto, C.: Three Phase Four Wire Shunt Active Power Filter from Theory to Industrial Facility Tests. In: 2011 11th International Conference on Electrical Power Quality and Utilisation (EPQU), pp. 1–5 (2011)
9. Gurrero, J.M., de Vicuna, L.G., Uceda, J.: Uninterruptible power supply systems provide protection. IEEE Industrial Electronics Magazine 1, 28–38 (2007)
10. Jiang, Z., Li, F., Qiao, W., Sun, H., Wan, H., Wang, J., Xia, Y., Xu, Z., Zhang, P.: A vision of smart transmission grids. In: Power Energy Society General Meeting, PES 2009, pp. 1–10. IEEE (2009)
11. Pitt, J., Bourazeri, A., Nowak, A., Roszczynska-Kurasinska, M., Rychwalska, A., Santiago, I.R., Sanchez, M.L., Florea, M., Sanduleac, M.: Transforming Big Data into Collective Awareness. Computer 46, 40–45 (2013)
12. Akagi, H., Watanabe, E.H., Aredes, M.: Instantaneous power theory and applications to power conditioning. Wiley-IEEE Press (2007)
13. Afonso, J.L., Couto, C., Martins, J.S.: Active filters with control based on the pq theory (2000)
14. Monteiro, L.F.C., Afonso, J.L., Pinto, J.G., Watanabe, E., Aredes, M., Akagi, H.: Compensation algorithms based on the pq and CPC theories for switching compensators in mi-crogrids. In: Power Electronics Conference, COBEP 2009, Brazilian, pp. 32–40 (2009)
15. Moschitta, A., Carbone, P., Muscas, C.: Performance Comparison of Advanced Techniques for Voltage Dip Detection. IEEE Transactions on Instrumentation and Measurement 61, 1494–1502 (2012)

AC Losses and Material Degradation Effects in a Superconducting Tape for SMES Applications

Nuno Amaro[1], Ján Šouc[2], Michal Vojenčiak[2], João Murta Pina[1],
João Martins[1], J.M. Ceballos[3], and Fedor Gömöry[2]

[1] Centre of Technology and Systems, Faculdade de Ciências e Tecnologia,
Universidade Nova de Lisboa, 2829-516 Caparica, Portugal
[2] Institute of Electrical Engineering, Slovak Academy of Sciences, Dúbravská cesta 9,
842 39 Bratislava, Slovak Republic
[3] "Benito Mahedero" Group of Electrical Applications of Superconductors,
Escuela de Ingenierías Industriales, Universidad de Extremadura, 06006 Badajoz, Spain
nma19730@campus.fct.unl.pt,
{eleksouc,michal.vojenciak,gomoelek}@savba.sk,
{jmmp,jf.martins}@fct.unl.pt, jmceba@unex.es

Abstract. Superconducting Magnetic Energy Storage (SMES) systems are one potential application of superconductivity in electric grids. The main element of such systems is a coil, made from superconducting tape. Although SMES systems work in DC conditions, due to highly dynamic working regimes required for some applications, AC currents can appear in the coil. It is then of utmost importance to verify the magnitude of these AC currents and take into account AC losses generated on the tape in the design phase of such system. To assure a proper operation, it is also necessary to know tape characteristics during the device lifetime, which in normal operation conditions can be of decades. Continuous thermal cycles and mechanical stresses to which the tape is subjected can change its characteristics, changing important quantities like critical current (I_C) and n-value. It is then also necessary to evaluate tape degradation due to these conditions. A study of AC losses will be here presented, for a short sample of BSCCO tape. I_C and n-value degradation due to consecutive thermal cycles will also be studied.

Keywords: Superconducting Magnetic Energy Storage, SMES, AC losses, HTS tape degradation.

1 Introduction

SMES systems are superconducting devices with several possible applications in electric grids and can contribute to the implementation of future's Smart Grids [1]. With such device it is possible to overcome several power quality issues, as voltage dips and swells, frequency oscillations or harmonic distortion [2]. This is possible because SMES systems store energy in a superconducting coil and can exchange active and reactive power, independently, with the electric grid where the system is placed. The main component of the system is a coil, made from high temperature

L.M. Camarinha-Matos et al. (Eds.): DoCEIS 2014, IFIP AICT 423, pp. 417–424, 2014.
© IFIP International Federation for Information Processing 2014

superconducting (HTS) tape. This particular work will be focused in first generation (1G) tape, built by BiSrCaCuO on its phase 2223 (BSCCO/Ag tape).

Although the coil operates in DC conditions, highly dynamic regimes to which it is subjected can lead to the appearance of an AC current. This means that in the project phase of an SMES system, it is necessary to take into account phenomena that usually occurs in AC conditions, like AC losses [3]. In fact, AC losses are one of the limiting factors for large-scale applications of superconductivity in power systems. For several years, AC losses have been an important object of study in superconductivity in both, single tape samples and coils. As a very small example one can see references [4–6]. The study of AC losses in BSCCO/Ag tapes in self-field conditions is then of utmost importance for applications that use this superconducting material, and must be addressed in the firsts steps of any project. This study is extremely important because in the design phase of an SMES cooling system, one has to consider the necessary power to extract from the system heat generated by those losses. Considering that there are several possible applications of superconductivity (and of SMES systems in particular), it is also necessary to verify the frequency dependence of such losses, in order to achieve an accurate value for total AC losses in the system [7].

Bearing in mind an envisaged application for SMES systems where a highly dynamic regime can introduce AC currents in the coil with frequencies up to few hundreds Hz, in this work, a study with a frequency range from 72 Hz to 576 Hz will be presented.

It is common to model high temperature superconductors using the power law [8]:

$$E = E_C \left(\frac{J}{J_C} \right)^n \tag{1}$$

Where E is electric field in the superconductor, E_C is the value of electric field in which the critical current density J_C is achieved (a usual criterion of $E_C = 1 \mu V / cm$ is used to define critical current). Parameter n is a material property and defines the shape of the E-J curve.

Operating conditions of superconducting devices in power systems subject HTS tape to various kinds of stress or strain. These can change tape characteristics, which will change important quantities like its critical current (I_C) or n-value. When subjected to bending and mechanical tension, the I_C value of a HTS tape will be degraded [9]. This fact should be taken into account in the design stage of the superconducting device, to assure a safe operation. To reduce mechanical vibrations in superconducting coils, it is common to impregnate HTS tape with epoxy. However, this impregnation may result in a higher tape heating, which can also decrease I_C [10].

In an SMES lifetime, the system can have many thermal cycles (cooling down and warming up the superconducting coil). These thermal cycles can also degrade the HTS tape, again leading to a lower I_c value [11] and possibly a different n-value. This effect will also be verified in this work.

2   Relationship to Collective Awareness Systems

Existing power grids are mainly unidirectional. However, in the electric grids of the future, energy is foreseen to flow up- and downstream. This "Smart Grid" concept involves both energy and information flows on the network [12]. This paradigm change transforms the power grid in a new grid, much more similar to information networks as internet. In the last years the concept of Internet of Things (IoT) in which every device is connected to the Web, is arising. Smart Grids represent a particular case of IoT, where reliability and security of supply are requirements of utmost importance, considering energy transmission and distribution. In this sense, power grids can become a Collective Awareness System, with distributed control and learning capabilities. Devices like SMES are foreseen to help the implementation of Smart Grids, thus being also part of these Collective Awareness Systems.

3   AC Losses in Superconducting Tapes

To verify the effects of AC losses in an SMES system it is first necessary to understand the origin of those losses and to classify them.

3.1   AC Losses Classification

Based on the physical mechanism that origins them and their frequency dependence, AC losses (power losses) in BSCCO/Ag tapes (and in HTS tapes in general) can be sorted into three different types [13]:

- *Superconducting hysteresis losses (P_h)*: linearly proportional to frequency, these appear due to variations in the superconductor magnetic state [14].

- *Eddy currents losses (P_{ed})*: due to the existence of silver in the BSCCO/Ag tape, there are Eddy currents through that metal, which also generates losses. These are quadratically proportional to frequency and, as demonstrated by Ishii et al. [15], can be expressed as:

$$P_{ed} = \frac{2\pi^2 \mu_0^2 f^2 I_p^2 d^3}{\rho l} \text{ (W/m)} \tag{2}$$

where μ_0 is vacuum permeability, f is frequency, I_p is current amplitude through the tape, d is sheath thickness (for this tape: 36 μm), ρ is resistivity of silver (here set as $3 \times 10^{-9} \, \Omega.m$) and l is perimeter of the outer superconducting filament layer (in this case: 6.5 mm).

- *Resistive losses (P_r)*: at high applied-to-critical current ratios (i), resistive losses appear in the superconductor. These are frequency independent and their value depends on the characteristics of the tape (mainly *n*-value). Usually they are negligible till i > 0.8 [13].

Total AC losses in the superconducting tape are then the sum of those three components. By dividing the power loss per meter (P) by frequency, one can achieve the energy loss per meter, per cycle (Q) as follows.

$$Q_t = \frac{P_t}{f} = \frac{P_h}{f} + \frac{P_{ed}}{f} + \frac{P_r}{f} = Q_h + Q_{ed} + Q_r \quad (\text{J/m}) \tag{3}$$

At frequencies below 200 Hz hysteresis energy losses (Q_h) are expected to have a main contribution. These are frequency independent. Eddy currents losses contribution for frequencies below this value are usually negligible, but they start to have a higher contribution to total losses with increasing frequencies. Resistive losses, as already stated, are frequency independent and depend on the used applied-to-critical current ratio. Their contribution to total losses is only visible for applied currents close to the tape I_C.

In this work, AC losses of a short sample of BSCCO/Ag tape, in which flows an AC current with frequencies ranging from 72 to 576 Hz, will be measured. The frequency dependence of these losses will be evaluated, especially concerning Eddy current losses, which are those most dependent of frequency.

To evaluate effects of thermal cycles in I_C and n-value, thus on operating conditions of the tape, after a first measurement of I_C, the sample is subjected to several of those cycles and critical current is measured again to obtain a comparison which can be very useful in real systems applications. By fitting all measured data to E-J power law (1) it will be possible to see if the n-value of the tape is also degraded.

4   Experimental Measurements of AC Losses

To evaluate tape degradation, the I_C of an HTS tape sample was measured prior and after several dozens of thermal cycles. AC losses were measured and compared to the well-known Norris ellipse model [14]. To evaluate frequency dependence, both total measured losses and calculated losses by taking out Eddy current component will be presented. It is expected that, after taking out the component related to Eddy currents, AC losses show no frequency dependence. InnoST insulated tape was used and its characteristics are shown in table 1.

Table 1. Characteristics of the SC tape, according to supplier

Characteristic	Value
Critical current @ 77 K, self-field (A)	85
Critical current density (SC) (A/mm^2)	8500
Minimum bending radius (mm)	30
Width (mm)	4.2
Thickness (mm)	0.25

4.1   Experimental Setup

To measure I_C (DC critical current) and AC losses, a four-point configuration was used. Figure 1 depicts the used sample with voltage taps 4 cm apart. AC losses were measured using an electrical method with a lock-in amplifier as main measuring device and a Rogowski coil. All measurements were made at liquid nitrogen (LN) temperature (77 K).

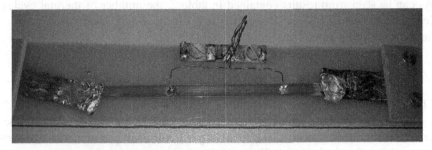

Fig. 1. HTS tape sample for measurements of critical current and AC losses

4.2   Measurements of Critical Current

DC critical current was measured for three different tape conditions:

- Unused sample (*case A*);
- After 25 thermal cycles (*case B*);
- After a total of 50 thermal cycles (*case C*).

For every thermal cycle the tape was cooled down by putting the tape directly into LN and then warmed up to room temperature (warm up by natural heating, just by taking the sample out of LN). All results are shown in figure 2.

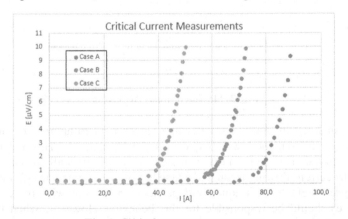

Fig. 2. Critical current measurements

To evaluate critical current and *n*-value degradation, experimental results obtained for the three previous cases were used to achieve fitting E-J curves, by using Matlab® cftool. Results are shown in table 2.

Table 2. Ic and n-value degradation due to thermal cycles

Number of cycles	Critical current [A]	n-value
0	77.02	15.35
25	59.6	11.73
50	37.9	8.44

As can be seen from results contained in table 2, the number of thermal cycles to which a superconducting tape is exposed degrade important characteristics like its critical current and n-value.

4.3   Measurements of AC Losses

AC losses were measured for four different values of frequency: 72 Hz, 144 Hz, 288 Hz, and 576 Hz. Figure 3 contains results for measurements made with the tape in situation B (after 25 thermal cycles). In this figure, Norris ellipse model is also shown for two values of critical current: 77 A and 60 A.

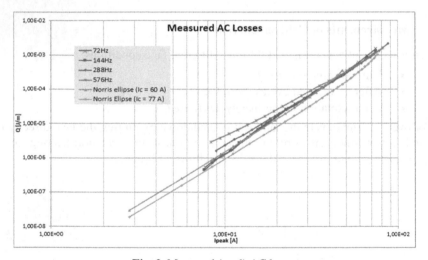

Fig. 3. Measured (total) AC losses

Results contained in figure 3 suggest that values tend to approach Norris ellipse model (for $I_c = 60$ A), for lower frequencies (72 Hz and 144 Hz) and for current values close to the critical current (for all frequencies). For small current amplitudes and when the current flowing through the tape has higher frequencies (288Hz and 576 Hz) AC losses seem to be frequency dependent, increasing with frequency. Eddy current losses should be responsible for this behavior. However, when the current increases resistive losses start to appear in the tape, making Eddy current losses negligible to the total losses value. So for currents whose peak is close to the critical current of the tape, there is no frequency dependence because resistive losses, which became the most important component in total losses (together with hysteresis losses) are frequency independent. As can be seen by the two Norris ellipse models depicted in the figure, for the same value of applied current, losses are higher if the tape has less critical current. This is an obvious result, since it means that for the same operating current, the tape is closer to its critical current if this value is lower. This also shows that AC losses in tapes with degraded critical current can still be modeled using Norris ellipse model, if the model is calculated considering the new value of critical current of that tape.

To verify if the frequency dependent behavior for small current amplitudes is due to Eddy current losses, this component was calculated using expressions (2), (3) and subtracted from the total value of losses. The results obtained are shown in figure 4.

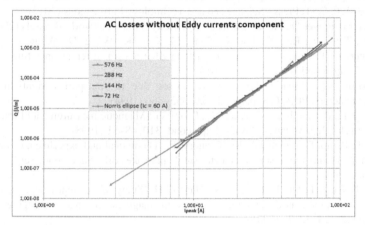

Fig. 4. AC losses after subtracting Eddy current losses component

As can be seen in figure 4, AC losses show a non-frequency dependent behavior, if Eddy current losses are not considered. Norris ellipse model for a critical current of 60 A is also shown in this figure, for easier comparison with the previous one.

5   Conclusions

Critical current and n-value degradation due to thermal cycles in an HTS BSCCO/Ag tape was studied. As shown by experimental results obtained, these important quantities in a superconducting tape are effectively degraded if the tape is subjected to a certain number of thermal cycles. In fact, by subjecting the tape to a total of 50 cycles, there is a reduction of 50% of critical current value and the n-value changes from 15.35 to 8.44. This phenomena is of extreme importance for superconducting devices like SMES, where operating current is calculated based on critical current of the tape. This means that during the lifetime of the device, due to tape degradation, the required power to extract heat losses is expected to increase (for the same operating current) and this aspect cannot be neglected in the design phase of an SMES project.

AC losses of a short sample of degraded BSCCO/Ag tape, in which there are currents flowing in a band of frequencies ranging from 72Hz to 576 Hz were also measured. Results indicate that AC losses in HTS degraded tapes can still be modeled using Norris ellipse model, if the critical current of the model is set to the effective critical current of the tape after degradation. AC losses seem to behave as frequency independent for the measured frequencies, if Eddy current losses are subtracted from the total value. As expected, Eddy current losses are frequency dependent and can be neglected for frequencies below 200 Hz.

References

1. Amaro, N., Murta Pina, J., Martins, J., Ceballos, J.M.: Superconducting Magnetic Energy Storage - A Technological Contribute to Smart Grid Concept Implementation. In: Proceedings of the 1st International Conference on Smart Grids and Green IT Systems, pp. 113–120. SciTePress - Science and and Technology Publications (2012)
2. Eurelectric: Power Quality in European Electricity Supply Network, Brussels (2003).
3. Xu, Y., Tang, Y., Ren, L., Jiao, F., Song, M., Cao, K., Wang, D., Wang, L., Dong, H.: Distribution of AC loss in a HTS magnet for SMES with different operating conditions. Phys. C Supercond. 494, 213–216 (2013)
4. Ashworth, S.P.: Measurements of AC losses due to transport currents in bismuth superconductors. Phys. C Supercond. 229, 355–360 (1994)
5. Gömöry, F., Bettinelli, D., Gherardi, L., Crotti, G.: Magnetic measurement of transport AC losses in Bi-2223/Ag tapes. Phys. C Supercond. 310, 48–51 (1998)
6. Šouc, J., Pardo, E., Vojenčiak, M., Gömöry, F.: Theoretical and experimental study of AC loss in high temperature superconductor single pancake coils. Supercond. Sci. Technol. 22, 15006 (2009)
7. Yuan, J., Fang, J., Qu, P., Shen, G.X., Han, Z.-H.: Study of frequency dependent AC loss in Bi-2223 tapes used for gradient coils in magnetic resonance imaging. Phys. C Supercond. 424, 72–78 (2005)
8. Rhyner, J.: Magnetic properties and AC-losses of superconductors with power law current—voltage characteristics. Phys. C Supercond. 212, 292–300 (1993)
9. Kim, H.-J., Kim, J.H., Cho, J.W., Sim, K.D., Kim, S., Oh, S.S., Kwag, D.S., Kim, H.J., Bae, J.H., Seong, K.C.: AC loss characteristics of Bi-2223 HTS tapes under bending. Phys. C Supercond. 445–448, 768–771 (2006)
10. Polák, M., Kvitkovic, J., Janšák, L., Mozola, P.: The effect of epoxy impregnation on the behaviour of a Bi-2223/Ag coil carrying DC or AC current. Supercond. Sci. Technol. 19, 256–262 (2006)
11. Chen, D.-X., Luo, X.-M., Fang, J.-G., Han, Z.-H.: Frequency dependent AC loss in degraded Bi-2223/Ag tape. Phys. C Supercond. 391, 75–78 (2003)
12. E.U., E. union: European SmartGrids Technology Platform - Vision and Strategy for Europe's Electricity Networks of the Future. Office for Official Publications of the European Communities, Brussels (2006).
13. Stavrev, S., Dutoit, B.: Frequency dependence of AC loss in Bi(2223)Ag-sheathed tapes. Phys. C Supercond. 310, 86–89 (1998)
14. Norris, W.T.: Calculation of hysteresis losses in hard superconductors carrying ac: isolated conductors and edges of thin sheets. J. Phys. D. Appl. Phys. 3, 489–507 (1970)
15. Ishii, H., Hirano, S., Hara, T., Fujikami, J., Sato, K.: The a.c. losses in (Bi,Pb)2Sr2Ca2Cu3Ox silver-sheathed superconducting wires. Cryogenics (Guildf) 36, 697–703 (1996)

Part XVI

Operation Issues in Energy - II

Active Power Filter with Relay Current Regulator and Common DC Link for Compensation of Harmonic Distortion in Power Grids

Maksim Maratovich Habibullin[1], Igor Sergeevich Pavlov[1],
Viktor Nikolaevich Mescheryakov[1], and Stanimir Valtchev[2]

[1] Lipetsk State Technical University, Moskovskaya 30, 398600 Lipetsk, Russia
maximuum@rambler.ru, mesherek@stu.lipetsk.ru
[2] Campus de Campolide, 1099-085 Lisboa, Portugal
ssv@fct.unl.pt

Abstract. The growing number of consumers representing non-linear loads has led to an increase in the level of harmonics and poor power quality. It is not news to apply active power filtering (APF), but the drawback is the lower efficiency and complexity. The described here active power filter control system is simpler and more efficient. It is based on a relay current regulator and its converter uses the existing DC bus of the rectifier. The theoretical basis of the APF is described. A model of the proposed APF (with its suggested control system and the common DC link) is simulated in Matlab. The result of this simulation demonstrates that the level of the Total Harmonic Distortion (THD) of the input current is less than 3 %. The input current matches the phase of the input (mains) voltage: i.e. the consumption of reactive power from the grid is minimized. The simulation is confirmed by experimental study of the relay current controller. The corresponding results are provided.

Keywords: Non-linear load, Quality of electric energy, Total harmonic distortion, Relay current control, DC-link, Active power filter.

1 Introduction

1.1 Power Quality

Less than twenty years ago the harmonic components of current and voltage in the electric alternating current were negligible and their impact on the grid could be ignored. Currently, in most countries the level of harmonic components in power grids has increased in several times. The tendency for further growth is here to stay [1]. The high level of harmonic components should be considered seriously because it has a negative impact both on the grid and its users.

The reason for the growth of the harmonic current and voltage is the growing number of electricity customers that use non-linear load, both in the industrial and domestic sectors. The main negative effect of the non-linear load consists in the specific nature of the harmonics in the consumed current from the grid. The harmonic components frequencies can be so high that they provoke electromagnetic disturbance

L.M. Camarinha-Matos et al. (Eds.): DoCEIS 2014, IFIP AICT 423, pp. 427–434, 2014.

but the low frequency harmonics are also a problem because their filtering requires large reactive elements. Currently, 60 % of the electrical consumption is due to the nonlinear load and now the requirements for power supply quality from the grid is one of the main tasks of the electrical energy industry [2].

The existence of higher harmonic contents in the grid current results in significant economic damages. These include the additional losses in rotating electrical machines, the loss in transformers, vibration in electric machine systems, single-phase ground short circuit, capacitor breakdowns, insulation aging, disruption and failure of high-precision measuring instruments, protective gears, control systems and microprocessor-based systems, interference and distortion of signals in telecommunication networks and systems. In addition, the nonlinear load has high level of reactive energy circulation; hence the losses in the energy transport system rise, as well as the overloading of generators, transformers and transmission lines. Oscillations of system voltage are then more probable. In general, the quality of the supplied energy goes downward [3].

1.2   Outlining the Research Question and Its Motivation

The harmonic components in the power network variables are produced in the consumer environment and they are originated by the load characteristics. This means that the consumers must themselves control the influence exercised by their equipment onto the grid.

The most characteristic behavior of the power grid working on active-inductive load has been modeled in Matlab in order to analyze the influence of a nonlinear load application. The above mentioned load is connected to a three-phase classical uncontrolled semiconductor bridge rectifier. The adopted for this simulation parameters, are: $U = 380$ V, $f = 50$ Hz, $R_L = 4.5$ Ohm, $L_L = 10$ mH, $C_d = 30$ mF. The simulation results are presented in Fig. 1 (a, b).

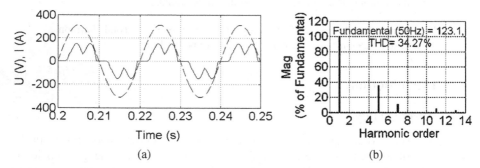

| (a) | (b) |

Fig. 1. (a) Waveforms of phase A current (solid line) and phase A voltage (dotted line); (b) Spectral contents of the phase A current, $THD_i = 34.27$.

As it is seen in Fig. 1 (a), the current and the voltage waveforms show a certain phase difference. The load requires reactive power from the supplying network and the THD_i is higher than normally is [4]. The usually applied equipment to correct the electric power quality parameters of the consumed power have been the passive power filters of high order harmonics, the compensating capacitor batteries and synchronous compensators. Based on the achievements in power electronics, it became possible to solve this issue by means of active power filters (APF) [5], [6], [7]. The APF that is currently applied consists of a voltage converter (with fully controlled power switches) and a capacitor battery in a dedicated DC link. However, the most non-linear consumers, having a rectifier incorporated, already possess a capacitor in their output DC link. It is suggested therefore that a new type of APF will use this already existing DC link: the converter of the filter is connected to the DC output voltage tank. This will reduce the number of circuit elements and simplify the control system of the APF [8]. The chosen control system is based on the relay (bang-bang) current controller. This control system is characterized by simplicity, stable performance, high speed and high accuracy [9], [10], [11].

The APF is supposed to operate as a step-up converter, i.e. the output voltage (the output filter's DC link voltage) will be always higher than the voltage applied to the APF converter (the damaged voltage that is necessary to correct). The main task in the development of this APF is the construction of the control system based on the balance of the power consumed by the nonlinear load from the AC supply and the power injected into the DC link. The other task is the control of the voltage level at the output of the APF converter in order to correspond to the consumed power.

2   Relationship to Collective Awareness Systems

The increase in the number and the power of non-linear loads is due to the rapid development of the modern electronics. From the point of view of the economy, the power electronics made the energy use more efficient. Unfortunately the electronic devices applied in the energy sector have led to quality problems of the power grid. Those problems our society needs to solve by the effort of both the energy producers and the electric power consumers.

In the industry, the modern static power converters and the frequency controllers for motor drives made possible by the high-power IGBT transistors and other emerging devices to facilitate the control of the induction motors, the cheapest and most common type of electric motors. The application of electronic devices brought certain effects in the general electric network: massively the controlled AC drives have been upgraded by introducing a frequency controlled converters in already existing power system and this distorted the consumed current.

In the domestic applications the switching power supplies increased dramatically in number. The switched power converter is highly efficient but the massive use of it created problems. For example, the millions of laptops, tablet PCs and mobile phones, the necessary servers, routers, etc., all of these electronic devices operate at a constant voltage supply, but they are connected to the AC mains. The converters from the AC to DC in the millions of devices and other equipment represent a non-linear load. As a final result, the quality parameters of the AC grid deteriorated.

In fact, it is not possible to avoid the non-linear load in the energy system. A promising solution to this problem is the use of APF in order to compensate the damage made to the grid parameters. This filtering is also a must for the interconnections between the numerous generators and consumers in the future smart grids. If a simple and reliable solution like the proposed one will be adopted, the manufacturers of electronic equipment will produce massively equipment that will be more easily made compact and cheap.

The consumers connected to the mains will be able to have their nonlinear influence compensated: the APF will be cheaper and more acceptable to be used in massive number to compensate for harmonic distortion. This will contribute to the development of Collective Awareness Systems, where the price of the nonlinear load compensation will be equally divided between the consumers and the producers of electric energy.

3   Theoretical Bases

The schematic diagram of the proposed APF is shown in Fig. 2 (a). A transformer is used as the matching element between the APF converter output (here it is an inverter) and the electric grid.

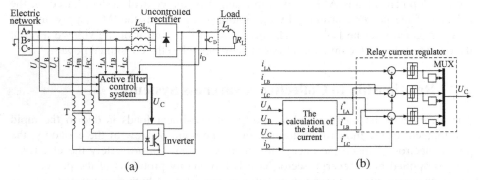

(a)   (b)

Fig. 2. APF: (a) its connection to the load and grid; (b) its control system

The operation of the suggested APF control system is planned to compensate for the difference ΔI_L between the real I_L and the ideal I_L^* load current that passes through the system in any instant of time [8]:

$$\Delta I_L = I_L^* - I_L . \tag{1}$$

The "Ideal" current, flowing from the mains, is defined by the power that the DC consumer requires.

The active power P_l at the input should correspond to that DC power, and P_l is:

$$P_l = \sum_{k=1}^{\infty} \sqrt{3} \cdot U_{k1} \cdot I_{kL} \cdot \cos \varphi_k , \tag{2}$$

where I_{kL} is the current drawn from the grid by the uncontrolled rectifier,

U_{k1} is the AC input voltage,

φ_k is the phase shift between the voltage and current (in fact, the first harmonic of the input current is the only harmonic contributing to the DC output power!).

The reactive power Q_l at the input is:

$$Q_l = \sum_{k=1}^{\infty} \sqrt{3} \cdot U_{k1} \cdot I_{kL} \cdot \sin \varphi_k . \tag{3}$$

The apparent power S_1 at the uncontrolled rectifier input is:

$$S_1 = \sqrt{P_1^2 + Q_1^2} \, . \tag{4}$$

According to the required (and expected) result from the filtering, it is assumed that the phase angle between the current and the voltage will be zero ($\varphi = 0$). Then the total power at the input of the scheme is:

$$\begin{aligned} Q_1 &= 0; \\ S_1 &= P_1 = \sqrt{3} \cdot U_1 \cdot I_L. \end{aligned} \tag{5}$$

The power P_2 that is expected to be consumed by the load is:

$$P_2 = k_u \cdot U_1 \cdot I_d \, , \tag{6}$$

where $k_u = 1.35$ is the DC/AC voltage ratio of the full bridge rectifier, and I_d is the DC current drawn by the load.

The power balance between P_1 and P_2 will be:

$$P_1 = P_2 + \Delta P \, , \tag{7}$$

where ΔP represents the total losses in the APF, the rectifier and the transformer.

In general, the lost part ΔP can be represented as a function of the current I_d multiplied by the coefficient $k_{\Delta P}$. Then the balance of power is:

$$\sqrt{3} \cdot U_1 \cdot I_L = k_u \cdot U_1 \cdot I_d + k_{\Delta P} \cdot I_d \, . \tag{8}$$

The peak value of the "ideal" current consumed from the grid is:

$$I_L^* = \frac{I_d}{\sqrt{3}} \left(k_u + \frac{k_{\Delta P}}{U_1} \right) . \tag{9}$$

The instantaneous values of the "ideal" phase currents are:

$$\begin{aligned} i_{LA}^*(t) &= \frac{I_d}{\sqrt{3}} \left(k_u + \frac{k_{\Delta P}}{U_1} \right) \cdot \sin(\omega t); \\ i_{LB}^*(t) &= \frac{I_d}{\sqrt{3}} \left(k_u + \frac{k_{\Delta P}}{U_1} \right) \cdot \sin(\omega t + 120°); \\ i_{LC}^*(t) &= \frac{I_d}{\sqrt{3}} \left(k_u + \frac{k_{\Delta P}}{U_1} \right) \cdot \sin(\omega t - 120°). \end{aligned} \tag{10}$$

As a general outlook, the APF control circuit is shown in Fig. 2 (b). The relay current control (RCC) is configured by adjusting the hysteresis value in order to determine the precision by which to process the signal from the current sensors. The difference between the "ideal" and real current is delivered to the RCC together with the signal from the DC current feedback i_D. The RCC block calculates and compares continuously the instantaneous "ideal" load currents and the real load currents. The control signals for the converter (it is an inverter in Fig. 2 are generated by RCC. These signals are aimed to form the APF output current instantaneous value. The instantaneous values of APF output currents are supplied to a step-up transformer. The secondary of the transformer supplies the missing instantaneous current continuously and thus compensates for the non-sinusoidal waveforms of the consumed current in the load. As a result, the current that the grid supplies to the whole circuit, becomes sinusoidal with a reasonable precision. The physical basis of this current waveform shaping, as in all the active filter circuits, is the temporary energy storage implemented by passive reactive devices. The particular difference in this method is the use of the output voltage: the filter capacitor or the battery, and not constructing a separate DC link as it is usually done.

4   Simulation Results

Following the described operation principle of this APF, a computer model has been developed in Matlab. The connection to the nonlinear load corresponds to the shown in Fig. 2 (a). The simulation parameters are: $U = 380$ V, $f = 50$ Hz, $R_L = 4.5$ Ohm, $L_L = 0.1$ mH, $C_d = 30$ mF, $L_{3L} = 0.1$ mH. The results shown in Fig. 3 demonstrate the current, supplied to the uncontrolled rectifier, having a sinusoidal form and a phase coincident with the mains voltage phase. As a result, no reactive power is consumed from the grid and the level of harmonic distortion of the current (THD$_i$) is close to ideal. This is illustrated by Fig. 3 (b) where THD$_i = 2.43\%$.

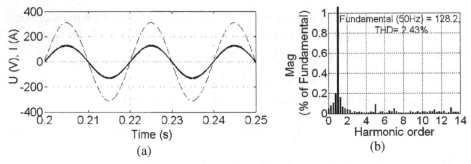

Fig. 3. Current in the phase A of the uncontrollable rectifier: (a) Graphically compared (solid line) to the phase A voltage (dashed line); (b) The spectral composition of the current

5   Experimental Results

The APF control system (RCC) was experimented in a real implementation. The parameters of the experimental study are: $U = 24$ V, $f = 50$ Hz, $R = 0 - 20$ Ohm, $L = 0.1$ mH. In Fig. 4 a general view of the construction is shown.

The power module is "Mitsubishi PS22054" connected to a "Mitsubishi 1200V DIP IPM (PS2205X) EVALUATION BOARD", used for debugging converter circuits. The module "Piccolo TMDX28069USB" of Texas Instruments" is a DSP controller, used here as the RCC. The module "CSLA1CH" of Honeywell is a linear current sensor (for measurement of the current in the load circuit). The waveforms were observed by oscilloscope "HANTEK DSO8060" and the values were measured by multimeter "MASTECH M890F". The power supply was a regulable block supplying 24 V at a maximum current of 2A) is connected to converter input. A single-phase load is experimented being a combination of a resistor and an inductive coil, connected to the power converter output.

Fig. 4. General view of set up to research RCC

The mathematical model of the RCC based APF control system as shown in Fig. 2 (b) was first implemented as a simulation in Matlab 8.1 and then was loaded to the DSP controller. The RCC control was supposed to maintain a sinusoidal current at a frequency 50 Hz and amplitude of 1.2 A. This task and the width of the hysteresis gap of the RCC were programmed in the DSP controller and the program was not changed during the experiment. The current feedback obtained from the current sensor placed in series with the load was introduced to the DSP controller too (analogue input). The signal processing speed is 3 MSPS. The "bang-bang" operation produced by the RCC control resulted in driving signals applied to the power converter of the APF. The signals depend on the programmed task, the width of the hysteresis gap and the signal from the current feedback. The driving signals from the DSP controller were sent to the power converter switches (through drivers). The correct waveforms from the experiments are shown in Fig. 5 (a, b).

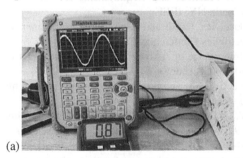
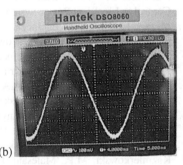

(a)   (b)

Fig. 5. Operation of the RCC: (a) Waveform and numerical value of the load current; (b) closer view of the signal of the load current sensor

As it is verified from the waveform in Fig. 5 (a, b) and the multimeter readings, the load current (DC) is corresponding to the sinusoidal input current from the grid at a correct magnitude, phase and frequency. The variation of the load (resistance of the variable resistor) across a range of values left the waveforms always correctly corresponding to the frequency, phase and amplitude required from the energy qualiy point of view.

6   Conclusions and Future Work

According to the results from the theoretical analysis, mathematical modelling, and experiments, the proposed APF based on the power rectifier filter capacitor as a DC link is designed and verified. The APF control based on RCC and limited calculation based on "ideal" load current was investigated. According to the simulation results, the uncontrolled rectifier combined with the new APF system is an effective means of compensation for harmonic current distortion and reactive power. For the experimental verification the APF set-up has been designed, and it confirmed the good performance of the simple "bang-bang" RCC control of the APF. The application of the proposed APF control system to non-linear power converters will improve the electromagnetic compatibility of the "naturally" nonlinear consumers and the power grids. It will have a positive influence on the power system and its customers. It provides increased reliability and energy efficiency. This will be especially important for the future smart grids construction. Further work on the APF will be in experimental studies of "uncontrolled rectifier – APF" to improve the construction and control system of the proposed APF.

References

1. Nejdawi, I.M., Emanuel, A.E., Pileggi, D.J.: Harmonic Trend in the USA: A Preliminary Survey. Transactions on Power Delivery 14(4), 1488–1494 (1999)
2. Bojarskaja, N.P., Dovgun, V.P.: The influence of the harmonic content of the currents and voltages on energy efficiency. Vestnik KrasGAU 4, 130–134 (2010)
3. Zhezhelenko, I.V.: The higher harmonics in power systems of industrial enterprises. 2nd ed. rev. and additional. Jenergoatomizdat, Moscow (1984)
4. IEEE Std 519–1992, IEEE Recommended Practices and Requirements for Harmonic Control in Electrical Power Systems, New York, NY: IEEE
5. Moran, L., Dixon, J., Espinoza, J., Wallace, R.: Using active power filters to improve power quality. In: 5th Brazilian Power Electronics Conference, COBEP (1999)
6. Sandeep, G.J.S.M., Rasoolahemmed, S.: Importance of Active Filters for Improvement of Power Quality. Int. J. of Engineering Trends and Technology 4(4), 1164–1171 (2013)
7. Meshherjakov, V.N., Koval', A.A.: The active filter-compensating devices for systems controlled dc drive. LSTU, Lipetsk (2008)
8. Meshherjakov, V.N., Bezdenezhnyh, D.V., Habibullin, M.M.: The active harmonic filter adapted to the alternating current. In: The 9th All-Russian Conference of Young Scientists. Managing Large Systems, Lipetsk, pp. 164–167 (2012)
9. Kazmierkowski, M.P., Malesani, L.: Current control techniques for three-phase voltage source PWM converters: A survey. IEEE Trans. Ind. Electron. 45(5), 691–703 (1998)
10. Chen, D., Xie, S.: Review of the control strategies applied to active power filters. In: IEEE International Conference on Electric Utility Deregulation, Restructuring and Power Technologies (DRPT2004), pp. 666–670 (2004)
11. Xiaobo, F., Dairun, Z., Qian, S.: Hysteresis Current Control Strategy for Three-phase Three-wire Active Power Filter. Automation of Electric Power Systems 31(18), 57–61 (2007)

Analysis of Power Quality Disturbances in Industry in the Centre Region of Portugal

Licínio Moreira[1,2], Sérgio Leitão[2], Zita Vale[3,4], João Galvão[1,5], and Pedro Marques[1]

[1] School of Technology and Management, Polytechnic Institute of Leiria,
P-2411-901 Leiria, Portugal
[2] UTAD-ECT Science and Technology School,
University of Trás-os-Montes and Alto Douro, Vila Real, Portugal
[3] Department of Electrical Engineering,
Polytechnic Institute of Porto (ISEP/IPP), Porto, Portugal
[4] GECAD (Knowledge Engineering and Decision Support Research Group),
Porto, Portugal
[5] INESC Coimbra - Institute for Systems Engineering and Computers at Coimbra,
Coimbra, Portugal
{licinio.moreira,jrgalvao,marques}@ipleiria.pt,
sleitao@utad.pt, zav@dee.isep.ipp.pt

Abstract. Power quality issues have taken a more prominent role in power systems over the last years. These issues are of major concern for energy customers, primarily for customers with a widespread use of electronic devices in their manufacturing processes. Even though the quality of service is increasing, customers are becoming more demanding of the energy provider. This research aims to provide some industrial managers the technical support in deciding of investments in the mitigation of power quality disturbances, such as the use of less sensitive devices or the use of interface devices (UPS, DVR ...) In order to recommend an appropriate solution, the problem is characterized. The technical and economic influences of the PQ disturbances in the manufacturing processes are assessed resorting to power quality audits in the customer facilities. This research covered a significant number of facilities in several industrial activities.

Keywords: Power Quality, Audit, Voltage Sag, Voltage Swell, Interruption, Wiring.

1 Introduction

Power quality (PQ) problems are inevitability, even in state-of-the-art power networks. The current worldwide economic competitiveness requires the organizations to be stricter in delivery times, following the just-in-time production model, and the profit margins tend to become smaller. Thus, any interference in production may lead to failure in delivery, sometimes implying large economic (direct and indirect) losses [1], [2] and [3].

The constant modernization of the manufacturing processes is based on the widespread use of electronic equipment. These electronic devices have an increased

L.M. Camarinha-Matos et al. (Eds.): DoCEIS 2014, IFIP AICT 423, pp. 435–442, 2014.
© IFIP International Federation for Information Processing 2014

sensitiveness to PQ disturbances and, at the same time are responsible for PQ degradation [4]. Several studies have been conducted to evaluate the costs of PQ problems for consumers. The assessment of an accurate value is nearly impossible; so all these studies are based on estimates. But all the studies point to losses valued in thousands of million Euros per year [5], [6] and [7]. In addition, the relevant studies point to the fact of investment made to minimize these disturbances is rather insignificant compared to the losses.

This research is centred in power quality audits conducted in several industrial facilities, such as plastic, moulds, wood, metallic construction and drywall manufacturing [8] and [9] in the centre of Portugal. Also some data from the quality of service reports issued by Portuguese energy regulator is used. The data is used to estimate losses (occurred and expected) and evaluate the economical feasibility of the use of mitigation techniques, including the use of uninterruptible power supplies (UPS), dynamic voltage restorers (DVR). The frequent contact with facility managers is used to encourage the option for devices that are less sensitive to short-duration PQ disturbances. Often, the PQ audits also deliver information used to the acting in the energy efficiency field.

2   Contribution to the Collective Awareness Systems

This work aims the rising of collective awareness on the mitigation of the problems arising from faulty power quality. Everyone in the electricity sector, from the generation to the consumption can and should be participative for the minimization of losses. The mission of the energy sector operators (generation, transmission and distribution) is to fulfil the requirements of the electricity consumers. The customers have become more aware and demand low-cost electricity of high reliability and quality, where the priorities are different for different customers. They are certainly no longer willing to accept their position as merely one parameter in a global optimisation [1].

On the other hand, the customers should take action, being proactive in the search for solutions that minimize the costs arising from PQ disturbances (mainly the ones with short duration) that are an inevitable. Standardization (both on supply and demand sides) plays an important role defining PQ parameters that are acceptable for the satisfaction of electricity consumers, but also defining standards for end-use equipments tolerance and interface devices. Fig. 1 resumes the actuation levels and strategies that could lead to a decrease in the magnitude of PQ events.

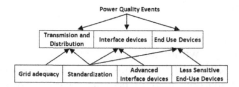

Fig. 1. Different levels of the electricity sector and actuation strategies

3   Data Collected from Distribution Network Operator

Every year, the operators of the transmission and distribution (T&D) networks are obliged to prepare and submit to the energy sector regulator reports stating the technical and commercial quality of the service provided. The technical quality of service is related to the analysis of the continuity of service, through the number and duration of supply interruptions, and the quality of the voltage wave (evaluation of its frequency values, amplitude, harmonic distortion, imbalance and others) [10].

To estimate the average annual number of voltage sags in each industrial facility, the values found in the quality of service report for medium voltage (MV) buses with permanent monitoring can be helpful, combined with the data collected in PQ audits. Table 1 shows the registered events categorized by time and amplitude.

Table 1. Average number of voltage sags in MV buses of the distribution network in 2012

Retained	Number / Duration (seconds)						% of events
Voltage (%U$_c$)	]0.01;0.1]	]0.1;0.25]	]0.25;0.5]	]0.5;1]	]1<;3]	]3;20]	
80%<U<90%	26.10	11.59	3.30	5.85	2.08	0.16	53.6
70%<U<80%	6.38	5.49	1.99	2.10	1.19	0.05	18.8
60%<U<70%	3.74	2.27	1.78	1.24	0.94	0.14	11.1
50%<U<60%	0.64	1.53	1.72	0.78	0.59	0.03	5.8
40%<U<50%	0.33	0.81	1.05	0.33	0.78		3.6
30%<U<40%	0.36	0.89	0.75	0.48	0.30		3.0
20%<U<30%	0.08	0.30	0.72	0.36	0.11	0.08	1.8
10%<U<20%	0.11	0.22	0.78	0.25			1.5
1%<U<10%	0.03	0.11	0.22	0.19	0.11		0.7
% of events	41.3	25.4	13.5	12.7	6.7	0.5	

The total number of sags recorded was 3843 over 42 buses. This results in an average of 91.5 sags per bus per year in 2012. This record is 15% lower than the one recorded in 2011. The shadowed cells on Table 1 represent the events that are under the lower limits of the ITIC curve, and are therefore likely to cause malfunction on electronic devices, such as computer numerical control (CNC) machines, widely used in manufacturing processes. The fraction of the occurrences under the ITIC curve is 18.8%, representing approximately 17 sags per year that can cause equipment malfunction.

This national average value is very similar to the value recorded in the substations located in the Leiria area, validated by the records of the distribution network operator and the data collected in the PQ audits conducted in this research.

4   Power Quality Audits

The referred several audits were performed using two Hioki Power Quality Analysers 3196. The PQ analysers were installed in low voltage buses with settings that are in accordance with EN 50160. These devices allow the continuous monitoring of electrical parameters and the recording of averaged values and the recording of transient phenomena, such as interruptions, voltage sags, swells and other.

Nearly all data collected was compliant with the norm. The only exception was registered on the 19th of January, 2013, when a gale occurred in Portugal, leading to the destruction of several power distribution network infrastructures, causing service interruptions that lasted for several days in some areas of the centre of Portugal. One can conclude that the values required by the norm are easily fulfilled (excluding acts of god) by the distribution operator, since the norm only applies to average voltage values of 10 minutes. Nevertheless, even when the quality of service complies with the norm, some disturbances which are detected are responsible for equipment malfunction and the disruption of manufacturing processes [11].

4.1   Voltage Sags

The voltage sags and swells recorded during a monitoring campaign in a metallic construction site, with several CNC machines (one week period) are represented in Fig. 2. These events are plotted in the ITIC curve, which defines the expected tolerance of electronic equipment to voltage variation events. Events plotted in the upper shaded area (swells) are dangerous for the equipment, susceptible of causing damages to it, while the events plotted in the lower shaded area may cause malfunction of the devices, but shouldn't be destructive. Any event in the non shaded area shouldn't interfere with normal function of the device.

The plot shows the occurrence of six voltage sags, with three of them under the tolerance limits defined in the ITIC curve.

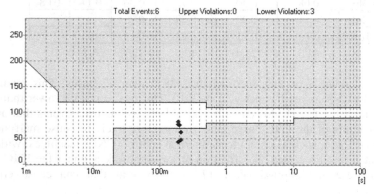

Fig. 2. Voltage sags and swells recorded over a week period (from the 14th to the 21st of March, 2013) and plotted over the ITIC curve

The average voltage values with 10 ms sampling time during a voltage sag is represented in Fig. 3. The voltage retained is 109 V and the duration is 201 ms.

The events recorded by the energy analyser were compared with the record of manufacturing process disturbances. The sag represented in Fig. 3 is responsible for a CNC machine malfunction causing the delay of the piece production of several hours. The CNC needed to be reprogrammed before continuing the piece machining.

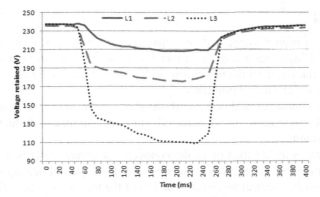

Fig. 3. Voltage sag recorded on the 18th of March, 2013

In some cases reported by facility managers, some malfunctions due to voltage sags or micro-interruptions cause damages to tools and / or machined piece. This can happen if the tool is performing a high speed operation and becomes uncontrolled due to the power failure in the controller.

4.2 Voltage Swells

Voltage swells are also transient phenomena susceptible of causing damage. These occur less frequently, but in some severe cases are more likely to cause direct damages on electrical equipment.

During the monitoring campaigns made, no damage could be directly related to the occurrence of a voltage swell, but it is widely known that voltage swells contribute to the lifetime shortening of electrical devices.

In Fig. 4, the most severe voltage swell (both in amplitude and duration) is plotted. In phase L3, the voltage reached 277.7 V (20.7% over rated voltage) and the duration was of about 20 seconds.

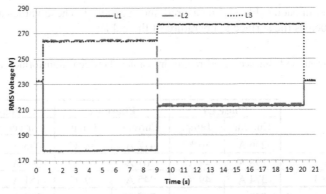

Fig. 4. Voltage swell recorded on the 15th of February, 2013

4.3   Wiring

The selection of proper wiring and protective devices is imperative for the operation of an electrical installation and the proper functioning of end use devices. Also, an incorrect installation, from the placement of the components to the correct tightening, can lead to the occurrence of problems.

During the conducted audits, several situations were detected, some before the destruction of equipment, others detected after the equipment failure because of improper tightening.

One example is the destruction of the connector in several ballasts in an industrial facility. The improper connection, combined with the expected heating of the ballast and lamp, led to an overheating and destruction of the terminals of the ballast.

With the use of a thermal camera (Fluke Ti45 Flexcam™), during a normal assessment, a difference of temperature was detected in a set of cables connecting the power transformer (in the customer substation) to the main electrical switchboard. The connection was made using three cables (H1XV-R 1x300 mm²) for each phase and for the neutral. The thermal image for each phase is shown in Fig. 5. It is easily found that the current distribution by parallel cables in every phase is unequal.

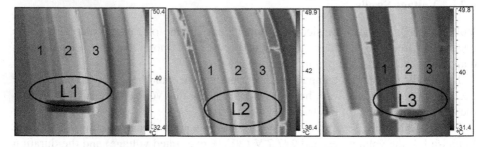

Fig. 5. Thermal imaging of electric cables

The current in each cable was measured and registered. These values are presented in Table 2, such as the surface temperature of the cables.

Table 2. Current and surface temperature in cables between transformer and the main switchboard

	Phase L1		Phase L2		Phase L3	
	Current	Temp.	Current	Temp.	Current	Temp.
Cable 1	236 A	36 °C	412 A	38 °C	483 A	46 °C
Cable 2	198 A	48 °C	201 A	42 °C	203 A	46 °C
Cable 3	374 A	49 °C	345 A	39 °C	181 A	37 °C
Phase rms current	**778 A**	-	**798 A**	-	**861 A**	-

The values show the expected imbalance of currents, of which the currents in phase L3stand out. In this phase, the most heavily loaded cable conducts 2.67 times the current of the least loaded cable. One can also easily conclude that the sum of the current in the three cables for each phase is greater than the rms current for the phase. For example, in phase L2, the algebraic sum of the current in the three cables is 958 A and the rms current is only of 798 A.

Considering that the admissible current of the conductors is of approximately 580 A, no cable is overloaded. Also, the temperature of every cable is within acceptable limits. The problem hasn't reached critical values, because the transformer (with 1600 kVA rated power) is underload, presenting an average load of 650 kVA and a maximum load of about 800 kVA. In the case of a load increase, the cables would suffer a lifetime reduction and, in an extreme event, could trigger a fire.

This imbalance is caused by an incorrect geometric distribution of the cables on the cable tray. The cables are arranged vertically with the sequence shown Fig. 7 (top). This arrangement promotes the increase of mutual inductances in some cables [12] and [13]. It was recommended the change of the arrangement for the one portrayed on Fig. 7 (bottom). The trefoil formation would represent a better solution, but it would be harder to implement, implying the enlargement of holes in a wall and the re-installation (or even replacement) of cable trays.

Fig. 6. Present arrangement (top) and proposed re-arrangement (bottom) of cables connecting the transformer and the main switchboard

5 Conclusions

This project has identified several issues related with power quality disturbances that effectively affected production processes, mainly in activity sectors with intensive use of CNC systems, whose methods of quality control are very strict in the manufacturing of steel (in the order of microns). In addition, some issues that could potentially become a problem were detected and solutions were proposed in order to take action anticipating possible damages.

The degradation of power quality can be a trouble resulting in large financial losses. With the increasing competitiveness of world economy, but also with the low financial capability, particularly of most of the Portuguese companies, every investment in power quality improvement must be analysed wisely. The implementation of a monitoring and recording smart system, where the PQ disturbances and their consequences are registered could support the taking of better, informed and sustainability-aware decisions. The linkage of objects, people and knowledge promotes the network effect for the cooperative awareness.

References

1. Bollen, M.H.J.: What is power quality? Electric Power Systems Research 66(1), 5–14 (2003), doi:10.1016/S0378-7796(03)00067-1, ISSN 0378-7796
2. Bayliss, C.R., Hardy, B.J.: Power Quality – Voltage Disturbances. In: Bayliss, C.R., Hardy, B.J. (eds.) Transmission and Distribution Electrical Engineering, 4th edn., Newnes, Oxford, ch. 25, pp. 1013–1026 (2012), doi:10.1016/B978-0-08-096912-1.00025-3, ISBN 9780080969121
3. Math, H.J.: Bollen, Understanding Power Quality Problems - Voltage Sags and Interruptions. Wiley-Interscience - IEEE Press, New York (2000)
4. Vairamohan, B., Komatsu, W., Galassi, M., Monteiro, T.C., de Oliveira, M.A., Ahn, S.U., Matakas Jr., L., Marafão, F.P., Bormio Jr., E., de Camargo, J., McGranaghan, M.F., Jardini, J.A.: Technology assessment for power quality mitigation devices – Micro-DVR case study. Electric Power Systems Research 81(6), 1215–1226 (2011), doi:10.1016/j.epsr.2011.01.014, ISSN 0378-7796
5. Chapman, D.: Costs of Poor Power Quality. Power Quality Application Guide – Copper Development Association (March 2001)
6. Targosz, R., Manson, J.: Pan-European Power Quality Survey – A study of the impact of power quality on electrical energy critical industrial sectors. In: Proceedings of the 9th Conference on Electrical Power Quality and Utilisation, Barcelona, Spain, October 9-11 (2007)
7. Coll-Mayor, D., Pardo, J., Perez-Donsion, M.: Methodology based on the value of lost load for evaluating economical losses due to disturbances in the power quality. Energy Policy 50, 407–418 (2012), doi:10.1016/j.enpol.2012.07.036, ISSN 0301-4215
8. Moreira, L., Leitão, S., Vale, Z.: Power Quality problems in the mould Industry. In: 11th Spanish Portuguese Congress on Electrical Engineering (11 CHLIE), Zaragoza, July 1-4 (2009)
9. [IYCE] Galvão, J., Moreira, L., Leitão, S., Silva, E., Neto, M.: Sustainable Energy for Plastic Industry Plant, Proceedings of 4th International Youth Conference on Energy, IYCE/IEEE, Siófok, Hungary (June 2013), doi: 10.1109/IYCE.2013.6604193
10. Entidade Reguladora dos Serviços Energéticos, Quality of Service in the Electric Sector 2012, Lisboa (October 2013)
11. Martinez, J.A., Martin-Arnedo, J.: Voltage sag studies in distribution Networks-part I: system modeling. IEEE Transactions on Power Delivery 21(3), 1670–1678 (2006), doi:10.1109/TPWRD.2006.874113
12. Du, P.Y., Wang, X.H.: Electrical and Thermal Analyses of Parallel Single-Conductor Cable Installations. IEEE Transactions on Industry Applications 46(4), 1534–1540 (2010), doi:10.1109/TIA.2010.2049819
13. Du, Y., Burnett, J.: Current distribution in single-core cables connected in parallel. Generation, Transmission and Distribution, IEE Proceedings 148(5), 406–412 (2001), doi:10.1049/ip-gtd:20010430

Sliding Mode Control of Unified Power Quality Conditioner for 3 Phase 4 Wire Systems

Nelson Santos[1], J. Fernando A. Silva[2], and João Santana[2]

[1] Instituto Superior de Engenharia de Lisboa
R. Conselheiro Emídio Navarro, 1959-007 Lisboa, Portugal
nsantos@deea.isel.ipl.pt
[2] Instituto Superior Técnico, Instituto Superior Técnico, INESC-id, DEEC, AC Energia
Av. Rovisco Pais, 1049-001 Lisboa, Portugal
jsantana@ist.utl.pt, fernandos@alfa.ist.utl.pt

Abstract. This paper presents the sliding mode control (SMC) of unified power quality conditioners (UPQC) intended to compensate power quality issues in three-phase four-wires systems. The SMC UPQC can be applied in electrical grids or isolated grids to mitigate power quality problems at the consumer facilities and also to minimize issues for the grid supplier. The UPQC is configured as a shunt-series filter. The shunt Active Power Filter (APF) uses a three-phase rectifier with SMC to enforce sinusoidal mains currents. The series APF uses three single-phase H-bridge inverters and SMC to improve the voltage quality at the point of common coupling. SMC design, analysis and simulation results are presented and discussed.

Keywords: Unified Power Quality Conditioner, Active Power Filter, Sliding Mode Control.

1 Introduction

In recent years, economic and environmental reasons entail the best possible use of energy resources. Recent technological developments enabled better usage of energy resources, mainly alternative energies [1]. However, due to the usually relatively low duty-cycle and randomness of these energies, their interconnection to the electrical grid might require the use of power electronics systems, and power quality mitigating devices, such APF, to increasing power quality [2], according to the standards recommended maximum power quality disturbances (IEEE 519, IEC 50160).

Most common power quality issues arise in the distribution grids, including voltage variations, harmonics, voltage sags and swells, voltage unbalances, harmonic currents, high reactive power burden and excessive neutral wire current. APFs can solve voltage and current waveform issues, reducing harmonics in a wide frequency band without resonance problems [3].

L.M. Camarinha-Matos et al. (Eds.): DoCEIS 2014, IFIP AICT 423, pp. 443–450, 2014.
© IFIP International Federation for Information Processing 2014

2   Contribution to Collective Awareness Systems

This paper proposes a UPQC system with the ability to increase the quality of distribution of electrical power to achieve better efficiency. It proposes a contribution towards the sustainability of global systems considering the economic aspect and the environmental, reaching an important advance in the sustainable growth systems while benefiting from future collective awareness system to take informed decisions, making effective the involvement of the customer and sustainable energy systems.

3   UPQC Topology

Conventional UPQC have two basic configurations [3]. The most common is the series-shunt which injects the voltage compensation at the source side and the current compensation at the load side. This configuration has the advantage of the current flowing through the series transformer containing only the fundamental frequency. The shunt-series UPQC injects currents at the source side and compensates voltage at the load side. This configuration has the advantage of avoiding the interference between the shunt inverter and the passive filters, caused by the high frequency switching of the UPQC [3].

This paper presents the study of a shunt-series configuration provided with a passive LC filter on the series side as depicted in Fig. 1.

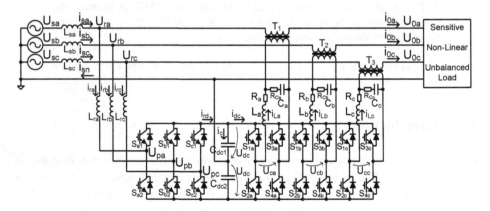

Fig. 1. Shunt-Series UPQC

To compensate the neutral current a split capacitor topology (C_{dc1}, C_{dc2}), with the midpoint connected to the neutral wire, enables return currents through the capacitors, being $i_{sa} + i_{sb} + i_{sc} = i_{sn}$. The return of the neutral current through the DC bus capacitors unbalances the voltage of the two capacitors and decreases the dynamic response of the shunt APF. To minimize this problem, an extra control system to equalize the voltage on the two capacitors is usually included [3], even for small neutral currents, with zero average values, flowing into the source due the high switching harmonics.

4   Control

4.1   Shunt APF

The shunt APF must enforce sinusoidal i_{sa}, i_{sb} and i_{sc} currents in the mains source, [4]. Additionally, the shunt APF must stabilize the DC bus voltage [3].

Sliding Mode Internal Current Control. The shunt APF injects the necessary currents to cancel unwanted harmonics produced by non-linear and unbalanced loads. A SMC technique is used to control the mains current. Considered one generic single phase k of the corresponding source, as represented in Fig. 1. Assuming ideal switches and ideal components, the dynamics of the i_{sk} current $k \in \{a, b, c\}$ can be described considering that the voltage U_{pk}, relative to the neutral voltage, is dependent on the switches states of the shunt inverter semiconductors, as in (1).

$$U_{pk} = \delta(t)U_{dc} = \begin{cases} U_{pk} = U_{dc} \text{ for } \delta(t) = +1 \ (S_{k1} \text{ ON and } S_{k2} \text{ OFF}) \\ U_{pk} = -U_{dc} \text{ for } \delta(t) = -1 \ (S_{k1} \text{ OFF and } S_{k2} \text{ ON}) \end{cases} \tag{1}$$

The first time derivative of the i_{rk} current is given as $\frac{di_{rk}}{dt} = \frac{U_{rk} - \delta(t)U_{dc}}{L_{rk}}$ considering that $i_{sk} = i_{rk} + i_{0k}$, and $\frac{di_{sk}}{dt} = \frac{di_{rk}}{dt} + \frac{di_{0k}}{dt}$, then, the phase canonical model of current i_{sk} is written as (2).

$$\frac{di_{sk}}{dt} = \frac{U_{rk}}{L_{rk}} - \frac{\delta(t)U_{dc}}{L_{rk}} + \frac{di_{0k}}{dt} \tag{2}$$

The first time derivative of i_{sk} contains the control action $\delta(t)U_{dc}$, thus the strong relative degree [5] of i_{sk} is 1, and a suitable sliding surface can be obtained as a linear combination of the control error $e_{i_{sk}} = i_{sk}^{ref} - i_{sk}r$, where i_{sk}^{ref} is the reference value to be tracked by the i_{sk} current. Considering a positive gain K_p, used to limit the semiconductors switching frequency, a suitable sliding surface $S(ei_{sk}, t)$ [5] is:

$$S\left(e_{i_{sk}}, t\right) = K_p \, e_{i_{sk}} = K_p\left(i_{sk}^{ref} - i_{sk}\right) = 0 \tag{3}$$

The switching strategy is obtained applying the sliding mode stability condition $S\left(e_{i_{sk}}, t\right) \dot{S}\left(e_{i_{sk}}, t\right) < 0$ [6], with $\dot{S}\left(e_{i_{sk}}, t\right) = K_p\left(\frac{di_{sk}^{ref}}{dt} - \frac{U_{rk}}{L_{rk}} + \frac{\delta(t)U_{dc}}{L_{rk}} - \frac{di_{0k}}{dt}\right)$. As $S\left(e_{i_{sk}}, t\right)$ is proportional to $e_{i_{sk}}$, the stability condition is written as:

$$\begin{cases} \text{if } e_{i_{sk}} > 0, \quad \text{then } \frac{de_{i_{sk}}}{dt} < 0 \Rightarrow \frac{di_{sk}^{ref}}{dt} - \frac{U_{rk}}{L_{rk}} + \frac{\delta(t)U_{dc}}{L_{rk}} - \frac{di_{0k}}{dt} < 0 \\ \text{if } e_{i_{sk}} < 0, \quad \text{then } \frac{de_{i_{sk}}}{dt} > 0 \Rightarrow \frac{di_{sk}^{ref}}{dt} - \frac{U_{rk}}{L_{rk}} + \frac{\delta(t)U_{dc}}{L_{rk}} - \frac{di_{0k}}{dt} > 0 \end{cases} \tag{4}$$

From (4), the sliding mode reaching condition can be obtained as $\frac{\delta(t)U_{dc}}{L_{rk}} >$ $\text{MAX}\left(-\frac{di_{sk}^{ref}}{dt} + \frac{U_{rk}}{L_{rk}} + \frac{di_{0k}}{dt}\right)$. Supposing that U_{dc} is high enough to satisfy this reaching condition and if $e_{i_{sk}} > 0$ then $\frac{de_{i_{sk}}}{dt} < 0$ the control action is $\delta(t) = -1$ or if $e_{i_{sk}} < 0$ then $\frac{de_{i_{sk}}}{dt} > 0$ the control action is $\delta(t) = 1$. To define a finite switching frequency, the i_{sk} current must have a non-zero ripple content Δi_{sk}. The switching law is then:

$$\delta(t) = \begin{cases} -1 \text{ for } S(e_{i_{sk}}, t) > +\Delta i_{sk}/2 \\ 1 \text{ for } S(e_{i_{sk}}, t) < -\Delta i_{sk}/2 \end{cases} \tag{5}$$

The sliding surface (3) and the non-linear switching law (5) do not depend on any load parameter, but only on the measured i_{sk} current error $e_{i_{sk}}$. To generate the sinusoidal and balanced i_{sk}^{ref} current references in the *abc* frame the inverse Park Transform (*dq0* to *abc*), is used where i_d, i_q and i_0 are the direct, quadrature and zero sequence component of the current reference. The i_{sk}^{ref} current references in *dq0* (i_d, i_q and i_0) will be obtained using a slow enough linear controller to minimize current total harmonic distortion (THD). The response time should be tailored to avoid DC bus voltage dropouts in the case where fast increments of load currents are required, which can affect the global dynamic of series APF.

Linear External DC Bus Voltage Control. The external loop to regulate the voltage in the DC bus capacitors uses a PI controller, $C_{v(s)} = \frac{1+sT_z}{sT_p}$, designed to present small load sensitivity. The PI output is the reference for the current i_d component of Park Transform [7], while the i_q and i_0 components of Park Transform are set to zero to have near unity power factor ($v_q = 0$) and to balance the three phases. The previously designed sliding mode internal current control closed loop behavior is modelled by a first order system represented as $\frac{i_{rk}}{i_{sk}^{ref}} = \frac{k_e}{T_d s+1}$ where $k_e = \frac{U_{sd}}{2U_{dc}}$ and T_d is estimated knowing that line currents should be sinusoidal to avoid line current distortion. The PI parameters T_z and T_p of the voltage controller $C_v(s)$ are calculated using the Symmetrical Optimum technique, [8]. Assuming $U_d = 2U_{dc}$ and $C = C_{dc1}/2 = C_{dc2}/2$, the differential equation of the DC bus voltage is $\frac{dU_d}{dt} = \frac{i_{rd}-i_{dc}}{C}$. Applying superposition with zero disturbances and supposing $\alpha_v = 1$ (sensor voltage gain), the closed-loop transfer function of the DC bus voltage control has a third order polynomial denominator, $\frac{U_d}{U_d^{ref}} = \frac{1+sT_z}{s^3\frac{T_pT_dC}{k_e}+s^2\frac{T_pC}{k_e}+sT_z+1}$, represented in (Fig. 2).

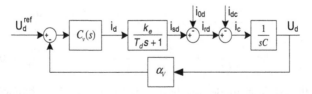

Fig. 2. DC bus voltage control

Applying Symmetrical Optimum criteria $b_k^2 = 2b_{k-1}b_{k+1}$, to ensure a low enough sensitivity to the UPQC disturbances. The value T_d=4ms was used to obtain the T_z and T_p, so that line currents are almost sinusoidal and DC bus voltage response is fast enough to avoid affecting the series APF performance [5]. The PI parameters are $T_z = 4T_d$ and $T_p = \frac{8T_d^2 k_e}{C}$.

4.2   Series APF

To compensate the voltage distortion, the three series transformer primaries are driven using three single-phase H-bridge inverters (Fig. 1) controlled using SMC. Considering the per-phase equivalent of the series APF, with $k \in \{a, b, c\}$, where the C_k is the filter capacitor and the load is reduced to the transformer primary.

The U_{ck} output voltage depends on the switches states S_{1k}, S_{2k}, S_{3k} and S_{4k}.

$$U_{ck} = \gamma_c(t)U_d = \begin{cases} U_{ck} = U_d \text{ for } \gamma_c(t) = +1 & (S_{1k}, S_{4k} \text{ } ON) \\ U_{ck} = 0 \quad \text{ for } \gamma_c(t) = 0 \text{ } (S_{1k}, S_{3k} \text{ } ON \text{ or } S_{2k}, S_{4k} \text{ } ON) \\ U_{ck} = -U_d \text{ for } \gamma_c(t) = - \text{ } (S_{2k}, S_{3k} \text{ } ON) \end{cases} \tag{6}$$

Assuming ideal components, the dynamic model with state variables u_{0k} and i_{Lk} is written as (7), were i_{0k} is the load current:

$$\frac{d}{dt}\begin{bmatrix} u_{0k} \\ i_{Lk} \end{bmatrix} = \begin{bmatrix} 0 & 1/C_k \\ -1/L_k & -R_k/L_k \end{bmatrix}\begin{bmatrix} u_{0k} \\ i_{Lk} \end{bmatrix} + \begin{bmatrix} -1/C_k & 0 \\ 0 & 1/L_k \end{bmatrix}\begin{bmatrix} i_{0k} \\ \delta_c(t)U_d \end{bmatrix} \tag{7}$$

The proposed of the series APF is compensate the voltage at the load side. According to [5], [6] and [9], to control the load voltage u_{0k} is necessary to obtain a equation which includes the dynamic of switching. The first order derivative of u_{0k} doesn't include the switch condition $\delta_c(t)$, for this reason we need to calculate the second order derivative for the u_{0k}. Thus, the strong relative degree of u_{0k} is 2 written as $\frac{d^2 u_{0k}}{dt^2} = -\frac{1}{C_k}\frac{d i_{0k}}{dt} - \frac{u_{0k}}{L_k C_k} - \frac{R_k i_{Lk}}{L_k C_k} + \frac{\delta_c(t)U_d}{L_k C_k}$.

The robust sliding surface (8) is a linear combination of the control error $e_{u_{0k}}$ and its derivative, $e_{u_{0k}} = u_{0k}^{ref} - u_{0k}$, where u_{0k}^{ref} is the voltage reference to be tracked and u_{0k} the voltage feedback.

$$S(e_{u_{0k}}, t) = k_1 e_{u_{0k}} + k_2 \frac{d e_{u_{0k}}}{dt} = 0 \tag{8}$$

The system response depends on the coefficients k_1 and k_2 which must be selected to ensure stability and fast response for operating conditions. Considering $\beta = k_2/k_1$ the sliding surface is written as:

$$S(e_{u_{0k}}, t) = \frac{C_k}{\beta}(u_{0k}^{ref} - u_{0k}) + C_k \frac{d u_{0k}^{ref}}{dt} + i_{0k} - i_{Lk} = 0 \tag{9}$$

The switching strategy is obtained applying the sliding mode stability condition $S(e_{u_{0k}}, t) \dot{S}(e_{u_{0k}}, t) < 0$, with $\dot{S}(e_{u_{0k}}, t) = \frac{i_{Lk}}{C_k} - \frac{i_{0k}}{C_k} + \beta\left(\frac{1}{C_k}\frac{di_{0k}}{dt} + \frac{u_{0k}}{L_k C_k} + \frac{R_k i_{Lk}}{L_k C_k} - \frac{\delta_c(t)U_d}{L_k C_k}\right)$. As $S(e_{u_{0k}}, t)$ is proportional to $e_{u_{0k}}$, the stability condition is written as:

$$\begin{cases} \text{if } e_{u_{0k}} > 0, \quad \text{then } \dfrac{de_{u_{0k}}}{dt} < 0 \Rightarrow \dfrac{i_{Lk}}{C_k} - \dfrac{i_{0k}}{C_k} + \beta\left(\dfrac{1}{C_k}\dfrac{di_{0k}}{dt} + \dfrac{u_{0k}}{L_k C_k} + \dfrac{R_k i_{Lk}}{L_k C_k} - \dfrac{\delta_c(t)U_d}{L_k C_k}\right) < 0 \\ \text{if } e_{u_{0k}} < 0, \quad \text{then } \dfrac{de_{u_{0k}}}{dt} > 0 \Rightarrow \dfrac{i_{Lk}}{C_k} - \dfrac{i_{0k}}{C_k} + \beta\left(\dfrac{1}{C_k}\dfrac{di_{0k}}{dt} + \dfrac{u_{0k}}{L_k C_k} + \dfrac{R_k i_{Lk}}{L_k C_k} - \dfrac{\delta_c(t)U_d}{L_k C_k}\right) > 0 \end{cases} \quad (10)$$

From (10), the sliding mode reaching condition is $\beta\frac{\delta_c(t)U_d}{L_k C_k} > \text{MAX}\left(\frac{i_{Lk}}{C_k} - \frac{i_{0k}}{C_k} + \right.$ $+\beta\left(\frac{1}{C_k}\frac{di_{0k}}{dt} + \frac{u_{0k}}{L_k C_k} + \frac{R_k i_{Lk}}{L_k C_k}\right)\bigg)$. Supposing that U_d is high enough to satisfy this reaching condition and using a hysteresis comparator, with Δu_{0k} as hysteresis, to avoid infinite switching frequency, the switching law is (11), being $\delta_c(t) \in \{-1; 0; 1\}$ to obtain three possible U_{ck} levels.

$$\begin{cases} \text{if } e_{u_{0k}} > \dfrac{\Delta u_{0k}}{2}, \quad \text{then } \dfrac{de_{u_{0k}}}{dt} < 0 \Rightarrow \delta_c(t) = 1 \\ \text{if } \dfrac{-\Delta u_{0k}}{2} < e u_{0k} < \dfrac{\Delta u_{0k}}{2}, \quad \text{then } \delta_c(t) = 0 \\ \text{if } e_{u_{0k}} < \dfrac{-\Delta u_{0k}}{2}, \quad \text{then } \dfrac{de_{u_{0k}}}{dt} > 0 \Rightarrow \delta_c(t) = -1 \end{cases} \quad (11)$$

This technique samples u_0, i_0 and i_L, but the closed loop dynamics does not depend on load parameters. Fast response with zero steady-state errors are also attained [10].

5   Simulation

The UPQC was simulated using Matlab/Simulink considering non-ideal mains voltage with 5% THD. Several loads, described in the Appendix, were tested.

To evaluate the UPQC response with wide operating conditions, different loads and conditions were tested successively. The test begins with a balanced R load. Then, an unbalanced RL load is switched on to add to the existing balanced R load. In a third step it is added a non linear load, made by three single-phase bridge rectifiers, with upstream inductance L and parallel RC type circuit downstream of the bridge. In the fourth step a voltage sag in the main source appears, being cleared in the fifth step.

The first simulation (Fig. 3) is performed without any compensation. High neutral current exists due to the switching on of unbalanced and non-linear loads. The load voltage has a THD of 5.4%. The THD of currents reach a maximum of 25,1% in the worse case.

Using the compensating UPQC, the delivered voltage (Fig. 4) has a maximum THD of 0.6%, and currents drops to a maximum THD of 2.37%. The UPQC clearly improves the quality power to the main source (current) and to the load side (voltage). The neutral current is mostly zero throughout the simulation (except for switching harmonics and switch on transients).

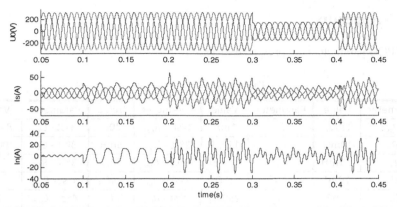

Fig. 3. Results under unbalanced and distorted voltage source with non-linear loads without compensation

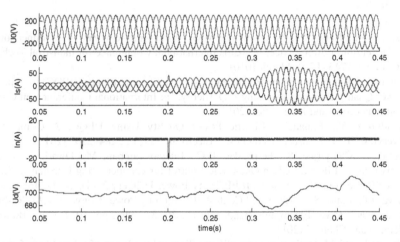

Fig. 4. Results under unbalanced and distorted voltage source with non-linear loads with compensation

6   Conclusion

This study demonstrates the sliding mode control of UPQC to improve power quality on three-phase four-wire systems. The UPQC can mitigate the harmonic content of voltage at the load side (88% reduction), and the harmonic current at the source side (90% reduction), while virtually eliminating the neutral current and improving the power factor. The measured values of the harmonic distortion of current source and the voltage delivered to the load were well within the recommended values according to standards IEEE-519 or IEC61000.

Acknowledgments. This work supported by Portuguese national funds through FCT – Fundação para a Ciência e a Tecnologia, under project PEst-OE/EEI/LA0021/2013.

A   Appendix

The UPQC parameters are presented in Tab. 1.

Table 1. System Parameters

Parameter	Value	Parameter	Value	Parameter	Value	Parameter	Value
U_{sk}	230V$_{ef}$	L_r	1mH	1ºst - R_{abc}	20Ω	3ºst - L_{NL}	8.4mH
R_s	10mΩ	C_k	112μF	2ºst - L_{abc}	5mH	3ºst - C_{NL}	1000μF
L_s	0.1mH	L_k	0,9mH	2ºst – R_{0a}	R_{abc}	3ºst - R_{NL}	50Ω
U_d	700V	R_{Ck}	15Ω	2ºst – R_{0b}	50R_{abc}	4ºst - R_{sag}	R_s
C_{dc}	5mF	U_{ref}	230V$_{ef}$	2ºst - R_{0c}	5R_{abc}	4ºst - L_{sag}	L_s

References

1. Farias, M.F., Battaiotto, P.E., Cendoya, M.G.: Wind Farm to Weak-Grid Connection using UPQC custom power device. In: IEEE International Conference on Industrial Technology (ICIT), pp. 1745–1750 (March 2010)
2. Babu, C., Dash, S.S.: Design of Unified Power Quality conditioner (UPQC) to improve the Power Quality Problems by Using P-Q Theory. In: International Conference on Computer Communication and Informatics (2012)
3. Khadkikar, V.: Enhancing Electric Power Quality Using UPQC: A Comprehensive Overview. IEEE Transactions on Power Electronics 27(5), 2284–2297 (2012)
4. Hirve, S., Chatterjee, K., Fernandes, B.G., Imayavaramban, M., Dwari, S.: PLL-Less Active Power Filter Based on One-Cycle Control for Compensating Unbalanced Loads in Three-Phase Four-Wire System. IEEE Trans. Power Delivery 22(4), 2457–2465 (2007)
5. Silva, J.F., Pinto, S.F.: Advanced Control of Switching Power Converters. In: Rashid, M., et al. (eds.) Power Electronics Handbook, ch. 36, 3rd edn., pp. 1037–1114. Butterworth Heinemann, Chennai (2011)
6. Silva, J.F.: PWM Audio Power Amplifiers: Sigma Delta Versus Sliding Mode Control. In: Proc. IEEE/ICECS 1998, Lisboa, Portugal, Set, vol. 1, pp. 359-362 (1998) ISBN 0-7803-5008-1
7. Bin, X.B.X., Ke, D.K.D., Yong, K.Y.K.: DC voltage control for the three-phase four-wire Shunt split-capacitor Active Power Filter. In: IEEE International Electric Machines and Drives Conference, pp. 1669–1673 (May 2009)
8. Bajracharya, C., Molinas, M., Ieee, M., Suul, J.A., Undeland, T.M., IEEE, F.: Understanding of tuning techniques of converter controllers for VSC-HVDC. In: Nordic Workshop on Power and Industrial Electronics (June 2008)
9. Silva, J.F., Rodrigues, N., Costa, J.: Space Vector Alpha-Beta Sliding Mode Current Controllers for Three-Phase Multilevel Inverters. In: Silva, J.F., Rodrigues, N., Costa, J. (eds.) IEEE Proc. PESC 2000, Galway, Ireland (June 2000) CD ROM ISBN 0-7803-5695-0
10. Martins, J.F., Pires, A.J., Fernando Silva, J.: A Novel and Simple Current Controller for Three-Phase PWM Power Inverters. IEEE Trans. on Industrial Electronics 45(5), 802–805 (1998)

Transformer and LCL Filter Design for DPFCs

Ivo M. Martins[1], J. Fernando A. Silva[2], Sónia Ferreira Pinto[2], and Isménio E. Martins[1]

[1] INESC-id, Department of Electrical Engineering, ISE, University of Algarve, Faro, Portugal
[2] INESC-id, Department of Electrical and Computer Engineering, IST, TU Lisbon, Portugal

Abstract. Flexible AC Transmission Systems (FACTS) can be used for power flow control in AC transmission grids, allowing simultaneous control of the bus voltage and line active and reactive power. However, due to high costs and reliability concerns, the application of this technology has been limited in such applications. Recently, the concept of Distributed FACTS (DFACTS) and Distributed Power Flow Controller (DPFC) has been introduced as a low cost high reliability alternative for power flow control.

This paper presents the design of a coupling transformer and a LCL filter for DPFC devices. To extract the electromagnetic energy from the transmission line a transformer with a single turn primary is designed and optimized. A third-order LCL filter is used to guarantee high order harmonics filtering. Simulations results are presented and discussed.

Keywords: FACTS, DFACTS, UPFC, DPFC.

1   Introduction

Nowadays the electrical network is facing increasing congestion and loss of reliability. Under this contingency, it is essential to improve the performance of existing power lines and optimize power flow. Flexible AC Transmission systems (FACTS) can be used for power flow control, both in static and dynamic conditions, making transmission systems more flexible [1]. Although FACTS devices offer several benefits, they have not seen widespread commercial acceptance due to a number of reasons [2]. As an alternative approach, the concept of distributed FACTS devices (DFACTS) has been proposed as a lower cost and higher reliability solution [2]. However, since the Distributed Static Series Compensator (DSSC) has no power source, it can only adjust the line impedance and is not as powerful as UPFC.

Using the concept of DFACTS devices, a new concept of distributed power flow controller (DPFC) has been proposed [3] to achieve the same functionality as the UPFC. The DPFC is derived from the UPFC but eliminates the common DC link between the shunt and series converters. As UPFC, DPFC devices give the possibility to control system parameters, such as line impedance and power angle.

After stating the Contribution to Collective Awareness System (section 2) this paper details the operation principle and configuration of DPFC devices (section 3). In section 4 the design of a coupling transformer that extracts electromagnetic energy from the transmission line is presented and in section 5 the design of a third-order LCL filter is shown. Simulation results are presented and discussed in section 6, using switching models of converters connected to the transformer secondary.

L.M. Camarinha-Matos et al. (Eds.): DoCEIS 2014, IFIP AICT 423, pp. 451–458, 2014.

2   Contribution to Collective Awareness Systems

This work follows previous research where sliding-mode controllers, based on switched state-space models, to achieve cross-decoupled (independent) control of active and reactive power flow were presented and two different DPFC series converter topologies were proposed [4], [5]. This paper proposes a power transformer and a third-order LCL filter to be part of a DPFC device. DPFC devices can contribute to sustainability of electrical power. This research work on energy management and smart grids might benefit from future collective awareness systems in order to implement cooperative control, or perform informed decision making or the effective involvement of the electrical or sustainable energy systems.

3   Distributed Power Flow Controller

DPFC devices can be used for power flow control in existing transmission lines. Multiple DPFC devices are distributed along the transmission line, cooperating together allowing cross-decoupled control of active and reactive power flow. Each DPFC device (Fig. 1) consists of a power IGBT full-bridge single-phase converter, with a DC capacitor capable to provide the required DC voltage, a series clamp-on transformer to be used in the connection to the transmission line and a low pass filter to reduce the switching frequency harmonics injected into the grid by the converter.

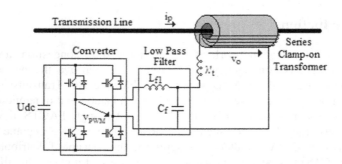

Fig. 1. DPFC configuration

From the conceptual viewpoint each DPFC device can be represented by two controllable voltage sources, connected in series with the transmission line. Each voltage source generates voltage at a different frequency, one at fundamental frequency, $v_{o,1h}$, and the other at the third-harmonic frequency, $v_{o,3h}$, so that the converter output voltage is $v_o = v_{o,1h} + v_{o,3h}$. The voltage source $v_{o,1h}$ injects a voltage vector with controllable magnitude and phase angle at fundamental frequency, allowing cross-decoupled control of the active and reactive power flow (P_{1h} and Q_{1h}). The voltage source $v_{o,3h}$ is responsible to maintain the DC bus voltage (U_{dc}) of the DPFC converter by using the third-harmonic frequency power. Therefore, using the third-harmonic line current, a controllable voltage vector is injected in series with the line, absorbing or generating active power from the third-harmonic current (P_{3h}).

4   Transformer Design

4.1   Operation Principle

A clamp-on transformer (COT) is used [6] to use the electromagnetic energy of the transmission line. The COT comprises two halves of a cylindrical torus magnetic core surrounding the power line, as shown in Fig. 1. The power line carrying the current i_p acts as the single-turn primary winding of the transformer ($n_p = 1$), being the secondary winding with n_s turns coiled around the magnetic core. The low pass filter and the AC/DC converter are connected to the transformer secondary terminals.

4.2   Transformer Design

Assuming linear operation and neglecting resistive and leakage voltage drops, the main transformer design equation, which relates the voltage V_p across the primary winding having n_p turns, given the maximum magnetic flux density B_{max}, frequency f_s and core section effective area A_{fe}, is [6]:

$$B_{max} A_{fe} \geq \frac{\sqrt{2}}{6\pi f_s n_p} (3V_{p,1h} + V_{p,3h})$$

(1)

Selecting the transformer magnetic material and establishing the allowed maximum transformer B_{max} value, knowing $V_{p,1h}$, $V_{p,3h}$ and frequency f_s, since $n_p = 1$, the core section effective area A_{fe} can be then calculated.

The transformer primary winding voltage $V_{p,1h}$, $V_{p,3h}$ is the voltage injected in series with the transmission line by the DPFC device and must be established according to specifications. For design purposes, consider a 220 kV, 300 MVA transmission line with a total of 4500 DPFC devices (1500 devices distributed along one phase) with 0.20 pu line power flow control capability. Each DPFC device must handle $S_{dpfc} = 13.3$ kVA. This means that the DPFC maximum output voltage at fundamental frequency is $V_{p,1h} = \sqrt{3} S_{dpfc} Z_{line} / V_n = 3.1$ V, where Z_{line} and V_n are the transmission line impedance and phase-to-phase voltage at fundamental frequency.

Since the maximum line current is $I_{line,1h} = 787$ A, the DPFC output apparent power is $S_{1h} = V_{p,1h} I_{line,1h} = 2.45$ kVA. Considering the phase angle δ of the output voltage $V_{p,1h}$ as $\delta = \pm \pi/2 \pm 10\%$, the maximum active power generated by the DPFC is $P_{1h} = S_{1h} \cos(\delta) = 383$ W. This active power generated at fundamental frequency must be equal (neglecting losses) to the active power P_{3h} absorbed at the third-harmonic frequency. Considering the transformer losses, low-pass filter and DPFC converter, it is assumed $P_{3h} = 1.5 \times P_{1h} = 574$ W, where $P_{3h} = V_{p,3h} I_{line,3h}$. To guarantee a low harmonic distortion of the transmission line current, the injected third-harmonic current $I_{line,3h}$ should not exceed 10% of the line nominal current ($I_{line,3h} = 0.1 \times I_{line,3h}$). Thus, from the above conditions, the DPFC maximum output voltage at third-harmonic frequency is $V_{p,3h} = 7.3$ V.

To start the transformer design from equation (1), the B_m value must be established according to the core magnetic material characteristics. Normally this value is chosen from the material magnetization curve as the highest B_m value before the saturation zone. Assuming a transformer core using M4 grade Grain-Orientation (GO) 3% Silicon Steel (Si-Fe) laminations, this value is estimated as $B_m = 1.8$ T. Thus, given the required output voltages $V_{p,1h}$ and $V_{p,3h}$, asthe primary number of turns is $n_p = 1$ and the fundamental frequency $f_s = 50$ Hz, the core section effective area can be calculated as $A_{fe} = 13.9 \times 10^{-3}$ m^2.

For the calculated A_{fe} value, the size and shape of the core is designed to minimize the total weight of the transformer (magnetic core and copper windings). To start the design, the secondary winding number of turns n_s is set according to the maximum current and voltage values in the secondary side of the transformer. Making $n_s = 18$ and considering the maximum current density $J_{cu} = 4$ A/mm^2, the section of the secondary winding wires is chosen as 10 mm^2. Therefore, the cross-sectional area of the secondary winding is $A_{w2} = 180$ mm^2. Taking into account the section of the power line cable $A_{w1} = 500$ mm^2, the total area of copper in the transformer window is $A_w = A_{w1} + A_{w2} = 680$ mm^2. Given the window space factor is nearly $K_w = 0.33$, calculated by the empirical formula $K_w = 10/(30 + KV_{hv})$, where KV_{hv} is the voltage of the secondary winding expressed in kV, the transformer window area is calculated as $W_A = A_w/K_w = 2.1 \times 10^{-3}$ m^2. This means that a transformer core with 5.1 cm inner diameter is needed. The core cross-section width around the power line cable and the transformer length can be now sized to optimize the total weight of the transformer. Representing the transformer core dimensions by the core average magnetic path length M_{gl}, the weight of the magnetic core and secondary winding as function of M_{gl} is presented in Fig. 2. As shown, the optimum value for M_{gl} is in the range 19-20 cm. Making $M_{gl} = 19.67 \times 10^{-2}$ m the core outer diameter is 7.4 cm. Given the calculated core section effective area, a 1.2 m long transformer is obtained.

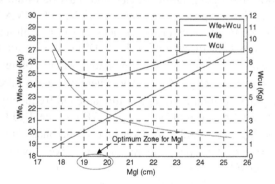

Fig. 2. Transformer weight optimization

5   LCL Filter

In grid-connected applications reduced levels of harmonic distortion are required to comply with IEEE 519-1992 standard. Therefore, a low pass output filter is used to connect the DPFC device to the electric power system (Fig. 1), to reduce the switching frequency harmonics injected to the grid by the DPFC converter.

Since between the filter and the grid a transformer is used, which inserts a leakage inductance seen by the grid, the output filter comprises an LC filter plus the transformer leakage inductance (λ'_t), which can be seen as an LCL filter but with constant leakage inductance L_{f2} on the output. Neglecting parasitic resistances and considering the output equivalent impedance $Z_o = R_o + sL_o$ seen from the transformer, for the low pass third-order order filter the transfer function is:

$$\frac{V_o(s)}{V_{PWM}(s)} = \frac{\dfrac{L_o}{C_f L_{f1}(L_{f2} + L_o)}(s + \dfrac{R_o}{L_o})}{s^3 + s^2 \dfrac{R_o}{L_{f2} + L_o} + s \dfrac{L_{f1} + L_{f2} + L_o}{C_f L_{f1}(L_{f2} + L_o)} + \dfrac{R_o}{C_f L_{f1}(L_{f2} + L_o)}} \quad (2)$$

While the numerator of the transfer function (2) has one real zero set by the output impedance $(z_1 = R_o/L_o)$, the denominator has one real pole (p_1) and two complex conjugate poles and can be represented by the polynomial $d(s) = (s + p_1)(s^2 + 2\xi \omega_p s + \omega_p^2)$, where ξ is the damping factor and ω_p the angular passband edge frequency. Equating the denominator coefficients from (2) with the polynomial $d(s)$, the filter parameters may be calculated from:

$$L_{f1} = \frac{2R_o(2\xi^2 \omega_p + \xi \omega_p^2 + p_1 \xi)}{p_1 \omega_p (p_1 + 2\xi \omega_p)}, \quad L_{f2} = \frac{R_o - L_o p_1 - 2L_o \xi \omega_p}{p_1 + 2\xi \omega_p},$$

$$C_f = \frac{(p_1 + 2\xi \omega_p)^2}{2R_o \omega_p (2\xi^2 \omega_p + \xi \omega_p^2 + p_1 \xi)} \quad (3)$$

Usually ξ and ω_p are set according to the desired filter characteristics, the pole p_1 is used to cancel z_1 and should be placed as near as possible from z_1 (ideally $p_1 = z_1$) to reduce the filter attenuation $(|A| = 20 \log_{10}(p_1/z_1))$ bellow ω_p. However, to fulfill the condition $L_{f2} > 0$ in (3), for a given ω_p, the values of ξ and p_1 are constrained by $\xi < (R_o - L_o p_1)/(2L_o \omega_p)$ and $p_1 < R_o/L_o$. Setting L_{f2} as the transformer leakage inductance $(\lambda'_t = 1.55 \,\mu H)$ then ξ and p_1 can be established from $\xi = (R_o - p_1(L_{f2} + L_o))/(2\omega_p(L_{f2} + L_o))$ and $p_1 < R_o/(L_{f2} + L_o)$.

Setting the passband edge frequency $f_p = 750$ Hz and pole $p_1 = 0.9R_o/(L_{f2} + L_o)$, the filter parameters are obtained $(R_o = 0.15 \,\Omega$ and $L_o = 4.9$ mH) as $L_{f1} = 0.55$ mH and $C_f = 91.6 \,\mu F$.

At the resonant frequency $\omega_r = \omega_p \sqrt{1 - 2\xi^2} = 4.7$ kHz the damping factor of the filter is $\xi = 3.2 \times 10^{-4}$ and the resonant peak $|M_r| = 20 \log_{10}(1/(2\xi\sqrt{1 - 2\xi^2})) = 63.8$ dB. The magnitude bode plot of the undamped filter is presented in Fig. 4.

5.1 Damped Filter Design

Since passive LCL filters have low damping characteristics at resonant frequency, they can cause instability. Therefore, the filter should be damped to avoid resonances without reducing attenuation at the switching frequency or affecting the fundamental. Several passive damping topologies can be used, each one having its particular properties [7]. Fig. 3 illustrates two practical approaches to damp the LCL low-pass filter.

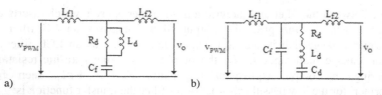

Fig. 3. Practical approaches to the damping of the LCL filter: a) Parallel R_d and L_d in series with the shunt capacitor. b) Series R_d, L_d and C_d in parallel with the shunt capacitor

5.1.1   Parallel R_d and L_d Damping in Series with the Shunt Capacitor

A damping resistor R_d can be added in series with the shunt capacitor C_f as shown in Fig. 3a. Since at the resonant frequency the impedance of the filter is zero, the aim of the damping is to insert impedance at this frequency to avoid oscillation. The main drawback of this damping method is that its transfer function contains a high-frequency zero ($z_2 = 1/(C_f R_d)$). The addition of R_d degrades the slope of the high-frequency asymptote, from -40 dB/decade to -20 dB/decade, reducing the filter attenuation above the resonant frequency. Hence, R_d must be chosen so that the value of z_2 is significantly greater than ω_r. This condition can be expressed as $R_d \ll 1/(C_f \omega_r)$. Setting the damping resistor impedance at a third of the capacitance at the resonant frequency then $R_d = 0.77\ \Omega$. The damping factor is now $\xi = 0.167$ and the resonant peak $|M_r| = 9.6$ dB. Fig. 4 illustrates how addition of the damping resistor modifies the magnitude of the transfer function, reducing oscillations in 54.2 dB, but also reducing the filter attenuation above the resonant frequency from -40 dB/decade to -20 dB/decade.

To avoid significant power dissipation in R_d, an inductor L_d can be placed in parallel with the damping resistor providing a low frequency bypass, as shown in Fig. 3a. To allow R_d to damp the filter, at the resonant frequency the inductor L_d should have an impedance magnitude sufficiently greater than R_d. However, increasing the inductance L_d increases weight and energy stored. Thus, the inductor is selected as $L_d = 4 R_d / \omega_r = 0.67$ mH.

5.1.2   Series R_d, L_d and C_d Damping in Parallel with the Shunt Capacitor

Another approach to damp the filter is to add resistor R_d in parallel with the shunt capacitor, as illustrated in Fig. 3b. The resistor results in increased power losses, therefore just by itself it is not a practical solution. To obtain the same damping factor as the previous method the resistor is calculated from:

$$R_d = \frac{L_{f2} + L_o}{C_f (\mathrm{p}_1 + 2\xi\omega_p)(L_{f2} + L_o) - C_f R_o} = 6.9\ \Omega \qquad (4)$$

Fig. 4 illustrates how the parallel damping resistor reduces filter oscillations at the resonant frequency without reducing attenuation above this frequency.

One practical solution to significantly reduce the power dissipation in R_d is to add a tuned L_d-C_d circuit in series with R_d, as illustrated in Fig. 3b. To allow R_d to damp the filter, the value of the high-frequency blocking inductor L_d and the DC blocking capacitor C_d are chosen such that, at the filter resonant frequency, the impedance of the damping branch is dominated by the resistor R_d. Therefore, the inductor is selected as $L_d = R_d/\omega_r = 1.5$ mH and the capacitor from $C_d = 1/(\omega_r^2 L_d) = 31$ μF.

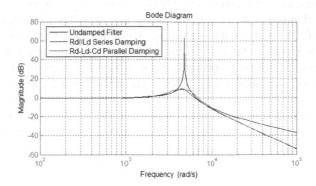

Fig. 4. Magnitude bode plot of the LCL filter

6   Simulation Results

The presented transformer and filter with parallel R_d and L_d damping in series with the shunt capacitor has been modeled and simulated in Matlab/Simulink environment, considering the implementation of the DPFC devices in a transmission network. The simulations values were obtained for a power system consisting of the sending and receiving end voltages V_s and V_R, connecting the load R_{load}, L_{load} through a transmission line R_{line}, L_{line}, with 4500 DPFC devices (1500 devices per phase).

Fig. 5a shows the PWM voltage v_{PWM} injected by the converter and its reference v_{oref}. The reference voltage is calculated according to the specified levels of active and reactive power. Fig. 5b shows the primary and secondary winding voltages v_p and v_s, where v_s is divided by the secondary winding number of turns n_s. As can be noted the effective transformer turns ratio is not exactly n_s/n_p, due to windings and leakage voltage drops.

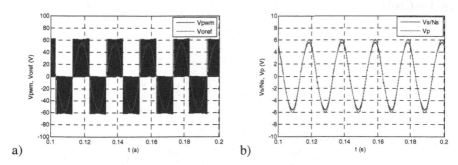

Fig. 5. a) Converter PWM output voltage. b) Transformer winding voltages.

7   Conclusions

In this paper a power transformer and a third-order LCL filter to be part of a DPFC device was presented. To couple the DPFC device to the transmission line, the transformer is clamped in series with the power line, avoiding galvanic contacts. The LCL low-pass filter interfaces the transformer with the single-phase full-bridge IGBT based converter, to reduce the high frequency switching harmonics. For the designed filter, two passive damping methods were presented. Simulation results were presented showing the effectiveness of the designed transformer and filter.

Acknowledgments. This work was supported by Portuguese national funds through FCT - Fundação para a Ciência e a Tecnologia, under project PEst-OE/EEI/LA0021/2013.

References

1. Gyugyi, L., Hingorani, N.G.: Understanding FACTS: Concepts and Technology of Flexible AC Transmission Systems. IEEE Press, New York (1999)
2. Divan, D., Johal, H.: Distributed FACTS – A New Concept for Realizing Grid Power Flow Control. IEEE Trans. Power Electronics 22, 2253–2260 (2007)
3. Yuan, Z., de Haan, S.W.H., Ferreira, B.: A New FACTS component – Distributed Power Flow Controller (DPFC). In: European Conference on Power Electronics and Applications, Aalborg, pp. 1–4 (2007)
4. Martins, I.M., Silva, F.A., Pinto, S.F., Martins, I.E.: Control of distributed power flow controllers using active power from homopolar line currents. In: IEEE 13th International Conference OPTIM 2012, Brasov, pp. 806–813 (2012)
5. Martins, I.M., Silva, F.A., Pinto, S.F., Martins, I.E.: Independent Active and Reactive Power Control in Distributed Power Flow Controllers (submitted for publication)
6. Silva, F.A., Lopes, D., Sequeira, J.: Designing Transformers for the Power Supply of a Transmission Line Inspection Robot. In: Congrès 2012 CIGRÉ Canada, Montréal, pp. 24–26 (2012)
7. Ahmed, K.H., Finney, S.J., Williams, B.W.: Passive Filter Design for Three-Phase Inverter Interfacing in Distributed Generation. In: Compatibility in Power Electronics 2007, Gdansk, pp. 1–9 (2007)

Part XVII

Power Conversion

Resonant Power Conversion through a Saturable Reactor

Luis Jorge, Stanimir Valtchev, and Fernando Coito

Faculdade de Ciências e Tecnologia, Universidade Nova de Lisboa
Monte de Caparica 2829-516 Caparica, Portugal

Abstract. The resonant converter control is usually implemented by electronic circuits that regulate the parameters of the necessarily produced energy pulses of circulating in the resonant circuit. In the same time the power circuit is characterized by fixed (or uncontrollably varying) parameters. In this article a research is described that shows a long time ago forgotten techniques: the magnetic amplifier that controls the power circuit parameters and as a result, changes the resonant frequency depending on the needs of control. The inductance that is made to vary its value in a controllable, continuous and linear mode would be a perfect (non-dissipative) regulator for the power converter. This can be achieved through a DC magnetization applied as a control command to the magnetic core. By varying this magnetization current and hence the magnetic parameters of the core, the inductance L is possible to be adjusted to a desired value. This operation is similar to the principle of the (long ago) well known "magnetic amplifier" or the "saturable core reactor" as it is often called. The magnetic amplifier usually has an AC source in series with the load and with its primary while in the secondary the DC control signal is applied. The application of a regulated inductance device in the Wireless Power Transfer (WPT) will guarantee a simple way to adjust the frequency of the transmitter and/or the receiver to achieve the required impedance adaptation and efficient energy transfer.

Keywords: Magnetic amplifier, mag amp, saturable reactor, wireless power transfer.

1   Introduction

The resonant converter control technique is more developed now compared to some decades ago, but still it is very complex and costly to implement. In search for new simple and reliable methods of control, this work is aimed to demonstrate the possible use of the magnetic amplifier principles in the control of the (power) resonant circuit.

The inductance that is capable to vary its value in a continuous and linear mode would be a perfect (non-dissipative) regulator for the power converter, especially if the switching frequency will be kept constant. This can be achieved through a DC current applied as a control. By varying the current and hence the magnetic flux, the value of the resonant inductor L will be regulated [1], [10].

This principle is similar to the functioning of the "magnetic amplifier" or the "saturable reactor core" as they are often called. This circuit was invented by E. F. W. Alexanderson in 1916, USA, and was used during long time for the theatre light dimming [1].

L.M. Camarinha-Matos et al. (Eds.): DoCEIS 2014, IFIP AICT 423, pp. 461–469, 2014.

During the World War II, the German scientists gave a big boost to those circuits, by using them in the V2 missiles and in the airplanes' navigation equipment.

The magnetic amplifier has the AC source and the load in series with the primary winding while in the secondary a DC control signal is applied. To avoid that the powerful AC signal appears in the secondary of that "transformer" and provoke problems in the control system, it is necessary to use a special "three legs transformer". This type of connection gives a possibility to isolate the two circuits: the control circuit and the power (source) circuit [1] [10].

The application of this type of circuit that provides regulation of the inductance, in the Wireless Power Transfer (WPT) will make possible to adjust the frequency of the transmitter and/or the receiver. This will permit to obtain impedance adaptation in order to achieve a maximum energy transfer.

The limit of the proposed method is set by the highest achievable frequency that is determined mostly by the eddy current losses in the core material.

2   Relationship to Collective Awareness Systems

The resonance based energy transfer is important solution for efficient energy application in electric and hybrid vehicles, very fast trains, embedded medical systems, home automation, biomedical systems and the emerging field of Internet of Things that includes also the energy transport. By the proposed resonant frequency adjustment through simple and robust regulation, it will be possible to achieve a widely spread and massive production of cheap wireless energy transmission equipment. The developed regulated inductive element would be useful also in the power quality compensation which is the main concern of the modern society and in particular, the emerging Smart Grids.

3   Saturable Reactor

The saturable reactor is a magnetic device that consists of a single magnetic core, having two windings. One is connected to AC source and the (power) load, normally designated by "load coil" and another is connected to a DC source namely "control coil". The idea is to control of the load current and voltage by the control coil, similarly to what is done in a transistor or another electronic device. In Fig. 1 (left) is shown a saturable reactor. In the left side of the magnetic core is the DC control

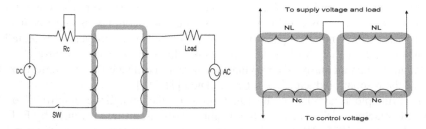

Fig. 1. Simple saturable reactor circuit (left); Saturable reactor having two cores (right)

winding (a variable resistance is representing the regulator). In the right side of the core are shown the AC power source and the load connected in series.

When the control current is varied, the impedance of the load winging is changing, once the permeability of the magnetic circuit varies, producing a change of the inductance. However, in this basic circuit, exists a problem: it is the mutual induction between the DC and AC windings that provokes interaction between both circuits. The power current in the AC winding will induce undesirable feedback to the DC side.

One way to solve this problem is the circuit presents in Fig. 1 (right), with two cores construction that separates the cores and the windings.

This circuit, having two separate cores with series control and series load windings, makes possible to cancel the induced voltages and to minimize the unwanted effects.

The saturable reactor is based on magnetization curve of the applied magnetic core material. A generic magnetization curve is shown in Fig. 2.

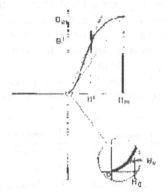

Fig. 2. Generic magnetization curve

The relation of B to H at any point on the curve is the *normal permeability* for that value of B. For the maximum flux density B_m the normal permeability is:

$$\mu = \frac{B_m}{H_m} \tag{1}$$

It is the slope of a straight line drawn starting in the origin.

A similar line drawn tangent to the curve at its "knee" is called the *maximum permeability* and its value is:

$$\mu_m = \frac{B'}{H'} \tag{2}$$

The slope B_0/H_0 of the normal induction at the origin (enlarge in Fig. 2) is the permeability for very low induction value B_0; it is called initial permeability and it is usually much lower than μ_m [2].

The most common saturable reactor consists of a three legged core as it will be shown in Fig. 5, where each coil is wound on separate magnetic "leg". The two AC coils in Fig. 6 should have relatively low and equal number of turns and the middle coil usually has much higher number of turns. This gives the possibility of amplification:

through lower control current to obtain control over powerful circuits. Because of this in many cases the (regulated) saturable reactor is used as a "Magnetic Amplifier".

In open circuit condition of the control circuit, the imaginary part of the complex impedance value is much higher than the real part of the impedance (the winding DC resistance). It must be recalled that a relatively small amount of direct current flowing into the control winding of the core reactor, because of the high number of turns and in case of no air gap, is easily capable to produce core saturation. Thus, the reactance of the core reactor may be varied by a small amount of DC power. In this way, the AC winding of the core regulates the amount of power delivered to the load by the AC source, controlled by a small amount of DC power flowing in the DC winding.

Usually, when one winding of the core reactor is used to control the power in the other winding through reactance variation, the magnetic device is called a saturable reactor [2], [3], [4]. The concept is the same for the magnetic amplifier: the lower DC component of the impedance permits regulation of the permeability. The main difference between a magnetic amplifier and a saturable core reactor is that the magnetic amplifier uses a half of the magnetic loop and the amplifier uses the complete magnetic loop. Because of this the term "Saturable Core" is often substituted by the term "Magnetic Amplifier".

Basically a saturable reactor consists of three essential elements (Fig. 3): Direct current source, Magnetic core with windings and Alternating current source.

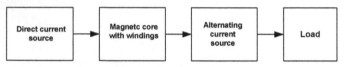

Fig. 3. Three essential elements of the saturable reactor

These three elements, when connected together, form the essentials of the Saturable reactor. Its operation is based upon the above mentioned principle: the current flowing in the coil wound on the magnetic core can be regulated by varying the core saturation [4], [5], [6]. In order to optimize the magnetic amplifiers, the choice of magnetic core is very important. It is also important to choose which part of the core is more convenient to saturate.

The main goal of any magnetic core is to supply an easy path for the magnetic flux in order to guarantee the flux linkage (magnetic coupling), between two or more elements. The energy storage in a magnetic core is an undesired parasitic effect. In order to minimize this problem, the choice of a magnetic material with high permeability and thin hysteresis loop is the right option.

Today, the wound magnetic cores used in practice are made by amorphous metal alloy. These cores are used in SMPS (Switch Mode Power Supply) application up to 100-200kHz, and especially in magnetic amplifiers [6], [7].

As it was shown, the magnetic amplifier is a relatively simple device. However, this simplicity is very misleading when mathematical analysis of its operation is attempted, because of the extremely nonlinear characteristics of the core material. The linear circuit theory may be applied only to carefully selected parts of its operation regimes. In order to simplify the analysis the common practice is to make the assumptions of perfect core material. As it was mentioned before, the AC winding voltage will induce undesired voltage in the DC winding. This influence may short circuit the DC winding if the impedance of the DC winding is low and by this, it would effectively short circuit the AC voltage in the power winding. This difficulty might be overcome by using high impedance in the DC control circuit [8], [9].

In Fig. 4 (a) a common solution is shown, in which two reactor cores are used. The DC current is applied in opposite direction in each winding. The AC current is applied in the same direction in both coils. A similar solution is shown in Fig. 4 (b).where the connection is parallel in order to reduce the necessary AC voltage [2], [3].

Fig. 4. Magnetic amplifier structures: (a) series; (b) parallel [2]

When there is zero direct current in the control winding of Fig. 4, both reactor impedances are large and prevent any load current except a small exciting current to flow in the AC loop in the lower parts of the voltage sinusoids. When non-zero direct current is applied to the control winding, the impedance remains large for the first part of a cycle, until the saturation flux density is reached. Then the reactor impedance is reduced and larger load current may flow.

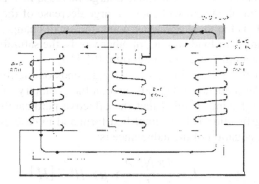

Fig. 5. Three legs transformer applied to magnetic amplifier [2]

In Fig. 5 a three-leg magnetic core is shown. The current in the DC coil will produce a change in the total flux and hence a change of inductance [9], [10]. This core has one DC coil and two AC coils and the total AC and DC flux is a changing sum: alternatively in each of the lateral legs it is either sum or difference of the fluxes. The figure shows the virtual paths for the AC and DC fluxes. The number of turns of the AC half-windings is equal and symmetric. The equal number of turns in the AC coils will provoke equal AC magneto-motive force, which will be cancelled in the centre leg and cause fluxes to flow as indicated by solid line. In consequence, no fundamental AC voltage is induced in the DC coil, but the DC flux is injected in both outer legs as indicated by the dotted lines.

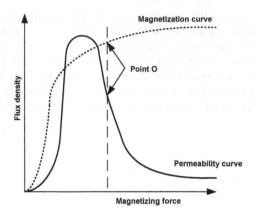

Fig. 6. Variation of flux and permeability with magnetization force

Fig. 6 shows the magnetization and permeability curves for the saturable core reactor with the ideal operating point indicated (point "O").

It is important to notice that this point is located on the "knee" of the magnetization curve. The "knee" of the curve corresponds to the maximum curvature. Saturable core reactor and magnetic amplifier should operate on this "knee" of the magnetization curve [9], [10].

When the saturable core reactor is set at the "knee" of the magnetization curve, any small increase in control current will cause a large increase of load current, and any small decrease in control current will cause a large decrease of the load current. This is why the point "O" is the ideal operating point: the small changes in control current will cause large changes of load current, in other words, the saturable core reactor can better and efficiently amplify the control current [9], [10].

The saturable reactor AC winding inductance depends on the flux provoked by the DC winding (-s). The inductance depends also on the geometry of the magnetic core and its permeability, having in mind that the effective permeability depends on the DC control winding function [1]. In simplified form and in case of flux concentrated only in the core, the equation of the inductance is:

$$L = \frac{4\pi N^2 A}{l} \times \mu \times 10^{-7} [H] \tag{3}$$

Once the geometry of the core is fixed by the manufacturing, the unique variable is the permeability μ. In this case the permeability will control the inductance L.

4   Results Presentation

In this section the results obtained from measurements during the experimentation are shown. The circuit that was used as a basis for different tests is the one shown in Fig. 5. The experiment was executed in circuit corresponding to the one in Fig. 7:

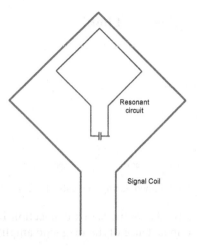

Fig. 7. The circuit base to the different tests

This circuit (Fig. 7) consists of two parts. The one in blue (external coil), is the signal coil with 10 turns, another is the resonant circuit, shown in red (internal coil). This coil has 30 turns and two capacitors 0,1µF each, connected in series.

The inductance of the resonant coil (shown in red) was measured and the (initial) value of $L=7,58mH$ has been obtained.

In these conditions the calculated resonant frequency is:

$$f_r = \frac{1}{2\pi\sqrt{L \times C}} = \frac{1}{2\pi\sqrt{7,58 \times 10^{-3} \times 50 \times 10^{-9}}} = 8,175kHz \qquad (4)$$

In fact, this value was confirmed when the experiments were executed, i.e. close to 8,2 kHz. The magnetic amplifier was constructed on a tree legs transformer, where the middle leg (control) have a coil of 30 turns and the two legs (one on each side) form two half-coils in series with 20 turns each.

In Fig. 8 the schematic diagram of the three legs transformer is represented. The "green" coils are the AC half-coils, A1 and A2, and the "red" coil, C, is the control coil, connected to the variable DC.

The graphic shown in Fig. 9 presents the values obtained during the tests of the magnetic amplifier.

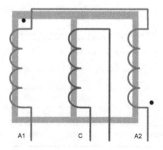

Fig. 8. Schematic of three legs transformer

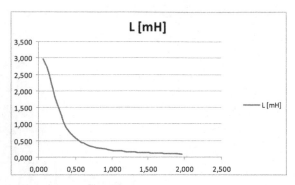

Fig. 9. The values obtained during the tests of the magnetic amplifier

The illustration in Fig. 10 shows the series connection between a fixed inductor (4,7 mH) and the regulated inductance of the magnetic amplifier. The values obtained are shown in Fig. 11.

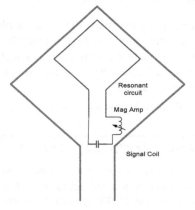

Fig. 10. Schematic of the tests on the saturable reactor in series with the resonant circuit inductor

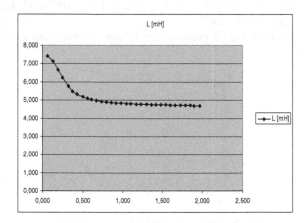

Fig. 11. The values obtained with Saturable reactor in series with resonant circuit

5   Conclusions and Future Work

A confirmation was found that the magnetic amplifier gives the possibility to adjust the inductance L, by a DC current control The correct starting point for the regulation (zero DC control current) of the power should be the in which the inductance guarantees the minimum output power. All the other points of operation will be achieved by the non-zero current of the DC coil.

By some value of a fixed (initial) and applying a regulated additional inductance, the resonant circuit will obtain the necessary capacity to regulate its resonant frequency and thus to regulate the power of the resonant converter.

The future work should involve the study of magnetic cores applied to optimize this circuit in terms of frequency and maximum allowed AC current along with the large spectrum of frequencies. The necessary electronic circuits should be constructed to control the completed system

References

1. Steffen Jr., E.W.: Lieutenant Commander, United States Navy– United States Naval Postgraduate School, Annapolis, Maryland (1948)
2. (November 6, 2013), http://www.vias.org/eltransformers/lee_electronic_transformers_04_17_01.html
3. Lufcy, C.W.: A survey of Magnetic Amplifier (January 7, 1955)
4. Woodson, H.H.: Doctor of Science Thesis. Magnetic Amplifier Analysis Using a generalized Model for the Saturable Reactor Core. MIT (1956)
5. Lee, I.-I., Chen, D.Y., Wu, Y.-P., Iamerson, C.: Modelling of Control Loop Behaviour of MagAmp Post Regulators. IEEE Trans. on Power Electronics 5(4) (1990)
6. Takashima, Y., Hata, S.: Variable Gain Magnetic Amplifier and Application for Power Invariance Control System. Osaka Prefecture University Education and Research Archives (March 31, 1971)
7. International Commission on Non-Ionizing Radiation Protection (ICNIRP) – Guidelines for Limiting Exposure to Time-varying Electric, Magnetic and Electromagnetic Fields (up to 300GHz) (1998)
8. Grilo, F.C.V.: O TRANSDUTOR (Amplificador Magnético) (1953)
9. Mali, P.: Magnetic Amplifiers - Principles and Applications. John F. Rider Publisher, Inc., New York
10. Platt, S.: Magnetic Amplifiers – Theory and Application. Prentice-Hall, Inc., Englewood Cliffs (1958)

Piezoelectric Energy Harvester
for a CMOS Wireless Sensor

Nuno Mancelos[2], Joana Correia[2], Luís Miguel Pires[1,2], Luís B. Oliveira[1,2],
and João P. Oliveira[1,2]

[1] Centre for Technologies and Systems (CTS) – UNINOVA
[2] Dept. of Electrical Engineering (DEE), Universidade Nova de Lisboa (UNL)
Campus FCT/UNL, 2829-516, Caparica, Portugal
{n.mancelos,jm.correia,l.pires}@campus.fct.unl.pt,
jpao@fct.unl.pt

Abstract. The emerging of collective awareness platforms opened a new range of driven forces that will converge to more sustainable systems. To achieve this task, these platforms have to support an increasing number of more sophisticated remote sensors and actuators that will need to cooperate smartly and strongly with each other in a mesh type of intelligent interconnectivity. These remote smart miniaturized nodes can add noninvasive intelligence but suffer from lifetime performance due to the small quantity of energy available in micro batteries. Therefore, harvesting energy from the environment is a promising technique. This work presents the study and experimental evaluation of a flexible piezoelectric material to validate the use of a piezoelectric harvester in a CMOS wireless actuator/sensor node.

Keywords: Energy harvesting, Piezoelectric transducers, Self-powered, Micro-systems, Smart systems, Wireless sensor nodes, MEMS, WSN.

1 Introduction

The recent advances in ultra-low-power device integration, communication electronics and Micro Electro-Mechanical Systems (MEMS) technology have fuelled the emerging technology of Wireless Sensor Networks (WSNs). The spatial distributed nature of WSNs often requires batteries to power the individual sensor nodes. One of the major limitations on performance and lifetime of WSNs is the limited capacity of these finite power sources, which must be manually replaced when they are depleted. Moreover, the embedded nature of some of the sensors and hazardous sensing environment make battery replacement very difficult and costly. The process of harnessing and converting ambient energy sources into usable electrical energy is called energy harvesting. Energy harvesting raises the possibility of self-powered systems, which are ubiquitous and truly autonomous, and without human intervention for energy replenishment. Among the ambient energy sources such as solar energy, heat, and wind, mechanical vibrations are an attractive ambient source mainly because they are widely available and are ideal for the use of

L.M. Camarinha-Matos et al. (Eds.): DoCEIS 2014, IFIP AICT 423, pp. 470–477, 2014.

piezoelectric materials, which have the ability to convert mechanical strain energy into electrical energy.

Contributions of this paper are summarized as follows. Section 2 presents the relationship to Collective Awareness Systems. In Section 3, state of the art of energy harvesting transducers is presented. In Section 4 the concept of synergy between piezoelectric energy harvesters and WSN is justified and a block diagram of the proposed self-powered Wireless Actuator/Sensor Node is presented. Section 5 presents experimental evaluation of the piezoelectric harvester. Finally some conclusions are drawn in section 6.

2   Relationship to Collective Awareness Systems

Nowadays, the emergence of collective awareness systems is pushing the performance of end user interactive objects. The support framework based in a interconnected objects and things (Internet of Things, IoT) is the bridge that combines technologies and components from micro-systems (miniaturized electric, mechanical, optical and fluid devices) with knowledge, technology and functionality from several areas of research.

However, Harbor Research [1] defines smart systems as a new generation of systems architecture (hardware, software, network technologies, and manage services) that provides real-time awareness based on inputs from machines, people, video streams, maps, new feeds, sensors and more that integrate people, process, and knowledge to enable collective awareness and decision making.

WSN provide endless opportunities, but at the same time pose formidable challenges, such as the fact that energy is a scarce and usually non-renewable resource. However, as part of WSN, micro-systems could provide advances in low power Very Large Scale Integration (VLSI), embedded computing, communication hardware, and in general, the convergence of computing and communications, are making this emerging technology a reality. Likewise, advances in nanotechnology and MEMS are pushing toward networks of tiny distributed sensors and actuators.

As mentioned in the previous section, energy harvesting can dramatically extend the operating lifetime of nodes on WSN. Finally, this technology enables battery less operation and reduces the operation costs of WSN, which are mainly due to battery replacement.

3   Energy Harvesting from the Surroundings

Energy harvesting techniques developed for micropower generators deal with the challenge of scavenging and making use of residual energy present in ambient sources, usually energy in the form of light, Radio Frequency (RF) electromagnetic radiation, thermal gradients and many sources of motion, namely rotation, vibration and fluid flow. Energy harvesting transducers that make use of energy in the form of motion are called Electromechanical and are separated in three different groups: electromagnetic, electrostatic and piezoelectric transducers. Piezoelectric transducers make use of the piezoelectric effect, which refers to the accumulation of an electrical charge in some solid materials, like crystals and certain ceramics, when a mechanical stress is applied to them. The effect is reversible, which means that movement, in the form of oscillation, can occur for the resonance frequency of the particular piezoelectric material to which an electrical charge is applied. These transducers are usually designed to harvest energy from vibration sources. Their harvesting optimization highly depends on the success of the characterization of the vibration from which the energy is to be harvested.

The three electromechanical transducer types are very different from each other. In general, power efficiency of a mechanical transducer could be considered as the ratio between the electrical power it delivers and the mechanical power it receives from the motion source. However, comparing different transducers is not trivial.

In [2], a variant of harvester effectiveness performance indicator is introduced, the Volume Figure of Merit (VFM), which is used to compare the performances of energy harvesting electromechanical transducers as a function of their size. The devices chosen for the actual comparison are electromechanical transducers of the three types mentioned above. Table 1 contains the two best results achieved for each of the transducer types. The main conclusion taken from the presented results is that piezoelectric transducers achieve reasonable values for power efficiency, when compared to the other electromechanical energy harvesting transducers.

Table 1. VFM for the Three Transducers Types (Extracted from [2])

Transducer type	Reference	VFM[%]
Electromagnetic	[3]	0.52
	[4]	0.64
Electrostatic	[5]	0.06
	[6]	0.68
Piezoelectric	[7]	1.39
	[8]	1.74

4   Synergy between Piezoelectric Energy Harvesters and Wireless Sensor Networks

Wireless Actuator/Sensor Nodes represent a wide range of devices with different functionalities and characteristics with an estimated power consumption level that can be retrieved from the data available in [9]. Some commercially available actuator sensor nodes, namely Crossbow MICAZ, Intel Mote 2 and Jennie JN5139, have power consumption levels of 2.8 mW, 12 mW and 3 mW, respectively.

In [10], the power consumption of a wireless sensor node based on the Nordic RF24L01 wireless transceiver is analysed. The conclusion reached is that the RF transceiver is responsible for 74% of the power consumption, the rest being associated with the power consumptions of the microcontroller, the power management module, sensors, actuators and Analog to Digital Converter (ADCs).

The power supplied by a piezoelectric energy harvester depends greatly on the piezoelectric transducer's characteristics and the vibration conditions it is submitted to. Furthermore, for each piezoelectric transducer setup there are optimal vibration conditions, which should match the resonance frequency of the piezoelectric material, and its optimal acceleration, which is directly related to the vibration amplitude. Considering an optimized energy harvesting setup regarding the transducer and the characterisation of the vibration frequency, the circuit responsible for rectification and eventual voltage supply regulation must be designed and implemented. It should be noted that this circuit has a power efficiency associated to it. As an optional component in the power management part of the system, and depending, once again, on the vibration predictability, an energy storage device might be considered.

Two examples of piezoelectric energy harvesters implemented with optimal vibration characteristics and resistive loads that maximize the power output are given in [9] and [11]. In the first case, a power output of 3 mW was achieved, while in the second work a value of 4.4 mW was also experimentally verified. For the first implementation, a Volture V22BL piezoelectric transducer was used, with vibration characteristics of 50 Hz and 1g. The second example makes use of a Piezo Systems T226-A4-503X piezoelectric transducer, with vibration characteristics of 50 Hz and 0.5 g. It should be noted that the power outputs are achieved using optimized resistive loads for each of the situations.

Although power consumption and power output levels do not match, energy harvesting solutions can be made possible using a duty cycling technique, making use of a battery or a supercapacitor. This technique consists in switching periodically the sensor node on and off. Since the wireless sensor node doesn't have to be constantly communicating, the energy harvested by the transducer is used in the communication time interval of the sensor node and stored in the battery (or capacitor) during the rest of the cycle (switched off time interval). In [2], a duty cycle of 1.6% is used for a solution using a piezoelectric energy harvester powering a custom designed radio transceiver that requires 12 mW when transmitting. The power consumption values previously presented, from [9], are based in operating conditions of 1% communication, 10% processing and 89% sleeping. The proposed block diagram of a self-powered actuator/sensor node is presented in Fig. 1.

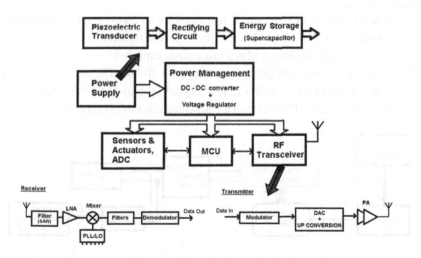

Fig. 1. Block diagram of a Piezoelectric-powered WSN

5   Experimental Evaluation of the Piezoelectric Energy Harvester

The chosen piezoelectric energy harvester Midé Volture™ V21bl [12], designed for vibration energy harvesting, is presented in Fig. 2. It uses the piezoelectric characteristics of its specific piezoceramic material to produce electrical charge when mechanically stimulated [13] and consists in a package of piezoelectric materials in a protective skin, with four pre-attached electrical leads.

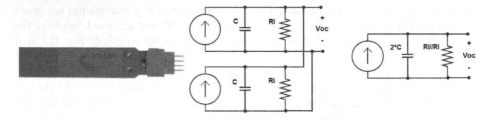

Fig. 2. Midé Volture™ V21bl [12]   **Fig. 3.** Parallel equivalent circuit

The piezoelectric harvester includes two electrically isolated piezo wafers, which may be used independently or in one of two possible combinations. These combinations can be optimised for increased output voltage (series configuration) or increased output current (parallel configuration). For this implementation, and since the desired output voltage is relatively low compared to the range of typical output voltages of the piezoelectric harvester, the parallel configuration is used, in order to maximize the output current needed to charge the capacitor. Fig. 3 presents the final equivalent circuit of the parallel configuration of the two wafers.

A block diagram of the test setup is shown in Fig. 4, which comprises three complementary parts:

1. a test vibration module which is formed by an audio amplifier (LM386) and vibration speaker coupled to the piezoelectric transducer;
2. a rectifying diode bridge plus a fast charge capacitor;
3. a data-acquisition subsystem, consisting of a 10-bit ADC and a Microcontroller (MCU) connected to both a Liquid Crystal Display (LCD) Module (HD44780 module, 16X2 LCD Panel) and a USB connection to a computer.

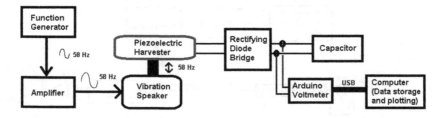

Fig. 4. Block diagram of the test setup

A picture taken of the test setup is shown in Figure 5.

The energy stored in the capacitor is calculated using the standard equation given by,

$$E = \left(\frac{1}{2}\right)cV^2 \tag{1}$$

For the Energy calculation, the actual capacitance of the parallel of capacitors was set to 2100 μF. In fact, this energy was calculated using the discharge time of the capacitor (from 1.3 V to 0 V), which is approximately 21 seconds, Fig. 6 a). Using this discharge time, the resistance value of the RC circuit (2 KΩ) and the RC time constant approximation, the actual capacitance of the capacitor can be calculated. Using this value, the estimated stored energy in the capacitor is 1.77 mJ, for 1.3 V.

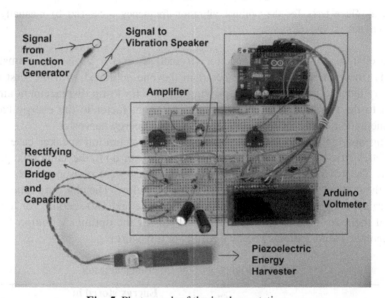

Fig. 5. Photograph of the implementation

Two different methods of vibration were used during the tests: the manually induced and the one produced by the small amplitude vibration speaker. The optimal vibration frequency was found for the second case.

For the manually induced movement test, a constant oscillation frequency was emulated. Given the equivalent circuit of the piezoelectric harvester, previously presented in Fig. 3, the expected behaviour of the charging circuit is described by,

$$V_C = V_{Final} * \left(1 - e^{-\frac{t}{RC}}\right) \tag{2}$$

where V_C is the instant voltage across the capacitor while V_{Final} depends on the open circuit voltage of the piezoelectric harvester. The charging time was approximately 68 seconds, Fig. 6 b). After some testing using a speaker, the optimal vibration frequency for energy harvesting purposes was found to be 58 Hz. Three tests were run with this vibration frequency, following the same procedure. The charging process was interrupted when the capacitor reached 1.3 V, after which the capacitor is connected to a 2 kΩ resistor. Fig. 6 a) shows the results of these three tests.

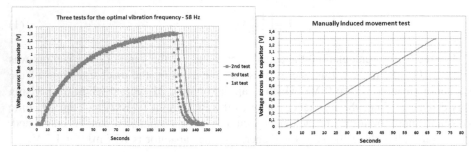

Fig. 6. a) Three tests for the optimal vibration frequency – 58 Hz; b) manually induced movement test

The charging time of the capacitor (from 0 V to 1.3 V) is approximately 125 seconds (medium value). It is greater than the one achieved with the first test (the manually induced movement test), proving that the tip-to-tip displacement amplitude, related to the vibration acceleration, is an important factor in the energy harvesting implementation, using this particular piezoelectric energy harvester.

Using non-optimal vibration frequencies for the vibration speaker case, the time the piezoelectric harvester takes to charge the capacitor up to the same voltage level is greater. To prove this, two non-optimal frequencies were chosen: 60 Hz, which is above the optimal frequency, and 56 Hz, which is below. The charging times for these tests are approximately 250 seconds for the first one and 180 seconds for the second. Both of them are greater than the one achieved for the optimal vibration frequency, 125 seconds. Table 2 shows practical results of this work.

Table 2. Summary of the experimental results

Vibration type	Vibration frequency	Charging time (from 0V to 1.3V)	Energy stored in the capacitor after charging time	Used capacitor
Vibration Speaker	58 Hz	125 s		
	56 Hz	180 s	1.77 mJ	2100 μF
	60 Hz	250 s		
Manual	-	68 s		

6   Conclusions

The main conclusion of the present work is that it is possible to implement an energy harvesting solution using the piezoelectric energy harvester Midé Volture™ V21bl for low voltage applications, even if the vibration conditions applied to the harvester are not optimal. Comparing the vibration speaker case and the test involving manually induced oscillation, it is possible to conclude that the tip-to-tip displacement amplitude is a very important factor in terms of energy generated by the piezoelectric device, more so than the actual selection of the right frequency for the oscillation.

For each configuration of the mechanical vibration method, there is an optimal vibration frequency, which is 58 Hz, in the described case.

Applying this study to a microsensor node makes energy-harvesting operation a possibility for microsensor networks [14] based on CMOS technology.

References

1. Harbor Research: Machine-to-Machine (M2M) an Smart Systems Forecast 2010-2014, Harbor Research (2010)
2. Mitcheson, P.D., Yeatman, E.M., Rao, G.K., Holmes, A.S., Green, T.C.: Energy Harvesting From Human and Machine Motion for Wireless Electronic Devices. In: Mitcheson, P.D. (ed.), vol. 96(9) (September 2008)
3. Ching, N.N.H., Wong, H.Y., Li, W.J., Leong, P.H.W., Wen, Z.: A laser-micromachined vibrational to electrical power transducer for wireless sensing systems. In: Proc. 11th Int. Conf. Solid-State Sensors Actuators, Munich, Germany (June 2001)
4. Ching, N.N.H., Wong, H.Y., Li, W.J., Leong, P.H.W., Wen, Z.: A laser-micromachined multi-modal resonating power transducer for wireless sensing systems. Sensors Actuators A, Phys. 97-98, 685–690 (2002)
5. Despesse, G., Chaillout, J., Jager, T., Léger, J.M., Vassilev, A., Basrour, S., Charlot, B.: High damping electrostatic system for vibration energy scavenging. In: Proc. 2005 Joint Conf. Smart Objects Ambient Intell. – Innov. Context-Aware Serv., Grenoble, France, pp. 283–286 (2005)
6. Arakawa, Y., Suzuki, Y., Kasagi, N.: Micro seismic electrets generator using electrets polymer film. In: Proc. 4th Int. Workshop Micro and Nanotechnology for Power Generation and Energy Conversion Applicat., Kyoto, Japan, pp. 187–190 (November 2004)
7. Fang, H.B., Liu, J.Q., Xu, Z.Y., Dong, L., Wang, L., Chen, D., Bing-Chu, C., Liu, Y.: Fabrication and performance of MEMS-based piezoelectric power generator for vibration energy harvesting. J. Microelectron 37(11), 1280–1284 (2006)
8. Roundy, S., Wright, P.K., Rabaey, J.M.: Energyscavenging for wireless sensor networks, 1st edn. Kluwer Academic, Boston (2003)
9. Yoon, Y.-J., Park, W.-T., Li, K.H.H., Ng, Y.Q., Song, Y.: A Study of Piezoelectric Harvesters for Low-Level Vibrations in Wireless Sensor Network. International Journal of Precision Engineering and Manufacturing 14(7) (July 2013)
10. Huang, L., Pop, V., Francisco, R.d., Vullers, R., Dolmans, G., de Groot, H., Imamura, K.: Ultra Low Power Wireless and Energy Harvesting Technologies – An Ideal Combination. In: IEEE, Eindhoven, The Netherlands and Osaka, Japan (2010)
11. Zhou, D., Kong, N., Ha, D.S., Inman, D.J.: A Self-powered Wireless Sensor Node for Structural Health Monitoring. In: Health Monitoring of Structural and Biological Systems, USA (2010)
12. Datasheet: Midé: PIEZOELECTRIC ENERGY HARVESTERS. Volture™
13. Datasheet: Volture Products - Material Properties. Volture Products - Material Properties, Volture™
14. Calhoun, B.H., Daly, D.C., Verma, N., Finchelstein, D., Wentzloff, D., Wang, A., Cho, S.-H., Chandrakasan, A.: Design considerations for ultra-low energy wireless microsensor nodes. IEEE Trans. Comput. 54(6), 727–740 (2005)

Bidirectional DC-DC Converter Using Modular Marx Power Switches and Series/Parallel Inductor for High-Voltage Applications

Ricardo Luís[1], J. Fernando A. Silva[2], José C. Quadrado[1], Sónia Ferreira Pinto[2], and Duarte de Mesquita e Sousa[2]

[1] Instituto Superior de Engenharia de Lisboa, R. Conselheiro Emídio Navarro, 1959-007 Lisboa, Portugal
rluis@deea.isel.pt, jcquadrado@isel.pt
[2] Instituto Superior Técnico, Instituto Superior Técnico, INESC-id, DEEC, AC Energia
Av. Rovisco Pais, 1049-001 Lisboa, Portugal
{Fernando.alves,duarte.sousa,soniafp}@tecnico.ulisboa.pt

Abstract. This paper presents the modelling and the numerical simulation results of a bidirectional DC-DC converter using modular Marx power electronic switches to be applicable in high-voltage converters. To achieve ample voltage ratio between high-voltage and low-voltage sides, the proposed DC-DC converter uses also a power electronic circuit, in the low-voltage converter side, that changes the connection between three inductors, as a series or parallel connection, to aid the energy transfer to the inductors.

Keywords: Bidirectional DC-DC converter, modular Marx switches, series-parallel inductor circuit.

1 Introduction

The recent industrial applications request energy storage systems (ESS) based on lithium-ion batteries banks and/or ultracapacitors cells for long-term energy storage or short-term energy storage, respectively. Examples of these applications can be seen in uninterruptible power supplies, hybrid powered traction systems, renewable energy generation, mobile energy generation and integrated active filters.

The control of the charge and discharge modes of the ESS need a bidirectional power electronic interface between the storage banks and the loads or to higher voltage DC-bus systems generally used with three-phase inverters, active front-end rectifiers and other power converters topologies.

Due the low nominal voltage of this batteries and ultracapacitors cells, it is required a large voltage conversion ratio for bidirectional DC-DC converter. In case of ultracapacitors their wide variation voltage in charging and discharging operation make difficult to design a bidirectional DC-DC converter with high voltage ratio.

In the last decade several works has been done to design solutions for bidirectional DC-DC converters. A good review can be found in [1][2], where the main isolated and non-isolated bidirectional converter topologies are deployed.

L.M. Camarinha-Matos et al. (Eds.): DoCEIS 2014, IFIP AICT 423, pp. 478–485, 2014.

In this paper, the proposed power electronic converter is a bidirectional non isolated half-bridge DC-DC converter based on buck-boost converter, [3]. To make the converter able to high-voltage applications, it uses a series-stacked semiconductors based on solid-state Marx generator concept, [4]. Using a Marx generator as a series switch, allows the use of low-voltage semiconductors and arrange the overall structure of the DC-DC converter with modular cells, which leads to compactness, low weight, low cost and portability of these power converters, [4].

The Fig. 1 presents the bidirectional DC-DC converter employing six modules of Marx cells and series-parallel inductor electronic circuit.

The modular Marx power electronic switches, SM_1 and SM_2, each one, uses three modular Marx cells. The HV side voltage, in steady-state operation, will be the sum of the capacitors voltage, v_{Ck}, for each three cells of SM_1 or SM_2.

To right operation of the bidirectional DC-DC converter, the DC voltage capacitor balancing should be taking account, using some strategies to equalize the v_{Ck}, which also depends of modular Marx cells quantity used, [5][6].

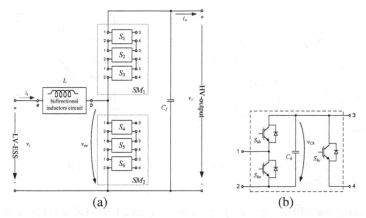

Fig. 1. Bidirectional DC-DC converter using Marx power switches and series-parallel inductor circuit: (a) the proposed bidirectional DC-DC power electronic converter; (b) the modular Marx power electronic cell

2   Contribution to Collective Awareness Systems

This paper proposes a bidirectional DC-DC power converter using modular Marx switches to achieve high voltage ratios. The bidirectional DC-DC power electronic converters together with the energy storage systems have significant importance to overcome the power availability issues of renewable energy sources. Also other applications for bidirectional DC-DC converters, as electric traction systems, have a remarkable position in sustainable growth. Due the application of modular cells in this DC-DC converter and also the proposed solution to achieve high voltage ratios, this work also allows the development of new technologies, which should be accounted in sustainability-aware decisions for collective awareness systems.

3   Series-Parallel Inductor Electronic Circuit

The design of the inductor L assumes considerable importance, since it is responsible for the power transfer, in both directions, between the low-voltage (LV) to high-voltage (HV) sides. The sizing of the inductor L determines also the LV side current ripple and the switching frequency of the DC-DC converter.

In this work, the inductor L can switch from a parallel to a series arrangement thought the use of bidirectional semiconductors circuit that changes between connections to obtain high voltages ratios from LV to HV sides.

Fig.2(a) presents the series-parallel inductor electronic circuit (SPIEC) with three non-coupled inductors, L_1, L_2, L_3, the semiconductors for parallel inductors connection, S_{P1}, S_{P2}, S_{P3}, S_{P4} and the semiconductors for series inductors connection, S_{S1}, S_{S2}. The terminals a and b represents the input and output of this electronic circuit.

The Fig.2(b) presents one cell of SPIEC, that can be used to extend this concept to higher number of non-coupled inductor connections, where $k = \{1, 2, ..., m+1\}$ and m the total number of SPIEC cells.

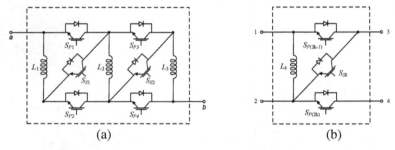

(a)   (b)

Fig. 2. The series-parallel inductor electronic circuit: (a) example of SPIEC with three inductors; (b) the SPIEC module cell

Considering the SPIEC operating in the bidirectional DC-DC converter the voltage ratio from HV to LV sides can be deduced taking account the duty cycle, δ, of switching period, T, of modular Marx switches, SM_1, SM_2 and the number of SPIEC cells, m.

The Fig. 3 presents an example of two SPIEC modular cells (three non-coupled inductors).

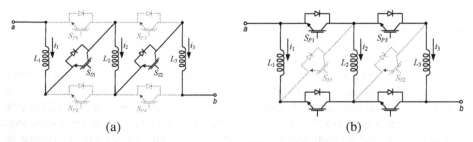

(a)   (b)

Fig. 3. The series-parallel inductor electronic circuit: (a) parallel mode operation; (b) series mode operation

Starting from parallel inductors connection, represented in Fig.3(a), the S_{S1} and S_{S2} semiconductors are synchronised with S_1 (which is turned Off) and S_{P1}, S_{P2}, S_{P3} and S_{P4} are tuned with S_2 (which is On) during time δT. Being v_{L1}, v_{L2}, v_{L3} the voltage across the inductors L_1, L_2 and L_3 respectively and v_i the voltage of LV source, then $v_{L1} = v_{L2} = v_{L3} = v_i$.

For series inductors connection, represented in Fig.3(b), the S_{S1} and S_{S2} semiconductors are synchronised with S_1 (which is turned On) and S_{P1}, S_{P2}, S_{P3}, S_{P4} tuned with S_2 (which is Off) during time $(1-\delta)T$. If v_o represents the voltage of HV side of DC-DC power converter, then $v_{L1} + v_{L2} + v_{L3} = v_i - v_o$. Considering equal inductors, where v_L is the voltage across each inductor, the previous equation can be rewritten as (1).

$$v_L = \frac{v_i - v_o}{3} \tag{1}$$

Considering the SPIEC steady-state operation with inductors current at continuous mode, the integral of the voltage across each inductor is zero over a cycle of switching T, which leads to (2).

$$\delta v_i + (1-\delta)\frac{v_i - v_o}{3} = 0 \tag{2}$$

From (2) the voltage ratio of DC-DC converter can be found as (3).

$$\frac{v_o}{v_i} = \frac{2\delta + 1}{1 - \delta} \tag{3}$$

Generalizing (3) for different number of non-coupled inductors, n, the voltage ratio of DC-DC converter becomes (4).

$$\frac{v_o}{v_i} = \frac{(n-1)\delta + 1}{1 - \delta} \tag{4}$$

Fig.4 presents the voltage ratio of (4) for two, three and four non-coupled inductors considering duty cycles varying from 0.4 to 0.9 and ideal circuit parameters.

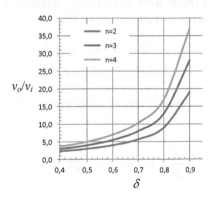

Fig. 4. Voltage ratios with different number of inductors and duty cycle values

From Fig.4 and considering acceptable values for duty cycle between 0.7 and 0.8 the voltage ratio of the DC-DC converter can be change from 6 to 17 times dependent of the non-coupled inductors number used in SPIEC.

4   Switching Control Loops

The switching control loops necessary to the bidirectional DC-DC converter operation are depicted in Fig. 5.

To control the output voltage from HV side, is used a proportional-integral controller (PI), that based on voltage error, $\left(v_{oref} - v_o\right)$, produces an input current reference, i_{ref}, to LV side in inner current control loop. On other hand, this inner control loop uses a current hysteresis controller that generates the switching state, γ, associated with SM_1 and SM_2.

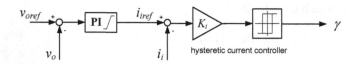

Fig. 5. Switching control loops of bidirectional DC-DC converter

The PI controller has an internal limiter, to avoid overcurrents in the inner control loop, with an anti-windup system to prevent the accumulation of integral error when that limiter is active. The tuning of PI controller can be made using the linear control theory or the well know Ziegler-Nichols method, [7], considering a linearized closed loop transfer function of current controller $i_i/i_{iref} \approx 1/(sT_d + 1)$, where T_d is the average current delay related to the switching frequency of bidirectional DC-DC converter, [8].

From Fig.1, analysing the turn-on and turn-off signals associated with variable γ, from the pulsed voltage across SM_2, v_{sw}, follows, (5).

$$v_{sw} = \begin{cases} v_o & \text{if } SM_1 \text{ is ON}; \left(\gamma = 1\right) \\ 0 & \text{if } SM_2 \text{ is ON}; \left(\gamma = 0\right) \end{cases} \tag{5}$$

The state-space model that describes the dynamic of input current, i_i, considering that $\left(v_{sw} = \gamma v_o\right)$ is (6).

$$L\frac{di_i}{dt} = u_i - u_{sw} = u_i - \gamma u_o \tag{6}$$

To control the SPIEC input current, i_i, its current error, e_{i_i}, should be zero. However, since the bidirectional DC-DC converter is operated at finite frequency, the current error is maintained with a small ripple, (7).

$$e_{i_i} = i_{iref} - i_i \approx 0 < \varepsilon \tag{7}$$

The hysteretic current controller, from Fig.5, uses a hysteresis band of width 2ε to maintain the input current error between: $-\varepsilon < e_{i_i} < +\varepsilon$. The command strategy that decides γ is given by (8).

$$\begin{cases} e_{i_i} > +\varepsilon \implies i_{iref} > i_i \implies i_i \uparrow \implies \dfrac{di_i}{dt} > 0 \implies v_{sw} = 0 \implies \gamma = 0 \\ e_{i_i} < -\varepsilon \implies i_{iref} < i_i \implies i_i \downarrow \implies \dfrac{di_i}{dt} < 0 \implies v_{sw} = v_o \implies \gamma = 1 \end{cases} \tag{8}$$

The hysteretic current controller uses also a gain, K_i, to adjust the variable switching frequency, f_s, within the semiconductors operating limits, [8], (9).

$$f_s = \frac{u_i (u_o - u_i)}{2\varepsilon K_i L u_o} \tag{9}$$

5   Numerical Simulation Results

The numerical simulation results of bidirectional DC-DC converter where performed with Matlab®/Simulink® and uses the SimPowerSystems™ toolbox to include the losses of semiconductors, inductors and capacitors.

The main circuit parameters are: $v_i = 1.5\,\text{kV}$; $v_{oref} = 15\,\text{kV}$; $L_k = 1\,\text{mH}$; $C_f = 10\,\mu\text{F}$; $2\varepsilon = 10\,\text{A}$. To simulate different load conditions, an output current source on HV side was considered, where the i_o values and simulation time, t, intervals are: $i_o = 0\,\text{A}$ (no load), $t \in [0; 0.02[$; $i_o = 15\,\text{A}$ (discharging LV side), $t \in [0.02; 0.08[$; $i_o = 15\,\text{A}$ (charging LV side), $t \in [0.08; 0.14]$. Since a voltage ratio of 10 is needed, for a $\delta \cong 0.8$ three non-coupled inductors are used (Fig.4).

Fig. 6 presents the results of bidirectional DC-DC converter with three non-coupled inductors in SPIEC and Fig.7 presents a zoom view in load transient at 0.08s.

From Fig.6, the DC-DC converter keeps a gain of 10 in voltage ratio. Due the load transients, the output voltage and the balance of voltage capacitors in Marx cells are tracked with variations about 3%.

In Fig.7 it is shown that the inductor currents have high ripple (210%), while the LV side current displays a ripple of 94% related with their reference values.

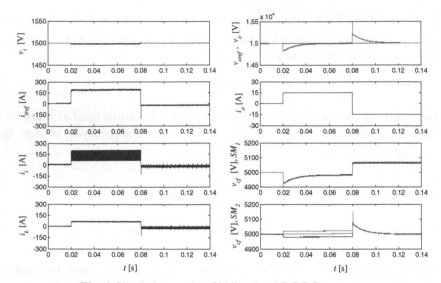

Fig. 6. Simulation results of bidirectional DC-DC converter

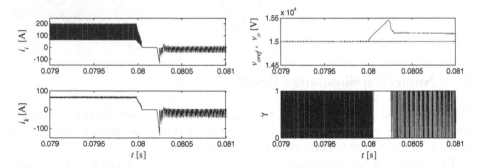

Fig. 7. The load transient in bidirectional DC-DC converter

6   Conclusions

This paper presents the application of modular Marx cells in a bidirectional DC-DC power converter for HV applications. The high voltage ratio depends of non-coupled inductors number used in SPIEC and the duty cycle.

The balance of the Marx capacitor of the need stack cells remaining under 3%, while the output voltage has a ripple below 0.1%.

The developed simulation model is valuable as a proof of concept and to mitigate some drawbacks as the high ripple current in LV side.

Acknowledgments. This work was supported by Portuguese national funds through FCT - Fundação para a Ciência e a Tecnologia, under project PEst-OE/EEI/LA0021/2013.

References

1. Kumar, S., Ikkurti, H.P.: Design and control of novel power electronics interface for battery-ultracapacitor hybrid energy storage system. In: Int. Conf. Sustain. Energy Intell. Syst., SEISCON, pp. 236–241 (2011)
2. Schupbach, R.M., Balda, J.C.: Comparing DC-DC converters for power management in hybrid electric vehicles. In: IEEE International Conference on Electric Machines and Drives (2003)
3. Karshenas, H.R., Daneshpajooh, H., Safaee, A., Jain, P., Bakhshai, A.: Bidirectional DC-DC Converters for Energy Storage Systems. In: Carbone, R. (ed.) Energy Storage in the Emerging Era of Smart Grids, p. 18. InTech (2011)
4. Redondo, L., Fernando Silva, J.: Solid State Pulsed Power Electronics. In: Power Electronics Handbook, 3rd edn., pp. 669–707. Elsevier Inc. (2011)
5. Encarnação, L., Silva, J.F., Pinto, S.F., Redondo, L.M.: Grid Integration of Offshore Wind Farms Using Modular Marx Multilevel Converters. In: Doctoral Conference on Computing, Electrical and Industrial Systems, DoCEIS 2012, pp. 311–320 (2012)
6. Encarnação, L., Silva, J.F., Pinto, S.F., Redondo, L.M.: A New Modular Marx Derived Multilevel Converter. In: 2nd Doctoral Conf. on Computing Electrical and Industrial Systems – DoCEIS Costa da Caparica, Portugal, pp. 573–580 (2011)
7. Ogata, K.: Modern Control Engineering, 5th edn. Prentice Hall (2010)
8. Fernando Silva, J., de Conversão Comutada, S.: Semicondutores e Conversores Comutados de Potência. Lisboa: ACEnergia/DEEC/IST (2012) (in Portuguese)

Experimental Study on Induction Heating Equipment Applied in Wireless Energy Transfer for Smart Grids

Rui Neves-Medeiros[1], Anastassia Krusteva[2],
Stanimir Valtchev[1], George Gigov[2], and Plamen Avramov[2]

[1] UNINOVA-CTS
[2] Dept. of Electrical Engineering, FCT/UNL, Portugal
[3] Research and Development Sector, TUS, Bulgaria

Abstract. This work is focused on the design of the contactless energy transmitters and testing of their electrical parameters, varying the working frequency in the kHz range. The intended application is related to the possibility to make most efficient wireless charging of different batteries from the grid, guaranteeing more acceptable use of the electric vehicles.

Keywords: transmitter, contactless, wireless, energy, power, converter, grid, battery, electric vehicles.

1   Introduction

Wireless On-Line Electric Vehicle Energy Transfer (WOLEVET) is a user-project within the Seventh Framework Programme (FP7) project Distributed Energy Resources Research Infrastructures (DERri) focused on battery charging when Wireless Energy Transfer (WET) is included. An exchange of energy is planned between the AC grid, the battery and the contactless energy converter. The wireless energy transfer is defined as "On-line" because of the varying position of the energy receiver in relation to the energy transmitter. To experiment this movement the energy receiver was displaced at different distance to the fixed charging station. The described experiments were aimed to prove the most efficient energy transfer conditions that will facilitate the integration of electric vehicles (EV) in micro-grids, adding storage capacity (EV batteries) to the grid. This integration will reduce the battery size as the necessary energy will be directly available from the nearby source of the grid. This is very important as the price of a propulsion battery is now roughly 80% of the EV price. Smaller batteries mean fewer cells and lower pollution.

The reported work is dedicated to the design and experimentation of a contactless energy transmitter/receiver set involving some available high frequency (HF) generators, most of them originally dedicated to induction and dielectric heating. The shape and construction of the inductors that guarantee a good magnetic coupling and best efficiency is presented. The magnetic coupling is tested at real power of one or more kW. The important choice of the power inverter parameters is limited to the type of resonance and the switching frequencies of the inverter.

L.M. Camarinha-Matos et al. (Eds.): DoCEIS 2014, IFIP AICT 423, pp. 486–493, 2014.

After the introduction, the Chapter 2 establishes a relation between this work and the scope of the conference. Chapter 3 explains the problem to be solved referencing previous works. Experimental achievements are revealed in Chapter 4 and followed by some conclusions in Chapter 5.

2   Relationship to Collective Awareness Systems

The recently adopted concept of Smart Grids is preparing our society to consider the inclusion of individual lower power generators into the energy system. Those individual generators will need a more complex control than the traditional grids, but some benefits will arise, e.g. better efficiency, more safety, more reliability and more power delivered to the system.

Wireless charging of batteries can be used to easily associate EV with the smart grid. In fact, if the bi-directional flux of energy will be made possible than every vehicle can be seen as a collective energy reserve and the grid will be able to manage all the energy portions resulting in peak shaving or longer time storage.

From the private users' point of view, the cars and other devices with significant amount of stored energy would need to be available to a more complex but profitable control system that will meet the social goals of reducing the price and ecological footprint and increasing the safety associated with the batteries maintenance.

Certainly this contribution of the EV in re-utilizing the distributed energy would not be enough to substitute a conventional power plant, but it would help tracing load profiles for the major cities helping to better plan the operation of the regional dam or coal power plant reducing their costs. For all these reasons, the EV charged wirelessly is seen as a Collective Awareness System.

3   State-of-the-Art

The EV is becoming a necessity due to environmental problems and growing prices of classical energy production. The batteries of the new vehicles will require a large number of charging in different places and moments. The knowledge about the resonant WET became very important. The study of the Series Loaded Series Resonant (SLSR) converter places it as suitable for the WET [1]. There are other possible resonant configurations too [2].

The operation of SLSR converter is analyzed in many articles, e.g. in [1] but to obtain a rapid and accurate reaction of this circuit remains a problem. The existence of stored energy in the resonant reactance elements (inductance and capacitance) makes the direct control of the power switches quite difficult, especially when the circuit elements are not ideal (WET). Many articles are published, aimed at resonant converters control, usually including calculation of normalized phase-plane trajectory as in [3]. A more complex calculation block (implemented as FPGA) is shown in [4].

All the known methods are not reacting immediately to the demand of the resonant tank as they measure and control the resonant current. The future Instant Energy Control (IEC) circuits that respond not only to the resonant current but more to the resonant capacitor voltage are presented in [1] and [5]. The IEC allows safer operation of the transmitter, but there is a lot to do for the bi-directional energy transfer. The charged vehicle is supposed to give back energy to the common grid.

The control of the bi-directional energy exchange is expected to be similar to the already known solutions but will be necessary to involve a new information grid, comparable to the mobile network although much faster. One of the main new functions will be to recognize and authorize the car (Fig.1), efficiently enough, to receive or to deliver energy passing near the transmitter cells and to continue this interchange with the next cell. This problem is similar to the requirements that other smart grid versions will ask from the information technology.

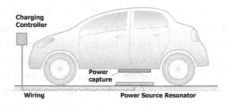

Fig. 1. Delphi Automotive's wireless electric-vehicle [6]

As referred in [6], the unit sales of wireless EV chargers in North America is expected to reach about 10,000 in 2014 and increase to more than 132,000 units by the end of the decade.

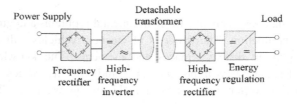

Fig. 2. The main blocks of the inductive charger [7]

Fig.2 illustrates the AC power (Power Supply) being supplied to the EV (Load) which concept is also reported in [8] and [9]. The circuit operates as following: the AC supply voltage is rectified and converted to a high frequency AC (tens or hundreds of kHz) within the charger station. Based on the resonant processes this high frequency power is transferred to the EV side by induction. Finally the receiver converts the high frequency AC power into a DC power for the battery charging.

4   Experimental Results

The main performed tasks were aimed to verify the efficient operation of the inductively coupled set of transmitter/receiver. This included the design of the transmitter, the proper choice of the resonant capacitor, the definition of the power for the experiments, the reconfiguration of the existing (from induction heating) system of coupling, the necessary measurements and analyses of the obtained parameters.

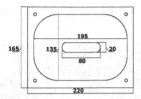

Fig. 3. Transmitter magnetic core design with dimensions in mm

The designed transceiver (transmitter/receiver), illustrated in Fig.3 consists in ferrite core (FLUXTROL 50) and coils of copper tube with d = 6 mm, N = 5 and water cooling. The thickness of the plate is 15 mm and the internal column is 9 mm thick. The parameters of the coil are: L = 40.6 µH and R = 82.0 mΩ.

Four different experiments were attempted with four different power sources.

In this text the, indexes "$_1$" and "$_2$" correspond respectively to the primary and the secondary sides of the transformer (magnetic link). The primary side of the transformer will be also referred as "the sender" or "the first coil" and the secondary side as "the receiver" or "the second coil". The index "$_r$" represents resonance. Upper case "R" and "D" correspond to electrical Resistance and Distance respectively. Lower case "r" and "d" correspond to radius and diameter.

4.1   First Experiment – HF Signal Generator

This experiment was aimed to determine the resonant frequency of the transceiver at low power.

Table 1. Circuit parameters for f_r = 145-175 kHz

D [cm]	f_{1r} [kHz]	U_{1max} [V]	I_{1max} [A]	f_{2r} [kHz]	U_{2max} [V]	I_{2max} [A]	k
5	145.8	1.99 (8)	+0.048-0.023		1.85 (6.46)	0.029 (0.27)	0.93
		1.88 (8)	0.2 (0.57)	191	1.5 (4.28)	0.05 (0.14)	0.79
10	173.5	2.438 (8)	0.049		1.64 (4.69)	0.0287 (0.07)	0.67
15	165.7	5.67	0.029/ 0.0208		2.08 (2.54)	0.0325	0.365
20	166	6.95	+0.019/ -0.016		1.16 (1.16)	0.020	0.167

The applied compensation (resonant) capacitor are $C_1 = C_2 = 0.2$ µF and the load was $R_{load} = 54$ Ω. Table.1 reveals the resonant frequencies in both sides of the transceiver and their measured values of voltage and current. The correlation k = U_1/U_2 for different distances is also presented. Values in parenthesis represent samples in the same conditions for different levels of power.

4.2   Second Experiment – HF Vacuum Tube Generator

The power for this experiment was higher. The same ferrite core was used (Fig.4).

Fig. 4. Transmitter with FLUXTROL 50 ferrite

The experimented distance between the coils was 120 mm and the compensation was made at the high voltage side of the transformer by capacitance $C_1 = 5000$ pF. The second coil is compensated by a capacitor $C_2 = 3 \times 6800$ pF = 20400 pF. The load is $R_{load} = 160$ Ω (1,2,3) or $R_{load} = 60$ Ω (4). The resonant frequency of the receiver is $f_{2r} = 509.7$ kHz and the operating frequency is f = 533 kHz.

Table 2. Circuit parameters for f = 533 kHz

U_{1ef} [V]	I_{1ef} [A]	φ_1 [°]	P_1 [W]	U_{2ef} [V]	I_{2ef} [A]	φ_2 [°]	P_2 [W]	η	R_{load} [Ω]
187.4	12.02	88.2	70.75	76.36	0.707	9	53.32	0.75	160
281.0	18.03	88.2	159.12	115.25	0.99	9	112.68	0.72	160
374.7	24	88.2	282.47	154	1.30	9	198.9	0.71	160
561.7	36.0	88.2	635.00	229	2.12	9	479.8	0.75	60

The primary circuit presents the expected inductive behavior, as shown in Table.2 by the phase shift $\varphi_1 = 88.2°$. The efficiency is high. A slightly inductive shift from the resonance is observed in the secondary ($\varphi_2 = 9°$). For the operation frequency f = 533 kHz and D = 12 cm the output power is $P_2 = 479.8$ W and the efficiency is 75%.

4.3   Third Experiment – MOSFET Inverter with Auto Generation

This experiment was prepared with the necessary rectifier and regulator circuit that permitted to charge a battery by WET. The equipment used in this case is a high frequency power converter implemented by MOSFET which is prepared for melting of gold. The resonant compensation is made both in primary and secondary sides by capacitors $C_1 = C_2 = 0.2 \ \mu F$. The input DC voltage is fixed at $U_d = 30$ V. For the distances higher than 10 cm another voltage source was joined in series.

Table 3. Circuit parameters with $f_r = 177$ kHz

D [cm]	f_1 [kHz]	U_0 [V]	U_{1max} [V]	I_{1rms} [A]	U_{2max} [V]	I_{2rms} [A]	φ [°]	U_{0ut} [V]
5	226	30	75.5	9.5	21.27	0.328	16.36	14.2
10	225.78		74.33	10	6.14	0.172	55.6	3.92
15	225.6		75.13	10.4	2.64	0.106	0.365	1.14
20	225.3		75.13	10.6	0.2	0.084	0.167	0.28

4.4   Fourth Experiment - MOSFET, Phase Shift Regulated and Full Bridge Inverter

The power converter used in this experiment was prepared for several kW. The operating frequency range was 100-200 kHz. The compensation is made by $C_1 = C_2 = 0.2 \ \mu F$, the experimented distance is D = 100 mm or D = 150 mm. The load was changed between 2.5 and 95 Ω. In this experiment the coils had to be water cooled, because of the higher currents. The resonant frequency is measured at $f_r = 184$ kHz.

Fig. 5. High power MOSFET inverter

The experimental results and the circuit parameters are presented in Table.4. The obtained efficiency is $\eta =70$ % for the D = 10 cm, $U_1 = 600$ V, $I_1 = 24$ A and $R_{load} = 11.5 \ \Omega$ or $R_{load} = 2.5 \ \Omega$. For the same input voltage and current and load $R_{load} = 16.7 \ \Omega$ the efficiency goes higher, $\eta = 75.1\%$ (Table.4).

Table 4. Inverter and sender side circuit parameters at frequency around 180 kHz

D [cm]	U$_1$ [V$_{max}$]	I$_1$ [A$_{max}$]	φ$_1$ [°]	P$_1$ [W]	R$_{load}$ [Ω]	U$_2$ [V$_{max}$]	I$_2$ [A$_{max}$]	φ$_2$ [°]	P$_2$ [W]	η [%]
	200	7.2	52.4	439	95	71.25	2.68	26.2	86	19.5
	350	12	75.6	522		88.75	3.1	26.6	123	23.6
	300	10.8	75.6	403	50	116	3.74	26.6	194	48.1
10	300	12	75.6	448	16.7	132.14	4.8	26.6	284	63.3
	600	24	80.0	1250		244.28	8.6	26.6	939	75.1
	600	24	80.2	1228	11.5	228	8.4	24.9	869	70.7
	600	24	78.8	1405	2.5	182	12.78	32.1	985	70.1
15	600	24	85.3	590	2.5	102.21	7.55	32.1	327	55.4

The waveforms of the primary voltage and current for the best efficiency case are presented in Fig.6. For the secondary circuit, the current waveform is presented in Fig.7.

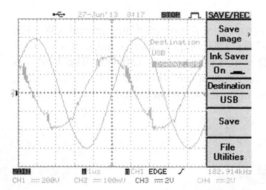

Fig. 6. Waveforms of the current I$_1$ and the voltage U$_1$

Fig. 7. Waveforms of the current I$_2$

5   Conclusions

The comparative analysis of the experimental results reveals the complexity of the problems and results in the following conclusions: the efficiency is related to the frequency and at the same frequency it is related to the load. The highest obtained output voltage depends on the input voltage and the distance transmitter/receiver. For these generators a cooling was needed at the highest transferred power.

In conclusion, the experiments have proven that the existing induction heating equipment can be used as a base for developing the new WET technology, especially for charging the batteries of the EV. It is a fast solution that can achieve a better cooperation between the electric vehicles and the grid.

Acknowledgments. The authors fully recognize the support from the European Commission FP7 project DERri GA No 228449 (http://www.der-ri.net). The authors are entirely responsible for the content of this publication. It does not represent the opinion of the European Community. We thank equally the enterprise Apronecs that allowed us to experiment at higher power.

References

1. Valtchev, S., Klaassens, J.B.: Efficient Resonant Power Conversion. IEEE Transactions on Industrial Electronics 37(6), 490–495 (1990)
2. Wang, C.S., Stielau, O.H., Covic, G.A.: Design Considerations for a Contactless Electric Vehicle Battery Charger. IEEE Transactions on Industrial. Electronics 52(5), 1308–1314 (2005)
3. Rossetto, L.: A Simple Control Technique for Series Resonant Converters. IEEE Transactions on Power Electronics 11(4), 554–560 (1996)
4. Moradewicz, A., Kazmierkowski, M.: FPGA Based Control of Series Resonant Converter for Contactless Power Supply. In: IEEE International Symposium on Industrial Electronics, pp. 245–250 (2008)
5. Valtchev, S., Brandisky, K., Borges, B., Klaassens, J.B.: Resonant Contactless Energy Transfer with Improved Efficiency. IEEE Transactions on Power Electronics 24(3), 685–699 (2009)
6. Delphi Media Releases, http://delphi.com/about/news/media/pressReleases/pr_2010_09_29_001/
7. Yuwei, Z., Xueliang, H., Linlin, T., Yang, B., Jianhua, Z.: Current Research Situation and Developing Tendency about Wireless Power Transmission. In: International Conference on Electrical and Control Engineering, Wuhan, pp. 3507–3511 (2010)
8. Lee, S., Huh, J., Park, C., Choi, N., Cho, G., Rim, C.: On-Line Electric Vehicle Using Inductive Power Transfer System. In: 2nd IEEE Energy Conversion Congress and Exposition, Atlanta, pp. 1598–1601 (2010)
9. Park, M., Shin, E., Lee, H., Suh, I.: Dynamic Model and Control Algorithm of HVAC System for OLEV® Application. In: International Conference on Control Automation and Systems, Gyeonggi-do, pp. 1312–1317 (2010)

Part XVIII

Telecommunications

On Quasi-Optimum Detection of Nonlinearly Distorted OFDM Signals

João Guerreiro[1,2], Rui Dinis[1,2], and Paulo Montezuma[1,3]

[1] DEE, FCT, Universidade Nova de Lisboa, Monte de Caparica, Portugal
[2] IT, Instituto de Telecomunicações, Lisboa, Portugal
[3] UNINOVA, Monte de Caparica, Portugal

Abstract. In this paper we considered OFDM schemes that somehow have a nonlinear operation on their transmission chain. Contrary to the conventional OFDM implementations, we consider the distortion term that come from that operation as something useful that has information on the transmitted signals. To efficiently harvest this information, we develop an optimum-based receiver that presents good performance improvements without introduce very high complexities. The performance of this sub-optimal receiver is investigated under different types of channel and considering receive diversity.[1]

Keywords: OFDM signals, nonlinear distortion effects, optimum receiver, Euclidean distance.

1 Introduction

OFDM (Orthogonal Frequency Division Multiplexing) [1] schemes support the physical layer of several wireless communications standards such as DVB [2], LTE [3] and WiMAX [4]. The reasons behind their popularity reside mainly on their facility to cope with frequency-selective channels without the need for complex equalization processes and their simplicity of implementation. However, OFDM signals have a critical issue: they present a very high Peak-to-Average Power Ratio (PAPR). If nothing is done, this will cause severe amplification difficulties and strong nonlinear distortion effects at the transmitter output. As the use of an high Input Back-Off (IBO) is not a good solution due to the inherent energy inefficiency, several techniques that aim to reduce the envelope fluctuations of OFDM signals have been proposed in the literature [5]. Between them, the simplest one involves the use of an envelope clipper followed by a frequency-domain filter (FDF), which allows an efficient PAPR reduction without introduce intolerable complexities and compromise the spectral efficiency [6]. However, although the clipped signals are linearly amplified with ease, they will also present nonlinear distortion effects since they are generated through a nonlinear operation.

---

[1] This work was supported by the FCT/MEC projects CTS PEst-OE/EEI/UI0066/ 2011, IT PEst-OE/EEI/LA0008/ 2013, OPPORTUNISTIC-CR PTDC/EEA-TEL/115981/ 2009, ADCOD PTDC/EEA-TEL/099973/2008, ADIN PTDC/EEI-TEL/2990/2012 and Femtocells PTDC/EEA-TEL/120666/2010 as well as by grant SFRH/BD/90997/2012.

L.M. Camarinha-Matos et al. (Eds.): DoCEIS 2014, IFIP AICT 423, pp. 497–506, 2014.

It is well known that the samples of an OFDM signal are well modeled by a Gaussian distribution, specially when a high number of subcarriers is considered. This approximation allows the use of the Bussgang theorem [7] that states that a nonlinearly distorted OFDM signal can be divided into two uncorrelated components: an useful term that is proportional to the transmitted signal and a second one that represents the nonlinear distortion. Considered as noise in the typical OFDM implementations, the distortion term is seen as something that degrades the performance. To avoid this, there are receivers that try to estimate and cancel this distortion [8], but their usefulness are very limited specially at high Signal-to-Noise Ratios (SNRs). In this work we considered not only to accept the nonlinear distortion but also to use it as something that can effectively improve the performance of the nonlinear OFDM schemes. In fact, the distortion term has useful information on the transmitted signals that can be used to improve the performance [9]. The best way to use this information is to consider an optimum receiver that makes an estimate of the transmitted signal based on the Euclidean Distance, which is evaluated using the two terms of the Bussgang Theorem. The research question can be stated as: can the nonlinear OFDM schemes perform better than the linear ones with the use of optimum receivers? To answer this question, we adopted a research approach based on the study of the relation between the Euclidean distance and the average bit energy with and without nonlinear distortion effects, which allows to conclude about the expected Bit Error Rate (BER). Surprisingly, it was verified that the nonlinear OFDM schemes can have better performances than the linear ones. As the complexity of the optimum receiver is very high even considering a small number of subcarriers, we propose a reduced-complexity receiver that in spite of not performing a full optimum detection can also achieve excellent performances.

The paper is divided in the following sections: In Sec.2 we present the relationship between this work and the collective awareness systems (CAS). Sec. 3 concerns about the communication scenario that is used as the basis of this work as well as the principal characteristics of nonlinearly distorted OFDM signals modeled with a Gaussian approximation. In Sec. 4 the optimum detection is introduced and the potential performance of the optimum receiver is presented. Sec. 5 presents a sub-optimal received and shows its performance results under different scenarios. Finally, Sec. 6 concludes this paper and presents some directions of future work.

2   Relationship to Collective Awareness Systems

CAS are today one of the biggest technological challenges in the digital world. The idea of everyone to communicate and use collective knowledge to decide in real-time presupposes solid communication mechanisms that can provide a good Quality of Service (QoS) without high energy consumption. Behind the real-time decisions, it is easily to see that wireless communications have a very important role. This work aims to present improvements in the quality of wireless communications that use OFDM modulations, which is the case of the most part of the standards, by reducing the energy consumption that is need to achieve a specific error rate and hence improving the energy efficiency of the equipments that compose the collective awareness systems.

3   Nonlinear OFDM Schemes

In this section we characterize the considered nonlinear OFDM scenario by describing the signals along the transmission chain depicted in Fig.2. We represent the baseband data signal with the vector $\mathbf{S} = [S_0\ S_1\ \dots\ S_{NM-1}]^T \in \mathbb{C}^{NM}$, where N is the number of subcarriers considered in each OFDM block and M is the oversampling factor. Each complex data symbol S_k is selected from an $\mathcal{M}$-order QAM constellation (e.g. 4-QAM). The time-domain version $\mathbf{S}$ is obtained through an Inverse Discrete Fourier Transform (IDFT), i.e., $\mathbf{s} = \mathbf{F}^{-1}\mathbf{S} = [s_0\ s_1\ \dots\ s_{NM-1}]^T \in \mathbb{C}^{NM}$, where $\mathbf{F}$ represents the Discrete Fourier Transform (DFT) matrix with the (n,n') element defined by

$$F_{n,n'} = \frac{1}{\sqrt{NM}} \exp\left(-\frac{j2\pi nn'}{NM}\right). \tag{1}$$

Fig. 1. Basic nonlinear OFDM scheme

If high values of N are considered (let's say $N \geq 32$), the imaginary and the real parts of the time-domain samples represented by $\mathbf{s}$ can be modeled by a Gaussian distribution. Representing $\bar{s}_I$ and $\bar{s}_Q$ as the considered Gaussian random variables, we can write

$$p_{\bar{s}_I}(s_I) = \frac{1}{\sigma\sqrt{2\pi}} \exp\left(-\frac{(s_I-\mu)^2}{2\sigma^2}\right), \tag{2}$$

and $p(s_I) = p(s_Q)$. The absolute value of $\mathbf{s}$ is represented by $\mathbf{R} = |\mathbf{s}| = [R_0\ R_1\ \dots\ R_{NM-1}]^T \in \mathbb{R}^{NM}$ and modeled by $\bar{R}$, which is a Rayleigh-distributed random variable with its PDF given by

$$p_{\bar{R}}(R) = \frac{R}{\sigma^2} \exp\left(-\frac{R^2}{2\sigma^2}\right) u(R), \tag{3}$$

with $u(\bar{R})$ denoting the unitary step function and $\sigma^2 = \text{var}(\bar{s}_I) = \text{var}(\bar{s}_Q)$, denoting the variance of the real and imaginary parts that is assumed to be equal. Under the Gaussian approximation, the Bussgang theorem [7] states that the nonlinearly distorted signal represented by $\mathbf{y} = f(|\mathbf{s}|) = f(\mathbf{R}) = [y_0\ y_1\ \dots\ y_{NM-1}]^T \in \mathbb{C}^{NM}$ can be decomposed in two terms, i.e.,

$$\mathbf{y} = \alpha\mathbf{s} + \mathbf{d}, \tag{4}$$

where $\mathbf{s}$ is the time-domain version of the modulated signal, $\mathbf{d} = [d_0\ d_1\ \dots\ d_{NM-1}]^T \in \mathbb{C}^{NM}$ is the distortion term and α is a scaling factor given by

$$\alpha = \frac{\mathbb{E}[R_n f(R_n)]}{\mathbb{E}[R_n{}^2]}. \tag{5}$$

The nonlinearity considered in our transmission chain is assumed to be memoryless. These nonlinearities can be modeled by the Saleh model [10] that characterizes the output of a nonlinear device using its AM-AM and AM-PM conversion functions, both receiving the absolute value of the signal as input (and represented by $A(\cdot)$ and $\Theta(\cdot)$, respectively). Using this model, we can also write the n-th output sample as

$$f(R_n) = A(R_n)\exp(j(\arg(s_n) + \Theta(R_n))), \tag{6}$$

with $\arg(s_n)$ being the original phase of the n-th time-domain sample. The considered memoryless nonlinearity is an ideal envelope clipper with normalized clipping level s_M/σ,

$$f(R_n) = \begin{cases} R_n, & R_n \leq s_M/\sigma \\ s_M\exp(j\arg(s_n)), & R_n > s_M/\sigma. \end{cases} \tag{7}$$

Note that in this case $f(R_n) = A(R_n)\exp(j\arg(s_n))$, since we don't have phase distortion. The use of a normalized clipping level is related to the random nature of our signals. A low s_M/σ means that the random samples of the signal will enter in the nonlinear region very often. Making use of the DFT definition of (1), we can express the frequency version of (4) by

$$\mathbf{Y} = \mathbf{Fy} = \alpha\mathbf{S} + \mathbf{D}, \tag{8}$$

where $\mathbf{D} = [D_0\ D_1\ \dots\ D_{NM-1}]^T \in \mathbb{C}^{NM}$ is the frequency-domain version of the distortion term that can be modeled by $\overline{D}$, which is a Gaussian random variable with zero mean as shown in [6]. In order to maintain a good spectral efficiency and the bandwidth of the original transmitted signal, a frequency-domain filter (FDF) must be used to eliminate the out-of-band radiation inherent to the nonlinear operation. In our signal processing scheme, the filtering operation is represented as a
multiplication by the diagonal matrix $\mathbf{G}$, that is defined as

$$\mathbf{G} = \text{diag}\left(\left[\underbrace{0\dots.0}_{(M-1)N/2}\ \underbrace{1\dots.1}_{N}\ \underbrace{0\dots.0}_{(M-1)N/2}\right]\right). \tag{9}$$

The filtered signal is given by

$$\mathbf{Y}_f = \mathbf{GY}. \tag{10}$$

At the reception side (after the channel effect), we have the time-domain block $\mathbf{z} = [z_0\ z_1\ \dots\ z_{NM-1}]^T \in \mathbb{C}^{NM}$ with the corresponding frequency-domain block $\mathbf{Z} = [Z_0\ Z_1\ \dots\ Z_{NM-1}]^T \in \mathbb{C}^{NM}$ defined by

$$\mathbf{Z} = \mathbf{HY}_f + \mathbf{N} = \mathbf{HGFf}(\mathbf{F}^{-1}\mathbf{S}) + \mathbf{N} = \alpha\mathbf{HGS} + \mathbf{HGD} + \mathbf{N}, \tag{11}$$

where $\mathbf{N} = [N_0 \ N_1 \ \dots \ N_{NM-1}]^T \in \mathbb{C}^{NM}$ represents the noise components and

$$\mathbf{H} = \text{diag}([H_0 \ H_1 \ \dots \ H_{NM-1}]^T), \tag{12}$$

represents the channel frequency responses.

4   Optimum Detection and Its Potential Performance

The presence of nonlinear distortion effects on the transmitted signals is typically seen as something undesirable. However, the optimum detection can take advantage of the distortion introduced by the nonlinearity for performance improvements. To understand this, let us first recall that the optimum receiver estimate is obtained by minimizing the least square error (LSE) between the received signal $\mathbf{Z}$ and a possible transmitted signal, $\mathbf{HY} = \mathbf{HF}f(\mathbf{F}^{-1}\mathbf{S})$. To evaluate the LSE between two vectors we can compute the squared Euclidean norm between them, which makes possible to write that the optimum estimate is given by

$$\hat{\mathbf{S}} = \min_{\mathbf{S}} \left\| \mathbf{Z} - \mathbf{HF}f(\mathbf{F}^{-1}\mathbf{S}) \right\|^2, \tag{13}$$

which, in fact, is the minimum Euclidean distance between $\mathbf{HY} = \mathbf{HF}f(\mathbf{F}^{-1}\mathbf{S})$ and $\mathbf{Z}$. Note that to choose the best sequence, all the possible transmitted signals must be evaluated. Here the term "evaluated" means that the possible transmitted sequence must be submitted to the same signal processing chain of the received signal. Let us also recall that what strongly conditions the BER of a communication system is the ratio between the average bit energy, E_b, and the minimum squared Euclidean Distance between the transmitted signals, $\mathcal{D}^2$. In the presence of nonlinear distortion effects, this ratio is higher and, consequently, we can have potential performance improvements. To verify this, let us consider two nonlinearly distorted OFDM signals that differ in one symbol, $\mathbf{Y}^{(2)} = \mathbf{F}f(\mathbf{F}^{-1}\mathbf{S}^{(2)})$ and $\mathbf{Y}^{(1)} = \mathbf{F}f(\mathbf{F}^{-1}\mathbf{S}^{(1)})$ associated to the modulated data signals $\mathbf{S}^{(2)}$ and $\mathbf{S}^{(1)}$, respectively. The Euclidean distance between $\mathbf{S}^{(2)}$ and $\mathbf{S}^{(1)}$ is

$$\mathcal{D}^2 = \left\| \mathbf{S}^{(2)} - \mathbf{S}^{(1)} \right\|^2 = \sum_{k=1}^{N} \left| S_k^{(2)} - S_k^{(1)} \right|^2. \tag{14}$$

With normalized QPSK constellations (i.e. $S_k = \pm 1 \pm j$), we have $\mathcal{D}^2 = 4E_b$, with

$$E_b = \frac{\|\mathbf{Y}\|^2}{\log_2(\mathcal{M})N}\bigg|_{\mathcal{M}=4} = 1. \tag{15}$$

Considering that the signals are nonlinearly distorted and using (8), the Euclidean Distance is given by

$$\mathcal{D}_{NL}^2 = \left\| \mathbf{Y}^{(2)} - \mathbf{Y}^{(1)} \right\|^2 = \sum_{k=1}^{MN} \left| \alpha \left(S_k^{(2)} - S_k^{(1)} \right) + D_k^{(2)} - D_k^{(1)} \right|^2, \tag{16}$$

and typically we have $\mathcal{D}_{NL}^2 > 4E_b$. In these conditions we can define an asymptotic gain given by

$$G = 10\log_{10}\left(\frac{\mathcal{D}_{NL}^2}{\mathcal{D}^2}\right). \tag{17}$$

In Fig.1 is shown the probability density function (PDF) of (17) without considering an FDF (i.e., the diagonal of **G** is zeros) and considering OFDM signals with $N = 64$ useful subcarriers distorted by an ideal envelope clipper with normalized clipping level s_M/σ. It is assumed an ideal AWGN channel (i.e., $|H_k| = 1 \; \forall \; k$) and μ different symbols between $\mathbf{S}^{(2)}$ and $\mathbf{S}^{(1)}$.

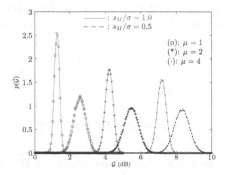

Fig. 2. Distribution of the asymptotic gain for different values of s_M/σ and μ

From Fig. 2, it is clear that gain is almost higher than one. Moreover, the stronger the nonlinear distortion effects the higher is the magnitude of the gain. For instance, with $\mu = 1$ and $s_M/\sigma = 1.0$, we have an asymptotic gain around 1.4 dB, but the gain can even reach 2.7 dB for $s_M/\sigma = 0.5$ and the same value of μ. From the figure it is also clear another important aspect: the average value of the gain is proportional to μ, i.e., $\mathbb{E}[G]_{\mu=a} = a\mathbb{E}[G]_{\mu=1}$, which unveils that even in a coded scenario where μ depends on the free distance of the code and is typically higher than one, we will also have potential performance improvements. To verify the asymptotic gains associated to the optimum detection in nonlinear OFDM schemes, we obtain the BER using the following approximation (that is valid considering that the minimum Euclidean distance is dominated by sequences that differ in only $\mu = 1$ symbols)

$$P_b = \Sigma_i \; P(G = G_i) Q\left(\sqrt{G_i \frac{2E_b}{N_0}}\right), \tag{18}$$

with G_i being the value that the gain can takes (measured in linear units) and $P(G = G_i)$ is the probability of the gain G be equal to G_i. In Fig.2 it is shown the BER obtained using (18) and considering OFDM signals with $N = 64$ subcarriers and different normalized clipping levels s_M/σ.

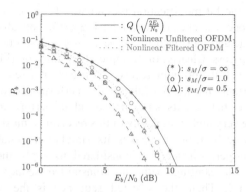

Fig. 3. Approximate BER for different values of s_M/σ with and without FDF

From the Fig. 3 it is clear that in the case of linear OFDM, $P(\mathcal{G} = 1) = 1$ which means that $\mathcal{D}_{NL}^2 = \mathcal{D}^2 = 4E_b$ and the approximate BER of (18) is equal to the theoretical linear OFDM BER given by $P_b = Q\left(\sqrt{\dfrac{2E_b}{N_0}}\right)$. As is expected from the gains distribution, when the nonlinear distortion effects are stronger the potential improvements in the BER are higher too. At $P_b = 10^{-3}$ and $s_M/\sigma = 0.5$ the asymptotic gain relative to the linear case is around 2.6 dB. When we consider a FDF to remove the out-of-band energy inherent to the nonlinear operation, the gains are lower, which is an expected result since less subcarriers are considered in the computation of the Euclidean distance, i.e., $\mathcal{D}_{NL}^2$ is computed only on the N in-band subcarriers. However, even considering filtered sequences, the asymptotic gain can reach a value around 1.3 dB.

5 Sub-optimal Receiver and Its Performance Results

The potential performance improvements unveiled in the previous section are inherent to the optimum detection, where the optimum receiver selects its estimate by comparing the received signal with all the possible transmitted signals. The computational load of this method is clearly too high, even considering a moderate number of subcarriers and small constellations. In this section we propose a sub-optimal receiver whose the main idea is to compare the received signal with only a set of all the possible transmitted signals. The main purpose of this receiver is achieve performance improvements closest the ones obtained by the optimum detection but, at the same time, reduce drastically the complexity associated with this type of detection. It makes uses of the fact that typically the transmitted signal differs in a few bits from the received signal and, thus, it is likely that we have the optimal sequence within the set of the tested ones without make all the possible comparisons. Our sub-optimal receiver starts its decision process by taking the hard decisions associated to the received signal. After that, it changes the first bit, modulates the resulting sequence and submits it to the same blocks whereby the received signal was submitted. Then, the Euclidean distance between the resultant signal and the received signal is evaluated. If this distance reduces, the bit modification is maintained, if not, the bit returns to its original value. This process is repeated for all the $\log_2(M)N$ bits that compose the data sequence. Moreover, it and can restart K times, since the sub-optimum estimate is modified during the process. Clearly, the complexity of this algorithm is very reduced when compared to the full optimum receiver, since it will test only $\log_2(\mathcal{M})NK$ sequences. An important metric to evaluate the performance of our sub-optimal receiver is the difference between its performance and the performance of the optimum receiver. However, the very high complexity of this receiver (it must analyze $2^{\log_2(\mathcal{M})N}$ sequences before taking a decision) makes its simulation inviable. To obtain an approximate optimum performance, we considered to apply the behavior of the sub-optimal receiver but starting with the transmitted sequence instead of the hard-decisions sequence. Then, the selected sequence is the sequence between: the hard-decisions sequence, the transmitted sequence, or one of their modifications that has the lower Euclidean distance relatively to the received signal. Fig.3 presents the BER performance of the sub-optimal receiver described above. The considered OFDM signal has $N = 64$ useful subcarriers with QPSK constelladions and an oversampling factor $M = 4$. The complex data symbols were mapped under a Gray mapping rule. The nonlinear device corresponds to an ideal envelope clipping with a normalized clipping level $s_M/\sigma = 1.0$. We considered an ideal AWGN channel and N_{Rx}-order diversity at reception.

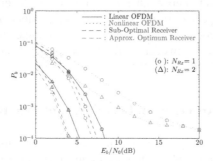

Fig. 4. Sub-Optimal receiver and approximate pptimum receiver BER in an ideal AWGN channel considering N_{Rx}-order diversity

From Fig.4 it is clear that the sub-optimal receiver improves the performance of conventional nonlinear OFDM schemes. Without diversity and at $P_b = 10^{-3}$, the sub-optimal receiver presents a gain of approximately 1.3 dB relatively to the linear OFDM. When $N_{Rx} = 2$ the gain is around 1.1 dB. It is also important to note the low difference between the sub-optimal receiver performance and the approximate optimum performance. This means that even testing only a small part of all the possible transmitted signals, the performance is still close to the one potentially obtained by the optimum receiver. In Fig.4 it is shown the BER of the sub-optimal receiver considering a XTAP channel with $N_{Ray} = 32$ rays of uncorrelated Rayleigh fading and N_{Rx}-order diversity at reception.

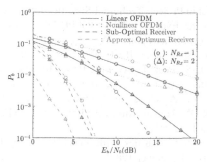

Fig. 5. Sub-Optimal receiver and Approximate Optimum receiver BER in a frequency-selective channel considering N_{Rx}-order diversity

From Fig. 5, it is clear that independently of the diversity order, in a frequency-selective channel the gains are higher than in a ideal AWGN channel. For instance, at $P_b = 10^{-2}$, the gain the sub-optimal receiver is around 5.4 dB relatively to a linear OFDM transmission and 10.3 dB relatively to the conventional nonlinear OFDM receiver. These higher gains can be explained due to the diversity associated to the selectivity of the channel that can be enhanced by the diversity inherent to the nonlinear distortion effects. It is also important to remark that the approximate performance of the optimum receiver is much better than the performance of our sub-optimal receiver which unveils that in a frequency-selective channel an increased complexity means better performances, however, even with a reduced complexity receiver, we have remarkable improvements relatively to the conventional nonlinear OFDM schemes.

6   Conclusions and Further Work

In this paper we considered the nonlinear distortion effects as useful information that can improve the performance of OFDM schemes. The adequate receiver to explore this information is the optimum receiver. However, its complexity is too high for practical applications. To overcome this problem, we present a sub-optimal receiver whose the performance is very close to the one that can be obtained by the optimum receiver but has much lower complexity. We include results with diversity and conclude that the sub-optimal detection also allows a performance improvement in these type of scenarios. The guidelines for future work can pass to theoretically derive the potential gains of the optimum (and sub-optimum) receivers under AWGN and frequency-selective channels. Another important open question is to theoretically quantify what is the impact of the filtering operation in the potential gains of the optimum-based detection.

References

1. Cimini Jr., L.: Analysis and Simulation of a Digital Mobile Channel Using Orthogonal Frequency Division Multiplexing. IEEE Trans. on Comm. 33(7), 665–675 (1985)
2. Digital Video Broadcasting (DVB); Framing structure, channel coding and modulation for digital terrestrial television, ETSI Standard: EN 300 744 V1.6.1 (January 2009)

3. 3rd Generation Partnership Project: Technical Specification Group Radio Access Network; Physical Layers Aspects for Evolved UTRA, 3GPPP TR 25.814 (2006)
4. IEEE Standard for Local and Metropolitan Area Networks - Part 16: Air Interface for Fixed Broadband Wireless Access Systems, IEEE 802.16- (October 2004)
5. Rahmatallah, Y., Mohan, S.: Peak-To-Average Power Ratio Reduction in OFDM Systems: A Survey And Taxonomy. IEEE Communications Surveys & Tutorials (99), 1–26 (2013)
6. Dinis, R., Gusmão, A.: A Class of Nonlinear Signal Processing Schemes for Bandwidth-Efficient OFDM Transmission with Low Envelope Fluctuation. IEEE Trans. on Comm. 52(11), 2009–2018 (2004)
7. Rowe, H.: Memoryless Nonlinearities with Gaussian Input: Elementary Results. Bell System Tech. Journal, 61 (September 1982)
8. Gusmão, A., Dinis, R.: Iterative Receiver Techniques for Cancellation of Deliberate Nonlinear Distortion in OFDM-Type Transmission. In: IEEE Int. OFDM Workshop 2004, Dresden, Germany (September 2004)
9. Montezuma, P., Dinis, R., Oliveira, R.: Should we Avoid Nonlinear Effects in a Digital Transmission System? In: IEEE WCNC 2010, Sydney, Australia (April 2010)
10. Saleh, A.: Frequency-Independent and Frequency-Dependent Nonlinear Models of TWT Amplifiers. IEEE Transactions on Communications Com-29(11) (November 1981)
11. Proakis, J.G.: Digital Communications, 4th edn. McGraw-Hill (2001)

UWB System Based on the M-OAM Modulation in IEEE.802.15.3a Channel

Khadija Hamidoun[1], Raja Elassali[1], Yassin Elhillali[3], Khalid Elbaamrani[1], A. Rivenq[3], and F. Elbahhar[2]

[1] ENSA, TIM, Marrakech, Morocco
[2] IFSTAR, LEOST, F-59666 Villeneuve d'Ascq,
[3] UVHC, IEMN-DOAE, F-59313 Valenciennes, France
k.hamidoun@gmail.com

Abstract. Known for many years but unexploited in the field of communications, ultra wideband systems (UWB) can be a solution to the saturation of the frequency bands. The advantage of such systems is that they are inexpensive in energy because of the limitations imposed by the standard. Our work is to design a high data rate IR-UWB system, through the use of modulation M-OAM (Orthogonal Amplitude Modulation), and the study of the system performances by adding the effect of the propagation channels dedicated to the UWB. The channel modeling is an important step in the implementation of any system, especially for those in development, as is the case of UWB systems. Although this model is one of the key elements of the overall chain system, it must take into account the distance between the two antennas, multipath and signal type. In this paper, we discover the existing models suitable for UWB channels and their implementation on our M-OAM system. The introduction of real channels in our chain of transmission reduces the quality of our system. Therefore, a RAKE receiver is proposed as a solution to improve performance.

Keywords: IR-UWB, channel effect, M-OAM, MGF, RAKE receiver.

1 Introduction

Recently, the main applications developed allowing the best use of the UWB spectrum, are: communication with high bit rate, low power, and short range, the second main applications are the communication with a low power and low bit rate [2]. As mentioned above, our study focuses on high-bit-rate systems, especially multimedia communication systems and intelligent transport such as vehicle to vehicle and vehicles to infrastructure applications [3].

The main purpose of our approach is to improve the quality of service of UWB communication systems, by using a high data rate modulation M-OAM (Orthogonal Amplitude Modulation). The IEEE.802.15.3a channel using M-OAM modulation is implemented and the perspective BER for each model is calculated. The performance is degraded with IEEE channels. By consequence, a RAKE receiver is proposed to circumvent this problem and improve performances.

L.M. Camarinha-Matos et al. (Eds.): DoCEIS 2014, IFIP AICT 423, pp. 507–514, 2014.

This paper is organized as follows. In the second section, we describe the orthogonal polynomials used in our UWB system. In this section, we focus on the properties of the modified Gegenbauer functions MGF. The third section presents an original high data rate modulation schemes called M-OAM based on orthogonal waveforms MGF. The main purpose of this paper is to evaluate the performances of our system in AWGN and IEEE.802.15.3a channels. Finally, the last section is dedicated to the study and implementation of RAKE receiver as a technique to improve the results. At the end, a conclusion is drawn with prospects.

2   Collective Awareness System

Our UWB M-OAM system is applied in the context of new information technologies related to transportation, in order to improve the quality of life of people by the collective awareness systems. Our application is included more precisely in the context of ITS (Intelligent Transportation System) that add some intelligence in transportation systems. Communication between vehicles would improve road safety, avoid traffic jams and reduce energy consumption, which is directly related to the release of gas (CO_2).

3   Orthogonal Polynomials and M-OAM Modulations

In this section, we present the Gegenbauer pulses as an example of orthogonal pulses that can be used for UWB communications. By definition, a family of polynomials $f_n(x)$ (where n is the degree of the polynomial) is called orthogonal on the interval of definition [a, b], if it satisfies the condition:

$$\int_a^b w(x)f_n(x)f_m(x)dx = 0 \tag{1}$$

Where m and n are non-negative integers unequal and $w(x)$ is the weight of generalized scalar product [4]. Assuming that $w(x)$ is strictly positive on the interval of definition. Equation (1) can be rewritten as a scalar product:

$$\int_a^b f_n(x)f_m(x)dx = 0 \tag{2}$$

Equation (2) ensures also that energy is normalized to unity, eg 1 J, for each order of polynomial. Functions f_n is then an orthonormal basis for a space of functions on the interval of definition [a, b]. A large number of families of polynomials can be considered, for example, the classical Hermite, Laguerre polynomials, Bessel and Gegenbauer polynomials.

3.1   Modified Gegenbauer Functions

The Modified Gegenbauer functions (MGF) are waveforms based on orthogonal polynomials. These waveforms have been developed in IEMN_DOAE and IFSTAR-LEOST laboratories for communication applications [4] [5].

Gegenbauer polynomials are defined on the interval [-1,1] and satisfies the differential equation of second order defined as follows:

$$G_{n,\beta}(x) = 2\left(1 + \frac{n + \beta - 1}{n}\right)xG_{n-1,\beta}(x) - \left(1 + \frac{n + 2\beta - 2}{n}\right)G_{n-2,\beta}(x) \qquad (3)$$

The first 4 orders of these functions are defined in following equations and presented in figure 1:

$$G_u(0,1,x) = 1 * (1 - x^2)^{1/4} \qquad (4a)$$

$$G_u(1,1,x) = 2x * (1 - x^2)^{1/4} \qquad (4b)$$

$$G_u(2,1,x) = (-1 + 4x^2) * (1 - x^2)^{1/4} \qquad (4c)$$

$$G_u(3,1,x) = (-4x + 8x^3) * (1 - x^2)^{1/4} \qquad (4d)$$

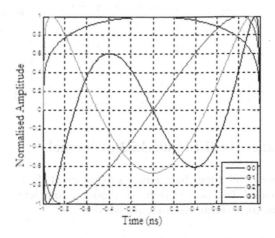

Fig. 1. The first 4 orders of Gegenbauer functions

These polynomials can be used in a UWB system to generate very short pulses. To do this, we propose to multiply $G_n(\boldsymbol{\beta},x)$ by the square root of the weight function of this family of polynomials $w(x,\beta)$. [7] [8]

3.2   M-OAM Modulation Schema

The Orthogonal Amplitude Modulation (OAM) is the combination between the Bi-Phase Modulation (BPM) and Pulse Position Modulation (PPM) modulation using the orthogonal signal. In this section, we will see how the OAM is constructed by combaining the both modulation.

The modulation 4-OAM transmits 2bits/symbol. This technique is the same as in the PPM bipolar; the first bit is represented by the position of the pulse and the second bit is represented by the phase (positive or negative) of the pulse. So, there are four possibilities combination of the pulse for this modulation. The symbol equation is:

$$x_i(t) = (2S_{iMSB} - 1)m(t + (2S_{iLSB} - 1)T) \tag{5}$$

With

$$m(t) = 2\alpha K t \exp(-\alpha t^2) \tag{6}$$

S_i is the symbol sent, m(t) the used waveform, T the time interval, LSB the Low Significant Bit and MSB the Most Significant Bit. Table 1 shows the 4-OAM pulse combination [7] [8].

Table 1. Symbols of 4-OAM modulation

MSB \ LSB	0	1
0	G1	G1
1	-G1	-G1

4   Performances in IEEE.802.15.3a Channel

At first, this section defines the characteristics of different IEEE 802.15.3a channels present in the standard. In a second step, we model a complete chain of our UWB system and evaluate its performances in terms of BER.

4.1   Standard IEEE.802.15.3a

IEEE 802.15.3a is based on the formalism of Saleh and Valenzuela; except that it is a real model and the amplitudes decay follows a law lognormal instead of an exponential decay, and also the first paths are not necessarily the strongest. [10] [11]

Parameters are defined to characterize the arrival of clusters Λ and rays rates λ as well as the coefficients of exponential decay inter-and intra-cluster (Γ and γ).

Four sets of model parameters are provided for four types of channels:
➢ The channel model CM 1: corresponds to a distance of 0-4 m in the LOS;
➢ The channel model CM 2: corresponds to a distance of 0-4 m in NLOS conditions;
➢ The channel model CM 3: corresponds to a distance of 4 to 10 m in NLOS conditions;
➢ The channel model CM 4: corresponds to a NLOS situation with a large spread delay τRMS = 25 ns.

This channel model has the merit of being a reference for studies of UWB systems. It applies in indoor environments at short range (0 to 10m). The following figures (Fig. 2 a-d) show the impulse responses of the four models.

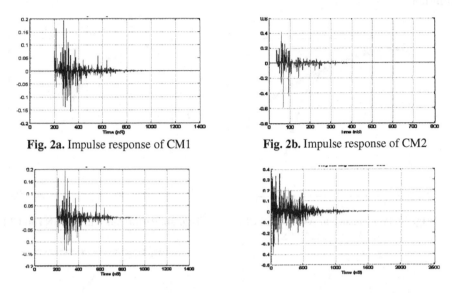

Fig. 2a. Impulse response of CM1   **Fig. 2b.** Impulse response of CM2

Fig. 2c. Impulse response of CM3   **Fig. 2d.** Impulse response of CM4

We note in the impulse response of CM1, that the delays have relatively low values, which makes sense since we're in a LOS (Line Of Sight) environment. For the last three models, we are in a NLOS (No Line Of Sight) environment, and we note here that the delays increase with the distance.

4.2   Simulation Results

To implement our UWB system, we simulated our transmission chain containing the transmitter using the 4-OAM modulations with MGF waveforms, the IEEE.802.15.3 channel according to the CM1-4 models, and finally, the demodulation at the reception. To examine the performance of our system, we calculated the BER versus SNR for each CM as well as for an AWGN channel, and then we compared the results. The Figures 3a-d show the BER versus SNR for each model.

Before the introduction of the channel, the BER was taking values less than 10^{-4} for a SNR greater than 10dB. While with the channel introduced, there is a degradation of the signal that can increase the value of BER up to 10^{-2} for the same SNR values chosen. By analyzing the graph of Figure 3, we see that the slope differs between AWGN and CM1-4, given the parameters used, namely the distance transmitter/receiver which is proportional to the BER, and the SNR that is inversely proportional to the AWGN and CM1-4.

According to figure 2 and the simulation results, we deduce that CM1 offers the best conditions. However, with the other channels, we have more degraded performances. For that, we focus in future work on performance improvement solutions.

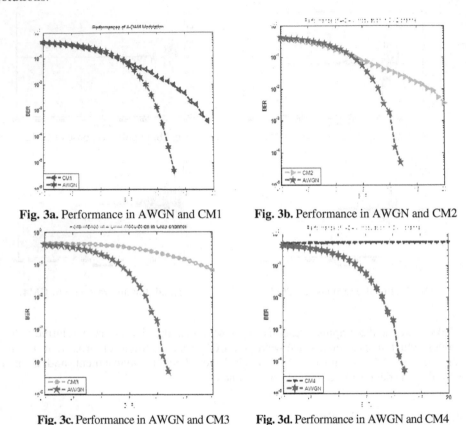

Fig. 3a. Performance in AWGN and CM1   **Fig. 3b.** Performance in AWGN and CM2

Fig. 3c. Performance in AWGN and CM3   **Fig. 3d.** Performance in AWGN and CM4

4.3   Performance Improvements

The transmitted signal takes different paths due to the propagation channel. These multiple paths have different physical lengths, thus the signal from each trip comes with amplitude and delay specific to each path. These echoes induce a temporal spreading of the signal giving rise to a phenomenon of interference between transmitted symbols (ISI), which increases the bit error rate.

Several solutions have been proposed in the literature to combat this phenomenon. Among, there are the equalization techniques introduced at the reception, as well as the Rake receiver using multipath constructively. In this section, we tried to improve the results obtained above, by implementing the Rake receiver as one of the solutions.

To verify the results obtained after introduction of the Rake receiver, we chose to draw for each CM: one graph for AWGN, the second one for CM channel only, then introducing the RAKE receiver. Figures 4a-d shows the simulation results.

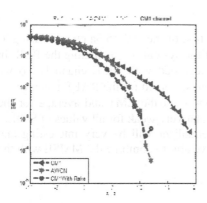

Fig. 4a. Comparison of CM1 and the Rake

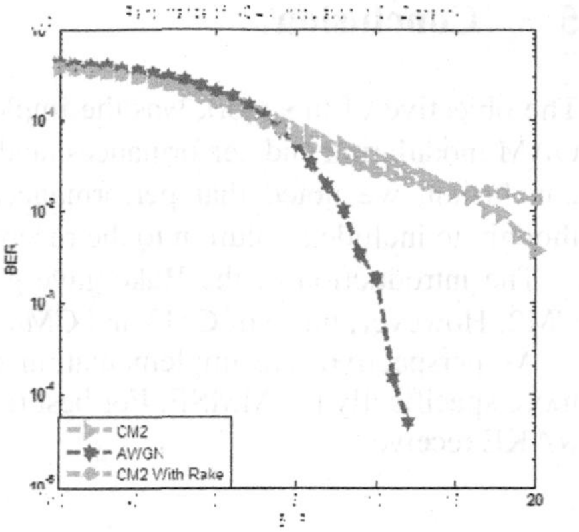

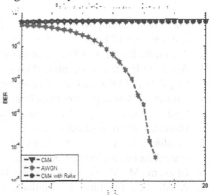

Fig. 4b. Comparison of CM2 and the Rake

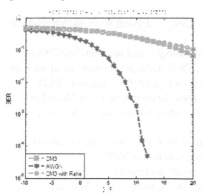

Fig. 4c. Comparison of CM3 and the Rake

Fig. 4d. Comparison of CM4 and the Rake

The above figures show that:

- The implementation of the RAKE receiver after CM1 gives better results throughout the axis of SNR. Since we reached a BER of 0.03% for 11dB.
- The contribution of the RAKE receiver with CM1 gives more improvement than the CM2. For example for SNR = 10dB, the BER has decreased from 3.9% to 2.7%. From figure b, we see that for a SNR greater or equal to 15dB, there is no more improvement.
- The introduction of the Rake receiver with CM3 and CM4 channels (Figures 4.c and 4.d) has not led to a noticeable improvement in system performance compared to its absence. For CM3, 38% of the BER were replaced by 36% for SNR = 3dB. While for the CM4, we spent 54.5% to 51% for SNR =-1dB. This can be explained by the extreme conditions of the channel, which is a NLOS environment with a transmitter / receiver distance that exceeds 4m for CM3 and 10m for CM4.

5   Conclusion

The objective of this work was the implementation of the 802.15.3a channel using 4-OAM modulation and performances analysis of the system by calculating the BER. In conclusion, we noted that performances are degraded with IEEE channels, so we thought to include a solution to the reception, and we opted for the RAKE receiver.

The introduction of the Rake gave good results for the CM1 and average for the CM2. However, those of CM3 and CM4 were relatively weak for all values of SNR.

As perspective, the implementation of an equalizer will be very interesting and more specifically the MMSE. For best results, we aim to combine the MMSE with the RAKE receiver.

References

1. Shen, X., Guizani, M., Qiu, R.C., Le–Ngoc, T.: Ultra-wideband wireless communications and networks (2006)
2. Aiello, R., Batra, A.: Ultra wideband systems technologies and applications (2006)
3. Abed, E.: Etude et conception d'un système IR-UWB dédié aux communications sans fils haut débit, PhDthesis, Univ Valenciennes and Hainaut Cambrésis (January 2012)
4. Lamari, A.: Conception et modélisation d'un système de communication Multi-Utilisateurs basé sur la technique Ultra Large Bande, PhDthesis, University of Valenciennes and Hainaut Cambrésis (January 2007)
5. Elbahhar, F., Rivenq, A., Rouvaen, J.M.: Multi-user ultra-wide band communication system based on modified gegenbauer and hermite functions (2005)
6. Ghavami, M., Michael, L.B., Kohno, R.: UWB signals and systems in engineering
7. Abed, E.L., Elbahhar, F., Elhillali, Y., Rivenq, A., Elassali, R.: UWB communication system Based on Bipolar PPM with orthogonal waveforms, Wireless Engineering and Technology (2012)
8. Faten SALEM BAHRI: Contribution à l'étude des systèmes ultra wide band différentiels, PhD thesis, the national school of Telecommunications of Britain with the national school of engineers of Tunisia (May 2009)
9. Hamidoun, K., Elassali, R., Elhillali, Y., Elbahharr, F., Rivenq, A.: New adaptive architectures of coding and modulation UWB for infrastructure vehicle communication. In: 6th International Symposium on Signal, Image, Video and Communications, ISIVC 2012 (July 2012)
10. Cassioli, D., Win, M.Z., Molisch, A.F.: The Ultra-Wide Bandwidth Indoor Channel From Statistical Model to Simulations. IEEE JSAC 20, 1247–1257 (2002)
11. Cassioli, D., Win, M.Z., Molisch, A.F.: A statistical model for the UWB indoor channel. In: Proc. 53rd IEEE Vehicular Technology Conf., vol. 2, pp. 1159–1163 (May 2001)

Distributed RSS-Based Localization in Wireless Sensor Networks with Asynchronous Node Communication

Slavisa Tomic[4], Marko Beko[1,3], Rui Dinis[2,5], and Miroslava Raspopovic[6]

[1] Universidade Lusófona de Humanidades e Tecnologias, Lisbon, Portugal
[2] DEE/FCT/UNL, Caparica, Portugal
[3] UNINOVA – Campus FCT/UNL, Caparica, Portugal
[4] Institute for Systems and Robotics / IST, Lisbon, Portugal
[5] Instituto de Telecomunicações, Lisbon, Portugal
[6] Faculty of Information Technology, Belgrade Metropolitan University, Serbia
s.tomic@campus.fct.unl.pt, mbeko@uninova.pt, rdinis@fct.unl.pt,
miroslava.raspopovic@metropolitan.ac.rs

Abstract. In this paper we address the node localization problem in large-scale wireless sensor networks (WSNs) by using the received signal strength (RSS) measurements. According to the conventional path loss model, we first pose the maximum likelihood (ML) problem. The ML-based solutions are of particular importance due to their asymptotically optimal performance (for large enough data records). However, the ML problem is highly non-linear and non-convex, which makes the search for the globally optimal solution difficult. To overcome the non-linearity and the non-convexity of the objective function, we propose an efficient second-order cone programming (SOCP) relaxation, which solves the node localization problem in a completely distributed manner. We investigate both synchronous and asynchronous node communication cases. Computer simulations show that the proposed approach works well in various scenarios, and efficiently solves the localization problem. Moreover, simulation results show that the performance of the proposed approach does not deteriorate when synchronous node communication is not feasible.

Keywords: Wireless localization, wireless sensor network (WSN), received signal strength (RSS), second-order cone programming problem (SOCP), cooperative localization, distributed localization.

1   Introduction

Wireless sensor network (WSN) consists of a large number of low-power sensor nodes that have some sensing, processing and communication capabilities. WSNs find application in various areas, and the capability to accurately locate all sensor nodes in the network is essential for many of them (e.g. monitoring, military operations, rescue missions, etc.). In general, sensor nodes can be classified as anchor and target (source) nodes [1]. The positions of anchor nodes are known *a priori* (usually measured by global positioning system (GPS) or manually), while the positions of target nodes are yet to be determined. For economic or other practical reasons, only a small fraction of

L.M. Camarinha-Matos et al. (Eds.): DoCEIS 2014, IFIP AICT 423, pp. 515–524, 2014.
© IFIP International Federation for Information Processing 2014

nodes are set to be anchor nodes, hence, an efficient algorithm for node localization is necessary for WSNs.

Instead of the use of GPS system, which is very expensive and limited to outdoor environments, algorithms that rely on distance measurements between neighboring nodes are emerging recently [2], [3], [4], [5], [6]. Depending on the available hardware, current distance-based algorithms extract the distance information from time-of-arrival (TOA), time-difference-of-arrival (TDOA), angle-of-arrival (AOA) or received signal strength (RSS) measurements [7]. Localization based on RSS measurements requires the least processing and communication (the least energy), and no specialized hardware [8], which makes it an attractive low-cost solution for the localization problem.

Localization algorithms can be executed in a centralized or a distributed fashion. The former approach assumes the existence of a fusion center which coordinates the network and performs all computational processing. This approach leads to large energy and bandwidth consumption, with a bottleneck around the fusion center and the computational complexity of such an approach grows with the increase of a number of nodes in the network. In many practical scenarios, it is not efficient or not possible for the nodes to share their private objective functions with a central processor or with each other [9], which makes the distributed localization a more preferable solution. Even though it is sensitive to error propagation, distributed concept is energy-efficient, has low-computational complexity and high-scalability [1]. In such an approach, the communication is possible only between the neighboring nodes, and the data associated with each node is always processed locally.

Recent RSS-based localization algorithms use a centralized concept to solve the sensor nodes localization problem [2], [3], [4], [5], [6]. A distributed approach for solving the RSS localization problem with synchronized node communication was introduced in [10], [11]. However, synchronized node communication requires the use of more sophisticated hardware in the network, which raises the cost of the network implementation.

In this work, we consider a large-scale RSS-based target localization problem, and we provide a solution that is completely based on a distributed approach. We investigate both the synchronous and asynchronous node communication scenarios, and propose a novel algorithm based on second-order cone programming (SOCP) relaxation. We first formulate the maximum likelihood (ML) optimization problem, which is highly non-linear and non-convex. To overcome these difficulties, we propose a SOCP relaxation approach to transform the original ML problem into a convex one, which can then be efficiently solved by interior-point algorithms [12].

2   Relationship to Collective Awareness Systems

Sensor nodes in WSNs are deployed over a monitored area in order to acquire the desired information (such as temperature, wind speed, pressure, etc.). Nowadays, the basic concept of WSNs is used for the Internet-of-Things (IoT) in which all devices, objects and environments are connected through Internet to form the so-called smart environments [13]. Such systems are capable of harnessing collective intelligence for promoting innovation and taking better-informed and sustainability-aware decisions.

Being able to accurately estimate object's position is a key factor in a number of sensor network applications (such as energy-efficient routing, target tracking and detection). Combining the location information with other information collected inside the network enables us to link objects, people and knowledge and develop intelligent systems (collective awareness systems). Such systems may improve safety and efficiency in everyday life, since each individual device can make better-informed and substantially-aware decisions to respond faster and better to the changes in dynamical environments (search and rescue missions, logistics in warehouses, etc.). Furthermore, such systems enhance sustainable growth, since they support new forms of social and business innovation.

3   Problem Formulation

We consider a p-dimensional ($p \geq 2$) WSN with $M + N$ nodes, where M and N are the number of target and anchor nodes, respectively. The locations of the nodes are denoted as $x_1, \dots, x_M, x_{M+1}, \dots, x_{M+N}$. The considered WSN can also be seen as a connected graph, $G = (V, E)$, where V and E represent the set of nodes (vertices) and the set of node connections (edges) in the graph, respectively. Due to lifetime of the network or other physical limitations, each node has a limited communication range, R. An edge exists between two nodes, i and j, if and only if they are within the communication range of each other, i.e. $E = \{(i, j) : \|x_i - x_j\| \leq R\}$. The set of neighbors of a target node i is defined as $\Omega_i = \{j : (i, j) \in E\}$, and each neighboring node j is seen as an anchor node in the localization process by the i-th target node.

We assume that each target node i is given an initial estimation of its position, $x_i^{(0)}$. For ease of expression, we define $X = [x_1, x_2, \dots, x_M]$ as the $2 \times M$ matrix of all target positions in the WSN; hence, $X^{(0)}$ contains all initial estimations of the target positions.

Node i measures the received power from the signal transmitted by its neighboring node j, P_{ij}, which, under the log-normal shadowing, can be modeled as (in dBm) [14], [15]

$$P_{ij} = P_0 - 10\gamma \log_{10} \frac{\|x_i - x_j\|}{d_0} + v_{ij}, \qquad \forall (i, j) \in E, \tag{1}$$

where P_0 is the power measured at a short reference distance d_0 ($\|x_i - x_j\| \geq d_0$), γ is the path loss exponent, and v_{ij} is the log-normal shadowing term between the i-th target node and its neighbor node j, modeled as a zero-mean Gaussian random variable with variance σ_{ij}^2, i.e. $v_{ij} \sim \mathcal{N}(0, \sigma_{ij}^2)$.

Based on the measurements from (1), we derive the maximum likelihood (ML) estimator as

$$\hat{X} = \underset{X}{\mathrm{argmin}} \sum_{(i,j)\in E} \frac{1}{\sigma_{ij}^2} \left[P_{ij} - P_0 + 10\gamma \log_{10} \frac{\|x_i - x_j\|}{d_0} \right]^2. \tag{2}$$

The least squares (LS) problem defined in (2) is non-linear and non-convex, hence, finding the globally optimal solution is difficult, since there may exist multiple local optima. Following [5], we show that the RSS measurement model in (1) can be approximated into a convex optimization problem, which can be solved by interior point algorithm [12], and obtain the global solution.

For the sake of simplicity, in the further text we assume that $\sigma_{ij}^2 = \sigma^2, \forall (i,j) \in E$. Furthermore, we assume that the transmit power, P_T, of each node in the network is equal, i.e. all nodes have identical P_0 and R.

4   Distributed Localization Using SOCP Relaxation

Note that the objective function in (2) depends only on the positions and pairwise measurements between the neighboring nodes. This means that we can portion the objective function in (2) and perform the minimization independently by each target node, using only local information gathered from its neighbors. Instead of having a sink, which collects and processes the information from all nodes in WSN, we can divide the optimization problem into smaller sub-problems which can be carried out locally by each target node. This kind of problem execution is particularly suitable for large scale networks, since the number of nodes inside the network has no major impact on the neighborhood fragments, and hence, the computational complexity remains the same (no significant changes) as more nodes are added in the network [1]. The price to pay for using this kind of approach is the increased energy consumption due to higher node communication, since distributive localization algorithms require repetition of the following phases:

- *Communication phase*: nodes in the network transmit their estimated position to their neighboring nodes.
- *Computation phase*: each target node computes its position estimation based on the information gathered in the communication phase.

If the information exchange between nodes is always performed at the beginning of each iteration, we say that the node communication is *synchronous* [16]. However, due to some physical imperfections (e.g. internal clock of the nodes), synchronous node communication may not be feasible in practice. This is the reason why we also investigate *asynchronous* node communication. In such communication, a randomly chosen target node performs an update of its position estimation based on the available information from its neighbors at that particular moment (computation phase) and transmits the updated information to its neighbors thereupon (communication phase). This process is then repeated until each node reaches the maximum number of iterations, K_{max}.

In the k-th iteration of the computation phase of our algorithm, each target node i solves a SOCP relaxation of the following problem:

$$\hat{x}_i^{(k)} = \underset{x_i}{\text{argmin}} \sum_{(i,j) \in E} \frac{1}{\sigma^2} \left[P_{ij} - P_0 + 10\gamma \log_{10} \frac{\left\| x_i - \hat{x}_j^{(k-1)} \right\|}{d_0} \right]^2, \tag{3}$$

where $\hat{x}_j^{(k-1)}$ denotes the estimated position of the neighboring node j in the $(k-1)$-the iteration. In the following text we will describe a SOCP relaxation method which approximates the problem in (3) into a convex one.

4.1   SOCP Relaxation

Approximating (1) as $P_{ij} \approx P_0 - 10\gamma \log_{10} \frac{\left\| x_i - x_j^{(k-1)} \right\|}{d_0}, \forall (i,j) \in E$, we get

$$\alpha_{ij} \left\| x_i - x_j^{(k-1)} \right\|^2 \approx d_0^2, \tag{4}$$

where $\alpha_{ij} = 10^{\frac{P_{ij} - P_0}{5\gamma}}$. According to (4), the following LS estimation problem can be formulated[1]:

$$x_i^{(k)} = \underset{x_i}{\text{argmin}} \sum_{(i,j) \in E} \left(\alpha_{ij} \left\| x_i - x_j^{(k-1)} \right\|^2 - d_0^2 \right)^2. \tag{5}$$

Define auxiliary variables $\lambda_{ij} = 10^{\frac{P_{ij}}{5\gamma}}$, $\rho = 10^{\frac{P_0}{5\gamma}}$, $y_i = \|x_i\|^2$, and $z = [z_{ij}]$, where $z_{ij} = \lambda_{ij} \left\| x_i - x_j^{(k-1)} \right\|^2 - \rho d_0^2$, $\forall (i,j) \in E$. Introduce an epigraph variable t, and apply second-order cone constraint (SOCC) to obtain a convex optimization problem:

$$\underset{x_i, y_i, z, t}{\text{minimize}} \; t$$

subject to

$$z_{ij} = \lambda_{ij} \left(y_i - 2{x_j^{(k-1)}}^T x_i + \left\| x_j^{(k-1)} \right\|^2 \right) - \rho d_0^2, \qquad \forall (i,j) \in E,$$

$$\| 2z; \; t - 1 \| \leq t + 1, \qquad \| 2x_i; \; y_i - 1 \| \leq y_i + 1. \tag{6}$$

---

[1] We can rewrite (1) as $\frac{P_{ij} - P_0}{5\gamma} + \log_{10} \frac{\left\| x_i - x_j^{(k-1)} \right\|^2}{d_0^2} = \frac{v_{ij}}{5\gamma}$, which corresponds to $\alpha_{ij} \left\| x_i - x_j^{(k-1)} \right\|^2 = d_0^2 10^{\frac{v_{ij}}{5\gamma}}$. For sufficiently small noise, we can apply the first-order Taylor series expansion to the right-hand side of the previous expression to obtain $\alpha_{ij} \left\| x_i - x_j^{(k-1)} \right\|^2 = d_0^2 \left(1 + \frac{\ln 10}{5\gamma} v_{ij} \right)$, i.e. $\alpha_{ij} \left\| x_i - x_j^{(k-1)} \right\|^2 = d_0^2 + \varepsilon_i$, where $\varepsilon_i = d_0^2 \frac{\ln 10}{5\gamma} v_{ij}$ is a zero-mean Gaussian random variable with variance $\frac{(\ln 10)^2 d_0^4 \sigma^2}{25\gamma^2}$, i.e., $\varepsilon_i \sim \mathcal{N} \left(0, \frac{(\ln 10)^2 d_0^4 \sigma^2}{25\gamma^2} \right)$. Clearly, the corresponding LS estimator is given by (5).

Problem in (6) is a SOCP problem, which can efficiently be solved by the CVX package [16] for specifying and solving convex programs. We will refer to (6) as "SOCP" in the further text.

5   Complexity Analysis

The trade-off between the computational complexity and the estimation accuracy is the most important feature of any localization algorithm, since it determines its applicability potential. Here, we consider only the worst case asymptotic complexity of an algorithm.

According to [18], the worst case complexity of the proposed "SOCP" approach is $M \cdot K_{max} \cdot \mathcal{O}((\max\{|\Omega_i|\})^3)$, where $|\Omega_i|$ represents the cardinality of the set Ω_i, for $i = 1, ..., M$.

As we can see from the above result, the worst case complexity of a distributive localization algorithm mainly depends on the neighborhood fragments (the biggest one). If the number of nodes in WSNs is increased, it will not significantly affect the size of the neighborhood fragments, which makes the distributed algorithms a desirable solution in highly-dense or large-scale networks.

6   Simulation Results

In this section, we present the computer simulation results in order to evaluate the performance of the proposed approach. The considered algorithms were solved by using the MATLAB package CVX [17], where the solver is SeDuMi [19].

Nowadays, flexibility and adaptability of a network are very important features in practical applications; hence we consider a random deployment of the nodes. All nodes were randomly deployed inside a square region of length $B = 30$ m in each Monte Carlo (M_c) run. In order to make the comparison fair, we first obtained $M_c = 500$ nodes positions, and we applied the proposed approaches for those scenarios. In each M_c run, we made sure that the network graph is connected. The path loss exponent is $\gamma = 3$, the reference distance $d_0 = 1$ m, the reference power $P_0 = -10$ dBm, and the communication range of a node $R = B/5$ m. We assumed that the initial estimation of the target positions, $X^{(0)}$, is in the intersection of the diagonals of the square area, and that $K_{max} = 200$. We assume that one iteration step is completed after M nodes compute and transmit their position estimations, for both synchronous and asynchronous node communication. As the performance metric we used the normalized root mean square error (NRMSE), defined as

$$\text{NRMSE} = \sqrt{\frac{1}{M M_c} \sum_{i=1}^{M_c} \sum_{j=1}^{M} \left\| x_{ij} - \hat{x}_{ij} \right\|^2}, \tag{6}$$

where $\hat{x}_{ij}$ denotes the estimate of the true location of the j-th target in the i-th Monte Carlo run, x_{ij}.

The NRMSE performance of the proposed method versus number of iterations for synchronous and asynchronous node communication, and variable N is depicted in Fig. 1. From Fig. 1 we can see improvement of NRMSE performance as the number of iterations is increased, as expected. Further, we can see that NRMSE performance improves as N is increased, as expected. Lower estimation accuracy is achieved with asynchronous than with synchronous node communication in the early phase of the algorithm. However, we can see that the asymptotical performance of the proposed approach is the same for both asynchronous and synchronous node communication. Furthermore, Fig. 1 exhibits that all major improvements in the performance occur until approximately 80 iterations, and we can conclude that our algorithm converges after this number of iterations, for the considered scenarios.

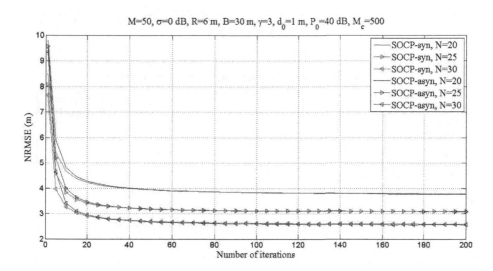

$M=50$, $\sigma=0$ dB, $R=6$ m, $B=30$ m, $\gamma=3$, $d_0=1$ m, $P_0=40$ dB, $M_c=500$

Fig. 1. Comparison of the performance of the proposed method for synchronous and asynchronous node communication: NRMSE versus number of iterations for variable N

The NRMSE performance of the proposed method versus number of iterations for synchronous and asynchronous node communication, and variable M is depicted in Fig. 2. We can see from Fig. 2 that NRMSE performance of the proposed approach does not degrade as more target nodes are added in the network. As it was anticipated, estimation accuracy is weaker in the early phase of the algorithm for higher M, and the proposed algorithm converges slower for this setting. Fig. 2 exhibits better NRMSE performance of the proposed approach with synchronized than with asynchronous node communication after the first few iterations. However, asymptotical performance is the same for both types of node communication.

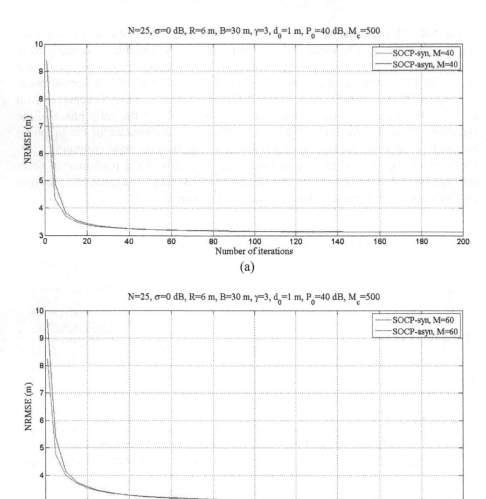

Fig. 2. Comparison of the performance of the proposed method for synchronous and asynchronous node communication: NRMSE versus number of iterations: (a) $M = 40$, (b) $M = 60$

7   Conclusions

In this work, we investigated the RSS-based sensor localization problem in large-scale WSNs, which we solved in a completely distributed fashion. We considered both synchronous and asynchronous node communication scenarios, and we proposed a distributed algorithm that is based on SOCP relaxation technique. Due to practical demands, such as flexibility and adaptability of the network, randomly generated WSN were taken into consideration. Simulation results show that the proposed approach efficiently solves the sensor localization problem for different settings. Moreover, simulation results show that, even though synchronous node communication is preferred over asynchronous when only few iteration steps are allowed, the asymptotical performance of the proposed approach does not suggest any preference between the mentioned types of node communication.

Acknowledgments. This work was partially supported by Fundação para a Ciência e Tecnologia under Projects PEst-OE/EEI/UI0066/2011, PTDC/EEATEL/099973/ 2008–ADCOD, PTDC/EEA-TEL/115981/2009–OPPORTUNISTIC-CR, PTDC/EEI-TEL/2990/2012–ADIN and PTDC/EEA-TEL/120666/2010–Femtocells, and Ciência 2008 Post-Doctoral Research grant. M. Beko is also a collaborative member of Instituto de Sistemas e Robótica – Instituto Superior Técnico, Av. Rovisco Pais, 1049-001 Lisboa, Portugal, and SITILabs/ILIND/COFAC: R&D Unit of Informatics Systems and Technologies, Instituto Lusófono de Investigação e Desenvolvimento (ILIND), COFAC, Lisboa, Portugal.

References

1. Destino, G.: Positioning in Wireless Networks: Noncooperative and Cooperative Algorithms. Thesis of Giuseppe Destino at University of Oulu, Finland (2012)
2. Ouyang, R.W., Wong, A.K.S., Lea, C.T.: Received Signal Strength-based Wireless Localization via Semidefinite Programming: Noncooperative and Cooperative schemes. IEEE Trans. Veh. Technol. 59(3), 1307–1318 (2010)
3. Wang, G., Yang, K.: A New Approach to Sensor Node Localization Using RSS Measurements in Wireless Sensor Networks. IEEE Trans. Wireless Commun. 10(5), 1389–1395 (2011)
4. Wang, G., Chen, H., Li, Y., Jin, M.: On Received-Signal-Strength Based Localization with Unknown Transmit Power and Path Loss Exponent. IEEE Wireless Commun. Letters (2012)
5. Tomic, S., Beko, M., Dinis, R., Lipovac, V.: RSS-based Localization in Wireless Sensor Networks using SOCP Relaxation. In: IEEE SPAWC (2013)
6. Vaghefi, R.M., Gholami, M.R., Buehrer, R.M., Strom, E.G.: Cooperative Received Signal Strength-Based Sensor Localization With Unknown Transmit Powers. IEEE Trans. Signal. Process. 61(6), 1389–1403 (2013)
7. Biswas, P., Ye, Y.: Semidefinite Programming for Ad Hoc Wireless Sensor Network Localization. In: IPSN 2004, Berkeley, California, USA (2004)
8. Patwari, N.: Location Estimation in Sensor Networks. Thesis of Neal Patwari at University of Michigan, Michigan, USA (2005)
9. Sundhar Ram, S., Nedic, A., Veeravalli, V.V.: Distributed Subgradient Projection Algorithm for Convex Optimization. In: IEEE ICASSP (2009)
10. Tomic, S., Beko, M., Dinis, R., Raspopovic, M.: Distributed RSS-basedLocalization in Wireless Sensor Networks Using Convex Relaxation. Accepted for publication in ICNC 2014, CNC Workshop, Honolulu, Hawaii, USA (2014)
11. Cota-Ruiz, J., Rosiles, J.G., Rivas-Perea, P., Sifuentes, E.: A Distributed Localization Algorithm for Wireless Sensor Networks Based on the Solution of Spatially-Constrained Local Problems. IEEE Sensors Journal 13(6), 2181–2191 (2013)

12. Boyd, S., Vandenberghe, L.: Convex Optimization. Cambridge University Press, New York (2004)
13. Jiang, J.A., Zheng, X.Y., Chen, Y.F., Wang, C.H., Chen, P.T., Chuang, C.L., Chen, C.P.: A Distributed RSS-Based Localization Using a Dynamic Circle Expanding Mechanism. IEEE Sensors Journal (2013)
14. Rappaport, T.S.: Wireless Communications: Principles and Practice. Prentice-Hall (1996)
15. Sichitiu, M.L., Ramadurai, V.: Localization of wireless sensor networks with a mobile beacon. In: Proc. IEEE International Conference on Mobile Ad-Hoc and Sensor Systems (2004)
16. Srirangarajan, S., Tewfik, A.H., Luo, Z.Q.: Distributed Sensor Network Localization Using SOCP Relaxation. IEEE Trans. Wireless Commun. 7(12), 4886–4894 (2008)
17. Grant, M., Boyd, S.: CVX: Matlab software for disciplined convex programming, version 1.21 (2010), http://cvxr.com/cvx
18. Pólik, I., Terlaky, T.: Interior Point Methods for Nonlinear Optimization. In: Di Pillo, G., Schoen, F. (eds.) Nonlinear Optimization, 1st edn., vol. 4. Springer (2010)
19. Sturm, J.F.: Using SeDuMi 1.02, a MATLAB toolbox for optimization over symmetric cones. Optim. Meth. Softw., 1–5 (2008)

Practical Assessment of Energy-Based Sensing through Software Defined Radio Devices

Miguel Duarte[1], Antonio Furtado[1,2], M. Luis[1,2], Luis Bernardo[1,2], Rui Dinis[1,2], and Rodolfo Oliveira[1,2]

[1] CTS, Uninova, Dep.º de Eng.ª Electrotécnica, Faculdade de Ciências e Tecnologia, FCT, Universidade Nova de Lisboa, 2829-516, Caparica, Portugal
[2] IT, Instituto de Telecomunicações, Portugal

Abstract. The Cognitive Radio is a solution proposed for the increasing demand of radio spectrum. Usually cognitive radios are adopted by the non-license wireless users, which have a certain degree of cognition in order to only access to a given frequency band when the band is sensed idle. However, these bands and their use must be assessed, to avoid interfering with licensed users (primary users). The way to assess band's occupancy is by discerning between just noise or noise plus signal. In this paper, energy-based sensing (EBS) is considered through the use of a classical energy detector. The work proposes and describes an implementation of an energy detector using a software defined radio (SDR) testbed and, after computing the probability of detection and false alarm from a real set of samples obtained with the SDR devices, we successfully validate a theoretical model for the probabilities. EBS' performance is validated for several points of operation, i.e. for different signal-to-noise-ratio (SNR) values. These findings may be useful for building an EBS detector that defines its own decision threshold in real time given the target probabilities, since the formal probabilities are successfully validated. Moreover, our contribution also includes a detailed description of the implemented blocks using GNU Radio's open-source software development toolkit.

Keywords: Software Defined Radio, Energy Sensing, System Performance.

1 Introduction

Cognitive Radio (CR) has been proposed as an effective answer to alleviate the increasing demand for radio spectrum [1]. CR nodes, usually denominated Secondary Users (SUs) due to its non-licensed operation, must be aware of the activity of the licensed users, denominated Primary Users (PUs), in order to dynamically access the spectrum without causing them harmful interference.

Spectrum Sensing (SS) aims at detecting the availability of vacant portions (holes) of spectrum and has been a topic of considerable research over the last years [1]. It plays a central role in CR systems. The traditional SS techniques include Waveform-based sensing (WBS) [2], a coherent technique that consists on correlating the received signal with *a priori* known set of different waveform patterns; Matched Filter-based sensing (MFBS) [3], an optimal sensing scheme where the received signal is also correlated with a copy of the transmitted one; and Cyclostationarity-based sensing (CBS) [4], a technique that exploits the periodic characteristics of the

L.M. Camarinha-Matos et al. (Eds.): DoCEIS 2014, IFIP AICT 423, pp. 525–532, 2014.

received signals, *i.e.*, carrier tones, pilot sequences, etc. MFBS assumes prior knowledge of the primary's signal, while WBS assumes that the received signal matches with one of the patterns previously known. This means that these sensing techniques are not feasible in some bands, where several communication technologies may operate without *a priori* knowledge. On the other hand, CBS is impracticable for signals that do not exhibit cyclostationarity properties.

Energy-based sensing (EBS) [5], [6] is the simplest spectrum sensing technique and its main advantage is related with the fact that it does not need any *a priori* knowledge of PU's signal. At the same time, it is well known that EBS can exhibit low performance in specific comparative scenarios [7], or when noise's variance is unknown or very large. EBS has been studied in several CR scenarios, namely on local and cooperative sensing schemes [1]. More recently, several EBS schemes adopting sub-Nyquist sampling have been proposed, which are advantageous in terms of the sensing duration [8].

This work evaluates the performance of the EBS technique in a real Software Defined Radio (SDR) platform. EBS' theoretical probabilities of detection and false alarm are compared with practical ones obtained with the SDR system. The comparison indicates that the theoretical probabilities are successfully validated.

2   Relationship to Collective Awareness Systems

As it is well known, collective systems massively rely on communication that is most of the times supported by wireless links. Wireless communication technologies allow high level of device's mobility, and are indicated for scenarios where mobility is required. Simultaneously, wireless technologies avoid the use of physical (wired) connections, being an effective solution for collective awareness systems relying on mobile players. As mentioned in the introduction, Cognitive Radio was proposed as a solution to alleviate the increasing demand for radio spectrum and, consequently, as more radio spectrum becomes available more wireless devices can be used. Consequently, CR will help the development of novel collective awareness systems by increasing the number of wireless devices that may access the network.

3   Energy-Based Sensing

This work considers a cognitive radio network with a pair of PUs accessing the channel and a pair of SUs that access the channel in an opportunistic way. The considered scheme is similar with the one presented in [11]. SUs are equipped with a single radio transceiver. However, because SUs are unable to distinguish SUs and PUs' transmissions, SU's operation cycle includes the sensing and transmission periods, which facilitates the synchronization of the sensing task. Sensing and transmission period durations are represented by T_S^{SU} and T_D^{SU} respectively, as illustrated in Fig. 1.

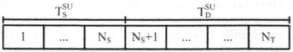

Fig. 1. SU's frame structure representing SU's operation cycle

The SU's frame, $T_F^{SU} = T_S^{SU} + T_D^{SU}$, contains N_T slots, where each slot duration is given by the channel sampling period adopted in the spectrum sensing task. The first N_S slots define the sensing period duration, and the remaining ones $(N_S + 1\ \text{to} N_T)$ represent the transmission period duration. To distinguish between occupied and vacant spectrum bands, SUs sample the channel during the sensing period T_S^{SU}, and for each sample k two hypotheses can be distinguished

$$
\begin{aligned}
\mathcal{H}_{00}&: x(k) = w(k) & k = 1,2,\dots,N_S \\
\mathcal{H}_{11}&: x(k) = w(k) + s(k) & k = 1,2,\dots,N_S,
\end{aligned} \tag{1}
$$

where $s(k)$ denotes the signal transmitted by the PUs, with distribution $\mathcal{N}(\mu_s, \sigma_s^2)$. $w(k)$ is assumed to be a zero-mean variable with variance $\sigma_n^2 = 1$, representing additive Gaussian white noise. The received signal is given by

$$
Y \sim \begin{cases} \mathcal{N}(\mu_n, \sigma_n^2), & \mathcal{H}_{00} \\ \mathcal{N}(\mu_n + \mu_s, \sigma_n^2 + \sigma_s^2), & \mathcal{H}_{11} \end{cases} \tag{2}
$$

Therefore, for a single SU, the probability of detection $P_D^{\mathcal{H}_{11}}$ and the probability of false alarm $P_{FA}^{\mathcal{H}_{00}}$ are represented by

$$
P_D^{\mathcal{H}_{11}} = Pr(Y > \gamma | \mathcal{H}_{11}) = \mathcal{Q}\left(\frac{\gamma - (\mu_n + \mu_s)}{\sigma_n^2 + \sigma_s^2}\right) \tag{3}
$$

$$
P_{FA}^{\mathcal{H}_{00}} = Pr(Y > \gamma | \mathcal{H}_{00}) = \mathcal{Q}\left(\frac{\gamma - \mu_n}{\sigma_n^2}\right), \tag{4}
$$

where $\mathcal{Q}(.)$ is the tail probability of the standard normal distribution.

In the testbed L time samples are used to determine the fast Fourier Transform (FFT). The FFT is represented in L frequency bins and Ns/L FFTs are computed during the sensing period (T_S^{SU}). During this period, the average energy received per sample, Y_n, is defined as

$$
Y_n = \frac{1}{Ns} \sum_{n=1}^{Ns/L} X_n, \tag{5}
$$

where X_n is the sum of the power of each individual FFT bin, given by

$$
X_n = \sum_{n=1}^{L} \left| x_n e^{-i2\pi k \frac{n}{L}} \right|^2, \quad k = 1,\dots L. \tag{6}
$$

The decisions performed in the testbed are according to the following conditions,

$$
\begin{aligned}
C_0 &= 0, & \mathcal{H}_{11} | Y_n < \gamma \\
C_1 &= 1, & \mathcal{H}_{11} | Y_n > \gamma \\
B_0 &= 0, & \mathcal{H}_{00} | Y_n < \gamma \\
B_1 &= 1, & \mathcal{H}_{00} | Y_n > \gamma,
\end{aligned} \tag{7}
$$

where C_1 contributes for the measured probability of detection, while B_1 represents the case of false alarm.

4   System Implementation

4.1   Testbed

The EBS was implemented using the GR (GNURadio) software architecture [9] and the USRP (Universal Software Radio Peripheral) hardware platform [10]. These tools are compliant with the notion of SDR (Software Defined Radio), mostly due to their flexibility when it comes to DSP (Digital Signal Processing). This flexibility opens up options when dealing with Digital Signal Processing. In this case, a range of FFT sizes can be chosen, instead of a fixed one like in most DSP devices. This allows for a variation in the N_S/N_T ratio.

The setup consisted of two USRP B100 devices connected to one computer. The USRPs are connected by a coaxial cable and two 30dB attenuators to achieve the desired signal-to-noise (SNR) values. One USRP implements a Primary User while the other behaves as the Secondary user, sensing for the presence of the first one. The basic blocks of the GR system are written in C++. These blocks link directly with the USRP and its FPGA. Python is used as a "glue" to connect these blocks and as a controller to the system flow (issuing start and stop commands). The script used to build the EBS was entirely written in python, also making use of python math library tools.

4.2   EBS Structure

A conventional EBS consists of a low pass filter, an A/D converter, a square-law device and an integrator (see Fig. 1). The detector model used in this work consists of an A/D converter followed by an L-Point FFT (6).The FFT provides the windowing needed to filter out unnecessary frequency bands. This data is then averaged (see Fig. 3), and the sensing output is determined according to C_0/C_1 or B_0/B_1 conditions (eq. (7)).

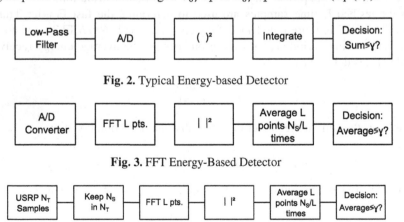

Fig. 2. Typical Energy-based Detector

Fig. 3. FFT Energy-Based Detector

Fig. 4. GNURadio Blocks

4.3   Primary User Signal

The Primary signal is assumed to obey a Normal Distribution with higher amplitude than the noise distribution for sake of simplicity. The signal generation is achieved using the blocks in Fig.5.

Fig. 5. PU signal generator

5   Comparison Results

5.1   Calibration and Parameter Calculation

This subsection validates the noise $w(k)$ and PU's signal $s(k)$. The USRP's parameters used in the validation are shown in Table 1.

Table 1. USRP Configuration

Freq. (GHz)	L	Sample Rate (Sps/s)	SU Gain (dB)	PU Gain (db)	SNR (dB)	Primary Amplitude
1.3	256	1,000,000	0	17	0.11	0.05

Noise ($w(k)$) and Signal plus Noise ($w(k) + s(k)$) signals were sampled during a 10 second interval. The histogram of the received energy is plotted in Fig. 6. In this figure the green curve approximates the noise distribution, while the red one approximates the noise plus signal distribution. Since the overlapping area of the approximate distributions is negligible, a decision threshold can be defined to identify the two conditions with low error.

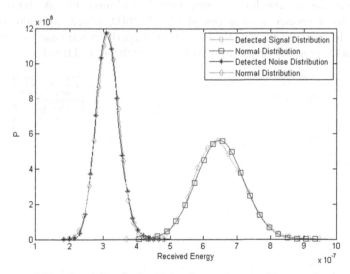

Fig. 6. Histogram of Signal and Signal plus Noise data and respective normal approximations. P is the number of occurrences.

5.2   Practical Probabilities

Taking the x-axis values of Fig. 6 as a hypothetical range for the decision threshold, the probabilities P_D and P_{FA} were computed from the conditions expressed in equation (7). Basically, P_D was computed as the ratio of the number of times the condition C_1 was observed over the total number of decisions. P_{FA} was similarly computed as the ratio of the number of times the condition B_1 was observed over the total number of decisions. These probabilities are shown in Fig. 7, where the achieved results are compared with the theoretical ones.

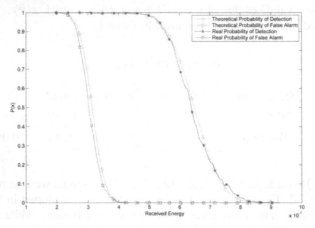

Fig. 7. Theoretical and practical probabilities P_D and P_{FA} ($T_S^{SU} = 0.512s$)

As can be seen, the registered practical values approach the theoretical ones. This is, however, for a very long sensing period (512 ms). Fig. 8 shows the results achieved for a shorter sensing period (N_S=512000, which corresponds to $T_S^{SU} = 51.2ms$). In this case the practical results underperform the case of longer sensing period, but the practical results roughly follow the theoretical trend.

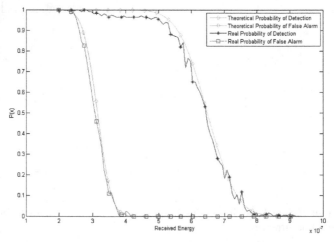

Fig. 8. Theoretical and Real Probabilities ($T_S^{SU} = 0.0512s$)

5.3   Performance Evaluation

With the knowledge of the signal and noise statistics, a threshold can be defined so that the desired conditions (a certain pair of P_D and P_{FA}) can be achieved. By inspecting the theoretical curves for the probabilities, a threshold was defined where the P_{FA} was set to 5%. Since P_{FA} only depends on noise power, and not on noise plus signal, its value should remain constant for any SNR value. P_D does not follow the same rationale, since it depends on SNR (eq. (3)). The following test evaluates the detection performance for a fixed decision threshold (3.7E-07) and varying the SNR from -10 to 2 dB. Different sensing periods were also tested, from 0.0125s to 0.512s. The obtained results are shown in Fig 9.

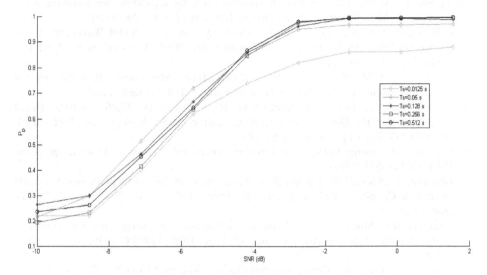

Fig. 9. P_D curve for SNR [-10,2] dB

As can be seen, the statistic test for sensing period durations longer than 0.05s start to be redundant, since they present almost the same performance. At 0dB we observe that the detector performance is high, and for $T_S^{SU} > 0.05s$ the miss-detection probability is lower than 10%. Due to lack of space we do not present false alarm probabilities for the same scenario, but for sensing period durations longer than 0.05s we have observed that the false alarm probability is always lower than 5%, which confirms the expected performance for the adopted decision threshold.

6   Conclusions

In this work we have addressed the assessment of energy detection for cognitive radio systems. We started to characterize the performance of an energy detector through practical USRP devices. The theoretical performance was successfully validated through practical results. For higher SNR values, the detection probability is high even for short sensing periods, achieving detection probabilities that approach 1. For lower SNR values a higher sensing time is required but even in these cases substantial performance can be achieved.

Acknowledgments. This work was partially supported by the Portuguese Science and Technology Foundation under the projects PTDC/EEA-TEL/115981/2009, PTDC/EEA-TEL/120666/2010, PTDC/EEI-TEL/2990/2012, PEst-OE/EEI/UI0066/2011, PEst-OE/EEI/LA0008/2011, SFRH/BD/68367/2010 and SFRH/BD/68367/2010.

References

1. Yucek, T., Arslan, H.: A survey of spectrum sensing algorithms for cognitive radio applications. IEEE Communications Surveys Tutorials 11, 116–130 (2009)
2. Zahedi-Ghasabeh, A.T., Daneshrad, B.: Spectrum Sensing of OFDM Waveforms Using Embedded Pilots in the Presence of Impairments. IEEE Transactions on Vehicular Technology 61, 1208–1221 (2012)
3. Bouzegzi, A., Ciblat, P., Jallon, P.: Matched Filter Based Algorithm for Blind Recognition of OFDM Systems. In: Proc. IEEE VTC 2008-Fall, pp. 1–5 (September 2008)
4. Al-Habashna, A., Dobre, O., Venkatesan, R., Popescu, D.: Cyclostationarity-Based Detection of LTE OFDM Signals for Cognitive Radio Systems. In: Proc. IEEE GLOBECOM 2010, pp. 1–6 (December 2010)
5. Urkowitz, H.: Energy Detection of Unknown Deterministic Signals. Proceedings of the IEEE 55, 523–531 (1967)
6. Ghasemi, A., Sousa, E.S.: Optimization of Spectrum Sensing for Opportunistic Spectrum Access in Cognitive Radio Networks. In: Proc. IEEE CCNC 2007, pp. 1022–1026 (January 2007)
7. Bhargavi, D., Murthy, C.: Performance comparison of energy, matched-filter and cyclostationarity-based spectrum sensing. In: Proc. IEEE SPAWC 2010, pp. 1–5 (June 2010)
8. Tian, Z., Giannakis, G.: Compressed Sensing for Wideband Cognitive Radios. In: Proc. IEEE ICASSP 2007, vol. 4, IV-1357–IV-1360 (2007)
9. GNURadio - Free Software Toolkit for SDR, http://www.gnuradio.org/
10. Ettus Research (December 2013), http://www.ettus.com/
11. Luis, M., Furtado, A., Oliveira, R., Dinis, R., Bernardo, L.: Towards a Realistic Primary Users' Behavior in Single Transceiver Cognitive Networks. IEEE Communications Letters 17(2), 309–312 (2013)

Part XIX

Electronics: Design

A Top-Down Optimization Methodology for SC Filter Circuit Design Using Varying Goal Specifications

Hugo Serra, Rui Santos-Tavares, and Nuno Paulino

Centre for Technologies and Systems (CTS) – UNINOVA
Dept. of Electrical Engineering (DEE), Universidade Nova de Lisboa (UNL)
2829-516 Caparica, Portugal
has14926@campus.fct.unl.pt, {rmt,nunop}@uninova.pt

Abstract. The design of Switched-Capacitor (SC) filters can be an arduous process, which becomes even more complex when the high gain amplifier is replaced by a low gain amplifier or a voltage follower. This eliminates the virtual ground node, requiring the compensation of the parasitic capacitances during the design phase. This paper proposes an automatic procedure for the design of SC filters using low gain amplifiers, based on a Genetic Algorithm (GA) using hybrid cost functions with varying goal specifications. The cost function first uses equations to estimate the filter transfer function, the gain and settling-time of the amplifier and the RC time constants of the switches. This reduces the computation time, thus allowing the use of large populations to cover the entire design space. Once all specifications are met, the GA uses transient electrical simulations of the circuit in the cost functions, resulting in the accurate determination of the filter's transfer function and allowing the accurate compensation of the parasitic capacitances, obtaining the final design solution within a reasonable computation time.

Keywords: Computer-aided design, genetic algorithm, filter design, switched-capacitor.

1   Introduction

Analog filters are very important blocks in several electronic systems such as RF transceivers or sigma delta modulators. They allow selecting signals with different frequencies and eliminating unwanted signals.

The scaling-down of transistors in advanced deep-submicron CMOS technologies results in the reduction of the intrinsic gain (g_m/g_{ds}) [1] and increases the variability, making the design of high gain high bandwidth opamps increasingly difficult. This limitation has large impact in the performance of filter circuits.

To avoid the difficulty of designing high gain amplifiers, SC filters can be implemented using low gain amplifiers [2]. However, this approach removes the circuit's virtual ground node. Without this node, parasitic insensitive SC branches cannot be used and the filter's transfer function becomes sensitive to the effects of parasitic s, which have to be compensated during the design phase of the filter.

This paper presents an automatic procedure for the compensation of the parasitic capacitances introduced by parasitic sensitive branches and a general design methodology for the design of SC filters, using low gain amplifiers, based on a GA using hybrid cost

L.M. Camarinha-Matos et al. (Eds.): DoCEIS 2014, IFIP AICT 423, pp. 535–542, 2014.
© IFIP International Federation for Information Processing 2014

functions with varying goal specifications. The exact transfer function of a SC filter can be obtained by simulating the circuit's impulse response. This, however, results in a large computation time that limits the maximum size of the population of the GA. An alternative is to estimate the filter's transfer function using equations, resulting in a low computation effort and allowing the evaluation of a larger population in a reasonable time. This approach allows the design space of the filter to be completely explored. Once a solution is found, it is then possible to reduce the population size in the GA, obtaining the final design solution using the more computation intensive and more accurate simulation based cost functions. During this step, the optimization will also calculate the final values of the capacitances taking into consideration the parasitic capacitance values. This hybrid cost function (equation based/simulation based) allows a larger design space to be explored and obtaining a design solution with a good accuracy, while still having a low computation time.

2   Relationship to Collective Awareness Systems

Collective awareness systems are information and communication technology systems that allow the interaction of people, knowledge, and objects through a network. The development of these systems, which require faster, smaller, and less power consuming circuits, while performing more complex operations, is only possible using advanced nanometer CMOS technologies. Filter circuits are essential blocks in front-end transceivers that allow the data collection and also support the communication between the agents involved in a network. This paper describes a top-down optimization methodology for the design of SC filter's using low gain amplifiers. Using this optimization methodology, it is possible to quickly study new SC filters in order to find new topologies capable of addressing the more demanding specifications required for collective awareness systems.

3   Proposed Design Methodology

A software add-on was developed to implement the proposed methodology. This add-on was integrated into an existing optimization software platform [3], [4], [5], based on the open-source circuit simulator, Ngspice [6]. Initially, a population consisting of N randomly created chromosomes is generated based on the maximum and minimum allowed values of each component, which defines the design space. The fitness of each chromosome is then calculated, based on the desired specifications, and sorted. A new population is created based on the genetic material of the best chromosomes and complemented with selection, cross-over and mutation operations among the remaining chromosomes. This process is repeated during a number of generations or until the desired specifications are met. The fitness is calculated using the exponential-based equations presented in [7].

The methodology uses a multi-step approach to obtain the optimized filter design. In the first step, the SC filter is optimized from an ideal standpoint using the equation that describes its transfer function. Once the desired specifications for the filter are met, the second step starts, where the tool focuses on optimizing the amplifier circuit at transistor level, in order to have enough gain and closed-loop bandwidth to replace the ideal gain obtained from the top-level optimization. After this step, it is necessary to compensate the effects of the amplifier's input and input to output parasitic capacitances in the SC filter. In the third step, the switches are optimized in order to have a RC time constant that is small enough to charge the capacitors in half the clock period. When these specifications are met the parasitic capacitances of the switches are compensated through a transient simulation, in Ngspice, to determine the fitness of each SC circuit at transistor level. During this step, the design space (the component's values allowed variations) is restricted to be around the best values obtained in the previous steps. After this final optimization step, the resulting component's values take into consideration all the parasitic effects in the circuit. In order to facilitate the convergence of the GA during the first step, the cost function uses variable goal specifications. By starting the optimization with an "easier" goal and then converging the goal to the desired specifications, it is possible speed up the convergence process and obtaining a solution in fewer generations.

4   A 2nd Order Low-Pass SC Filter

A 2nd order low-pass biquad filter (Fig. 1) is presented as a proof-of-concept for the proposed design method. This architecture is based on the continuous low-pass Sallen-Key topology replacing the resistors with parallel SC branches. Capacitors C_2, C_B and C_{B2} were added to the architecture to facilitate the compensation of the parasitic capacitances in each node.

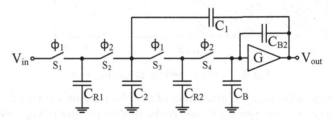

Fig. 1. Low-pass SC biquad filter in single-ended configuration

4.1   First Step – Top-Level Optimization

In the top level optimization, all the filter components are considered ideal (capacitors, switches and amplifier). The chromosome used in this step, which is shown in Table 1, contains the values of the capacitors and the desired gain for the amplifier. Once the values of the capacitors are generated, they are scaled down based on the smallest capacitor and the desired minimum value. The optimization will then try to find a solution that best fits the design specifications, while minimizing the gain value, because this will make the optimization of the amplifier circuit, in the second step, easier, as this value will be used as a design specification. The optimization will also try to minimize the total value of the capacitors.

Table 1. Chromosomes used in each optimization step by the GA

Top level	C_{R1}	C_{R2}	C_1	C_2	C_B	C_{B2}	G					
Amplifier	W_1	L_1	W_2	L_2	W_3	L_3	W_4	L_4	W_5	L_5	I_D	V_{cm}
Switches	W_{s1}	W_{s2}	W_{s3}	W_{s4}	L_s							

The design specifications in this step are given in the frequency domain (Fig. 2), and are evaluated based on the ideal transfer function of the filter, which was obtained from a charge conservation perspective, considering that the filter's output is sampled at the end of phase Φ_1. The passband specification is controlled by the indicators $F_{passlow}$ and F_{cutoff}, in which the signals attenuation must be delimited between A_{max1} and A_{max2}, i.e., these four indicators are used to obtain the desired low frequency gain of the filter and move the poles to frequencies around or above F_{cutoff}. To place the poles as close as possible to the cutoff frequency, two additional indicators are used ($F_{specific}$ and A_{min1}). By indicating the desired attenuation at a specific frequency we are forcing the poles to move to the cutoff frequency. The indicators $F_s/2$ and A_{min2} are used to reinforce the pole placement at the cutoff frequency.

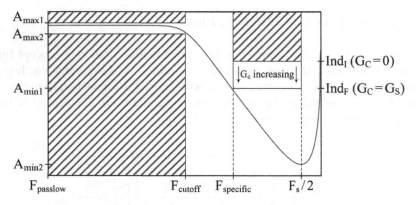

Fig. 2. Filter's variable goal specifications

In this optimization step, the evaluation of the chromosomes is based on a variable goal equation (1), i.e., the specifications (goals) of the indicators (A_{min1} and A_{min2}) will start with an initial value (Ind_I) and converge as the generations progress, following an exponential evolution, to the final value (Ind_F). The speed at which the indicator converges to the final value is controlled by the variable w. The variables G_C and G_S represent the current generation and the generation at which the indicator reaches its final value at the latest, respectively. Fig. 2 shows how the indicator A_{min1} behaves when using the variable goal equation. While converging from Ind_I to Ind_F, the optimizer will find the chromosome that satisfies the passband specification and that has an attenuation at $F_{specific}$ closest to the variable goal equation, at the current generation. Once the variable goal equation reaches the final value ($G_C = G_S$), the optimizer will find the chromosome that satisfies the passband specification and that has the smallest A_{min1} at $F_{specific}$. By temporarily softening the specifications we are increasing the probability of converging to the best solution.

$$\text{Variable Goal Specification} = Ind_F + (Ind_I - Ind_F)e^{-5\,w(G_C/G_S)} \tag{1}$$

4.2   Second Step – Amplifier Optimization

In this step, the amplifier circuit is designed to have the gain value obtained in the previous step and optimized following the time-domain methodology that was presented in [5], which ensures that if a given settling error is reached within the desired settling time, then the amplifier has enough open-loop gain, closed-loop bandwidth, and slew-rate. This methodology simplifies the circuit's evaluation and helps the optimization process to converge faster. To calculate the settling time it is necessary to obtain the closed-loop circuit transfer functions during phase Φ_2 and Φ_1 (Fig. 3a and Fig. 3b), and apply the inverse Laplace transform multiplied by a unity input step (1/s). The closed-loop transfer function can be obtained by replacing the amplifier block G(s) with its equivalent medium frequency small signal model. Both transfer functions need to be calculated because charge flows to the filter's output in both clock phases.

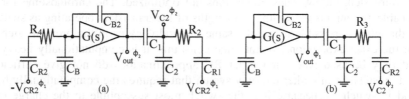

Fig. 3. Equivalent closed-loop circuit during clock phase (a) Φ_2 and (b) Φ_1

$V_{CR2}{}^{\Phi_1}$ and $V_{C2}{}^{\Phi_2}$ represent the charge stored in capacitor C_{R2} during phase Φ_1 and in capacitor C_2 during phase Φ_2, respectively, and can be calculated from (2), where V_{amp} represents the amplitude of the input signal.

$$V_{C2}{}^{\Phi_2} = C_{R1}V_{amp}/(C_1 + C_2 + C_{R1})$$
$$V_{CR2}{}^{\Phi_1} = (C_1 + C_2)V_{C2}{}^{\Phi_2}/(C_1 + C_2 + C_{R2})$$

$$(2)$$

The amplifier circuit used is shown in Fig. 4. Due to its differential configuration, this amplifier is capable of achieving gains larger than one.

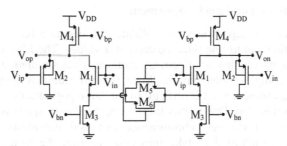

Fig. 4. Voltage-combiner amplifier with source degeneration

The chromosome for this step (Table 1) contains the width and length of the amplifier's transistors, the biasing currents and the input common-mode voltage. The design specifications which are evaluated during this step are the average between the amplifier's input and output common mode voltages, which is evaluated based on DC Operating Point (OP) simulation; the closed-loop gain, which is calculated from the step response with s = 0; the settling time during phase Φ_2 and Φ_1, also calculated from the step response; and the amplifier's current and area.

During this optimization step, the amplifier's input and the input to output parasitic capacitances are calculated based on a DC OP simulation. Once the specifications are met the corresponding parasitic capacitances are stored, to be later compensated into capacitor C_B and C_{B2}. Considering the amplifier circuit from Fig. 4, the compensated capacitor values can be obtained from (3), where G_1 and G_2 represent the voltage gain between the gate and drain and the voltage gain between the gate and the source of transistor M_1, respectively.

$$C_{B2}' = C_{B2} - C_{gb2} - C_{gs2}$$
$$C_B' = C_B - C_{gb1} - (1 + G_1)C_{gd1} - (1 + G_2)C_{gs1} - C_{gd2} - C_{gg5}$$
(3)

4.3   Third Step – Switch Sizing Optimization

In the third step, the switches dimensions are optimized. The chromosome for this step (Table 1) contains the widths and lengths of the transistors operating as switches. Even though it is possible to use the same methodology as in the previous step and obtain the exact expressions for the time constants, it is computationally heavy and should only be used when first order RC approximations do not give sufficiently accurate results. In this filter the only switch that requires the computationally heavy method is switch S_2, because it is the switch most susceptible to the charge effect from the amplifier at transistor level. The settling time of this switch can be obtained from the closed-loop circuit shown in Fig. 3a considering node $V_{c2}^{\Phi_2}$ as the output instead of $V_{out}^{\Phi_2}$. The remaining RC time constants are calculated using first order approximations. The fitness equation of this step is based on the RC time constants/settling time and the overall switch area.

Clock-boosted phases were used to reduce the switches conductance non-linearity and overall value. During this step the value of the switches parasitic capacitances are calculated based on a DC OP simulation and stored, so that they can be used in the next step to compensate their effect on the filter.

4.4   Fourth Step – Top-Level Adjustment

The last optimization step is used to validate the equation-based design through electrical simulation and to fine-tune the capacitor values. The filter transfer function is now evaluated by running an electrical transient time simulation of the impulse response of the complete filter circuit. Since this procedure is very computational intensive and takes some time, it is difficult to use large populations or a large number of generations in this step. Due to this limitation, the design space is restricted to be around the values of the best chromosome obtained previously. Moreover, the capacitor values are adjusted to take into consideration the parasitic capacitances previously calculated. Unlike the calculation of the amplifier's parasitic capacitances which is relatively straightforward, due to their value being constant for a fixed voltage, the value of the switch's parasitic capacitances can have a variation of over 5 times depending if the switch is ON or OFF, making the compensation process more complex. Equation (4) was implemented to overcome this problem. The first generation chromosomes are all identical apart from a random compensation factor

(CF) which varies between 0 and 5. From this initial population, it is possible to converge to an optimized solution very rapidly, i.e., within a few generations.

$$C_{comp} = C_{nom} - CF\,C_{par}, \qquad 0 < CF < 5 \tag{4}$$

5   Simulation Results

The biquadratic low-pass filter shown in Section 4 was designed in a standard 1.2 V 130 nm CMOS technology. The filter specifications used in this section are shown in Table 3. All simulations obtained from the optimizer were validated in Spectre using BSIM3V3.2.4 models of the transistors and MIMCAPS capacitor models (including parasitics).

In the first step, several optimizations were performed with and without the use of the varying goal specifications, on average, without using the varying goal specifications; it took the optimizer 96 generations to find a chromosome with fitness above 90%, and only 58 generations when using the varying goal specifications during the first 30 generations. In both cases, each generation was composed of 2500 chromosomes. The values obtained in this step were $C_{R1} = 54.44$ fF, $C_{R2} = 50.00$ fF, $C_1 = 4530.80$ fF, $C_2 = 99.24$ fF, $C_B = 473.95$ fF, $C_{B2} = 131.64$ fF, and $G = 1$.

Table 2 shows the amplifier specifications, the simulation results and the best chromosome optimized in the second step, obtained after only 10 generations of 2500 chromosomes each. The results show that the continuous time equation-based methodology gives similar results to the ones obtained through Cadence. Based on the DC OP simulation of the amplifier using the best chromosome, the values of the parasitic capacitances were calculated and compensated into capacitors C_B and C_{B2}.

Table 2. Amplifier specifications, simulation results, and best chromosome

	Norm. closed-loop gain	Settling-Time Φ_2 (%)	Settling-Time Φ_1 (%)
Specifications	3.8808	< 75.0	< 75.0
Optimizer	3.8808	65.13	59.92
Cadence	3.8808	65.20	55.20

W_1 (µm)	W_2 (µm)	W_3 (µm)	W_4 (µm)	W_5 (µm)	Power (µW)	C_{Bcomp} (fF)
0.37	62.29	39.19	90.5	0.38	261.38	468.67
L_1 (nm)	L_2 (nm)	L_3 (nm)	L_4 (nm)	L_5 (nm)	V_{cmi} (mV)	C_{B2comp} (fF)
520	240	350	450	220	450	26.99

The filter's switches were also optimized in order to have time constants small enough to charge the capacitors in less than half a clock period. The length chosen for the switches was fixed at L = 120 nm and the optimized widths obtained were W1 = W3 = W4 = 240 nm and W2 = 300 nm. The optimizer took 15 generations of 2500 chromosomes to find the smallest widths for the switches while maintaining times constants small enough to charge the capacitors in less than half a clock period.

Table 3 shows the results obtained during each step of the optimization process. From rows (B) and (C), it can be concluded that the equation method for the automatic compensation of the amplifier's parasitic capacitances is very accurate. From (E) and (F), it can be seen that the open source simulator Ngspice gives very similar results to Cadence.

Table 3. Specifications and simulation results of the low-pass SC filter

		$F_{passlow}$	F_{cutoff}	$F_{specific}$	$F_s / 2$
		10 kHz	1 MHz	6 MHz	50 MHz
(A)	Specifications	≤ 0.01	≥ -3.01	≤ -30	≤ -59
(B)	Equations/Cadence using ideal components	0.00	-2.95	-30.78	-59.87
(C)	Cadence using real amplifier and ideal switches	0.01	-2.94	-30.78	-59.87
(D)	Ngspice real circuit before parasitic compensation	0.00	-2.91	-30.73	-59.89
(E)	Ngspice real circuit after optimization complete	0.00	-3.00	-31.47	-60.63
(F)	Cadence real circuit after optimization complete	0.00	-3.03	-31.50	-60.61

In the last step, after 12 generations, the capacitor values obtained were $C_{R1} = 53.14$ fF, $C_{R2} = 47.78$ fF, $C_1 = 4754.51$ fF, $C_2 = 97.29$ fF, $C_B = 459.30$ fF, $C_{B2} = 27.71$ fF. Although a solution that satisfies all the specifications can be found after a single generation in the top-level adjustment, using equation (4), the optimization process continued for another 11 generations in order to try to find an even better solution than the one obtained after the first generation.

6   Conclusion

A design methodology for SC filters based on a GA using hybrid cost functions with varying goal specifications was proposed. The cost functions first use equations to obtain approximate solutions, and once all specifications are met, the GA uses the results of transient electrical simulations of the circuit in the cost functions, resulting in accurate transfer functions and allowing the compensation of parasitic capacitances within a reasonable amount of generations. Furthermore, it was concluded that using variable goal specifications reduces even further the amount of generations needed to obtain a solution within the design specifications.

References

1. Perez, A.P., Maloberti, F.: Performance enhanced op-amp for 65nm CMOS technologies and below. In: IEEE Int. Symp. Circuits and Systems (ISCAS), pp. 201–204 (May 2012)
2. Serra, H., Paulino, N., Goes, J.: A switched-capacitor biquad using a simple quasi-unity gain amplifier. In: Proc. IEEE Int. Symp. Circuits Systems (ISCAS), pp. 1841–1844 (May 2013)
3. Santos-Tavares, R., Paulino, N., Goes, J.: Time-Domain Optimization of CMOS Amplifiers - Based on Distributed Genetic Algorithms. LAP LAMBERT Academic Publishing (2012)
4. Figueiredo, M., Santos-Tavares, R., Santin, E., Ferreira, J., Evans, G., Goes, J.: A two-stage fully differential inverter-based self-biased CMOS amplifier with high efficiency. IEEE Trans. Circuits Syst. I, Reg. Papers 58, 1591–1603 (2011)
5. Santos-Tavares, R., Paulino, N., Goes, J., Oliveira, J.: Optimum sizing and compensation of two-stage CMOS amplifiers based on a time-domain approach. In: Proc. IEEE Int. Conf. on Electronics, Circuits Systems (ICECS), pp. 533–536 (December 2006)
6. Nenzi, P., Vogt, H.: Ngspice users manual version 25, http://ngspice.sourceforge.net/ (accessed October 8, 2013)
7. de Melo, J.L.A., Nowacki, B., Paulino, N., Goes, J.: Design methodology for sigma-delta modulators based on a genetic algorithm using hybrid cost functions. In: Proc. IEEE Int. Symp. Circuits Systems (ISCAS), pp. 301–304 (May 2012)

Parallel Algorithm of SOI Layout Decomposition for Double Patterning Lithography on High-Performance Computer Platforms

Vladimir Verstov, Vadim Shakhnov, and Lyudmila Zinchenko

Ul.Baumanskays 2-ya, 5, 105005, Moscow, Russia
v.verstov@gmail.com, {shakhnov,lyudmillaa}@mail.ru

Abstract. In the paper silicon on insulator layout decomposition algorithms for the double patterning lithography on high performance computing platforms are discussed. Our approach is based on the use of a contradiction graph and a modified concurrent breadth-first search algorithm. We evaluate our technique on both real-world and artificial test cases including non-Manhattan geometry. Experimental results show that our soft computing algorithms decompose layout successfully and a minimal distance between polygons in layout is increased.

Keywords: VLSI, Layout, Double Pattering, Parallel Algorithms, High-Performance Computing, Radiation Hardening.

1 Introduction

Collective awareness systems for sustainability and social innovation are information and communications technology (ICT) platforms using electronic components and networks to drive social innovation. Chernobyl (1986) and Fukushima (2011) nuclear incidents demonstrate an increasing role of radiation hardening for emergency systems, especially for very-large-scale integration (VLSI) chips. Design of electronic devices resistant to damage caused by radiation requires new approaches. Reliability is the main silicon on insulator (SOI) technology advantage in a comparison with bulk-silicon technology [1]. However, floating body effect is the major parasitic effect in SOI technology. It results in circuit instabilities. In order to overcome this negative effect multi-gate transistors were proposed [2]. The use of pi-gate, sigma –gate transistors etc. results in non-Manhattan layout topology.

Currently, double and multiple patterning technologies play an important role because the extreme ultraviolet lithography (EUV) has been delayed for the emerging technology nodes. For the multiple patterning technologies layouts are decomposed into two or more masks. Fig. 1 demonstrates a layout decomposition case for double patterning. In [3] the sequential algorithms based on a conflict graph and integer linear programming were discussed. In [4] the sequential algorithms based on a contradiction graph and graph coloring were proposed. The contradiction graph

L.M. Camarinha-Matos et al. (Eds.): DoCEIS 2014, IFIP AICT 423, pp. 543–550, 2014.

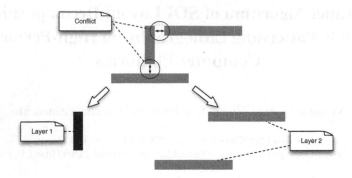

Fig. 1. An example of the layout decomposition for double patterning technology. Layer 1 and Layer 2 are layout layers after decomposition.

accumulates information about contradictions between requirements of minimal distance between polygons and negative diffraction effects for adjacent objects. The approach proposed in [4] was implemented in the TPLConverter program [5]. However, high-performance computing platforms are requires for SOI layout decomposition for double patterning technology because of non-Manhattan layout and huge size of layout data file. However, the application of sequential algorithms for high-performance computing platforms results in the decrease of computing productivity [6]. Parallel algorithms are required for these computing systems.

In [7] concurrent algorithms of the VLSI layout decomposition for double pattering were proposed. However, these algorithms are valid for Manhattan layouts only.

In this article, we propose methods of SOI VLSI layout decomposition for double patterning. In comparison with the algorithms [7] these methods can be applied for both Manhattan and non-Manhattan layouts including H-transistors, O-transistors etc. The proposed algorithms were implemented in ParallelDPLayout Migrator software for high-performance computing systems. Preliminary experimental results for artificial and real-world tests are outlined.

The rest of paper is structured as follows. The next Section reviews collective awareness platforms features. Section 3 presents our methods for layout decomposition for double patterning. We discuss our experimental results in Section 4. Finally, conclusions are derived in Section 5.

2   Relationship to Collective Awareness Systems

The coming revolution in personal and community risk awareness is based on collective awareness platforms. Chernobyl and Fukushima nuclear incidents showed a danger of radiation for a person and a community. In addition, many radioactive isotopes are used in industry and health care. A radiation level can be monitored by individual devices and data about radiation are collected. Neris [8] project has started in 2008. Safecast project [9] has started in 2011. In 2012 more 3 million radiation data points were collected using individual devices.

However, commercial devices can be damaged or malfunctioned because of radiation. Special radiation hardened electronic devices have to be used in emerging collective awareness platforms. SOI [1] including silicon on sapphire (SOS) technologies were developed for radiation-hardened applications. SOI devices are more reliable for the applications in emerging systems. Another way to do radiation hardened circuits is redundancy. But this technique increases area of a chip design. However, some negative effects hinder their wide applications.

In particular, floating body effect results in variation of delay time, circuit parameters etc. In order to overcome these deficiencies multi-gate transistors are used. Many multi-gate transistors have non-Manhattan layouts that realize in a non-Manhattan layout of chips. In addition, layout data file size is increased that result in huge computational efforts to manipulate by the data. Some files to be processed have to be divided and special methods to verify, decompose or to migrate have to be used for these layouts.

High performance computing is one of promising solutions for the mentioned above problem. However, new methods to process layout and then modify them for a target technology are required.

In this paper our focus is on layout decomposition for double patterning technology while the proposed layout processing methods can be applied for VLSI routing, layout migration, verification etc.

3 Concurrent Algorithms for VLSI Layout Decomposition for Double Patterning

3.1 Concurrency during Layout Processing

Special techniques are requiring handling a non-Manhattan SOI layout including multi-gate transistors with circular structure. We propose a special data structure for the concurrent VLSI layout processing. Fig. 2 illustrates our approach for a layer of a circular structure transistor (O-transistor).

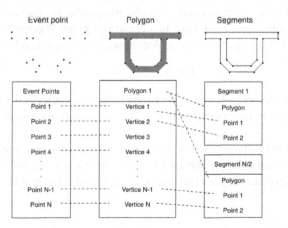

Fig. 2. An example of our data structure for the SOI VLSI layout processing

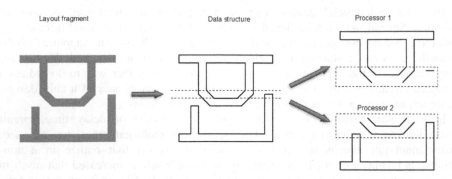

Fig. 3. An example for the two processors case

The VLSI layout represents a set of polygons. Each polygon consists of a set of segments. Each segment is stored as two references to boundary points. Every point is stored in the "events points" array. This data structure allows us to describe non-Manhattan layout.

Our algorithm assumes the high-performance computer (HPC) platform architecture to be a one-dimensional mesh of processors. Each layer splits to horizontal bars. The number of horizontal bars is equal to the number of processors. To avoid conflicts in adjacent bars they overlay each other. Therefore, one geometric object splits to two features and two overlapping ones are assigned to two different processors to be decomposed. Fig. 3 illustrates our approach for the test layer case including F- and O-transistors for two processors. We apply our approach for layout decomposition for double patterning. However, this approach can be expanded for compaction and verification of SOI VLSI layouts as well.

3.2   Constraint and Contradiction Graphs Construction

We propose a modified sweep line algorithm to construct constraint and contradiction graphs. Each vertex in constraint and contradiction graphs is crisp and corresponds to a polygon in a layout layer. Each edge in a constraint graph is crisp and depends from spacing constraints for objects in layers. A contradiction graph is a fuzzy graph [10]

$$\sim G=(V, \sim E) \tag{1}$$

where V is the node set, $\sim E$ is the fuzzy edge set, characterized by the matrix of membership functions depending from the double patterning technology parameter $\mu_E(x)$.

If a distance x between two segments related to different polygons is smaller than the double pattern technology parameter the correspondent edge and its value defined by the membership function $\mu_E(x)$ are added to a contradiction graph.

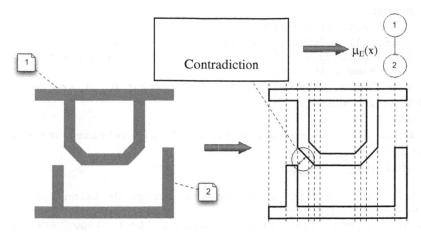

Fig. 4. An example of the contradiction graph construction flow

The contradiction graph accumulates an information about all contradictions in a given layout and a level of the contradiction according to the membership function for double patterning technology. Fig. 4 illustrates our approach for our test layer case including F- and O-transistors.

3.3   Parallel Layout Decomposition Algorithm

The parallel algorithm for the layout layer decomposition is discussed below. This algorithm based on testing bipartiteness for α-cuts family of the contradiction graph

$$\mu_E (x) \geq \alpha, \tag{2}$$

where the α-cut of a contradiction graph is the crisp graph.

If the corresponding crisp graph is bipartite this layer could be decomposed into two new ones for the current α-cut.

We propose a modified concurrent breadth-first search algorithm to test bipartiteness of the α-cut of a contradiction graph. For the current α we can find either two-coloring for the graph or an odd cycle for non-bipartite graph in linear time. If the algorithm did not find an odd cycle, then the α-cut of the graph is bipartite according to [11].

Our algorithm is given below.

```
Input: A graph ~G, α_min, α_max.
Output: All design alternatives, a coloring of the α-cut of ~G
and α.
```

Begin
```
Step 1. α=α_min.
  Repeat
    Step 2. The crisp graph G is constructed for the current α.
    The current color is equal to 1. Assign 1 color to the
    source vertex.
```

Repeat
Color all the neighbors with 2 color.
Color all neighbor's neighbor with 1 color.
If an edge from *i* to *j* exists and destination *j* is colored with the same color as *i*
    Then design alternatives for the current α-cut are collected and α is increased.
    Else
    Until all vertices are colored. *If* all vertices are colored *Then* Step 3.
    Until $\alpha < \alpha_{max}$

Step 3. Our algorithm stops when a design decision for double patterning technology is found or $\alpha = \alpha_{max}$. In the last case, recommendations for multi-patterning technology are given according the found minimal number of the required colors.
End

Fig. 5 illustrates our approach for one design alternative. The metal layer for the 4AND cell layout and the corresponding crisp graph are shown on Fig. 5, a. Fig. 5, b shows the corresponding crisp graph that has been colored using two colors. Fig. 5, c shows the metal layer of the 4AND element layout after decomposition. The first layer polygons are shown as grey objects, while the second layer polygons are shown as black objects.

4   Experimental Results

The proposed approaches were implemented using C++ programming language on Cent OS 5.5 operating system. Additionally, we use Boost libraries and OpenMP framework.

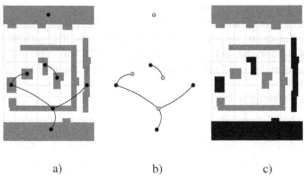

a)   b)   c)

Fig. 5. The 4AND element metal layer decomposition for double patterning: a – the layout layer and the α-cut of the contradiction graph; b – the α-cut of the contradiction graph after coloring; c – two new layout layers (the first layer is shown as grey objects, the second layer is given as black objects).

Table 1 summarizes our simulation results for our artificial and real-world tests. We evaluate our solutions with respect with solution quality. We use the relative minimal distance between the polygons after the decomposition that is calculated as the ratio of the minimal distance between the polygons of the layer i ($i = 1, 2$) after the decomposition to the minimal distance in the critical layer before the decomposition [7].

Table 1. Test cases parameters

Layout	Polygons, total/ Layer 1/Layer 2	Relative minimal distance, Layer 1	Relative minimal distance, Layer 2
Multiplexer	18 / 13 / 5	1,15	3,68
4AND	8 / 4 / 4	1,65	2,18
4NOR	7 / 5 / 2	1,27	6,35
XNOR	9 / 7 / 2	1,15	1,18
Memory Cell	278 / 249 / 29	1,10	1,24
4xAdder	79 / 58 / 21	1,80	2,05
Total	67 / 56 / 11	1,35	2,78

It should be mentioned that one of the simplest DFM rules is an increase of the minimal distance between polygons [12]. In average, the minimal distance between the polygons in the layer for all our tests is increased above 35%. It is obvious that the reproducibility of the critical layout layer will be better in a comparison with the initial design. However, this improvement strongly varies for several test cases, from the minimum of 10% for the memory cell to 80% for the adder. In addition, this parameter varies for layers (Layer 1 and Layer 2) that were created after layout decomposition for double patterning. This deficiency has to be overcome to increase yield.

5   Conclusion

We have proposed a novel layout decomposition algorithm to address design needs for radiation hardened layout design. Our approach practically and effectively improves layout quality. It has been implemented using the contradiction graph and our modified concurrent breadth-first search algorithm. Our soft computing approach to manage contradictions in the layouts is novel and has advantage in terms of adaptability. Another aspect of our research is the use of high performance computing platforms for our design iterations. Our experiments using artificial and real-world test cases indicate that minimal distance is increased for all test cases.

Our ongoing research is in the following directions. The irregularity of layouts after decomposition results in lower yield. In order to overcome this deficiency optimization techniques will be used.

Our results can be used in design flow of radiation hardened electronics for several applications including space industry, collective awareness systems, emerging systems etc.

Acknowledgments. This study was partially supported by the Program of the President of the Russian Federation for the Support of Leading Scientific Schools, project LS-2903.2014.9 and Russian Fund for Basic Research, project 13-07-0073-a.

References

1. Bernstein, K., Rohrer, N.J.: SOI Circuit Design Concepts. Kluwer Academic Publishers, London (2003)
2. Colinge, J.: Multi-gate SOI MOSFETs. Solid-State Electronics 48, 897–905 (2004)
3. Kahng, A.B., Park, C.-H., et al.: Layout Decomposition for Double Patterning Lithography. In: Proc. IEEE Intl. Conf. on Computer-Aided Design (2008)
4. Shakhnov, V.A., Zinchenko, L.A., Rezchikova, E.V., Averyanikhin, A.E.: Algorithms of the VLSI layout decomposition. Vestn. Mosk. Gos. Tekh. Univ. 1, 76–87 (2011)
5. Zinchenko, L.A., Averyanikhin, A.E.: The Software for VLSI Layout Decomposition for Double Patterning. Program. Produkty Sist. 1, 7–10 (2011)
6. Shakhnov, V.A., Zinchenko, L.A.: Features of Application of Computing Systems in Nanoengineering CADs. Vestn. Mosk. Gos. Tekh. Univ., spec. issue "Nanoinzheneriya" (Nanoengineering), 100–109 (2010)
7. Shakhnov, V.A., Zinchenko, L.A., Verstov, V.A.: Topological Transformation of Submicron VLSIs for Double Lithographical Mask Technology. Russian Microelectronics 42(6), 427–439 (2013)
8. Neris Project, http://www.eu-neris.net
9. Safecast Project, http://safecast.org
10. Zinchenko, L.A., Kureichik, V.M., Redko, V.G., et al.: Bionic Information Systems and their Practical Applications. Fizmatlit, Moscow (2011)
11. Kleinberg, J., Tardos, É.: Algorithm Design. Addison Wesley (2006)
12. De Dood, P.: Impact of DFM and RET on Standard Cell Design Methodology. In: Proc. EDP Workshop (2003)

Assessment of Switch Mode Current Sources for Current Fed LED Drivers

Olegs Tetervenoks and Ilya Galkin

Riga Technical University, Faculty of Power and Electrical Engineering,
Kronvalda Boulevard 1, Riga, Latvia
olegs.tetervenoks@rtu.lv, gia@eef.rtu.lv

Abstract. Today solid state lighting is one of the most rapidly growing industries. Unfortunately light-emitting diodes require additional electronics (ballasts, drivers) for proper operation with conventional energy sources. Therefore this paper summarizes the previous studies of LED luminous flux regulation techniques, as well as extends previous studies of current fed (CF) converters. Requirements for the switch mode constant current source are discussed in this paper. Two suitable circuits as well as control algorithm are considered here. PSIM models of the converter were created and the operation with CF converters was approved. The considerations regarding the practical implementation are given in the end of this paper.

Keywords: Light-emitting diode, ballast, LED driver, DC-DC converter.

1   Introduction

Today the lighting industry is one of the fastest growing and developing branches in the production of electrical devices. The ground of this is the continued development and increase of efficacy of white light-emitting diodes (LEDs) [1], which are reasonably called the light sources of the future. LEDs gives opportunity to improve lighting system parameters such as efficiency and light quality, therefore also to reduce electrical power consumption. Even greater benefits can be achieved by implementation of smart lighting systems.

2   Relationship to Collective Awareness Systems

Today's development should lead to smart lighting systems with self-learning capabilities. In the future LED lighting systems will be equipped with communication modules (such as Bluetooth) for data exchange with personal mobile devices (smartphones) of the user. For the user it will be capable to fit parameters through mobile device using a special application, which also will be connected to a special server. The preferences of the users will be collected and processed on this server and the operation profile for the lighting system can be generated on this server based on the collected data. The data about simultaneous changes of the electricity price and

L.M. Camarinha-Matos et al. (Eds.): DoCEIS 2014, IFIP AICT 423, pp. 551–558, 2014.
© IFIP International Federation for Information Processing 2014

availability also can be considered for the generation of profile of lighting system. User will be capable to accept or modify generated profiles and to upload them in lighting system. Therefore such a complex control system interacting with people and taking into account their preferences as well as external factors in future may form a collective awareness lighting system.

Convenient and energy efficient light (luminous flux) regulation is necessary for the implementation of smart lighting systems.

3   LED Luminous Flux Regulation Methods

Besides the efficacy of LEDs themselves the performance of the ballast (utilized light regulation technique, efficiency) also plays significant role on the overall efficiency of LED luminaire. Therefore different luminous flux regulation techniques for LEDs and appropriate converters were studied and discussed in previous article [2]. Considered light regulation techniques are summarized in Fig. 1. Their benefits and drawbacks are listed in Table 1. Pulse mode flux regulation is the most appropriate for high performance devices where stable light color temperature is critical (backlit of LCD panels, displays) [3], [4]. This method might suffer from stroboscopic effect (because the luminous flux of LED follows the forward current at very high speed [5]), which is unwanted phenomenon in general lighting. Stroboscopic effect is especially dangerous for industrial lighting, where spinning mechanisms under certain conditions may seem motionless. Therefore step mode and amplitude (fluent) luminous flux regulation methods are the most appropriate in general lighting applications, however, fluent mode regulation technique allows utilizing LED in more efficient way (approximately by 7% in case of dimming at 50%) [5].

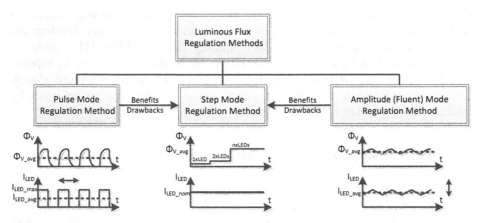

Fig. 1. Luminous flux regulation methods for LED lamps as well as typical waveforms of LED forward current and luminous flux

Table 1. Benefits and drawbacks of luminous flux regulation methods

Pulse Mode	Step Mode	Amplitude (Fluent) Mode
+ high accuracy and resolution	+ no stroboscopic effect	+ higher efficacy of LEDs
+ stable color temperature	+ stable color temperature	+ no stroboscopic effect
+ simple control system	+ simple control system	+ longer life span
− undesirable stroboscopic effect	− shorter life span	− relatively complex control system
− shorter life span	− low resolution (small number of regulation steps)	− unstable color temperature
− worse efficacy of LEDs	− worse efficacy of LEDs	− accuracy and resolution depends on complexity of control system

In Fig. 2 the ampere-lumen (A-Lm) curve of the common high power LED is shown. In case of pulse mode or step mode luminous flux regulation the LED operates at fixed point, usually maximal allowable or nominal (test) current. This maximal or nominal current is applied to LED periodically at high frequency, but the luminous flux is proportional to the duty cycle D (ratio of the time when the current is applied to the time of whole period.). It can be imagined that change of duty cycle moves the operation point of LED in a straight line which is connected between crossing point of the axes and previously mentioned fixed point (thick dashed lines in Fig. 2).

In case of fluent luminous flux regulation the operation point of LED moves along A-Lm curve in this way achieving higher efficacy, especially at smaller input current (power) as it shown in Fig. 2. Therefore amplitude mode regulation is the most suitable for general lighting applications. A lot of different kind of converters and approaches for amplitude mode regulation were described in scientific papers during last years. This paper deals with current fed (CF) converters.

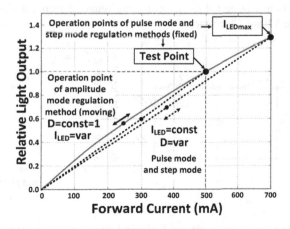

Fig. 2. Ampere-lumen curve of a common high power LED and operation points for different luminous flux regulation methods

4   Current Fed LED Drivers

LEDs are current consumers rather than voltage consumers. At the same time the most of the today's LED drivers are based on voltage fed (VF) converters, which means LED current (luminous flux) is controlled indirectly through applied voltage (volt-ampere V-A stage in Fig. 3). In previous studies it has been hypothesized that current fed (CF) converters are more suitable for driving LEDs, as the LED current is regulated in direct way [6], [7]. Also the circuits for three basic topologies of the current fed converters have been derived [7]. CF Buck topology (Fig. 4 a) is the most appropriate for step down applications. Also it has convenient controllability curve (Fig. 4 b).

CF drivers require constant current source at the input (Fig. 4). Current sources based on transistors operating in active region are energy inefficient, therefore the possibilities to build switch mode constant current source are considered in the following section.

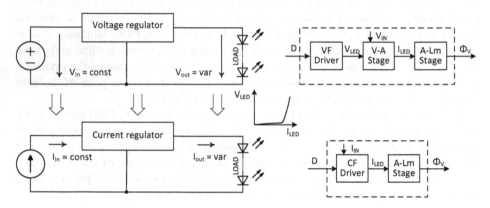

Fig. 3. Transition from voltage fed to current fed converters. Functional diagram of the light regulation loops for VF and CF chains.

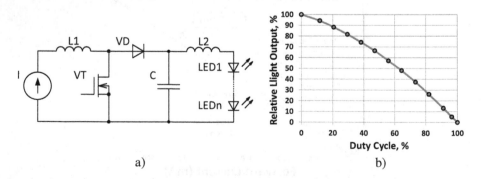

Fig. 4. CF buck converter: a) electrical principal circuit; b) controllability curve [6]

5   Switch Mode Constant Current Sources

The primary energy source almost always is a voltage source. This means constant current source must be voltage fed converter operating in constant output current mode. The requirements for the constant current source are:

1. uninterruptible current flow at the output
2. the capacitors (voltage source) at the output are not allowed in order to prevent high short circuit current.

The inductor is the only element, which meet the first requirement. Therefore the inductor current must be input current of the CF driver (inductor must be connected in series with the input of CF driver).

In [8] non-inverting buck-boost converter is described (at least two controllable switches). In fact it is combination of VF buck converter, which operates as constant current source and CF buck converter, which operates as current regulator. The constant current is formed in inductor, which is between VF and CF parts of the converter. In [8] it has been hypothesized that constant current in this inductor can be achieved by synchronized operation of the switches. The main feature of control law for the switches in this case: the sum of duty cycles of both switches must be approximately 100% [8]. Non-inverting buck-boost converter is capable to provide output voltage less or equal to the input voltage, but the efficiency decreases at higher difference between input and output voltage. It becomes a serious problem at high input voltage (rectified mains voltage). Also galvanic isolation is necessary for safety reasons. Therefore the converter with pulse transformer is necessary.

Special attention should be paid to the possibility of saturation of the core. Transformer isolated converter with bidirectional flow of current in primary winding should be used to prevent this. The circuit shown in Fig. 5 is based on half-bridge converter. It has been hypothesized that the similar control law as in [8] can be used for circuit shown in Fig. 5. For half-bridge converter the maximal allowable duty cycle of switches VT1 and VT2 is 50%, then control law for this circuit can be written as:

$$D_{VT1} + D_{VT2} + D_{VT3} = 2 \cdot D_{VT1} + D_{VT3} = 100\% . \qquad (1)$$

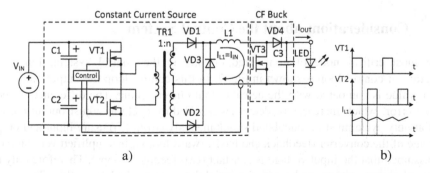

a)   b)

Fig. 5. Constant current source based on half-bridge converter: a) configuration; b) typical waveforms

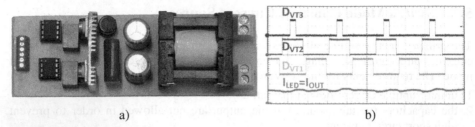

a)   b)

Fig. 6. The prototype of half-bridge based constant current source for CF buck converter: a) photo of the prototype; b) typical control signals ($D_{VT3} \approx 12\%$) and the waveform of the output current

Table 2. Results of the experiments with the prototype shown in Fig. 6 (at constant input voltage 260V and 7 connected in series high power LEDs as a load)

D_{VT3}, %	P_{IN}, W	P_{OUT}, %	Efficiency, %
12.1	29.9	27.5	91.9
49.8	6.4	5.4	85.2
87.4	2.6	1.7	63.7

But the relationship between the input voltage V_{IN}, output voltage V_{OUT}, duty cycle D and transformer turns ratio n is given in the following expression

$$V_{OUT} = 0.5 \cdot n \cdot D \cdot V_{IN} \text{ or } D = \frac{2 \cdot V_{OUT}}{n \cdot V_{IN}}. \tag{2}$$

The prototype shown in Fig. 6 was built to verify hypothesis (1) and the results of several experiments are summarized in Table 2. The conclusion from these preliminary experiments may be the same as in [8]: the duty cycle itself (D_{VT3}) is the main regulation parameter while the balance (1) plays additional tuning role. Synchronized operation of the switches is not enough to ensure stable operation of the whole system: closed loop regulation is necessary.

6   Considerations about the Control System

The most critical node of considered circuits is the control system. Functional diagram of control loop is shown in Fig. 7. In this control loop the output current set point value is compared with the actual output current value. The difference of them is the error, which increases or decreases initial duty cycle to adjust output current. Initial duty cycle must be calculated simultaneously (from (2)) using input and output voltage of the converter (feedback and feedforward loops). In simplified version it can be assumed that the input voltage is constant (two feedback loops). Therefore only the value of output voltage can be used for initial duty cycle calculations (Fig. 7).

To verify the functionality the model was simulated in PSIM software. PSIM model with control system discussed above is shown in Fig. 8.

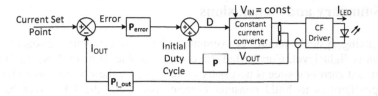

Fig. 7. Functional diagram of control loop for considered constant current converter

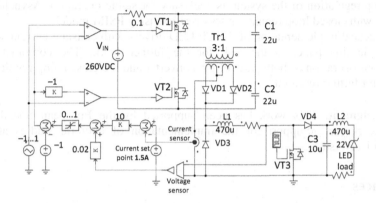

Fig. 8. PSIM model for considered constant current converter

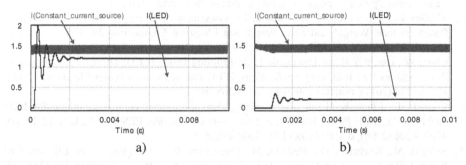

a)   b)

Fig. 9. Simulation results of the considered constant current sources operating with CF buck converter: a) startup process at 85% of maximal LED current (D=15% for CF buck); b) steady state operation at 15% of maximal LED current (D=85% for CF buck).

The simulation results for the considered circuit operating with the CF buck are shown in Fig. 9. It should be noted that this model operate switches of the constant current source and CF buck asynchronously. The simulation results Fig. 9 show that this configuration of control system is capable to operate properly.

7   Summary and Conclusions

Smart lighting systems require convenient and energy efficient approaches for the regulation of light level. Current fed drivers are capable to meet these requirements, but a constant current source is necessary at the input for their proper operation.

The possibilities to build constant current source for the CF driver have been discussed in this paper. Half-bridge based constant current source prototype was built and experimentally verified for synchronized operation. It was found that additional closed loop regulation of the system is necessary for stable operation. Asynchronous operation with closed loop regulation was verified using PSIM model.

The practical implementation of the CF LED drivers with constant current sources described in this paper is the task for the further work. The combination of synchronous operation of both parts of the converter and the closed loop regulation is the topic for further research.

Acknowledgment. This work has been supported by the European Social Fund within the project «Support for the implementation of doctoral studies at Riga Technical University».

References

1. Cree, Cree Reaches LED Industry Milestone with 200 Lumen-Per-Watt LED, Press Releases (December 18, 2012), http://www.cree.com/news-and-events/cree-news/press-releases2012/december/mkr-intro
2. Milaševski, I., Tetervenoks, O., Galkin, I.: Assessment of Energy Efficient LED Ballasts Based on their Weight and Size. Power and Electrical Engineering 29, 105–112 (2011) ISSN 1407-7345, doi:10.2478/v10144-011-0018-6
3. Chen, W., Hui, S.Y.R.: A Dimmable Light-Emitting Diode (LED) Driver With Mag-Amp Postregulators for Multistring Applications. IEEE Transactions on Power Electronics 26(6), 1714–1722 (2011), doi:10.1109/TPEL.2010.2082565
4. Chiu, C.-L., Chen, K.-H.: A high accuracy current-balanced control technique for LED backlight. In: IEEE Power Electronics Specialists Conference, PESC 2008, June 15-19, pp. 4202–4206 (2008), doi:10.1109/PESC.2008.4592615
5. Schmid, M., Kuebrich, D., Weiland, M., Duerbaum, T.: Evaluation on the Efficiency of Power LEDs Driven with Currents Typical to Switch Mode Power Supplies. In: 42nd IAS Annual Meeting Industry Applications Conference. Conference Record of the 2007 IEEE, September 23-27, pp. 1135–1140 (2007), doi:10.1109/07IAS.2007.178
6. Milashevski, I., Galkin, I., Tetervenok, O.: Assessment of buck converter powered by current or voltage sources for LEDs luminary. In: 2012 13th Biennial Baltic Electronics Conference (BEC), October 3-5, pp. 239–242 (2012), doi:10.1109/BEC.2012.6376861
7. Galkin, I., Tetervjonok, O., Milashevski, I.: Comparative study of steady-state performance of voltage and current fed dimmable LED drivers. In: 2013 8th International Conference on Compatibility and Power Electronics (CPE), June 5-7, pp. 292–297 (2013), doi:10.1109/CPE.2013.6601172
8. Galkin, I., Tetervenoks, O.: Validation of Direct Current Control in LED Lamp with Non-Inverting Buck-Boost Converter. In: 2013 39th Annual Conference of the IEEE Industrial Electronics Society (IECON), November 10-13, pp. 6019–6024 (2013) ISBN 9781479902231

Part XX
Electronics: RF Applications

RF Synthesizer Loop Filter Design for Minimal OFDM Inter-carrier Interference

Vitor Fialho[1,2], Fernando Azevedo[2,3], Fernando Fortes[2,3], and Manuela Vieira[1,2]

[1] Universidade Nova de Lisboa – Faculdade de Ciências e Tecnologia– DEE.
Monte da Caparica 2829-516, Lisbon, Portugal
[2] Instituto Superior de Engenharia de Lisboa – ISEL.
Rua Conselheiro Emídio Navarro,1 1959-007-Lisbon, Portugal
[3] Instituto das Telecomunicações
Av. Rovisco Pais, 1 1049 - 001 Lisbon - Portugal
{vfialho,fazevedo,ffortes}@deetc.isel.ipl.pt, mv@isel.pt

Abstract. This paper describes the influence of Radio-Frequency synthesizer loop filter design on OFDM inter-carrier interference. OFDM modulation scheme is very sensitive on carrier phase and frequency variations, resulting on inter-carrier interference leading to incorrect message decoding and the increase of system bit error rate. Radio-Frequency synthesizer has the main role on modulation and demodulation processes of base band IQ signals. The loop filter is one of its building blocks designed for loop stability. It also shapes the phase noise frequency response at the output of the frequency synthesizer. The achieved results are supported by simulation scenarios based on OFDM architecture, including radio-frequency synthesizer topology. The loop filter frequency response, local oscillator phase noise power and bandwidth are configurable. These simulations features allow the design trade-off between synthesizer stability and output phase noise power.

Keywords: Synthesizer, loop filter, phase noise, inter-carrier interference.

1 Introduction

Orthogonal frequency division multiplexing (OFDM) is a widely known modulation technique used in several wireless communication standards [1], [2]. OFDM specificity, based on slow sub-channel rate, minimizes the inter-symbolic interference (ISI) and channel multi-path interference [3], [4]. The sub-channel orthogonality is ensured by the inverse fast-Fourier transform (IFFT), where each input element represents the data to be carried on the correspondent sub-channel. However, OFDM is highly sensitive to phase and frequency variations which may cause the loss of orthogonality, leading to incorrect message decoding and, consequently, the increase of system bit error rate (BER) [4]. These effects are mainly caused by the radio-frequency (RF) front-end impairments during frequency conversion.

RF front-end is the main block between the base-band (BB) processor and the antenna, and its building blocks are responsible for the up and down conversion of

L.M. Camarinha-Matos et al. (Eds.): DoCEIS 2014, IFIP AICT 423, pp. 561–568, 2014.
© IFIP International Federation for Information Processing 2014

in-phase (I) and quadrature (Q) BB signals. These operations are performed by the mixer and oscillator. The oscillator used in RF transceivers is embedded in a synthesizer topology to achieve an accurate output frequency, with small and precise variations steps [5]. The need for an output signal with high accuracy implies the use of phase lock loop (PLL)/ synthesizer architecture to obtain a stable local oscillator (LO) signal for the frequency conversion. However, the oscillator in the synthesizer is responsible for the injection of phase noise into it and consequently, through the entire frequency conversion process. Its main effect on OFDM is the loss of channel orthogonality, leading to the interference between the adjacent sub-channels [4]. In OFDM transmission, besides the transmitted data, it is also included carrier pilots which provide the estimation of channel propagation conditions. This is done before the RF frequency conversion, therefore, these pilots will be also affected with phase noise.

Phase noise is a widely studied topic in free run oscillators [6], [7]. When the oscillator is inserted in a synthesizer to obtain the LO, the phase noise spectral behavior is modified [5], [8].

There is some literature addressing OFDM under phase noise influence [8], [9], [10], however these works are based on a classical synthesizer topology, where the presented results are based on synthesizer noise characterization, not addressing the effects of phase noise on the up-converted channel and how to minimize it.

Due to the constant demand for transceiver miniaturization and low power consumptions, the actual RF front-end, namely RF synthesizer, is totally based on monolithic implementations, which minimizes the margin for improvement of a given circuit inserted on RF transceiver. Loop filter is one of synthesizer building blocks, which is designed for loop stability, settling time and locking range. The obtained values for the loop components may lead to a solution external to the synthesizer monolithic implementation. This enables the study, characterization and possible optimization of phase noise under different filter topologies.

The objective of this work is the characterization of phase noise in RF synthesizer based on the obtained modulated channel, minimizing its influence with loop filter topology optimization. The main contribution and novelty of this work consists on performing phase noise shaping without changing the LO characteristics. The validation of the proposed work is based on OFDM simulation scenario with configurable RF parameters [11], which enables the extraction of inter-carrier interference (ICI) for several loop filter topologies.

2   Technological Innovation for Collective Awareness Systems

Since every collective awareness system needs to be capable of linking several entities such objects, people or enterprises, it needs to connect to a physical device that enables the network interconnectivity between such elements. The communication system that supports collective awareness platforms must be connected to Internet in order to fulfill the information transaction. Assuming that every equipment establishes a wireless link, they need to have a RF unit to enable such communication. Therefore the study of optimization of radio systems contributes, indirectly, for the technological innovation in this area of study.

3   RF Charge-Pump Synthesizer Analysis

Several synthesizer topologies can be found in literature [5], [8]. In this work the adopted topology is based on a charge-pump PLL with a divider in the feedback loop, as presented in Fig. 1.

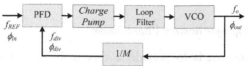

Fig. 1. Charge-pump synthesizer topology

Typical charge-pump synthesizer architecture uses a low frequency reference signal which phase ϕ_{in} and frequency f_{REF} are compared with ϕ_{div} and f_{div}, obtained from the M division of the output signal. The phase-frequency detector (PFD), together with the charge pump, produces an output voltage proportional to the phase and frequency difference between the reference signal and the feedback signal. The loop filter suppresses high frequency components from the output of the charge-pump and ensures loop stability, as presented in 3.1.1. The M divider is configurable to obtain an output frequency multiple of the reference input signal. The output frequency (f_{out}) is expressed by (1).

$$f_{out}=M\,f_{ref}.$$

(1)

Depending on the standards, the output signal must change in small frequency steps, typically hundreds of kHz in channels centered in the GHz range. These requirements make the synthesizer very sensitive to the building blocks impairments, namely LO phase noise.

Synthesizer transient response is a nonlinear process and is not addressed in this work. However, a linear approximation can be used, in order to obtain the trade-off building blocks design and their contribution to the loop stability [5]. Fig. 2a) shows the linear model of a synthesizer in lock conditions, with the transfer function for each block [5]. The obtained close loop transfer function is presented in expression (2).

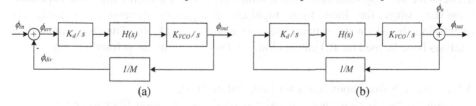

(a)   (b)

Fig. 2. Synthesizer linear model: (a) output to input; (b) output to VCO phase noise

$$\frac{\phi_{out}(s)}{\phi_{in}(s)}=\frac{K_d\cdot K_{VCO}\cdot H(s)}{s^2+K_d\cdot K_{VCO}\cdot H(s)\Big/M}$$

(2)

If H(s) = 1, the closed loop reveals two conjugated imaginary poles, showing loop instability and its dependency on the loop filter design.

Classical approaches suggest that a zero must be added in the transfer function to avoid instability [5], [8]. However, these studies are not addressed on how the loop filter will influence the synthesizer phase noise.

3.1   Charge-Pump Synthesizer Phase Noise

Since the synthesizer needs to provide a very stable signal makes this topology susceptible to phase noise, which can be caused by the input reference signal and VCO [5]. These two influences must be studied independently, since their impact on the synthesizer is different [5].

The study of phase noise at input is presented in [5], [8]. The transfer function is similar to (2), where ϕ_{in} and ϕ_{out} reflects the phase variation of the input and output signal, respectively. The synthesizer itself, due to its negative feedback characteristic, can compensate the phase error. If the phase of the input signal has a slow variation with time, the synthesizer can track the signal. However, if the input phase has a fast variation with time, the synthesizer fails to track with the reference signal.

When VCO phase noise is taking into account, for simplicity of the loop analysis, it is assumed that the input excess phase is zero. The phase noise inherent to the VCO can be modulated as an additive noise [5]. Therefore, the loop is redrawn as depicted in Fig. 2b). Assuming that ϕ_{VCO} and ϕ_{in} are uncorrelated, the superposition theorem can be applied, obtaining the transfer function (3), which presents the same poles of (2). The transfer function also contains two zeros at the origin, showing that synthesizer phase noise transfer function has a high pass characteristic.

$$\frac{\phi_{out}(s)}{\phi_{VCO}(s)} = \frac{s^2}{s^2 + \dfrac{K_d \cdot K_{VCO} \cdot H(s)}{M}} \tag{3}$$

Classical synthesizer analysis addresses transfer function denominator into control theory form, where the loop natural frequency and damping factor is taking into account for the loop bandwidth and stability [5]. This is a useful and valid approach, however, when the loop filter topology complexity increases, an analytical expression, based on these parameters, is difficult to obtain. Therefore, numerical analysis is performed for ICI calculation as a function of the loop filter topology.

3.1.1 Phase Noise Dependency on Loop Filter Design

As depicted by (3) phase noise transfer depends on the loop filter topology, which will be shaped by the transfer function.

Table 1 shows the zero-pole values for the several loop filters used in this work. Filter $H_A(s)$ corresponds to the particular case when there is no loop filter in the synthesizer. The remaining loop parameters are given by $K_{VCO} = 1\text{MHz/V}$, $K_d = 1\text{krad/s}$ and $N = 1000$.

Table 1. Loop filter zero and pole values for each topology

Loop Filter	Zero [Hz]	Pole [Hz]	Laplace Transfer Function
$H_A(s)$	-	-	1
$H_B(s)$	-	10000	$\dfrac{\omega_p}{s+\omega_p}$
$H_C(s)$	50	10000	$\dfrac{\omega_p}{\omega_z} \cdot \dfrac{s+\omega_z}{s+\omega_p}$
$H_D(s)$	50	100000	

To minimize the synthesizer phase noise effect the different loop filter topologies presented in Table 1 are applied in expression (3). For each topology it is possible to infer the one that shows the minor overshoot, and consequently, evaluate the system stability and bandwidth.

Fig. 3a) and b) shows the transfer functions of the synthesizer and phase noise, respectively, for each loop filter presented in Table 1.

The obtained results shows that the synthesizer instability decreases with the increase of the loop filter complexity. The synthesizer transfer function and the respective phase noise transfer function, without the loop filter, presents an instability characteristic, since the poles are located on the imaginary axis.

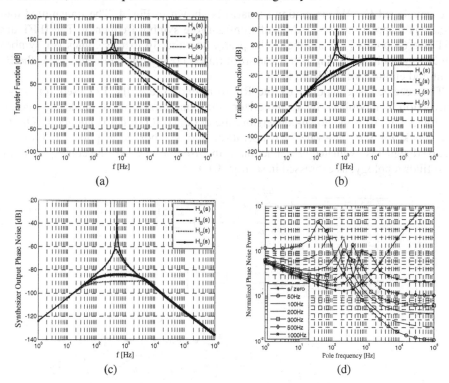

(a)

(b)

(c)

(d)

Fig. 3. (a) synthesizer transfer function; (b) phase noise transfer function; (c) synthesizer output phase noise (d) normalized integrated phase noise for frequency pole sweep

Synthesizer output phase noise is shown in Fig. 3c). These values are obtained by multiplying (3) by the VCO integration function. As depicted, for each filter topology, the synthesizer transfer function presents different behaviors. To simplify the filter analysis, a numeric simulation is performed, based on frequency pole sweep. For each sweep, the integrated phase noise power is obtained and shown in Fig. 3d), regarding different zero values of the loop filter $H_C(s)$.

As expected, due to the loop instability, a filter with no zero is undesirable, since the integrated phase noise power will increase as the frequency pole increases. However, by increasing the frequency of the zero, for low pole values, the integrated phase noise power decreases, though these values do not allow the synthesizer loop to be stable. Increasing the pole frequency, the integrated phase noise increases until the frequency of the zero approximately, and then decreases. For the performed numeric simulations, the minimal integrated phase noise power corresponds to the zero at 50Hz. This evaluation will be taken in to account for the frequency conversion of the OFDM simulation scenario.

4   Numerical Simulation of OFDM Inter-carrier Interference

The developed OFDM simulation scenario is shown in Fig. 4. The transmitted data stream is mapped into 16-QAM with configurable symbol frequency. The IFFT and FFT use 64 points [11] and the cyclic prefix contains 16 extra samples per symbol. OFDM symbol windowing is a square-root raised cosine filter. For this work, the RF front-end emitter is considered ideal.

The receiver is composed by the RF front-end where, the LO used for down-conversion, is modulated by a synthesizer with configurable noise mask presented in section 3.1.1. After the down-conversion, the obtained IQ signal is applied to OFDM RX block for CP removal, FFT operation and symbol de-mapping. ICI on the received signal is evaluated as a function of the used noise mask and consequently the loop filter topology, as expressed in section 3.1.1.

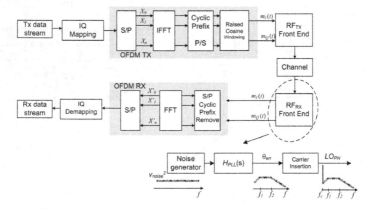

Fig. 4. OFDM block diagram MatLab/Simulink simulation

The LO model, presented in Fig. 4, allows the configuration of the integrated phase noise power and bandwidth. Previous works show how phase noise power distribution affects single carrier [12] and OFDM [11] modulation schemes. In this work, it is possible to predict the noise power within an OFDM sub-channel under several phase noise masks for a specific synthesizer loop filter topology.

4.1   Numerical Results

Unlike single carrier systems [11], [12], OFDM spectral content does not present a direct relation with phase noise, due to information split through the several sub-channels, the CP insertion and raised-cosine windowing. To obtain a significant result based on spectrum analysis, it is necessary to cancel the analyzed sub-channel, extracting the noise mean power within its bandwidth. This methodology allows the analysis of the adjacent sub-channels influence on the analyzed sub-channel.

Fig. 3c) and d) shows that loop filters $H_C(s)$ and $H_D(s)$ minimizes the synthesizer phase noise. Based on this results, $H_{PLL}(s)$ is modulated according to both phase noise mask depicted in Fig. 3c). Table 2 presents the synthesizer noise masks based on zero-pole mapping, where $H_{PLLc}(s)$ $H_{PLLd}(s)$ correspond to the output noise mask when the loop filters $H_C(s)$ and $H_D(s)$ are used.

Table 2. Zero-pole values for LO modulated noise

Noise Mask	Zero [Hz]	Pole 1 [Hz]	Pole 2 [Hz]
H_{PLLc} (s)	0	50	5000
H_{PLLc} (s)	0	100	2000

As depicted in Fig. 5, OFDM ICI for the same number of active sub-channels, presents a lower value for the phase noise mask $H_{PLLc}(s)$ comparing with $H_{PLLd}(s)$, which confirms the results presented in Fig. 3c) and d).

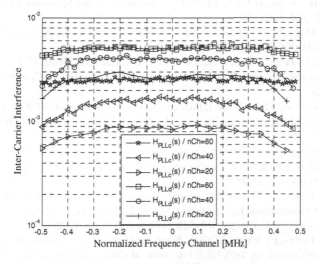

Fig. 5. ICI for different loop filter topologies

5   Conclusions

The influence of the synthesizer loop filter design on the phase noise of a charge pump RF synthesizer was described. Different passive filter designs were studied in order to obtain a topology that minimizes the synthesizer phase noise. The validation method was performed with a developed OFDM simulation model that allows the synthesizer modeling based on a specific noise mask.

The obtained results show that is possible to minimize OFDM ICI by changing the synthesizer loop filter topology, namely the poles and zeros values. This is an interactive process since the minimal noise value may drive synthesizer loop into instability. OFDM ICI, besides the loop filter topology, also depends on the number of active sub-channels. Therefore, besides the loop filter design, ICI can be optimized in function of the active sub-channels. This feature revels to be and important trade off factor when the optimization of phase noise is take into account.

As future work, the proposed method will be experimentally validated with a radio transceiver.

References

1. IEEE Standard for Wireless LAN Medium Access Control (MAC) and Physical Layer (PHY) Specifications: High Speed Physical Layer Extension in the 2.4GHz, IEEE Standard 802.11b-1999
2. IEEE 802.16. IEEE Recommended Practice for Local and metropolitan area networks Coexistence of Fixed Broadband Wireless Access Systems
3. Armstrong, J., et al.: OFDM for Optical Communications. Journal of Lightwave Technology 27(3) (February 2009)
4. Armanda, A.G.: Understanding the effects of phase noise in orthogonal frequency division multiplexing (OFDM). IEEE Transactions on Broadcasting 47(2) (June 2001)
5. Razzavi, B.: RF Microelectronics. Prentice Hall (1998)
6. Leeson, D.B.: A Simple Model of Feedback Oscillator Noise Spectrum. Proceedings of the IEEE 54(2), 329–330 (1966)
7. Lee, T.H., Hajimiri, A.: Oscillator Phase Noise: A Tutorial. IEEE Journal of Solid-State Circuits 35(3), 326–336 (2000)
8. Mathecken, P., et al.: Characterization of OFDM Radio Link Under PLL-Based Oscillator Phase Noise and Multipath Fading Channel. IEEE Transactions on Communications 60(6) (June 2012)
9. Petrovic, D., et al.: Phase Noise Influence on Bit Error Rate, Cut-off Rate and Capacity of M-QAM OFDM Signaling. In: Proc. Intl. OFDM Workshop of M-QAM OFDM Signaling, InOWo (2002)
10. Tchamov, N.N., et al.: VCO phase noise trade-offs in PLL design for DVB-T/H receivers. In: 16th IEEE International Conference on Electronics, Circuits, and Systems-ICECS, pp. 527–530 (December 2009)
11. Fialho, V., Fortes, F., Vieira, M.: Local Oscillator Phase Noise Influence on Single Carrier and OFDM Modulations. In: Camarinha-Matos, L.M., Tomic, S., Graça, P. (eds.) DoCEIS 2013. IFIP AICT, vol. 394, pp. 513–520. Springer, Heidelberg (2013)
12. Fialho, V., Fortes, F., Vieira, M.: Test Setup for Error Vector Magnitude Measurement on WLAN Transceivers. In: 19th IEEE International Conference on Electronics, Circuits, and Systems-ICECS, pp. 917–920 (December 2012)

Single-Objective Optimization Methodology for the Design of RF Integrated Inductors

Fábio Passos[1], Maria Helena Fino[1], and Elisenda Roca[2]

[1] Faculdade de Ciências e Tecnologia, Universidade Nova de Lisboa
2829-516 Caparica, Portugal
f.passos@campus.fct.unl.pt, hfino@fct.unl.pt
[2] Instituto de Microelctrónica de Sevilla, CSIC and Universidade de Sevilla,
Seville, Spain
eli@imse-cnm.csic.es

Abstract. Designing integrated inductors for RF applications is quite a challenging task due to the necessity of minimizing the parasitic effects arising from using today's technologies. The multiplicity of non-ideal effects to be minimized makes imperious the use of optimization-based design methodologies. In this paper a model-based optimization methodology is considered as a way of offering the designer the possibility to obtain inductors with maximum quality factor. The inductor model accounts for square, hexagonal or octagonal topologies. Furthermore tapered inductors are also accounted for in the proposed model. The use of the inductor model reduces the optimization time significantly. The validity of the results obtained is checked against electromagnetic (EM) simulations. As an application example, the particular case for the design of inductors for 2.4 GHz is illustrated.

Keywords: Integrated Inductors, RF, Design, Optimization etc.

1 Introduction

Integrated Inductors are a very important element on modern radio frequency (RF) design [1]. During the past few years, the design of integrated inductors has attracted much interest in both IC design and electronic design automation communities. In order to design an integrated inductor three main parameters must be considered by the designer [2]. One of the most critical parameter, if not the most critical one, is inductance value. Afterwards, quality factor (Q) and self-resonance frequency (SRF) [3] are to be taken into consideration.

Given an inductor design, different approaches may be followed in order to evaluate these parameters. The most common and widely used are: electromagnetic (EM) simulations, surrogate models and lumped element models.

EM simulations are time consuming, i.e. a simulation of a single inductor in a visual environment with ADS Momentum [4] can last up to eight hours. However this is the most accurate simulation possible when designing inductors. Surrogate models i.e. [5] are mathematically extensive, computing exhausting and they are not

L.M. Camarinha-Matos et al. (Eds.): DoCEIS 2014, IFIP AICT 423, pp. 569–574, 2014.
© IFIP International Federation for Information Processing 2014

physically understandable, and they introduce errors due to the mathematical expressions. On the other hand, lumped element models are based on physical expressions, which allow a much easier interpretation of the models [6]. Furthermore, as they rely on geometrical and technological parameters, design migration to new technologies is a straightforward process. In the end, these lumped element models are the option that allows the best trade-off between accuracy and simulation complexity. In order to obtain an inductor with desired impedance or the maximum quality factor possible, optimization based simulations must be used. It is possible to understand that if EM simulations and surrogate models are time prohibitive for only one inductor, to integrate these evaluation tools into optimization loops, is not a practicable solution. These leave designers with only one solution that is the usage of lumped element models. In this work the common π-model is used to model integrated inductors. After presenting the main contribution of the work to the implementation of collective Awareness systems, the inductor model is briefly described. Then, the integration of the inductor model into an optimization tools is presented. For the optimization a selection based differential evolution algorithm (SDBE) is used. Validity of the proposed methodology is performed by comparing the results of the optimization process for the design of an integrated inductor for 2.4 GHz against EM simulations. Finally conclusions are driven.

2   Relationship to Collective Awareness Systems

Collective Awareness systems link objects, people and knowledge in order to foster new forms of social and business innovation. The implementation of such systems relies on networks supporting the interconnections of a large number of heterogeneous cooperating devices. The development of systems has been made possible due to the rapid evolution of electronic technologies, which enable the implementation of ever more complex functions, in smaller and more rapid and less consuming circuits. To cope with the necessity of minimizing the power consumption of such systems, new design methodologies must be adopted so that the ever more stringent specifications may be attained. In the particular case of communications services, e.g. wireless communications devices, integrated inductors are becoming widely used in either Voltage Controlled Oscillators, LNAs and other RF blocks.

The mains contribution of this paper is the definition of a single-objective model-based optimization methodology for the design of integrated inductors. The main added value relies on the possibility for designing square, hexagonal or octagonal, tapered/no tapered inductors.

3   Integrated Inductor Model

The simple p-model is the most widely used for characterizing integrated inductor. Its applicability, however is limited to frequencies in the order of 1GHz. F The model used in this work is based on the segment model approach [7]. The inductor is divided into segments and then each segment is characterized with the π lumped-element circuit shown in Fig. 1. This figure illustrates a typical square integrated inductor, where n is the number of turns, w, the width of the metal turn, s, the spacing between metal turns, D_{out}, the outer diameter and, finally, D_{in}, the inner diameter.

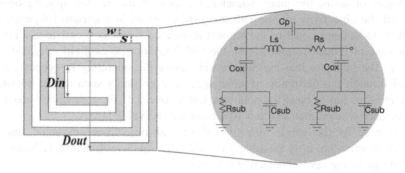

Fig. 1. Square inductor, its physical parameters and the π model

The series branch of this model consists of L_S, R_S and C_S. The series resistance, R_S arises from metal resistivity of the inductor and is closely related to the quality factor, being a key issue for inductor modeling. The series feed-forward capacitance, C_p, is usually considered as the overlap capacitance between the spirals and the underpass metal lines, also called C_O [8]. However, as the minimum feature size of CMOS process continues to shrink, the spacing between metal spirals, s, can be reduced to a value similar to the distance to the underpass. Therefore the coupling capacitance between metal lines, C_S, can increase its significance in the total capacitance of the device, and therefore in the self-resonance frequency (SRF). C_p is then given by the sum of C_S and C_O. The capacitance C_{ox} represents the oxide capacitance between the spiral and the substrate. The silicon substrate is modeled with C_{sub} and R_{sub} [9]. A detailed explanation on how to evaluate L_S may is presented in [10]. For the remaining elements analytical expressions may be obtained in [11].

For the sake of simplicity some physical aspects were not taken into account in the analytical characterization of the model elements. For example, the model does not take into account *Eddy* currents [8], thus for small inner diameters, the model is less accurate. Also, with a wider metal, the parasitic capacitances and fringing capacitances tend to increase and again, induce higher errors. Since this model is to be integrated into an optimization based design tool, the validity of the model along the overall design space was performed through a statistical analysis [11].

4   Optimization of Integrated Inductors

This section discusses the optimization of integrated inductors in a 0.35 μm CMOS technology for 2.4 GHz. In this work an octagonal layout is simulated because it uses less metal to draw turns than square or hexagonal layouts, and can therefore achieve higher quality factors. This section presents a single-objective optimization using the selection based differential evolution (SBDE) algorithm is presented. The results of the optimization processes are validated against EM simulations.

For inductor optimization, common design variables are the geometric parameters (i.e. number of turns, the inner diameter, the turn width and the spacing between turns) and the objectives are the inductor performances at a certain frequency (i.e. equivalent inductance, quality factor, self-resonance frequency and area). For the specific case of inductor optimization the SRF may not be explicitly defined as an objective. The guaranty that the SRF is well above the operating frequency may be obtained through the definition of some specific constraints such as the imposition that the inductor has to be in the plan bandwidth zone, i.e., the frequency band where the inductance value remains within 10% from the DC value [15].

In the case of the single objective, if several objectives envisaged into one optimization process, it is possible to build an objective function reflecting the weighted sum of the several objective functions.

The general constraints imposed in the optimization processes are:

- The inductance value is limited to only 5% error from the inductance desired at 2.4 GHz,
- The inductance curve has to be in the plain bandwidth zone,
- The quality factor has to be before the peak value, which means that the quality factor has to be in the positive slope region of the quality factor.

The single-objective optimization was made with the objective of finding an inductor with a given inductance value, while maximizing the quality factor and minimizing the area. This was possible through the weighted objective function shown in (1)

$$Objective = Min(-(0.7 \cdot Q) + 0.3 \cdot Area) \tag{1}$$

Several simulations were made in order to obtain inductors with a desired inductance for 1nH to 5nH. All optimizations were done with 300 individuals and 300 iterations and the simulation time was about 300 seconds per simulation. The inductors obtained were then simulated with ADS Momentum, in order to check the validity of the results. The comparison values and curves are shown for each of the inductors obtained. The physical parameters of the inductors are given in Table 2, as well as the values for inductance and quality factor at 2.4 GHz. All the physical parameters are given in μm, and the inductance in nH. The relative error (ε) is given in %. The physical parameters also had to be limited so that the model is accurate over the design space considered. So, the geometric parameters were limited according to the values shown in Table 1. Layout and fabrication processes impose limits on the precision of the physical dimensions, and so not all values in a range can be used, only some discrete values. These values are defined by the grid of the optimization process and are related to each technology.

Table 1. Restrictions on the inductor physical parameters

Parameter	Minimum	Maximum
n	2	7
D_{in} (μm)	100	250
w (μm)	5	10

Table 2. Comparison between inductance and quality factor for optimized inductors

L_D	n	w	D_{in}	s	L_{Model}	L_{EM}	ε	Q_{Model}	Q_{EM}	ε
1 nH	2	10	123	2.5	1.04	1.01	**2.97**	8.80	9.07	**2.98**
2 nH	3	10	106	2.5	2.10	1.87	**12.3**	11.17	9.85	**13.4**
3 nH	3	10	144	2.5	2.80	2.66	**5.26**	11.32	10.35	**9.37**
4 nH	3	8.03	185	2.5	3.77	3.71	**1.62**	10.24	1.10	**1.39**
5 nH	4	5.63	143	2.5	5.03	4.90	**2.65**	9.21	9.55	**3.56**

It is possible to conclude that the model can be used in RF frequency range in the design space where the optimization process was developed. The errors observed in the optimization process are usually quite acceptable. It is also possible to observe that the inductor that has the highest error has the Din and the w close to its boundaries, which increases the error values.

5   Conclusions and Future Work

In this work an efficient lumped-element model to characterize integrated inductors is presented. The model uses analytical expressions to evaluate the lumped elements values. The model is able to predict the inductance and the quality factor values with relative accuracy. The model was also integrated into an optimization process and the results validated against EM simulation. It has been proved that the model is suitable for RF circuit design and its best feature is the ability to save time in the design process. As future work, the model can be fitted to smaller frequency ranges in order to obtain higher accuracy. The complexity of the model can also be increased, adding elements to the equivalent circuit.

References

1. Niknekad, A.M.: Electromagnetics for High-Speed Analog and Digital Communication Circuits (2007)
2. Zhan, Y., Sapatnekar, S.S.: Optimization of integrated spiral inductors using sequential quadratic programming. In: Proceedings of the Design, Automation and Test in Europe Conference and Exhibition, February 16-20, vol. 1, pp. 622–627 (2004)
3. Mohan, S.S.: The Design, Modeling and Optimization of On-chip Inductor and Transformer Circuits. Stanford University, Dept. of Electrical Engineering (1999)
4. ADS Momentum. Agilent technologies. EEsof division, Santa Rosa, CA (2006)

5. Nieuwoudt, A., Massoud, Y.: Variability-Aware Multilevel Integrated Spiral Inductor Synthesis. IEEE Transactions on Computer-Aided Design of Integrated Circuits and Systems 25(12), 2613–2625 (2006)
6. Passos, F., Helena Fino, M., Roca, E.: A Wideband Lumped-Element Model for Arbitrarily Shaped Integrated Inductors. In: IEEE ECCTD Conference Proceedings (2013)
7. Koutsoyannopoulos, Y., Papananos, Y., Alemanni, C., Bantas, S.: A generic CAD model for arbitrarily shaped and multi-layer integrated inductors on silicon substrates. In: Proceedings of the 23rd European Solid-State Circuits Conference, ESSCIRC 1997, pp. 320–323 (September 1997)
8. Yue, C., Wong, S.: Physical modeling of spiral inductors on silicon. IEEE J. Transactions on Electron Devices 47, 560–568 (2000)
9. Lee, T.H.: The Design of CMOS Radio-Frequency Integrated Circuits. Cambridge University Press (2004)
10. Passos, F., Helena Fino, M., Roca, E.: Analythical Characterization of Variable Width Integrated Spiral Inductors. In: IEEE MIXDES Conference Proceedings (2013)
11. Passos, F., Helena Fino, M., Roca, E.: Lumped Element Model for Arbitrarily Shaped Integrated Inductors – A Statistical Analysis. In: IEEE COMCAS Conference Proceedings (2013)
12. Greenhouse, H.: Design of planar rectangular microelectronic inductors. IEEE J. Trans. PHP 10(2), 101–109 (1974)
13. Grover, F.W.: Inductance calculations: Working formulas and tables. Courier Dover Publications (1929)
14. Yue, C., Wong, S.: On-chip spiral inductors with patterned ground shields for Si-based RF ICs. IEEE Journal of Solid-State Circuits 33(5), 743–752 (1998)
15. Sieiro, J., Lopez-Villegas, J.M., Osorio, J.A., Carrasco, T., Vidal, M.N., Ahyoune, S.: Synthesis of compact planar inductors in LTCC technology. In: 2012 International Conference on Synthesis, Modeling, Analysis and Simulation Methods and Applications to Circuit Design (SMACD), pp. 45–48 (2012)

A 0.5 V Ultra-low Power Quadrature Ring Oscillator

João Eusébio[2], Luís B. Oliveira[1,2], Luís Miguel Pires[1,2], and João P. Oliveira[1,2]

[1] Centre for Technologies and Systems (CTS) – UNINOVA
[2] Dept. of Electrical Engineering (DEE), Universidade Nova de Lisboa (UNL)
Campus FCT/UNL, 2829-516, Caparica, Portugal
{jje19386,l.pires}@campus.fct.unl.pt,
{l.oliveira,jpao}@fct.unl.pt

Abstract. In this paper we present a CMOS quadrature ring oscillator operating at 0.5 V. Due to this very low voltage conditions, new project technique using the available terminal of the transistors (bulk) is used in order to reduce the threshold voltage of the transistors, thus improving the voltage headroom. The technique is applied in a conventional inverter-based ring oscillator with a feedback topology capable to generate quadrature signals. Simulations results in a 130 nm CMOS technology shows that a very simple VCO in the GHz range can be obtained, by changing the bulk voltage of transistors (NMOS or PMOS). The circuit operates with less than 50 µW achieving a FoM of about -115 dBc/Hz at 10 MHz offset.

Keywords: Quadrature RC oscillator, CMOS circuit, Low voltage, Ultra-low power.

1 Introduction

Nowadays there is a huge demand on ultra-low power circuits for portable equipment, in wireless communication applications. Moreover, designers try to reduce the production cost, reducing the circuits die area (avoiding the use of on-chip inductors) and using standard CMOS technology. Thus, they try to obtain high frequency performance, with low supply voltage and low power.

This is the motivation for use modern receivers receiver architectures such as Low-IF Intermediate Frequency (IF) or Zero-IF, which can be used to further reduce the cost since they do not require eternal image reject filters and can be designed in a single chip using standard CMOS technology without inductor [1]–[3]. Typical applications are in biomedical bands, such as Wireless Medical Telemetry Services (WMTS) and Industrial, Scientific, and Medical (ISM), since the main requirements are low power consumption and low cost [4]. The oscillator that is a key block in these systems will be investigated in this paper.

Quadrature Ring oscillators are key blocks on the design of these architectures since they can be fully integrated, they do not have inductors and thus occupying very low area. They have higher tuning range (useful to cover several bands) and are able to provide quadrature signals with low quadrature error [5].

L.M. Camarinha-Matos et al. (Eds.): DoCEIS 2014, IFIP AICT 423, pp. 575–581, 2014.

In this paper, we present a CMOS ring oscillator [6] capable to achieve quadrature outputs for very low voltage operation (in our case 0.5 V). A circuit design in a 130 nm CMOS technology proves that changing the voltage of the bulk node in NMOS and PMOS transistors allows the reduction of the transistor threshold voltage (V_{th}) [7]. With that change we are able to increase the transistor transconductance (g_m) (in comparison with the original), and therefore, designing a Voltage Controlled Oscillator (VCO) for the GHz range with very low power and high Figure Of Merit (FoM). The tuning of the oscillator in the GHz range can be achieved by changing the bulk voltage, while keeping a low supply voltage, with a FoM of -114.7 dBc/Hz at 10 MHz offset.

In section 2, we describe the contributions of this paper for the conference theme. In section 3, we describe low voltage techniques for CMOS inverter. In section 4, we present the proposed circuit and we present the simulation results, and finally, in section 5 we draw the conclusions.

2   Contributions for Technological Innovation for Collective Awareness Systems

Nowadays, the emergence of collective awareness systems is pushing the performance of end user interactive objects. These devices are embedded with sensors and actuators with the ability to communicate and exchange information, in a global cloud environment "Internet of the future" or "Internet of Things". The physical communication plays a critical role in portable wireless devices equipped with very low power modern receivers with very low noise, and accurate and stable frequencies signals to ensure a reliable and efficient communication in a crowded channel environment. With this goal in mind the design of Radio Frequency (RF) front-end blocks for low power applications, in CMOS technology, will contribute towards the achievement of more cheap and robust devices. In this paper we will focus on the design of low power wideband CMOS quadrature oscillators.

3   Low Voltage Technique for a CMOS Inverter

A basic ring oscillator is composed by and odd number of single-ended blocks in a feedback topology [1]-[2]. In the work proposed in [6] they combine 4 blocks of 3 inverter stages in a quadrature ring oscillator. In this work we use the circuit topology proposed in [6], but with the inverter operating in ultra-low power voltage (0.5 V) for very low power consumption. The basic inverter cell is shown in Figure 1.

This cell is composed by two MOSFETs. The top MOSFET is a PMOS transistor and the bottom is a NMOS transistor. The bulk is linked to the respective sources, PMOS source is linked to V_{DD}, NMOS source is connected to Vss, reducing the value

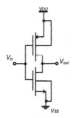

Fig. 1. Basic CMOS Inverter

of the threshold voltage, V_{th}, to V_{T0} (V_{th} voltage at $V_{BS}=0V$). Interestingly, this threshold voltage, can be further reduced by directly biasing the bulk-source junction, as shown in [1]-[2]

$$V_{th} = V_{T0} - \gamma \cdot \left(\sqrt{\phi_o} - \sqrt{\phi_o - V_{BS}} \right) \tag{1}$$

where ϕ_F is the Fermi potential, γ is the body-effect coefficient, $\Delta \phi \approx 6\ kT/q$ and $\phi_0 = 2\phi_F + \Delta \phi$.

The first modification for the basic CMOS inverter is to add a voltage different than zero to the bulk of transistor NMOS, V_{CTRL}, as shown in Figure 2-a). This is possible since we are using CMOS triple well process.

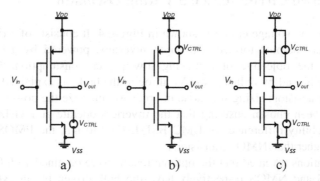

a)   b)   c)

Fig. 2. Voltage source added to: a) bulk of the NMOS b) bulk of the PMOS c) both NMOS and PMOS

The second change is to perform the same modification to the transistor PMOS, as shown in Figure 2-b). Thus, the final circuit is shown in Figure 2-c), with the objective of changing simultaneously both bulk voltages (in the range of 0 to 0.5 V) in order to reduce the transistor's V_{th}. The final configuration of the inverter represented in Figure 2-c), is formed by PMOS transistor with width of 3 μm and a length (L_{min}) of 120 μm while for the NMOS transistor the width is set to 1 μm

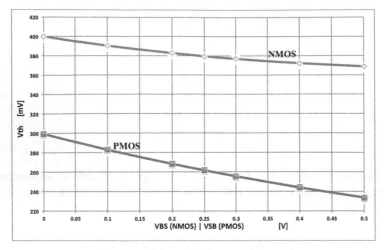

Fig. 3. Vth reduction for the NMOS and the PMOS transistors

maintaining L_{min} in 120 µm. In every single schematic the results were similar in the sense of reducing the values of V_{th}. But more effective changes are achieved by changing simultaneously the bulk voltages of both devices. In Figure 3 it is shown that the reduction curve of V_{th} varying the bulk voltage between 0 and 0.5 V.

4   Proposed Circuit for a 0.5 V Ring Oscillator

The proposed low-voltage circuit is shown in Figure 4. It consists of a ring oscillator with quadrature outputs formed with eight inverters, powered by a 0.5 V source voltage with the objective of achieving low power consumption. Low voltage operation is achieved by reducing the V_{th} through the bulk, as explained in section 3. The inverter transistors sizing was done in order to minimize the power consumption and L_{min} has been chosen ensuring that the inverters operate at 1 GHz. Due to the transistors mobility differences in UMC BSIM3V3 130 nm, the PMOS have width three times higher than NMOS transistors.

The simulations revealed that the optimal results were obtained with 0.5 V in bulk V_{CTRL}. PMOS and NMOS respectively have the bulk driven by an external source voltage in a similar process like the work proposed in [7].

To evaluate the performance of the overall circuit, it is used the accepted FoM given by,

$$\text{FoM} = L_{measured} + 10 \cdot \log\left[\left(\frac{\Delta f}{f_{OSC}}\right)^2 \cdot \frac{I_{core} \cdot V_{DD}}{1mW}\right], \tag{2}$$

where, $L_{measured}$ is the value of the phase noise measured at 10 MHz, Δf is 10 MHz and I_{core} is the current consumption of the circuit core.

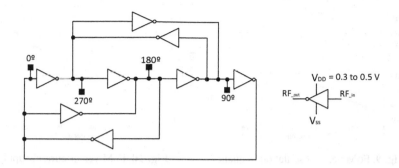

Fig. 4. Quadrature four-stage ring oscillator

Using the traditional inverter in the quadrature ring oscillator, represented in Figure 1, and varying the supply voltage between 0.3 and 0.5 V, it is possible to see that the FoM has a variation smaller than 1.4 dB, as shown in Figure 5. Other important result is the current consumption that rises when the FoM reach higher values, as shown in Figure 6.

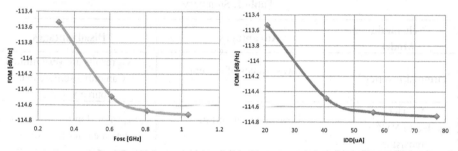

Fig. 5. FoM versus oscillation frequency   **Fig. 6.** FoM versus current supply

The result in Figure 7 was chosen from the range of FoM variations, a range over 3 dB, for the case of schematic in Figure 2-c). Clearly the best situation in Figure 8 is the first one, were the FoM has the best value and the current is the smallest one.

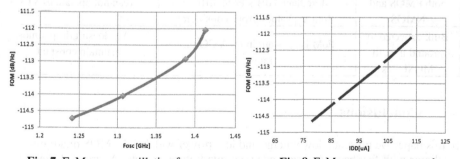

Fig. 7. FoM versus oscillation frequency   **Fig. 8.** FoM versus current supply

Figures 9 and 10 show the results obtained when changing the threshold voltage of both transistors.

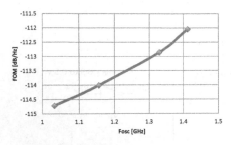

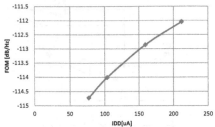

Fig. 9. FoM versus oscillation frequency   **Fig. 10.** FoM versus current supply

In Table 1 is possible to visualize the advantages and disadvantages from each schematic presented previously (Figure 2-a), b) and c)). The best results are obtained using the inverter with both reduced V_{th}, were the FoM varies up to 3 dB. Moreover, this simultaneous V_{th} reduction in both transistors enables the operation of the oscillator at very low voltage supply, with reduced current supply.

Table 1. Summary

Schematic Model	Advantages	Disadvantages
Original Inverter	Lowest values of current	V_{th} results are very limited Can only oscillate main source voltage
VCTRL added to NMOS transistor	Low values of current	V_{th} value don't lower enough to get an earlier functioning
VCTRL added to PMOS	Low value of current; the best V_{th} values 3 dB's FoM difference The circuit works with less voltage	Only V_{th} drops considerably
VCTRL added to both PMOS and NMOS	Low values of current More than 3 dB's FoM difference Both V_{th} values drops considerably	Since both bulks are used (can not operate as VCO)
Bulk conected to ground in both MOSFETs	FoM is constant in the VCO tuning range	V_{th} doesn't drop almost maintain constant

5   Conclusions

In this paper we present a low voltage and low power wideband CMOS quadrature ring oscillator with feedback to ensure accurate quadrature outputs. A circuit design at 0.5 V is presented in a 130 nm CMOS technology, which validates the proposed methodology. Simulation results show that changing the bulk voltage named V_{CTRL} can control the oscillator frequency. Therefore, this circuit can be used as VCO in Phased

Locked Loop (PLL) block, under low voltage supply. V_{CTRL} can be reduced to 0.4 V or 0.3 V with very low power consumption, less than 50 µW, with a FoM of -114.7 dBc/Hz at 10 MHz offset. The proposed circuit is especially useful for low power and low voltage operation in biomedical applications.

References

[1] Razavi, B.: RF Microelectronics, Prentice-Hall (1998)
[2] Lee, T.H.: The Design of CMOS Radio Frequency Integrated Circuits, 2nd edn. Cambridge University Press (2004)
[3] Crols, J., Steyaert, M.: CMOS Wireless Transceiver Design. Kluwer (1997)
[4] Iniewski, K.: VLSI Circuits for Biomedical Applications. Artech House (2008)
[5] Oliveira, L.B., Fernandes, J.R., Filanovsky, I.M., Verhoeven, C.J., Silva, M.M.: Analysis and Design of Quadrature Oscillators. Springer (2008)
[6] Grozing, M., Philipp, B., Berroth, M.: CMOS Ring Oscillator with Quadrature Outputs and 100 MHz to 3.5 GHz Tuning Range. In: 28th Int. IEEE European Solid-State Circuits Conference 2003, ESSCCIR 2003, pp. 679–682 (September 2003)
[7] Nose, K., Sakurai, T.: Optimization of VDD and VTH for Low-Power and High-Speed Applications. In: Proceedings of the ASPDAC 2000, pp. 469–474 (2000)

Part XXI

Electronics: Devices

Stability Improvements in a Rail-to-Rail Input/Output, Constant G$_m$ Operational Amplifier, at 0.4 V Operation, Using the Low-Voltage DTMOS Technique

Joana Correia[2], Nuno Mancelos[2], and João Goes[1,2]

[1] Centre for Technologies and Systems (CTS) – UNINOVA
[2] Dept. of Electrical Engineering (DEE), Universidade Nova de Lisboa (UNL)
Campus FCT/UNL, 2829-516, Caparica, Portugal
{jm.correia,n.mancelos}@campus.fct.unl.pt, goes@fct.unl.pt

Abstract. The use of the dynamic threshold MOS (DTMOS) technique is evaluated in a two-stage rail-to-rail Input/Output, constant G_m amplifier. The proper choice of specific transistors in which the technique should be used is presented, as well as the resulting improvements, mainly regarding stability of the circuit at low voltage operation. The DTMOS technique is used in the NMOS transistors of the folded-cascode input stage, allowing the circuit to be stable at V_{DD} = 0.4 V, with equivalent gain and gain-and-bandwidth product (GBW) values achieved with the same V_{DD} value, for the initial circuit operating at 0.8 V. The implemented changes allow the circuit to be stable at low voltage operations without requiring any increase in the cascoded-Miller compensation capacitors, saving circuit area and, consequently, cost.

Keywords: Low-voltage DTMOS technique, Low supply voltage OPAMP, Low-voltage OPAMP stability issues.

1 Introduction

The general tendency for reducing the supply voltage in ICs represents a challenge for analog amplifier designers, in terms of maintaining required gain levels, bandwidth (BW) and stability. In terms of stability and frequency compensation, several techniques can be implemented, but the solutions represent an increase of the circuit area, since they often include the need of additional capacitors.

As a starting point for the testing of the DTMOS technique, an operational transconductance amplifier (OTA) was chosen. The choice is based on an original, rail-to-rail Input/Output, constant G_m OTA, described in [1], and operating at a 0.8 V supply. The technology used for the initial amplifier was the purely-digital 180 nm CMOS, whereas the present implementation, fully supported in the initial study, [1], was simulated using the CADENCE, with an UMC 130 nm, purely-digital, CMOS technology. The change in technology doesn't represent a relevant factor in the overall behavior of the circuit.

L.M. Camarinha-Matos et al. (Eds.): DoCEIS 2014, IFIP AICT 423, pp. 585–591, 2014.
© IFIP International Federation for Information Processing 2014

The main objective of this work is to improve the gain, GBW and stability of the original amplifier, and to guarantee that the OTA still works, with reasonable performance, when operating at much lower supply voltages.

The DTMOS technique was the technique chosen to reach the objectives, and consists in connecting the bulk of a transistor to its gate terminal. This allows the bulk voltage to be variable, instead of being fixed at ground or V_{DD} (NMOS or PMOS, respectively) [2]. The transistor suffers from body-effect, in this configuration, since the voltage between the source and bulk terminals is not zero. However, the transcondutance that is a result of the body-effect contributes, positively, to the total transcondutance of the transistor, being this contribution an increase of approximately 20-to-30% of the transcondutance of the transistor. This means that in the DTMOS configuration, the body-effect of a transistor is not a degrading factor in its G_m, but rather an improvement.

The disadvantages of using the DTMOS technique are the increase of the parasitic capacitances in the transistor, which can ultimately lead to a loss of circuit BW, and the possibility of a latch-up problem. In the case of this particular application and circuit, the latch-up situation is not a real problem, because there won't be sufficient voltage to trigger the effect, since the intended test supply voltages will be lower than 0.7 V, which is the typical problematic threshold that allows that effect to be a problem. In terms of the increase of the parasitic capacitances, and consequent decrease in BW, the situation can be dealt with separately, although the changes in the parasitic capacitances are not relevant enough to reflect some problematic difference in terms of the BW.

2   Relationship to Collective Awareness Systems

Collective Awareness Systems that are capable of harnessing collective intelligence, allowing the creation of a network of distributed knowledge and data from real environments, include a multitude of individual subsystems that must be able to collect and transmit the actual information that is shared in the system. These subsystems may include a wide range of different devices and applications, but they usually consist in wireless and portable devices [3] for which concerns like power consumption and autonomous lifetime are very pertinent.

The DTMOS low voltage technique used in this work is one of the techniques that allow reduction in circuit supply voltage, and consequently power consumption. Moreover, as it will be demonstrated in this paper, the use of DTMOS in certain specific devices can improve the stability of the circuit, which translates into smaller area (there is no need for larger compensating capacitors). The significant supply voltage reductions that can be achieved with this technique are promising for the hardware implementations of Collective Awareness Systems or any other system where power consumption and cost (silicon area) are key factors.

3   Amplifier Description

3.1   Initial Amplifier

The 0.8 V, rail-to-rail Input/Output, constant G_m OTA, whose schematic is presented in Fig. 1, consists of two stages.

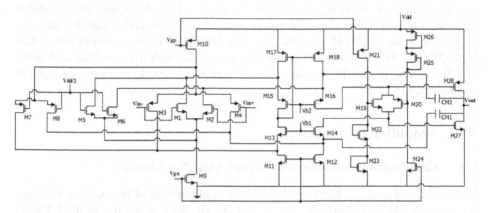

Fig. 1. Initial operational amplifier

The rail-to-rail folded-cascode input stage includes two differential pairs in parallel, consisting of the transistors M1, M2 (NMOS differential pair), M3 and M4 (PMOS). This configuration allows rail-to-rail operation at the input. For high common-mode voltages at the input, the NMOS differential pair is on, while for low common-mode voltages it is the PMOS differential pair that is on. For medium levels of the common-mode voltages, both differential pairs are on, increasing the current flowing to the summing circuit and consequently increasing the G_m. To avoid this increase in the transconductance, since it is desirable to have it constant for the whole range of common-mode voltages, the current switches M5 to M8 are added. The following folded-cascode configuration is responsible for the summing of the signals (current) coming from the complementary differential pairs.

The class AB output stage features, as stated above, the cascaded-Miller frequency compensation technique, which includes the capacitors CM1 and CM2. Class AB operations are made possible by M19 and M20. These transistors are driven by the signal currents coming from transistors M14 and M16 and are polarized by the diode-connected transistors M22, M23, M25 and M26. The output transistors M27 and M28 are in common-source configuration to allow rail-to-rail operations at the output. The M27 transistor functions at the positive swing, while M28 takes care of the negative swing. The frequency of non-dominant pole is shifted to higher frequencies by the cascoded-Miller frequency compensation technique.

3.2   Proposed Modifications in the Original Amplifier

The low-voltage DTMOS technique was chosen as the approach to use because of the simplicity of the implementation, since it basically consists in disconnecting the bulks of the PMOS and NMOS devices, respectively from V_{DD} and V_{SS}, and re-connect them to the gate terminal [4, 5].

Many possible configurations have been simulated and evaluated, using the DTMOS technique applied to different sets of transistors. The best results, taking into account simplicity and gain results, were achieved using the DTMOS technique only in the NMOS cascode devices M13 and M14. Notice that the DTMOS technique also has the benefit of reducing the threshold voltage of the transistor. Since NMOS transistors have a higher threshold voltage than PMOS, it is understandable that it is preferable to use the DTMOS technique in the NMOS cascode transistors rather than in the PMOS. This requires fabrication processes either with triple-well or with deep-Nwell but this is clearly the case, for all state-of-the-art deep nano-scale CMOS technologies.

4   Amplifier Performance

4.1   Electrical Simulations of the Original Amplifier Circuit

The simulation results achieved with the initial configuration of the circuit ported into the 130 nm CMOS technology (the original circuit, without the use of the DTMOS technique) are presented in Table 1. The presented results include simulations run under different supply voltages, focusing on the values of DC gain, GBW and Phase Margin (PM).

Table 1. Simulated results for the initial configuration of the amplifier circuit

V_{DD} (V)	Gain (dB)	GBW (kHz)	Phase Margin (°)
1.2	82.62	857	> 56
0.8	82.58	864	> 56
0.7	81.16	824	> 56
0.6	81.44	833	52
0.5	78.5	818	56
0.4	72.15	878	53
0.3	57.7	794	51

The simulated PM is evaluated at a typical closed-loop gain of 6 dB, for each V_{DD} value. The presented PM results are lower than 55 °, which leads to the conclusion that the circuit is not stable for any of these supply voltages. In terms of DC gain and GBW, the values decrease as a consequence of the lowering of the V_{DD} value, as it is expected. However, the decrease in those values is not as substantial as was estimated, revealing that the topology of the circuit is capable of maintaining both the gain and the GBW performance parameters even at lower supply voltages (i.e., below 0.8 V).

4.2   Re-Design of the Amplifier Using DTMOS

For the circuit amplifier including the proposed changes, the same simulations have been carried-out, under the same conditions (common-mode voltages at 55 % of V_{DD}, the PM measured with a closed-loop gain of 6 dB). The results are summarized in Table 2.

Table 2. Results for the new circuit configuration (DTMOS)

V_{DD} (V)	Gain (dB)	GBW (kHz)	Phase Margin (^{o})
1.2	80.37	887	> 60 (higher)
0.8	79.54	844	> 60 (higher)
0.7	79.14	902 (higher)	> 60 (higher)
0.6	79.5	872 (higher)	57 (higher)
0.5	77.14	909 (higher)	61 (higher)
0.4	72.45 (higher)	924 (higher)	60 (higher)
0.3	58.82 (higher)	794 (same)	59 (higher)

As it can be observed in Table 2, there are just minor improvements in the DC gain and in the GBW. The most significant improvement is the PM parameter, particularly at very low supply voltages below 0.6 V. Since the values are superior to 55°, the OTA circuit is now stable for those V_{DD} values and can operate down to 0.3 V.

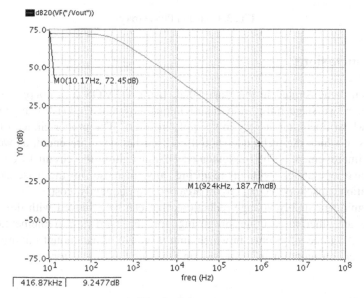

Fig. 2. Gain trace

The gain and phase plots are presented in Fig. 2 and in Fig. 3, for a 0.4 V supply voltage. This V_{DD} value was chosen as a reference example because it is half the value of the supply voltage for which the original circuit was designed (which constitutes, by itself, a significant reduction in the supply voltage) and because it presents good gain, GBW and PM results, in comparison to the results achieved by the original circuit, for that same V_{DD}.

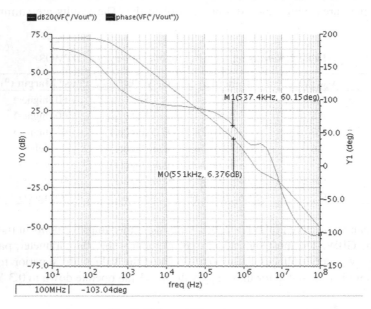

Fig. 3. Gain and Phase traces

5   Conclusions

It has been demonstrated in this paper that, using DTMOS in a couple of specific devices of a two-stage rail-to-rain Input/Output, constant G_m OTA can improve its stability whilst operating at very low supply voltages. The proper choice of specific transistors in which the technique should be used has been presented, as well as the resulting improvements. The implemented changes allow the circuit to be stable at low voltage operations without requiring any increase in the cascoded-Miller compensation capacitors, saving circuit area and, consequently, cost.

The significant supply voltage reductions that can be achieved with this DTMOS technique are, therefore, quite promising for the hardware implementations of Collective Awareness Systems, in which amplifiers are of paramount importance, and where power consumption and cost (silicon area) are limiting factors.

References

1. Citakovic, J., Riis Nielsen, I., Hammel Nielsen, J., Asbeck, P., Andreani, P.: A 0.8V, 7μA, Rail-to-Rail Input/Output, Constant G_m Operational Amplifier in Standard Digital 0.18μm CMOS. In: IEEE Xplore, Denmark, pp. 54–57 (2005)

2. Chatterjee, S., Pun, K.P., Stanić, N., Tsividis, Y., Kinget, P.: Analog Circuit Design Techniques at 0.5V. Springer (2007)
3. Bicocchi, N., Fontana, D., Mamei, M., Zambonelli, F.: Collective Awareness and Action in Urban Superorganisms, Italy (2013)
4. Abdollahvand, S., Gomes, A., Rodrigues, D., Januário, F., Goes, J.: Design of Robust CMOS Amplifiers Combining Advanced Low-Voltage and Feedback Techniques. In: Camarinha-Matos, L.M., Shahamatnia, E., Nunes, G. (eds.) DoCEIS 2012. IFIP AICT, vol. 372, pp. 421–428. Springer, Heidelberg (2012)
5. Abdollahvand, S., Santos-Tavares, R., Goes, J.: A Low-Voltage CMOS Buffer for RF Applications Based on a Fully-Differential Voltage-Combiner. In: Camarinha-Matos, L.M., Tomic, S., Graça, P. (eds.) DoCEIS 2013. IFIP AICT, vol. 394, pp. 611–618. Springer, Heidelberg (2013)

Simple and Complex Logical Functions in a SiC Tandem Device

Vitor Silva[1,2], Manuel A. Vieira[1,2], Paula Louro[1,2],
Manuel Barata[1,2], and Manuela Vieira[1,2,3]

[1] Electronics Telecommunication and Computer Dept. ISEL, Lisboa, Portugal
[2] CTS-UNINOVA, Quinta da Torre, Monte da Caparica, 2829-516, Caparica, Portugal
[3] DEE-FCT-UNL, Quinta da Torre, Monte da Caparica, 2829-516, Caparica, Portugal

Abstract. In this study we demonstrate the basic AND, OR and NOT logical functions based on SiC technology. The device consists of a p-i'(a-SiC:H)-n/p-i(a-Si:H)-n heterostructure with low conductivity doped layers. Experimental optoelectronic characterization of the fabricated device shows the feasibility of tailoring channel bandwidth and wavelength by optical bias through illumination at the back and front sides. Results show that, front background enhances the light-to-dark sensitivity of the long and medium wavelength range and strongly quenches the others. Back violet background has the opposite behavior; it enhances the magnitude in short wavelength range and reduces it in the long ones. This nonlinearity provides the possibility for selective removal or addition of wavelengths.

Keywords: Optoelectronics, Digital light signal, Logical functions, SiC Technology.

1 Introduction

Boolean algebra is the underlying mathematical theory that holds present computing systems. Since there is research to improve the materials that are used to build computational platforms there is also a need to guarantee that the basic logical functions are possible to be built and identified with such innovations.

Current CMOS technology is going to approach a scaling limitation in deep nanometer technologies. Quantum-dot cellular automata (QCA) is one of the promising new technologies for future generation ICs that overcome the limitation of CMOS. The fundamental unit of QCA-based design is majority gate; hence, efficient construction of QCA circuits using majority gates has attracted a lot of attention. Since every QCA circuit can be implemented by using only majority and inverter gates, the inverter becomes another important component in constructing QCA circuits [1].

Majority logic [2] is a way of implementing digital operations in a manner different from that of Boolean logic. The logic process of majority logic is more sophisticated than that of Boolean logic; consequently, majority logic is more powerful for implementing a given digital function with a smaller number of logic gates. One of the most important component in any arithmetic and digital circuits in QCA and VLSI is the full adder [3], [4].

L.M. Camarinha-Matos et al. (Eds.): DoCEIS 2014, IFIP AICT 423, pp. 592–601, 2014.

The majority gate can also be used for error correction in data transmission where several bits hold the pair functions of some selected bits (majority logic decoding) [5].

Research question answered in this article: Is it possible to use the SiC:H sensor for basic and complex logical functions?

Since previous work of the SiC:H sensor has proven its use as a multiplexer / demultiplexer device, and a multiplexer function allows for the construction of different logical functions, it is expected that by using this multiplexing capability that other logical functions can be identified with a SiC:H sensor.

2   Relationship to Collective Awareness Systems

A global earth picture of the present moment owes its possibility to a very remote past in which the global earth picture then would show reduced communities holding their bindings by being aware of everything that could endanger their survival or give them advantage over their everyday living. Communication is the basis for the collective awareness of a community. Collective awareness occurs when two or more people are aware of the same context and each is aware that the others are aware of it [6]. Visibility, awareness and accountability are properties of social translucent systems. Too see is not to be aware, to be aware includes rules, some social, that govern our actions and through which we are responsible. For example, opening a door which has a sign "Please open slowly" does not force whoever opens it to do it with care, being aware that it may knock someone down. The person inside the room relies that the door will effectively be opened with care. This is a form of collective awareness. Privacy and visibility are different views, as it is for example a transparent door into a room. Collective awareness systems must be translucent. The action / interaction theory in sociology [7] is present when users interact with interfaces [8]. There is a cyclic action theory, in which the user acts by recognizing the effects that lead to his goal. Present society is covered with different objects that allow for collective awareness used for social sensing and social mining [9]. The sensor device that is presented in this manuscript allows for light sensing. Used as a social sensing device it may be used to distinguish colors [10], for people that have difficulty in this type of visualization. During daylight the sky color can be used as a collective sensing sensor and aid in weather forecasts. Used in social events, different lighting may become informative of special happenings and gather people's attention without using sound. Even though we can be aware of other people's constraints, this too is a collective awareness of constraints [6] that shape the way as we all participate in the process of organization.

3   Sensor Characterization and Operation

The sensor is a two stacked p-i-n structures (p(a-SiC:H)- í'(a-SiC:H)-n(a-SiC:H)-p(a-SiC:H)-i(a-Si:H)-n(a-Si:H)) sandwiched between two transparent contacts one at each end. The thicknesses and optical gap of the í'- (200nm; 2.1 eV) and i- (1000nm; 1.8eV) layers are optimized for light absorption in the blue and red ranges [11]. Based in silicon carbon technology [12] this structure can be seen in Fig. 1, where the wavelength arrows indicate the absorption depths during operation and λV, λB, λG, λR the digital light signals within the visible spectrum.

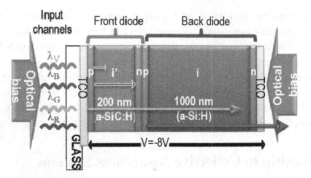

Fig. 1. Device structure and operation

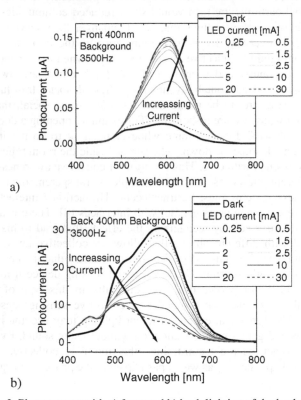

Fig. 2. Photocurrent with a) front and b) back lighting of the background

General purposes LEDs are used as light sources in two different ways: as digital signals and as background lighting. The digital signals are impinged on the front side of the sensor. The background lighting is either at the back or at the front side. The intensity of the signal sources is very low compared to the background intensity. Different wavelength signal sources are used: violet (400nm), blue (470nm), green (524nm) and red (626 nm). For background lighting the same violet wavelength is applied in a continuous and steady flux. When the background side changes the digital signals are sampled after transient influence has diminished.

The sensor is electrically biased with -8V. Readings were accomplished with a monochromator in 10nm steps from 400 to 800nm wavelengths.

Photocurrent results are presented in Fig. 2.

The experimental results of Fig. 2a show the photocurrent's increase in the 470-700nm bandwidth. There is a significant increase just by the presence of the violet light; the fivefold increase from no LED current to 0.25mA is outstanding when compared to the increase from 0.25 to 30mA. In Fig. 2b the thick black curve is the same of the previous figure and represents the dark level. With increasing LED current the photocurrent in the 470-700nm bandwidth gradually decreases and there is an almost fixed increase of the photocurrent in the 400-470nm bandwidth. The photocurrent gain is the ratio between the photocurrent output and the value of the dark curve when there is no background lighting. This gain is shown in Fig. 3.

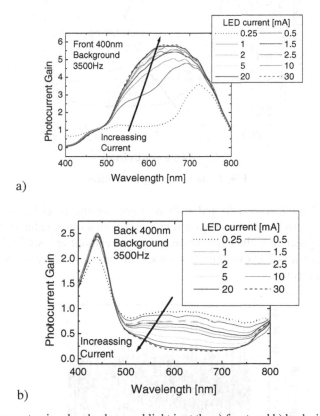

Fig. 3. Photocurrent gain when background light is at the a) front and b) back side of the device

The spectral gain shown in Fig. 3a, with front background, reduces the short wavelengths (<470nm) and increases the long wavelengths (>470nm). This behavior is that of a selective filter centered in 650nm. The opposite happens when the background lighting is at the back side, Fig. 3b, the short wavelengths increase while the long wavelengths decrease. This is also a selective filter centered in 440nm. Thus the sensor can act as a selective filter, where the gain of the short and long pass wavelengths is controlled by optical bias at either one of the sides. The gains of both filters suffer almost no changes with LED currents above 10mA. These experimental results were made with the front Led positioned at 6.5cm from the sensor and the back LED at 2.5cm.

Normalizing the photocurrent gains shown in Fig. 3 and plotting them in the same graph, Fig. 4, enables a view of the two filters and the eye figure assures the effectiveness of the filtering capabilities of the sensor. A notch filter around 500nm is also perceived.

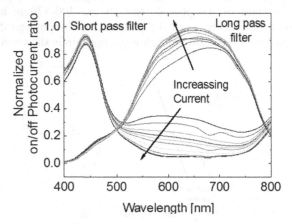

Fig. 4. Short and long pass filters

The sensor biased at -8.00V and with no light shining in either of its surfaces presents a noise current which is depicted in Fig. 5 with a mean value of 0.89nA. The values were registered with a low-noise current preamplifier (SRS-SR570). Static power consumption is 7.15nW. The low values are due to the reverse polarization of the two pi'npin junctions.

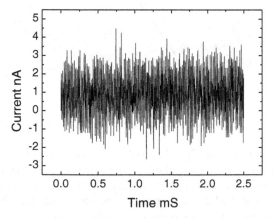

Fig. 5. Dark current

4   Logical Functions

Logical functions are commonly used as logic gates [13] and by combining them, other logical functions are created, and some, due to their special function, are named. One of them is the multiplexer which is a combinational function. Due to its behavior the multiplexer can also be used as a basic circuit which in turn can produce results as simple as the basic logic gates. This functionality is presented in Fig. 6.

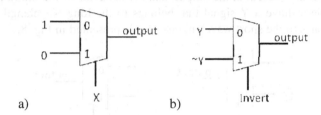

Fig. 6. A multiplexer used as an invert function

The 2x1 multiplexer shown in Fig. 6a) is composed by a selector input, X, which chooses one of the input signals to be set as the output. When X hold the 0 value, the 0 input is set at the output, in this case the output presents the value 1. Accordingly, when X is 1 the output is 0. This means that the output is the inverse of X.

In Fig. 6b the Invert signal is the selector that chooses between inputs Y and ~Y (NOT Y). The output follows the Y signal when the Invert selector is 0 and follows the ~Y signal when Invert is 1. By this setup, the Invert selector can choose between a Y signal or its inverse. This is the basis of the following work using the sensor's capability of being a multiplexer [14] to act as an inverter.

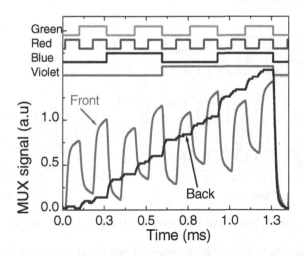

Fig. 7. The multiplex function

Presented in Fig. 7 is the multiplex/demultiplex function of the SiC sensor. By choosing the front background the output signal follows the influence of the red and green digital light signals, and the back side lighting allows the output to follow the blue and violet signals. The patterns in the top of the figure are displayed to guide the eyes into the output signal. According to Fig. 4 the red and green wavelengths belong to the long wavelengths whereas the blue and violet wavelengths belong to the short wavelengths. Having in mind that choosing the background side chooses the wavelengths that are at the output, and the behavior of the multiplexer in Fig. 6b it is possible to have a Y signal that belongs to the long wavelengths and a ~Y signal that belongs to the short wavelengths. This is presented in Fig. 8.

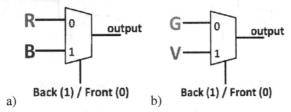

a)   b)

Fig. 8. Digital light multiplexer

The digital light multiplexer presented in Fig. 8 works in the following way: the selector illuminates the background either at the front (0) or at the back (1), thus choosing a long wavelength (R or G) to be presented as output or the short wavelength (B or V). According to Fig. 6b, the R signal must have its inverse equal to ~B, and the G signal its inverse equal to ~V. The long, short wavelength pair forms a digital light signal and is represented by *Signal[Long, Short]*. In the experimental results that follow two different digital light signals will be used, signal D [Red, Blue] and signal P [Green, Violet].

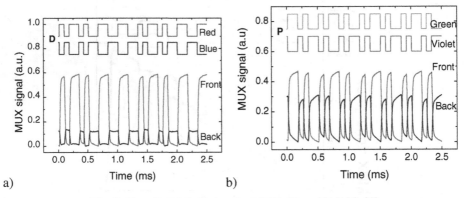

a)   b)

Fig. 9. Two digital light signals a) D [R, B] and b) P [G, V]

Plotted in Fig. 9a is the resultant output signal under front and back illumination for the same input digital light signal D[R, B]. The waveforms at the top of the figure represent the on/off input sequence. The front output follows the Red component of the digital signal and the back follows the Blue component. Fig. 9b is identical in its behavior, were the output signal with back lighting follows the Green component and the back waveform follows the Violet component. This shows the inverse function with two different examples. The simultaneous lighting of the digital signals D and P is presented in Fig. 10.

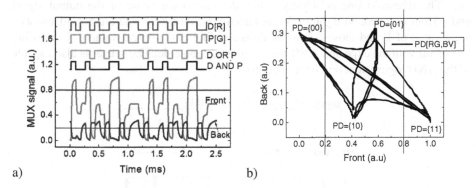

a)   b)

Fig. 10. a) Interaction of digital light signals D [R, B] and P [G, V]. b) Front and back signal correlation.

The two digital signals D and P applied simultaneously to the sensor are depicted in Fig. 10a. There are four different combinations. The shaded part in the figure holds these combinations that are shown with zoom in the next figures. The front and back signals can be plotted simultaneously showing the correlation between them in Fig. 10b. This allows the selection of threshold values that are used to determine the logical functions in Fig. 11 and Fig. 12. By adding a threshold line (not shown) at 0.15V on the back signal, all four combinations can be extracted. Noise will surround the vertexes of Fig. 10b creating areas of uncertainty which can be improved with hysteresis resulting in wider noise margins.

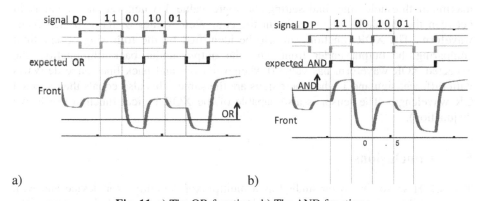

a)   b)

Fig. 11. a) The OR function. b) The AND function.

The waveforms at the top of Fig. 11a show the four different combinations of the digital signal D with P, only the long wavelength of each digital signal is shown. The output signal under front illumination is also depicted. By setting a threshold value above the minimum value of the output signal and assuming that values above the threshold have a logic value of 1 and the values below the line the logic value 0, the result is equal to the expected OR waveform. The sensor is thus capable of the OR logical function (disjunction).

Depicted in Fig. 11b is the same figure of Fig. 11a, but with a different threshold line. The threshold line is slightly below the maximum value of the output signal under front background lighting. Values above the threshold line are considered as a logic value of 1, and those below the threshold line have logical value 0. Comparing this result with the expected AND waveform both coincide. The sensor is also capable of the AND logical function (conjunction).

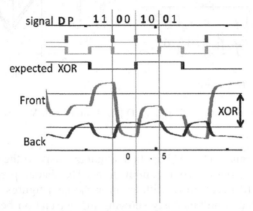

Fig. 12. The XOR function

The same description of Fig. 11a is applied to Fig. 12 which is completed with the threshold line of Fig. 11b and also displays the output under back illumination of the background. Using both threshold lines and setting the logic value 0 whenever the output signal under front illumination is below the minimum threshold and above the maximum threshold line, and setting the logic value 1 whenever the signal is in between the threshold lines, results in the equivalent values of the expected XOR waveform. The XOR function can also be identified by the following observation: comparing the output under front with the output under back illumination, the expected XOR waveform has value 0 whenever front and back lines have derivates with different sign and 1 when their signs are the same. This also equals the expected OR wavelength. The sensor is also capable of the XOR logical function (exclusive disjunction).

5   Conclusions

The SiC:H sensor has been studied as a multiplexor/demultiplexor device and with the possibility of determining the full 16 combinations of four digital inputs. The SiC:H sensor has the characteristic of a tunable filter in two distinct bandwidths, the long and short wavelengths. By defining a digital light signal as the combination of two wavelengths, one from the long and other from the short bandwidth it is possible to identify the inverse function using the tunable filter characteristic. The interaction of two digital light signals allow for the identification of other basic logical functions: AND, OR. The more complex digital function, the XOR is also a possible outcome of the sensor. The work that is under development aims for the identification of complex logical and arithmetic functions as the majority and the adder. The SiC:H sensor presented in this article is capable of logic functions and that is a good expectation that it can be used for optical computational systems.

References

1. Azghadi, M.R., Kavehei, O., Navi, K.: A Novel Design for Quantum-dot Cellular Automata Cells and Full Adders. J. Appl. Sci. 7(22), 3460–3468 (2007)
2. Pacuit, E., Salame, S.: Majority Logic Majority Logic: Syntax, pp. 598–605 (2004)
3. Farazkish, R., Sayedsalehi, S., Navi, K.: Novel Design for Quantum Dots Cellular Automata to Obtain Fault-Tolerant Majority Gate. J. Nanotechnol. 2012, 1–7 (2012)
4. Azghadi, M.R., Kavehei, O., Navi, K.: A Novel Design for Quantum-dot Cellular Automata Cells and Full Adders. J. Appl. Sci. 7(22), 3460–3468 (2007)
5. Hauck, P., Huber, M., Bertram, J., Brauchle, D., Ziesche, S.: Efficient Majority-Logic Decoding of Short-Length Reed – Muller Codes at Information Positions, pp. 3–10
6. Kellogg, W.A., Erickson, T.: Social Translucence, Collective Awareness, and the Emergence of Place, pp. 1–6 (2002)
7. Goodwin, C.: Action and embodiment within situated human interaction. J. Pragmat. 32(10), 1489–1522 (2000)
8. Ryu, H., Monk, A.: Analysing interaction problems with cyclic interaction theory: Low-level interaction walkthrough 2(3), 304–330 (2004)
9. Giannotti, F., Pedreschi, D., Pentland, A., Lukowicz, P., Kossmann, D., Crowley, J., Helbing, D.: A planetary nervous system for social mining and collective awareness. Eur. Phys. J. Spec. Top. 214(1), 49–75 (2012)
10. Louro, P., Vygranenko, Y., Martins, J., Fernandes, M., Vieira, M.: Colour sensitive devices based on double p-i-n-i-p stacked photodiodes 515, 7526–7529 (2007)
11. Vieira, M., Louro, P., Fernandes, M., Vieira, M.A., Torre, Q., Caparica, M.: Three Transducers Embedded into One Single SiC Photodetector: LSP Direct Image Sensor, Optical Amplifier and Demux Device, ch. 19, pp. 403–426 (March 2011)
12. De Cesare, G., Irrera, F., Lemmi, F., Palma, F., Tucci, M.: a-Si:H/a-SiC:H Heterostructure for Bias-Controlled Photodetectors. In: MRS Proc., vol. 336, p. 885 (February 2011)
13. Stallings, W.: Computer Organization And Architecture Designing For Performance, 8th edn., pp. 694–730. Prentice Hall
14. Vieira, M., Vieira, M.A., Silva, V., Louro, P., Costa, J.: SiC monolithically integrated wavelength selector with 4 channels. In: MRS Proc., vol. 1536, pp. 79–84 (June 2013)

Simulation in Amorphous Silicon and Amorphous Silicon Carbide Pin Diodes

Dora Gonçalves[1], Miguel Fernandes[1,2], Paula Louro[1,2], Alessandro Fantoni[1,2], and Manuela Vieira[1,2,3]

[1] ADEETC, ISEL, Lisbon, Portugal
[2] CTS-UNINOVA, Caparica, Portugal
[3] DEE-FCT-UNL, Caparica, Portugal
{dgoncalves,mfernandes,plouro,afantoni}@deetc.isel.ipl.pt,
mv@isel.pt

Abstract. Photodiodes are devices used as image sensors, reactive to polychromatic light and subsequently color detecting, and they are also used in optical communication applications. To improve these devices performance it is essential to study and control their characteristics, in fact their capacitance and spectral and transient responses. This study considers two types of diodes, an amorphous silicon pin and an amorphous silicon carbide pin, whose major characteristics are simulated, using the AFORS-HET program .The pin diode structure can be defined using contacts, interfaces and optical layers and then common measurements can be simulated by a numerical model, the AFORS-HET program. I-V, C-V characteristics, spectral response are simulated for both devices, without and under different illumination wavelengths. The results will allow a comparison between the main properties of amorphous silicon and amorphous silicon carbide diodes. We can conclude that sinusoidal frequency varies capacitance values as well as incident light wavelength. And when carbon is included in an amorphous silicon diode structure, its electrical and optical properties change.

Keywords: Photodiode, amorphous silicon, background illumination.

1   Introduction

This paper concerns the following research question: "By superimposing background illumination the device behaves as a filter, producing signal attenuation, or as an amplifier, producing signal gain, depending on wavelength combination. Is it possible (and if yes, how) to improve the frequency response of a a-Si:H photodiode for optoelectronic applications?", taking into consideration the next research hypothesis: "It is possible to improve the device performance by controlling the material properties (defect density, conductivity and light absorption) and layer thicknesses in order to get to an optimal definition of the device photo-capacitance".

Photodiodes are widely used for a variety of optoelectronic applications; one of the most important is their use in the context of optical communications. The state of

L.M. Camarinha-Matos et al. (Eds.): DoCEIS 2014, IFIP AICT 423, pp. 602–609, 2014.
© IFIP International Federation for Information Processing 2014

the art of information transmission over optical fibers systems is well established over the Infra-Red technologies for light emitting, transmission and receiving, allowing communication with large (and always increasing) bandwidth characteristics. Nevertheless recent advances in the last two decades have produced an increasing market growth for solid state technology illumination with Light Emitting Diodes (LED), because it reduces considerably power consumption and expands architectural capabilities. Visible light wireless communication efforts, specifically with Light Emitting Diodes (LED), for indoor spaces have gained great interest lately and have already been included in standardization activities (IEEE 802.15) [1].

Amorphous Silicon pin structures, well studied during the 90's as solar cells devices for terrestrial photovoltaic applications and low power electronic systems [2], are getting a renewed interest as photodiodes devices working in the visible light range for communication applications [3]. The literature about the physics of a-Si:H devices is mainly centered on device characteristics and effects which are important for development and optimization of solar cell devices. Less attention has been paid to frequency response and, generally speaking, to the transient characteristics of the a-Si:H multilayer structures. The characterization of the device under pulsed light illumination has highlighted the presence of transient effects which are not measured under steady-state condition [4], [5]. At a first approach, these effects have been related to a photo-capacitive behavior of the a-Si:H structures, revealing a need for a more detailed analysis, able to relate the light controlled device capacitance to the physical characteristics of the device, like interface quality and/or defect density of the materials used. Our simulation program ASCA [6] has been found not able to reproduce the device transient characteristics under pulsed light conditions. This paper report our initial study based on simulation performed with the simulation program AFORS-HET [7]. We report here the simulation results about Current-Voltage characteristics, Spectral Responses, and Capacitance Voltage characteristic. The simulations have been performed under different conditions of illumination and applied bias. Two different kind of photodiodes have been selected for our simulations: one pin structure of a-Si:H and one pin structure of a-SiC:H. The intent of this study is the comparison of the optoelectronic properties of this two kind of structures.

2   Collective Awareness Platforms for Sustainability and Social Innovation

Collective Awareness Platforms for Sustainability and Social Innovation (C.A.P.S.) are systems that combine data from real environments ("Internet of Things"), open online social media, and disseminated knowledge creation in order to generate new forms of social innovation by means of awareness of problems and possible solutions using collective efforts. The work hereby presented is motivated by a clear research question: Is it possible (and if yes, how) to improve the frequency response of a a-Si:H photodiode for WDM applications? The answer to this question is directly related to a future development of a network based on visible range optical communications. This technology will certainly contribute to the future development of C.A.P.S.

3   Simulation

The AFORS-HET simulations have been performed under different conditions of illumination and applied bias. Two different kind of photodiodes have been selected one pin structure of a-Si:H and one pin structure of a-SiC:H. The intent of this study is the comparison of the optoelectronic properties for these two kinds of structures. Varying the i-layer thickness of the photodiodes we have simulated the I-V and C-V characteristics and the spectral response. The material parameters used in the simulator are reported in Table 1. Our study is mainly directed to study the influence of the intrinsic layer. The adding of carbon into the a-Si:H base is assumed to provoke a lager optical gap (Eg), an increasing of the defect density (Ntr) and a lowering of the carriers mobility ($\mu n, \mu p$).

Table 1. Material parameters used in the AFORS-HET simulations (i-layer)

Parameter	Description	a-Si:H	a-SiC:H
E_g (eV)	Energy gap	1.72	2.1
μ_n (cm^{2}V^{-1}s^{-1})	Electrons mobility	20	2
μ_p (cm^{2}V^{-1}s^{-1})	Holes mobility	5	0.5
$E_{U,v}$ (eV)	Urbach energy (valence tail)	0.05	0.05
$E_{U,c}$ (eV)	Urbach energy (conduction tail)	0.035	0.035
N_{tr} (cm^{-3})	Defects density	5×10^{15}	5×10^{16}
N_{tr} (cm^{-3})	Defects density	5×10^{15}	5×10^{16}

In Fig. 1 and 2 are reported, respectively, the short circuit current and the open circuit voltage produced by a Si:H pin structure as a function of the i-layer thickness. The simulation has been performed under illumination with different wavelength: red (650nm), green (560nm), blue (440 nm) and violet (410nm). The intrinsic layer thickness varies between 200 and 1100 nm. The blue radiation is absorbed in the front part of the device, so an increasing of the i-layer thickness produces a bottom region subject to a low illumination level. The series resistance of the photodiode increases and the overall photocurrent falls down. When illuminated with a red light the increasing of the i-layer thickness produces an enhancement of the photocurrent. This is due to the lower absorption coefficient in the red part of the spectrum and the higher penetration of the red radiation within the pin structure. With a value of 1100 nm for the i-layer thickness, the short circuit current saturates. A further increasing of the i-layer thickness would not produce any enhancement and it would end in a photo current degradation. A thickness of 400 nm appears to be an optimal choice for absorption of a green radiation. This would be the optimal choice for a solar cell device. For all the radiations, the open circuit voltage suffers a smooth lowering with the i-layer thickness. As matter of fact a photodiode for communication application would not work in photovoltaic mode, but will be always reversed biased. The photocurrent is the most important parameter to be extracted by these simulation runs.

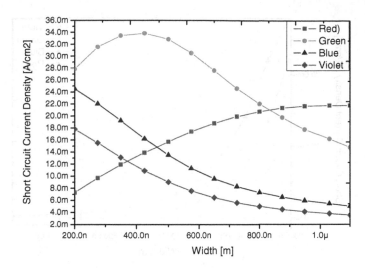

Fig. 1. Short circuit current produced under different illumination conditions by a a-Si:H pin structure as a function of the i-layer thickness

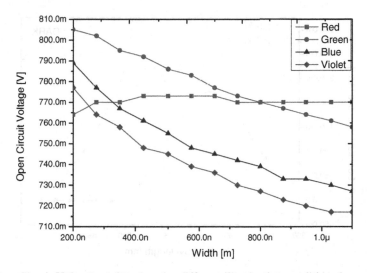

Fig. 2. Open Circuit Voltage produced under different illumination conditions by a a-Si:H pin structure as a function of the i-layer thickness

In Fig. 3 is reported the short circuit current produced by a a-SiC:H pin structure as a function of the i-layer thickness. Like in the precedent case, the simulation has been performed under illumination with different wavelength: red (650nm), green (560nm), blue (440 nm) and violet (410nm). Again, the intrinsic layer thickness varies between 200 and 1100 nm. The larger optical gap produces a very low absorption in the red range and the photocurrent under red illumination is almost zero. The photodiode has a good response in the blue spectrum, while under green irradiation

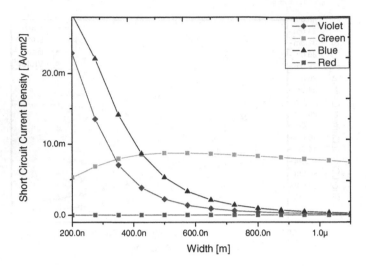

Fig. 3. Short circuit current produced under different illumination conditions by a a-SiC:H pin structure as a function of the i-layer thickness

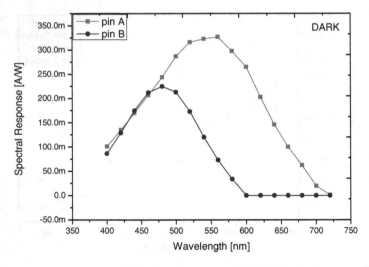

Fig. 4. Spectral response of a a-Si:H (Pin A) and a a-SiC:H (pin B) structure. The i-layer thickness is 500 nm for both cases.

the photocurrent is a lot lower than in the a-Si:H case. This is caused by the inferior value of the free carrier mobility. The increasing of the i-layer thickness produces a fast degradation of the photocurrent. This is the effect of the augmented density of defects and the consequentially increased trapped charge density and recombination centers. A similar behavior was already reported for a-Si:H pin solar cell with increasing density of defect in the intrinsic layer [8].

In Fig. 4 is depicted the spectral response of a a-Si:H and a a-SiC:H structure. The i-layer thickness is 500 nm for both cases. In both structures the curves exhibit a similar trend. The spectral sensitivity increases with the wavelength until reaching a maximum, and then exhibits a decrease reaching very low values. It is also observed that for the a-SiC:H structure the sensitivity is confined to the range 400 - 600 nm, while the a-Si:H device exhibits sensitivity in the whole visible range. This difference is related mainly to the energy gap of both materials. It is also observed that the maximum of each curve occurs at different wavelengths depending on the nature of the material. For the a-Si:H photodiode it is located around 560 nm while for the a-SiC:H device at 480 nm. This is related mainly to the energy gap and absorption coefficients of each absorber layer. For the a-Si:H structure the magnitude of the spectral response is higher than the observed for the a-SiC:H structure, except in the short wavelength range (up to 470 nm) where the curves overlap.

In Fig. 5 it is displayed, respectively, the capacitance-voltage characteristic (under reverse bias) for the device a-SiC:H at different illumination conditions (dark, red, green, blue and violet light). The C-V measurement is simulated by considering an applied AC voltage of 0.02 V with frequency of 1 kHz. It is observed that under red and green light the capacitance remains low and it is independent on the applied bias, while under blue and violet light it decreases with the increase of the reverse bias. Maximum values are observed with violet illumination at short circuit voltage (230 nF/cm^2).

In Fig. 6 it is displayed the C-V characteristic of the structure a-SiC:H under violet illumination simulated at different AC frequencies in the range 1 kHz up to 100 kHz. Results show that the capacitance trend with the reverse bias is identical for every frequency. Higher values of capacitance correspond to low frequency values (1 kHz).

Practical measurements are shown in Fig. 7, for comparison.

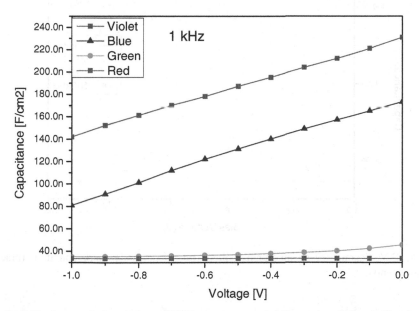

Fig. 5. C-V characteristic of the a-SiC:H structure at 1 kHz under different illumination conditions

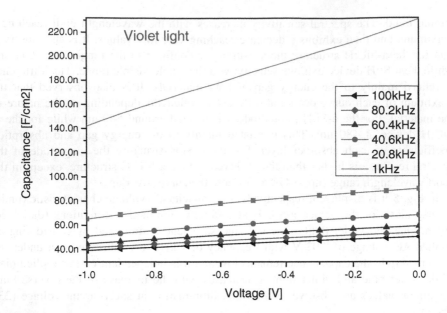

Fig. 6. C-V characteristic of the a-SiC:H structure under violet light simulated at different AC frequencies

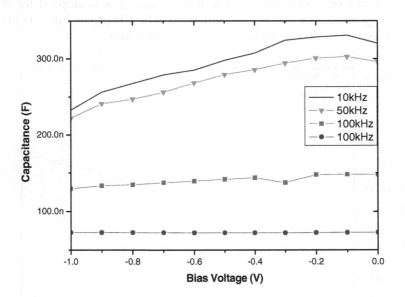

Fig. 7. C-V diagram for an amorphous silicon pin diode, under dark condition (practical measurements)

4   Conclusions and Future Work

We have performed a simulation study on the electrical characteristics of an a-Si:H pin photodiode and compared to a a-SiC:H pin photodiode.

I-V characteristic, for an amorphous silicon sample, shows that short circuit density current has a maximum for a pin intrinsic layer width around 400 nm, when illuminated with green light. When using blue or violet illumination, values of short circuit current reduce with thickness, although when using red illumination it increases. C-V characteristic depends on frequency and illumination, the peak values corresponding to 1 kHz and violet light.

For a better understanding of these characteristics the transient response under different wavelength pulsed light must to be simulated and related to capacitive effects. The AFORS-HET simulator is very powerful simulation software when applied to pin structures, but it does not allow for simulation of multi-junction devices. Our focus in future work will be on the development of a transient model able to simulate the photo capacitive effects in a-Si:H/a-SiC:H double junction photodiodes.

References

1. Rajagopal, S., Roberts, R.D., Lim, S.-K.: IEEE 802.15.7 visible light communication: modulation schemes and dimming support. IEEE Communic. Magazine 50 (2010)
2. Deng, X., Schiff, E.A.: Amorphous Silicon–based Solar Cells. In: Handbook of Photovoltaic Science and Engineering. John Wiley & Sons (2003)
3. Vieira, M., Vieira, M.A., Louro, P., Fernandes, M., Fantoni, A., Silva, V.: SiC multilayer photonic structures with self optical bias amplification. In: MRS Proceedings, vol. 1426 (2012)
4. Fantoni, A., Fernandes, M., Louro, P., Vieira, M.A., Vieira, M.: Capacitive effects in pinpin photodiodes. Microelectronic Engineering 108, 195–199 (2013)
5. Gonçalves, D., Fernandes, L.M., Louro, P., Vieira, M., Fantoni, A.: Measurement of Photo Capacitance in Amorphous Silicon Photodiodes. In: Camarinha-Matos, L.M., Tomic, S., Graça, P. (eds.) DoCEIS 2013. IFIP AICT, vol. 394, pp. 547–554. Springer, Heidelberg (2013)
6. Fantoni, A., Vieira, M., Cruz, J., Schwarz, R., Martins, R.: A two-dimensional numerical simulation of a non-uniformly illuminated amorphous silicon solar cell. Journal of Physics D: Applied Physics 29, 3154 (1996)
7. Stangl, R., Leendertz, C., Haschke, J.: Numerical simulation of solar cells and solar cell characterization methods: the open-source on demand program AFORS-HET. In: Rugescu, R.D. (ed.) Solar Energy, pp. 319–352. INTECH (2010)
8. Fantoni, A., Viera, M., Martins, R.: Influence of the intrinsic layer characteristics on a-Si: H p–i–n solar cell performance analysed by means of a computer simulation. Solar Energy Materials and Solar Cells 73, 151–162 (2002)

Touch Interactive Matrix LED Display
for the Collective Awareness Ecosystem

Fábio Querido[2] and João P. Oliveira[1, 2]

[1] Centre for Technologies and Systems (CTS) – UNINOVA
[2] Dept. of Electrical Engineering (DEE), Universidade Nova de Lisboa (UNL)
Campus FCT/UNL, 2829-516, Caparica, Portugal
f.querido@campus.fct.unl.pt, jpao@fct.unl.pt

Abstract. Nowadays, the devices have the role to be user-friendly with the touch interfaces. In this paper it is presented an interactive touching device fully supported in LED technology. This is achieved through the use of an array of LEDs where each one works as a light emitter and as a sensor. The implementation uses a microcontroller to control the LED operation. Also it is proposed a new LED unit matrix configuration that avoids the use of an external light source to detect any touch. This new architecture only needs two microcontroller input/output to each row plus column which provides a user-friendly interactive contactless interface. Also it is explained how to aggregate these unit modules into a bigger LED display panel, reducing implementation costs.

Keywords: Light Emitting Diode, LED Matrix, Bidirectional LED, Air Touch, Remote Touch, Multi Touch, Interactive LED, Touch Sensor, Display Devices.

1 Introduction

Light Emitting Diodes, or LEDs, are nowadays a very common source of light due to their higher energy efficiency, and easy dimming light control through standard Pulse With Modulation (PWM). In addition, this device can also be used as a photosensor [1], which can operate as a light emitter in one phase, and as a light sensor in other phase, all without hardware modifications. This coordinated operation can transform a large LED display, like advertising panels or queuing information panels, into a bidirectional interface with, basically, the same hardware. Moreover, the LED sensing capability can also be explored as a wireless two-way communication port [1].

Considering a matrix of LEDs, by quickly switching them between forward-biased (light-emitting) and reverse-biased (light-sensing) modes, it is possible to create dynamic images while measuring the lighting level, at the same time. In [2], some techniques to sense a touch finger with an LED matrix are presented, which can measure the reflected light through finger proximity. The operation can be managed by a master controller which controls each desired pixel intensity, while a slave module is dedicated to forward and reverse bias the LEDs.

In this paper, techniques will be studied with the objective to implement a display that can produce an image, while measuring the ambient environment. A master controller, using a dedicated protocol, controls its operation but at the same time

L.M. Camarinha-Matos et al. (Eds.): DoCEIS 2014, IFIP AICT 423, pp. 610–617, 2014.
© IFIP International Federation for Information Processing 2014

transform it into an interactive contactless input interface. For instance, each frame defines all intensity to each pixel followed by the setup of one of three modes, only light emission, sense, or at the same time sense and light emission.

Each module only completes the cycle when all LEDs provide the desired operation requested by master controller. For the light emission stage, the microcontroller (μC) can provide at the same time the complete LED row, and finishes the cycle when all rows are completed. In case of sensing only the ambient lighting level, the microcontroller read all columns in the row at the same time, and after finishes all rows, send each value of each to the master controller, and wait for other cycle mode. In the sense and light emission mode, all LED in the same row in odd columns are emitting. In the neighbor row, sensing light in even-numbered columns are accomplished. The cycle is completed when all pixels generated a proper value of intensity light, received the feedback light of each one and sent the feedback values to the master controller.

With this methods, the touch display implementation have low cost and low complexity hardware. It can also be used to sense multiple lasers with different wavelengths which are pointed by distinct users at the same time allowing the differentiation of the remote touch of each user, being similar to [3] in remote touch parameter.

2   Contribution to Collective Awareness Systems

Nowadays, the emergence of collective awareness systems is pushing the performance of end user interactive objects. The support framework based in interconnected objects and things (IoT) is the bridge that combines technologies and components from micro-systems (miniaturized electric, mechanical, optical and fluid devices) with knowledge, technology and functionality from several areas of research. In other words, the devices have to have embedded capacity to integrate a system, to be capable to interact dynamically having into account proprieties like lower resources and multi-functionality.

This paper will focus on the functionality of the smart touching devices for a wide range of dynamic and creative interactive applications, combining low cost and, in some cases, the usage of the same hardware with only a software update.

3   Related Works

Normally, the touch screens have separated technologies for imaging and touching functionality, [4]. The latter is can be embedded in the display using capacitive or resistive membranes. Alternatively, a lateral infrared strip can be used [5] or standard external cameras with sophisticated image recognition can detect the fingertips.

The novelty in this paper is the possibility to transform a standard LED display into a multi-touch, contactless touch and distance touch (e.g., using lasers) with almost no additional costs beyond the display itself, while maintaining a low system complexity, since it needs software modifications and only a few ones in hardware.

4   Using an LED as an Emitting/Sensing Light Device

Light emitting diodes have a similar exponential current-voltage curve when compared with a normal diode. The main property of an LED is the emission of light with a certain wavelength profile. Interestingly, this device also reacts to incoming photons with the same wavelength, enabling the use of the LED to measure the intensity of that radiation. One of the techniques is to measure a tinny voltage across the LED that is proportional to the intensity radiation. The other one, used in this paper, is based on the behavior of reverse biased pn junction under radiation conditions. When reverse biased, the radiation charges the LED intrinsic capacitor inverse to the radiation, this means that the equivalent capacitor is smaller in high light conditions, and bigger in low light conditions.

To measure the charge in the capacitor after the inverse biased phase, it is important that the measuring current is small. One method mentioned in [1] is using a digital comparison input, counting from the moment that the I/O_1 changes to High-impedance until a threshold that the microcontroller assumes a LOW value. That count is related to the incoming light intensity on the LED.

Figure 1 shows how to drive a simple LED with two input/output of the microcontroller to enable the switching between emitting and sensing light.

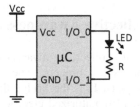

Fig. 1. Schematic of a bidirectional LED

To understanding the full behavior of LED when reverse biased, the charge reading process is equivalent to a Resistor-Capacitor (RC) circuit, that the R is the input impedance of the microcontroller, and C the intrinsic capacitor of the LED, and including a parallel of a current source, equivalent to a small leakage current. The electric equivalent in each state is shown in Fig. 2, and in c) shows a resistor that can be added (in grey) to have shorter time constant. This technique has the same power consumption, because the main contribution comes from the forward biasing phase.

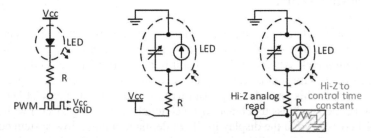

Fig. 2. Different stages of biasing LED with equivalent electric behavior. a) as light emitting variable. b) reverse bias to charge capacitor. c) read received light.

Figures 2 a) to c) show the sequence of emitting and measuring light steps. The case in Fig. 2 b) is needed to charge the LED junction that is related with the incident light.

The operation can be analyzed as a first order RC circuit with a time constant given by,

$$\tau = RC, \tag{1}$$

where R is the equivalent of the current limiting resistor in series with the parallel of input impedance of microcontroller and the pull-down high-impedance resistor (if needed) to allow the adjustment of R, and consecutively τ due to equation (1), and C the intrinsic LED capacitance.

The implementation using an RGB LED has to enter, firstly, into a pre-calibration step to adjust the sensitivity in relation to each color. The lenses in some types of LEDs are used to mix colors, which introduce an additional degradation factor to sense light. The pre-calibration step will adapt the firmware of the microcontroller to adjust the sensing sensitivity when red, green or blue lights are within the detection range, or the combination of them.

5   Bidirectional LED Array for an Interactive Interface

When the goal is to have the largest possible number of LEDs with just one microcontroller, a non-inverter driver can be used drive at the same time an entire row. This is needed because the microcontroller outputs are not capable to drive all LEDs in a row. However, to allow emitting and sensing light at the same instant, only two rows are used, while the others must be in high-impedance to not interfere in the measurements, meaning that the driver can be a tri-state type (Fig. 3).

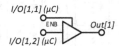

Fig. 3. Schematic of driver for multiple LED array (Tri-State)

An issue to take into account is the refresh rate of the total LED matrix, which has to be higher than 100 Hz to be undetectable by the human eye. On lower light conditions (bigger time constant), it should be defined a fixed maximum time to each operation and if the time constant is still too high a proper resistance value has be selected to obtain shorter discharges (grey in Fig.2. c)). In higher light conditions, the precision depends on the comparison rate of the microcontroller.

The matrix in Fig. 4 is a model that can be configured depending on inputs/outputs available on microcontroller, wherein two outputs are used for each row and an I/O for each column. The time for each cycle goes from the beginning of the frame received from master controller, Fig. 5, until light emission mode, receive mode, or both be completely performed for each LED.

For the first case, light emission mode, the controller output that controls the row through the tri-state driver puts the logical HIGH value, while the outputs directly connected to current limiting resistors generate a PWM with the desired Duty-Cycle for a certain time. The cycle ends when all rows are covered.

In the case that only is needed to measure the ambient light, the process is similar, but now it is necessary to impose the LOW value in the row, and HIGH value in each column to charge the intrinsic capacitor. Only after, the measurement of the charge accumulated in the junction of each LED in that row is taken. Note that the measuring light occurs from all LEDs in the row at the same time, and when some finishes, the microcontroller waits a programmed guard time. The cycle finishes when all LEDs are measured and that values are sent to a master controller.

When the microcontroller enters into a emit and read modes, first it has send the frame containing all PWM values for each LED, and after that, the first row is taken HIGH, while the odd columns are generating PWM. The sensing is made in the LED in the diagonal immediately at lower right position, while the other rows are in high-impedance state. After that, the other rows should have the same behavior, until the last one does not have any LED around in a sensing state, because that row is emitting (odd columns), and the first one is sensing (even-numbered columns)., This is made keeping in mind that the LED matrix can be surrounded by another one, in large display structure. Yet it is necessary to maintain the same behavior, but now emitting in even-numbered columns and sensing in odd columns. Finally, each value sensed should be sent to a master controller.

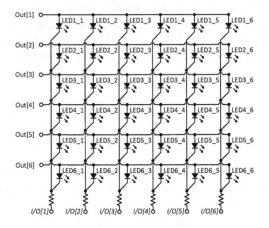

Fig. 4. Example model of an LED matrix

Figure 5 shows how to connect multiple modules, creating an interactive display controlled by a master controller or a simple computer, all this through a serial communication link.

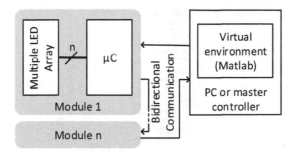

Fig. 5. Interconnection between the LED and master controller through a serial communication

6   Experimental Evaluation of the Proposed Matrix

To validate the proposed LED matrix, a single array of six basic red LEDs of three millimeters were designed and implemented. The green LED also was tested, and have basically the same behavior.

The microcontroller used was the ATmega328, from Atmel. Figure 6 shows the schematic assembly of the LED array prototype, in which 'Out_1' is driven by a non-inverter similar to Fig. 3, but without ENABLE since an Hi-Z state is not need for a one row configuration.

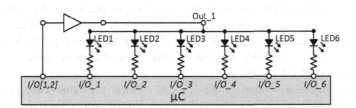

Fig. 6. LED array setup with MCU

The setup used is similar to Fig. 5, but having just one module. A serial communication to a virtual environment, created in Matlab, was used to demonstrate the user interactivity.

Remember that the microcontroller only finishes the phase of sensing when all LEDs reaches the threshold voltage, meaning than if one LED arrives at the threshold value the microcontroller waits until the last one, allowing a measured with an oscilloscope than the voltage across the LED measured drops much faster because the ten megohm input impedance.

To determine the time constant, measurements where performed with a digital oscilloscope. Figure 7 shows the measurement results of each phase.

In Fig. 7 a), the time constant is the time that the voltage across LED drops from Vcc to $Vcc \cdot e^{-1}$, but the threshold voltage of microcontroller is around $2V$.

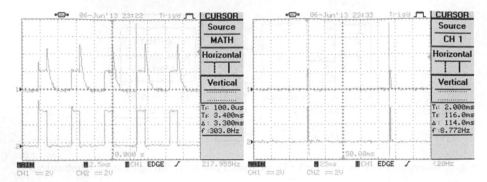

Fig. 7. Measure of reading time of received light $\varDelta$ (CH1 cathode of LED, CH2: anode of LED). a) case with high light conditions b) case with low light conditions.

Through the proportionality rule,

$$\frac{\varDelta}{\tau} \leftrightarrow \frac{Vcc - 2}{Vcc - Vcc \cdot e^{-1}},$$

which $\varDelta$ is the value obtained in each measure of Fig. 7, and time constant τ is 3.5ms and 120ms for high and low light conditions, respectively.

Since the input impedance of microcontroller is in the range of some megohm, these experimental results show that the capacitor is in the range of a few picofarad and near nanofarad in high and low light conditions, respectively. We cannot change this intrinsic values, but adjusting a proper resistor like in Fig. 2 c) in grey, the time constant can be manipulated to boost higher refresh rates. However, for lower refresh times, a faster microcontroller may be needed. Note that these values depends on LED, and microcontroller features.

A real time interactive interface was created to view the values determined by the microcontroller from which a synthetized a musical tone proportional to the proximity distance of a finger, was generated. For example, Fig. 8 b) shows a finger proximity in the third LED.

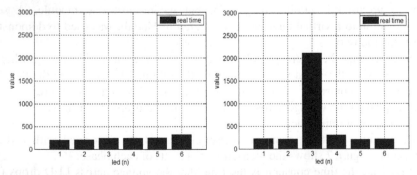

Fig. 8. Measure ambient light (lower value means higher light). a) without light obstruction (ground level). b) with a finger on third LED.

The Matlab demo application consists of a piano synthesizer, in which each LED corresponds to a musical tone, and giving an indication to the user of the tone touched by changing the corresponding LED light intensity (amplitude of the sound level is proportional to the light intensity).

7 Conclusion

The techniques presented in this paper allows the implementation of contactless LED based matrix smart interactive interface that can operate in a robust way for each application. The information processing may be performed by a computer or a master controller, which enables the setup of large area displays. In each module, each slave microcontroller can be replaced by a System on Chip (SoC) capable of producing the desired outputs for driving each LED, read ambient light and transm. In smaller applications, only one microcontroller can be used to generate their own interactivity.

With this architecture of touch displays, the global cost is almost the same as a normal LED matrix display, with practically the same complexity and a software modification, not forgetting that the consumption remains practically equal due to the small intrinsic capacitors that have to be charged and uncharged to measure light.

Through this system, we expect a touch interface as the important component in these days to provide the high quality of services to users for applying to a variety of applications easily and giving the intuitive interface.

References

1. Dietz, P., Yerazunis, W., Leigh, D.: Very low-cost sensing and communication using bidirectional LEDs. In: Dey, A.K., Schmidt, A., McCarthy, J.F. (eds.) UbiComp 2003. LNCS, vol. 2864, pp. 175–191. Springer, Heidelberg (2003)
2. Scott, E.: Hudson: Using light emitting diode arrays as touch-sensitive input and output devices. In: Proceedings of the 17th Annual ACM Symposium on User Interface Software and Technology (UIST 2004), pp. 287–290. ACM, New York (2004)
3. Pasquariello, D., Vissenberg, M.C.J.M., Destura, G.J.: Remote-Touch: A Laser Input User–Display Interaction Technology. Journal of Display Technology 4(1), 39–46 (2008)
4. Nichols, S.J.V.: New Interfaces at the Touch of a Fingertip. Computer 40(8), 12–15 (2007)
5. Lee, B., Hong, I., Uhm, Y., Park, S.: The multi-touch system with high applicability using tri-axial coordinate infrared LEDs. IEEE Transactions on Consumer Electronics 55(4), 2416–2424 (2009)

Author Index

Printed in the United States
By Bookmasters

Printed in the United States
By Bookmasters